International Conference on Civil and Architecture Engineering (ICCAE-14)

IOP Conference Series: Earth and Environmental Science
Volume 1056

Cairo, Egypt
29-31 March 2022

ISBN: 978-1-7138-6019-8
ISSN: 1755-1307

Printed from e-media with permission by:

Curran Associates, Inc.
57 Morehouse Lane
Red Hook, NY 12571

CURRAN ASSOCIATES INC.
proceedings
.com

Some format issues inherent in the e-media version may also appear in this print version.

Institute of Physics
Dirac House, Temple Back
Bristol BS1 6BE UK

Phone: 44 1 17 929 7481
Fax: 44 1 17 920 0979

techtracking@iop.org

Additional copies of this publication are available from:

Curran Associates, Inc.
57 Morehouse Lane
Red Hook, NY 12571 USA
Phone: 845-758-0400
Fax: 845-758-2633
Email: curran@proceedings.com
Web: www.proceedings.com

TABLE OF CONTENTS

ARCHITECTURE ENGINEERING

BUILDING TECHNOLOGY

CONCRETE STRUCTURES

ENVIRONMENTAL ENGINEERING

GEO-TECHNICAL ENGINEERING

METALLIC STRUCTURES

TRANSPORTATION ENGINEERING

URBAN PLANNING

WATER RESOURCES

IOP Conf. Series: Earth and Environmental Science 1056 (2022) 011001 doi:10.1088/1755-1315/1056/1/011001

Arab Republic of Egypt
Ministry of Defense
Military Technical College

BOOKLET OF THE 14TH INTERNATIONAL CONFERENCE ON CIVIL AND ARCHITECTURE ENGINEERING

ICCAE-14

Military Technical College
The 14th International Conference on
Civil and Architecture Engineering
29 - 31 March, 2022
https://iccae.conferences.ekb.eg/

PREFACE

The Military Technical College is pleased to organize the 14th International Conference on Civil and Architecture Engineering, sponsored by the Ministry of Defence in the period 29 - 31 March 2022. The conference is a chance to bring together all scientists and engineers from Egyptian armed forces and their counterparts in the universities, research center, academic and technical institutions from Egypt and other countries. It is our hope that the conference will provide a good opportunity to exchange knowledge, and to find appropriate solutions to present progress in Civil and Architectural Engineering fields. The conference has the aim to provide an environment of submitting and discussion of modern advances and experience on various Civil and Architectural Engineering activities. The order of conference sessions has covered the following themes:

1- Concrete Structures.
2- Geotechnical Engineering.
3- Metallic Structures.
4- Project Management.
5- Transportation Engineering.
6- Water Resources Engineering.
7- Architectural Engineering.
8- Building Technology.
9- Environmental Engineering.
10- Urban Planning.

From (126) full paper submitted to the conference, number of (47) papers were accepted and the rest were rejected/withdrawn. All submitted manuscripts were peer reviewed and checked against similarity.
These selected papers have been presented during the conference interval, 29 - 31 March 2022 in different (13) scientific sessions. In addition to different (4) workshops and seminars. The list of workshops are:

Sponsor Organizer IOP Publishing Publisher Egyptian Knowledge Bank بنك المعرفة المصري Partner

Military Technical College
The 14th International Conference on
Civil and Architecture Engineering
29 - 31 March, 2022
https://iccae.conferences.ekb.eg/

W1 - The 5th Generation of Egyptian Sustainable Cities;

a) The 5th Generation of Egyptian Cities; Reporting for AGA urban sustainable upgrade: by (Principle Investigator) Brig. Gen. Prof. Dr. Mohammad Fahmy, Military Technical College, Egypt.

b) Urban development policies and strategies from the ISDF to UDF: by Maj. Gen. Eng. Khaled Sedik, Eng. Hesham Gohar, Dr. Marwa Ahmed, Urban Development Fund, Egyptian Cabinet of Ministries.

W2 - Achievements towards Egypt Vision 2030;

a) Efforts of the Ministry of Environment towards Egypt Vision 2030: by Dr. Ali Abo-Senna, CEO of Environmental Affairs Agency, Egyptian Ministry of Environment.

b) Cities after COVID-19: by Dr. Rania Adham, Housing & Building Research Center, Egyptian Ministry of Housing and Urban Communities.

W3 - Behavior of Substructure and Superstructure Subjected to Blast Loads by using ABAQUS: by Dr. Marwa Nabil, Zagazig University, Egypt.

W4 - Role of Geological Methods and Techniques in Solving Construction Engineering Problems: by Prof. Zakaria Hamimi, Benha University, Egypt.

Finally, the conference high committee would like to thank and acknowledge all contributors, committees' members and boards of the conference session for achieving all planned missions.

BEST REGARDS,
BRIG. GEN. PROF. DR. MOHAMMAD FAHMY
ICCAE-14 EDITOR IN CHIEF
https://www.scopus.com/authid/detail.uri?authorId=55972043400

Sponsor Organizer IOP Publishing Publisher Egyptian Knowledge Bank بنك المعرفة المصري Partner

Military Technical College
The 14th International Conference on
Civil and Architecture Engineering
29 - 31 March, 2022
https://iccae.conferences.ekb.eg/

COMMITTEES

Sponsor Organizer Publisher Partner

IOP Conf. Series: Earth and Environmental Science 1056 (2022) 011001 doi:10.1088/1755-1315/1056/1/011001

Military Technical College
The 14th International Conference on Civil and Architecture Engineering
29 - 31 March, 2022
https://iccae.conferences.ekb.eg/

Conference High Committee

Maj. Gen. Prof. Dr. Ismail Mohamed Kamal	Chairman.
Maj. Gen. Assoc. Prof. Dr. Moataz Abo Elnoor	Vice – Chairman.
Maj. Gen. (R) Prof. Dr. Ahmed Elsaid Abdallah	General Rapporteur.
Brig. Gen. Dr. Prof. Dr. Mohammad Fahmy	Editor in Chief.

Editorial Committee

Professor Dr. Mohammad Fahmy - Editor in Chief.
Professor (Asist.) Dr. Mohamed Zahran - Associate Editor.
Professor (Asist.) Dr. Hany Mokhtar - Associate Editor.

Scientific Committee

Prof. Adel Yaseen, Ain Shams Uni., Egypt.
Prof. Ahmed Abdin, Cairo Uni., Egypt.
Prof. Morad Abdel-Kader, Ain Shams Uni., Egypt.
Prof. Ahmed Sherif, AUC, Egypt.
Prof. Osama M. Moussa, MTC, Egypt.
Prof. Neveen Hamza, Newcastle Uni., UK.
Prof. Islam Abo-Hela, Staffordshire Uni., UK.
Prof. Mohamed Salheen, Ain Shams Uni., Egypt.
Prof. Marialena Nikoloupoulo, Kent Uni., UK.
Prof. Alaa Mandour, Helwan Uni., Egypt.
Prof. Abd El-Reheam Khalel, Ain Shams Uni., Egypt.
Prof. Ahmed Abdelkhaleak, Future Uni., Egypt.
Prof. Adel H. Salem, Ain Shams Uni., Egypt.
Prof. Ahmed A. Hassan, Ain Shams, Uni., Egypt.
Prof. Ahmed El-Desoky, Canadian Uni., Egypt.
Prof. Nabil Amer, AASTMT, Egypt.
Prof. (Assoc.) Sherif Ezz El-Din, AASTMT, Egypt.
Prof. Suzette M. Aziz, AASTMT, Egypt.
Prof. Ashraf A. El-Ashaal, Irrigation Ministry, Egypt.

Sponsor Organizer IOP Publishing Publisher Partner

Steering Committee

IOP Conf. Series: Earth and Environmental Science 1056 (2022) 011001 doi:10.1088/1755-1315/1056/1/011001

Military Technical College
The 14th International Conference on
Civil and Architecture Engineering
29 - 31 March, 2022
https://iccae.conferences.ekb.eg/

CONTRIBUTORS

Armed Forces

Military Technical College.
Armed Forces Units.

Egyptian Universities and Organizations

Cairo University.
Ain Shams University.
Helwan University.
Menoufia University.
Benha University.
Tanta University.
Alexandria University.
American University in Egypt.
Modern Academy for Engineering and Technology.
National Research Center for Sciences.
British University in Egypt.
Suez Canal University.
Al-Azhar University.
Damietta University.
Future University.
Farous University.
Kafr El-Shaikh University.
Arab Academy for Science Technology & Maritime.
Menoufia Higher Institute of Engineering.
Tenth of Ramadan Higher Institute of Engineering.
Housing and Building Research Center.

Arab and Foreign Universities and Organizations

Lund University, Sweden.
Concordia University, Canada.
Hadhramout University, Yemen.
Prince Sultan University, KSA.
University of Nyala, Sudan.

Sponsor Organizer Publisher Partner

IOP Conf. Series: Earth and Environmental Science 1056 (2022) 011002 doi:10.1088/1755-1315/1056/1/011002

Peer Review Statement

All papers published in this volume have been reviewed through processes administered by the Editors. Reviews were conducted by expert referees to the professional and scientific standards expected of a proceedings journal published by IOP Publishing.

- **Type of peer review:** Double Anonymous
- **Conference submission management system:** https://iccae.conferences.ekb.eg/
- **Number of submissions received:** 126
- **Number of submissions sent for review:** 105
- **Number of submissions accepted:** 47
- **Acceptance Rate (Submissions Accepted / Submissions Received × 100):** 37.3
- **Average number of reviews per paper:** 2
- **Total number of reviewers involved:** 73
- **Contact person for queries:**
  Name: Prof. Dr. Mohammad Fahmy
  Email: md.fahmy@mtc.edu.eg
  Affiliation: Conference Rapporteur

Prevention and Control of Airborne Infections; investigating the efficiency of hospital design using (AIC) evaluation tool

Marianne Nabil Guirguis[1, *], **Rania Rushdy Moussa**[1]

[1]Architectural Department, Faculty of Engineering, The British University in Egypt, El-Shorouk City, Cairo, Egypt

[*]Corresponding Author Email: marianne.nabil@bue.edu.eg

Abstract. Our world is resisting the new pandemic "severe acute respiratory syndrome Coronavirus 2" (SARS-CoV-2) causing the disease known as COVID-19. To date, more than two hundred and three million cases were confirmed out of who more than four million died. Sharing data that will help the community to intervene with measures that will decrease the spread of the virus and protect the population is an obligation. This will help the world cope with this pandemic. This research aims to highlight the different criteria that will determine that the building of a health facility is ready to control the infection of this virus and similar airborne viruses. The research developed an evaluation tool that can be used by hospital administration to assess the hospital building readiness to prevent and control airborne infection from the viewpoint of architecture if it is an existing one or alternatively it can assess the design in case of a new hospital building, determining required roles and responsibilities.

Keywords: Hospital design, Contamination control, Infection control checklist, Disease transmission prevention, Infection mitigation, Infection evaluation tool

Abbreviations: (AIC) airborne infection control; (AIIR) airborne infectious isolation room; (GUV) germicidal ultraviolet; (HAI) Hospital acquired infection; (ICU) intensive care unit; (IPC) Infection prevention control; (PPE) Personal protection equipment.

1. Introduction

Coronavirus outbreak in 2020 raised questions about the preparedness of hospitals for pandemics, especially airborne diseases. Infection control within a hospital is not simple and is a multidisciplinary task, on one hand, comprising architecture and electromechanical works, in addition to administrative and personal measures. On the other hand, it comprises the science of medicine and the respective procedures. Creating sustainable hospital that can stand in front of airborne diseases is the main scope of this research. Previous research tackled different ways to create sustainable building [1], but this research intends to create evaluation tool to measure the efficiency of medical facilities against airborne infections. All stakeholders should be playing harmoniously on site to eliminate HAI. The aim of this article is to highlight the main criteria and practical design solutions that could serve as interventions to limit disease spread within hospitals from the viewpoint of airborne infection control (AIC) principally by developing an evaluation tool for AIC in hospitals.

2. Materials and Methods

The research adopts a theoretical method, by collecting the data from different standards and references. Data has been categorized according to the nature of the different aspects and arranged to result an evaluation tool in the form of a checklist assessing the existence of each aspect while evaluating its condition.

IOP Conf. Series: Earth and Environmental Science 1056 (2022) 012001 doi:10.1088/1755-1315/1056/1/012001

Figure 1: Research Methodology (developed by author).

3. Disease transmission

Diseases transmit via direct or indirect contact. Transmission by direct contact may occur by person to person contact through different body fluids and secretions, or by droplet spread, principally through speaking, coughing or sneezing, when they reach the mouth and nasal mucosa, or conjunctiva of another person [2]. Indirect contact comprises more types amongst which are contaminated things, food and drink, animal to person contact, and airborne transmission [3] causing infection to the susceptible persons. WHO determined that transmission of Covid-19 is via different modes of transmission by direct contact, droplet, airborne, faecal-oral, blood borne, mother-to-child, and animal-to-human transmission complicating the task of (IPC) It comprises hygiene, proper use of PPE and environmental sanitation, and the evaluation of the readiness of the building itself. The last task is critical, as it has to mitigate transmission of all types. However, one of the most difficult disease transmission modes to be controlled is the indirect contact by airborne transmission [4] that occurs when the particulates of microbes from the infection source remain suspended in the air carrying these pathogens to other persons who become infected upon inhaling them if they are susceptible [5].

4. Design considerations in medical facilities addressing infection control

4.1. Functional zoning and space considerations

Each kind of space within a hospital has its own function generated from the occupants, durations, and activities done in it. Separate zoning is essential to allow different approaches to the reduction of aerosols of airborne infection. Transitions between different spaces should be well considered[6], and separation between the blocks of the building is a higher level of zoning, by introducing gaps between blocks of different function; hence, achieving maximum access control and separation between functions, consequently a higher level of infection control [7]. To facilitate correct functional zoning, modular design is highly recommended providing clarity and order. Hospital design integrates a multitude of infrastructure systems. Also, there is a need to consider future improvements to the different systems to accommodate technological advances and future extensions, thus, putting modular design as a viable solution leading to adaptive hospital architecture with better performance [5]. Also, functions of certain zones or spaces within zones are changed to accommodate arousing circumstances such as viral outbreaks or activities expansion [8], hence, the design should allow this kind of flexibility, such as changing certain zones to isolation rooms upon need[7]. This may be accomplished by dividing an open space into smaller spaces by installing partitions, making it easier to adapt a current hospital design in case of pandemics that are airborne without much work in the HVAC system. Airflow should be reviewed to check for the modifications that will result from the added partitions[9]. One of the most successful frameworks for zoning is the "three zones and two channels". The three zones (cleaning zone, semi-contaminated zone, and contaminated zone) are separated from one another in a strict way without any intersection between the three zones. Warning signage should be installed in a extremely noticeable places at the doorways and entrances of the zones; The three channels (the medical personnel channel, the patient channel, and the sample delivery channel) are also separated without any crossing circulation [7].

4.2. Circulation

Circulation within a hospital is a critical topic within mitigating infection. The most important aspect is a non-overlapping circulation with low intensity traffic[6]. The circulations that must not overlap are that of patients with hospital staff. Physical mobility between different spaces was the only means of communication previously; however, with technological advances in different installations for interactive communication and monitoring systems, traffic flow intensity is minimized eliminating infection by transmission. Different circulation routes start by a good study of the quantity of entrances' types and locations, and visibility of certain entrances in a hospital is important such as the main entrance, the entrance to the emergency department and the outpatient clinic[5]. Stairways present a potential of infections transmission via stack effect, hence, they are located far from the operation theatres and the ICU preferably[5]. Enclosed pressurized stairwells can be adopted.

4.3. Natural ventilation:

The basic elements that determine effect of natural ventilation are the ventilation rate, airflow direction, and pattern. The ventilation rate is directly affected by following factors: area of the window, climatic and weather conditions, and relative position of the opening[10]. There are standards that present the minimum accepted levels of hourly averaged natural ventilation rates, to mitigate airborne infection transmission [9]. At times of airborne pandemics, natural ventilation is a key solution, as neither infection methods nor efficient pharmaceutical treatments are scientifically determined conclusively, as they are still under scientific investigation and research. Hence, it is strongly recommended to spread awareness to the importance of substitution of re-circulated air in ventilation by natural air ventilation through existing openings, whenever possible [10]. If natural ventilation is insufficient, then mixed mode ventilation can be used. Exhaust fans present a possible complementary solution in this case to improve the ventilation rate used[3]. During SARS outbreak it was determined that natural ventilation reduced the intensity of airborne infection sources through the high ventilation rates with outdoor fresh air[5]. This comes in accordance to the guidelines set by "WHO" which recommends minimizing the need for air-conditioning, with windows that are designed effectively. Cross ventilation is also recommended, where possible[8]. Hence, design solutions of the square or U-shape corridor that depend totally on air conditioning are not the best solution as they withdraw the possibility of natural ventilation[5]. Although natural ventilation has a multitude of advantages yet, it should be understood that it is not the permanent substitute to mechanical ventilation. Natural ventilation should be available, but mechanical ventilation is necessary due to its reliability in the first place, in addition to ease of control of the indoor air temperature, humidity, and airflow path. Installation of filtration systems is an advantage of mechanical ventilation. Hence, when designing a hospital, the worst case scenario for natural ventilation is the absence of wind, which means that natural ventilation by itself cannot satisfy the requirements, to underpin that mixed-mode or alternatively mechanical ventilation are obligatory from the viewpoint of infection control and thermal comfort as well[9].

4.4. Finishing materials.

Sterile non-porous materials are recommended to minimize contamination and facilitate cleaning including harsh scrubbing and cleaning with strong disinfecting chemicals. For finishing of walls and ceilings, special paints should be applied that are washable in accordance to the standards. Also, high solidity plaster should be used[5]. Hence, porous paints that cannot withstand any attack from any organism should be avoided. As cleaning of disinfection materials of different chemical bases do floors, hence, floor-finishing materials should be resistant to acids, alkali and salts. Additionally, they must be easy to clean and designed with the minimum number of joints.

4.5. Partitions,

Metal panels in partitioning are a good solution in terms of flexibility, and sterility. However, these are not the only factors in material choice of partitions, but their strength should be considered. At pandemic outbreaks separate sinks can be added to be fixed on partitions outside the patients' room to enhance hands hygiene procedures and avoid direct transmission by knobs [5].

4.6. Openings,

Designing the used door system and accessories in a hospital is be done from the viewpoint of infection control. To achieve more sterility, the hands-free door operation is essential throughout a medical facility. In patient rooms, using self-closing doors with stainless steel or polymer kick plates and push plates will mitigate transmission of contaminated air and will minimize using hands. Details as the gap under a standard door should be considered as in AIIR, this gap is approximately 1cm[11]. For entrances as emergency egress, automatic sliding doors that can be swung open providing an adequate solution. Sliding doors without tracks with top-hung roller system provide a practical solution that has many advantages; in addition to easy access, these doors reduce bumping of stretchers, patient beds and different wheeled equipment thus decreases the joints in the floor. Material choice for the door should be sterile such as doors with polymer panels cladding that provide cleanable surfaces. There are many spaces in a health facility which tightness the room envelop is of due importance such as operation theatres, ICUs, AIIRs. Hence, joints and penetrations should be designed critically in order to maintain the airflow rate and pressure difference in these spaces [11].

4.7. Furniture.

The patient rooms use for quarantine must adequately ventilated, spacious single rooms with en suite facilities and that means the toilet facility and the hand hygiene. Sometimes single rooms are unavailable, hence, beds should space at least 1m in accordance to WHO. Also, the airflow path must not be blocked with furniture. The furniture arrangement should consider social distance.

4.8. Signage.

The whole arrangement of the hospital should be set is clarity with a multitude of measures to communicate certain message to the users. These messages will warn different types of users to go from "clean" zones to "dirty" ones. Using signage, colors of the ground and/or walls, and automatic counting sensors that give alert to administration if crowdedness occurs in areas as the waiting rooms.

5. Electromechanical considerations for infection control in hospitals

5.1. HVAC:

The ventilation systems are a means of infection transmission. Inadequate maintenance and low air change rates put the health facility at risk of outbreaks [12]. Hence, in hospital HVAC system design, ventilation of infected areas must be separated with specific filtration methods for AIC [13].

5.1.1. Supply air quality: The ventilation system should allow an option of disabling re-circulated air, where outdoor air dampers introduce fresh air instead. The conditions of ventilation in terms of the pressure relation to the other areas are set by "WHO"[8]. However, if the system used does not allow this option like the split air-conditioning units, in this case, they will have to be turned off, as if operated in a contaminated space, they will spread the airborne infection. But, this will need a way out to achieve thermal comfort without increasing contamination risk such as natural ventilation through the existing window openings [10].

5.1.2. Air filtration: It is another method to improve supply indoor air quality. Certain filters can be added to the existing ventilation system as an intervention such as: HEPA filters, where HEPA is "High Efficiency Particulate Air (or absorption)" [14] or electrostatic precipitator (ESP) filters, where appropriate with an advantage of negligible pressure drop as they do not resist the airflow [13].

5.1.3. Ventilation rate: The role of the ventilation system from the perspective of disease transmission control is to dilute the air from different pathogens. Previous research showed that room air should be changed 10 to 12 times per hour when there is an infected patient with an airborne infectious disease as opposed to a value of 4 to 8 times per hour for general wards [7].

5.1.4. Airflow distribution and pattern: The relative positions of supply and exhaust openings determines the airflow distribution. For instance in AIIR ceiling-level exhaust was found to be better in removing air impurities of different fine particles and aerosols than floor-level exhaust. furthermore, it was determined that due to the thermal plume of the patient, these air contaminations flow upwards more in an effortless way more than downwards [11]. In addition, the internal arrangements and installations within each space should be taken care of as the curtains that are drawn for privacy or the partitions, as they will hinder the air distribution pattern within the space. Moreover, no furniture must be located within the path of the airflow from the patient to the exhaust terminal [11].

5.1.5. Air pressure in different spaces: There are rooms that require positive, or negative, or neutral air pressure according to the function [15]. Some spaces will require maintaining negative air pressure such as one of the design scenarios for the AIIR. The direction of the airflow should take place from the clean area towards the infection source and velocity of air should be less than 0.25 m/s. A self-closing door is essential to maintain the pressure difference and a permanent pressure monitor is essential [15]. .However, this may attract secondary infections from adjacent spaces even though the design in case of quarantine isolation rooms comprises an anteroom [16]. Awareness to this by the hospital management affects decisions within the daily operations. Negatively pressurized anterooms leading to negatively pressurized isolation rooms need to be entered by healthcare workers while wearing PPE for respiration. Amongst such spaces that require negative air pressure are toilets [6].

5.2. Disinfection units and air purifiers

Using disinfection units in hospitals is another complementary solution in severe times of pandemics as there is a diversity of technologies as mobile disinfections units that utilize UV light or hydrogen peroxide vapor. Some of them use disinfectant fogging systems[17]. GUV was proved to be effective with many microorganisms amongst which is Coronavirus, by mitigating aerosol transmission or UVGI that removes or deactivates potential viral contamination from the re-circulated air[18]. In addition, separate air filtration devices with the designated filters can be introduced to the different spaces. For instance in areas where there are COVID-19, running air purifiers in the quarantine room may will mitigate infection risk for caregivers and healthcare workers[19]. In addition to antivirus sterilization and isolation equipment, that is important in isolation rooms to avoid cross infection through the medical staff. These units come in a host of designs to suit the context.

6. Human factors:

Human factors encompass three levels; the state, the administrative, and the personal levels. The first level comprises governmental strategies and policies dealing with the clear concise sovereign measures on the scale of the population, whereas the second level is the administrative regulations and bylaws in health facilities. It provides localized essential roles and responsibilities whose implementation necessitates spreading awareness, providing personnel training plans and clear procedures for the medical staff, caregivers, administrative staff, for patients and visitors. Finally, the last level is the personal measures comprising awareness and behavioral training for the public to protective measures including proper hygienic [4], and awareness of social distancing measures.

7. Results & Discussion: AIC (airborne infection control) evaluation tool

This research developed an evaluation tool to determine the deficiencies within any existing hospital design in controlling airborne infection such as COVID-19. AIC evaluation tool can be considered as an initial hospital design guideline checklist that new hospitals should consider its different aspects during the design and construction phases to prevent spreading the airborne infections such as Coronavirus. At the same time, when used by hospital management team, it stimulates increased preparedness for such pandemics. The evaluation is based on yes or no; 'N' represents absence of an aspect and 'Y' represents the existence of the aspect, knowing that the tool assesses the condition of the existing aspect. The aspects condition range between 1 to 3; 1 is poor condition till 3 good condition. The tool is divided into three sections; the first section is presented in **Error! Reference source not found.**evaluating the design aspects assessing the hospital architectural design toward decreasing the airborne infection transmission through five sub-aspects; zonings, modular design, design flexibility, hospital circulation and the airflow. In this section of the tool, the assessment is conducted on the whole hospital building.

Table 1. AIC evaluation tool section 1, Checklist for designing aspects in each space.

Aspects		Sub-aspects		Y 1	N 2	3	Action	By who
1.1	Zoning	1.1.1	Separate functions for zones					
		1.1.2	Clean air supply for each zone					
		1.1.3	Separation between building blocks					
1.2	Modular design	1.2.1	Possibility of future extension					
1.3	Design flexibility	1.3.1	Possibility of changing function					
		1.3.2	Possibility of adding partitions					
1.4	Circulation	1.4.1	Non overlapping circulation					
		1.4.2	Low traffic					
		1.4.3	Separation of entrances of different functions					
		1.4.4	Visibility of the main entrance to patients					
		1.4.5	Visibility the entrance to the outpatient clinic to patients					
		1.4.6	Visibility of emergency entrance to patients					
1.5	Airflow	1.5.1	Airflow direction in building from clean to dirty zones					
		1.5.2	Separate functional zones have separate HVAC systems					
Further comments:								

Table 2. AIC evaluation tool section 2, Checklist for interior design aspects of each individual space.

Zone Name		…………………		Room Number	…………………			
Aspects		Sub-aspects		Y 1	2	N 3	Action	By who
2.1	Wall/ partitions	2.1.1	Non-porous					
		2.1.2	Solid (can withstand harsh scrubbing & impact resistant)					
		2.1.3	Washable (can withstand cleaning with disinfecting chemicals)					
2.2	Availability of Signage	2.2.1	labelling for departments to show restricted access signs for certain users					
2.3	Floor	2.3.1	Resistant to acids, alkali and salts.					
		2.3.2	Minimum number of joints					
		2.3.3	General condition: it has cracks or tear					
2.4	Doors	2.4.1	Hinged door					
		2.4.2	Sterile , non-porous material					
		2.4.3	Washable withstanding disinfecting chemicals					
		2.4.4	Hands-free door operation					
		2.4.5	Self-closing doors					
		2.4.6	Stainless steel or polymer kick plates					
		2.4.7	Stainless steel or polymer push plates					
		2.4.8	Gap under a standard for airborne infectious isolation room (AIIR) is approximately 1 cm					
		2.4.9	Door tightness					
		2.4.10	Sliding door on floor					
		2.4.11	Sterile , non-porous material					
		2.4.12	Washable withstanding strong disinfecting chemicals					
		2.4.13	Sliding doors without tracks with top-hung roller system					
		2.4.14	Self-closing doors via IR					
2.5	Furniture	2.5.1	The minimum distance between beds is 1m					
Further comments								

The second section in AIC evaluation tool is presented in **Error! Reference source not found.**, evaluating interior design and finishing materials of different hospital zones to prevent the infection of airborne viruses. For this section of the AIC evaluation tool, the assessment is conducted on each distinct space.

The third section as presented in **Error! Reference source not found.**, evaluates for each distinct space the ventilation system and the air quality toward preventing airborne infection transmission.

Table 3: AIC evaluation tool section 3-Checklist for hospital ventilation & air quality aspects for spaces

Zone Name			....	Room Number						
Aspects			Sub-aspects	Y 1	2	3	N	Action	By who
3.1	Natural ventilation	3.1.1	Is there a possibility of natural ventilation?						
		3.1.2	Is there cross ventilation?						
3.2	Mixed mode ventilation	3.2	Are there Exhaust fans?						
3.3	HVAC	3.3.1	Is there a AC Split system?						
		3.3.2	Does Air Changes per Hour meet standards?						
		3.3.3	Is the system efficient in delivering the outdoor air to each location in the room? (to judge, check the internal furniture, curtains for privacy, partitioning , & location of supply and exhaust)						
		3.3.4	Disabling re-circulated air option available						
3.4	Air filtration	3.4.1	Re-circulated air is mandatory in the system						
		3.4.2	Are there filters for fine particulates						
		3.4.3	Are there ceiling-level exhaust?						
		3.4.4	Are there above floor level exhaust?						
3.5	Room air pressure	3.5.1	Positive pressure						
		3.5.2	Negative pressure						
		3.5.3	Neutral pressure						
		3.5.4	Does it meet the standards?						
3.6	Disinfection units & air filtration	3.6.1	Is there a GUV?						
		3.6.2	Is there a UVGI?						
		3.6.3	Are there Disinfection units that use H_2O_2?						
Further comments									

8. Conclusion:

Due to COVID -19 pandemic outbreaks, the airborne infection spread in hospitals with a significantly negative effect on the medical team and other patients due to the inappropriate hospital design. The research develops an evaluation tool in the form of a checklist, to assess the efficiency of existing hospitals toward preventing the spreading of airborne infection, thus helping the medical institution to assess its efficiency toward facing the pandemic and other airborne viruses' attacks in the future. This is achieved by identifying the weakness points that need to be enhanced and developed to improve the efficiency of the design by reduction of spreading the infection within the medical facility.

References

[1] R.R. Moussa, Zero Energy Educational Building: A Case Study of The Energy and Environmental Engineering Building in The British University in Egypt, in: 2021 IEEE 48th Photovolt. Spec. Conf. (PVSC), 2021: pp. 1913–1915. doi:https://doi.org/10.1109/PVSC43889.2021.9518995.

[2] B. Blocken, T. van Druenen, T. van Hooff, P.A. Verstappen, T. Marchal, L.C. Marr, Can indoor sports centers be allowed to re-open during the COVID-19 pandemic based on a certificate of equivalence?, Build. Environ. 180 (2020) 107022. doi:10.1016/j.buildenv.2020.107022.

[3] WHO Regional Office for the Eastern Mediterranean, Hospital readiness checklist for COVID-19 Interim document-Version 1, 2020.

[4] I. Eames, J.W. Tang, Y. Li, P. Wilson, Airborne transmission of disease in hospitals, J. R. Soc. Interface. 6

(2009) S697–S702. doi:10.1098/rsif.2009.0407.focus.

[5] T. van Khai, Adaptive Architecture and the Prevention of Infections in Hospitals, Trans. VŠB – Tech. Univ. Ostrava, Civ. Eng. Ser. 16 (2016) 165–172. doi:10.1515/tvsb-2016-0028.

[6] C.H. Cheong, B. Park, S. Lee, Design method to prevent airborne infection in an emergency department, J. Asian Archit. Build. Eng. 17 (2018) 581–588. doi:10.3130/jaabe.17.581.

[7] Q. Zhang, S. Cheng, Q. Cheng, Experience summary of a COVID-19 designated community hospital and its operation model, Panminerva Med. (2020). doi:10.23736/S0031-0808.20.03908-7.

[8] WHO, District hospitals: guidelines for development, 2nd ed., World Health Organization Regional Publications. Western Pacific Series, 1996. apps.who.int › iris › bitstream › handle › 9290611170_eng.

[9] J. Atkinson, Y. Chartier, C. Lúcia Pessoa-Silva, P. Jensen, Y. Li, H. Seto, Natural Ventilation for Infection Control in Health-Care Settings, WHO Publication, 2009.

[10] L. Morawska, J.W. Tang, W. Bahnfleth, P.M. Bluyssen, A. Boerstra, G. Buonanno, J. Cao, S. Dancer, A. Floto, F. Franchimon, C. Haworth, J. Hogeling, C. Isaxon, J.L. Jimenez, J. Kurnitski, Y. Li, M. Loomans, G. Marks, L.C. Marr, L. Mazzarella, A.K. Melikov, S. Miller, D.K. Milton, W. Nazaroff, P. V. Nielsen, C. Noakes, J. Peccia, X. Querol, C. Sekhar, O. Seppänen, S. ichi Tanabe, R. Tellier, K.W. Tham, P. Wargocki, A. Wierzbicka, M. Yao, How can airborne transmission of COVID-19 indoors be minimised?, Environ. Int. 142 (2020). doi:10.1016/j.envint.2020.105832.

[11] M. Hyttinen, A. Rautio, P. Pasanen, T. Reponen, G.S. Earnest, A. Streifel, P. Kalliokoski, Airborne infection isolation rooms - A review of experimental studies, Indoor Built Environ. 20 (2011) 584–594. doi:10.1177/1420326X11409452.

[12] G. Correia, L. Rodrigues, M. Gameiro da Silva, T. Gonçalves, Airborne route and bad use of ventilation systems as non-negligible factors in SARS-CoV-2 transmission, Med. Hypotheses. 141 (2020). doi:10.1016/j.mehy.2020.109781.

[13] M. Holt, Health care facilities, EC M Electr. Constr. Maint. 108 (2009). doi:10.1007/978-3-642-01999-9_17.

[14] J.L. Perry, J.H. Agui, R. Vijayakumar, G.C. Marshall, Submicron and Nanoparticulate Matter Removal by HEPA-Rated Media Filters and Packed Beds of Granular Materials, (2016). https://ntrs.nasa.gov/search.jsp?R=20170005166.

[15] © Tahpi, Isolation Rooms-Guideline Section-International Health Facility Guidelines, Int. Heal. Facil. Guidel. (2017) Part D 19-28. http://healthfacilityguidelines.com/Guidelines/ViewPDF/iHFG/iHFG_part_d_isolation_rooms.

[16] E.S. Mousavi, K.R. Grosskopf, Directional airflow and ventilation in hospitals: A case study of secondary airborne infection, Energy Procedia. 78 (2015) 1201–1206. doi:10.1016/j.egypro.2015.11.184.

[17] Automated disinfection systems | HFM, (2017). https://www.hfmmagazine.com/articles/1516-automated-disinfection-systems (accessed July 3, 2020).

[18] C.M. Walker, K. Gwangpyo, Effect of Ultraviolet Germicidal Irradiation on Viral Aerosols, Environ. Sci. Technol. 41 (2007) 5460–5465. doi:10.1021/es070056u.

[19] P. Santanachote, What to Know About Air Purifiers and Coronavirus - Consumer Reports, Consum. Reports, Inc. (2020). https://www.consumerreports.org/air-purifiers/what-to-know-about-air-purifiers-and-coronavirus/.

Achieving sustainability through reducing risks during architecture design process: a lean management perspective

Sherouk M Abdelrehim[1,*], Ashraf A Nessim[2], Ayman A E Othman[3]

[1] Teaching assistant, Architecture Engineering department, Faculty of Engineering, the British University in Egypt, AinShams University, Cairo, Egypt

[2] Architectural Engineering Department, Faculty of Engineering, Ain Shams University

[3] Head of Architectural Engineering Department, Faculty of Engineering, the British University in Egypt, Cairo, Egypt

*Corresponding Author Email: sherouk.mohamed@bue.edu.eg

Abstract. The construction industry, as one of the largest industries in the world, plays a critical role in achieving national and international social and economic sustainable development goals. Socially, it is concerned with delivering long-term projects that turn client and end-user requirements into designs that specify technical aspects, functional specific requirements, and standards of quality. Economically, the construction industry contributes to a country's gross domestic product (GDP) by creating jobs and providing the country's infrastructure. the construction industry's nature is more vulnerable than other industries. This is believed to be due to the complex and dynamic nature of construction projects, as well as the involvement of a large number of participants and organisations with varying goals and priorities. The architectural design process is a crucial stage in the life cycle of a construction project. This is due to the fact that decisions made during this phase have an impact on the project throughout its life cycle. Failure to consider risk factors during this phase will have an impact on the project's sustainability. Although Lean Management has been successfully applied in many sectors of the construction industry, its application in the design process has received far less attention than in construction literature. Accordingly, this research aims to investigate the role of lean management towards achieving sustainability through reducing risks during the architectural design process.

Keywords: Sustainability, Architecture Design, Lean management.

1. Introduction

The construction industry is one of the biggest industries worldwide which has a great influence on society and the economy. The contribution of the construction industry in the development of countries and economic growth is widely acknowledged [1]. The construction industry takes part in achieving social development goals for nations [2]. In addition, construction projects socially cope with the continuous growth of population and the way they live, work, play and relate to one another as a part of the society [3]. Economically, the construction industry has a great influence on the gross domestic product (GDP) as it acts as a factor of growth for other sectors of industry [2]. On the other hand, the construction industry is criticized for its unsustainable practices in respect of the amount of waste generated, pollution, resources, and energy consumption. The construction industry is one of the most complex industries which involve many parties with different objectives, in addition to its dynamic nature which subject it to several risks that can affect the value of the project. The life cycle of a construction project is widely affected by the decisions made during the architecture design phase.

IOP Conf. Series: Earth and Environmental Science 1056 (2022) 012002 doi:10.1088/1755-1315/1056/1/012002

Therefore, it is important to identify and consider the risks facing the construction process in the early stages of the project [4]. Lean project management is an approach developed to enhance the performance of industries. Thus, applying lean management principals throughout the architecture design phase will enhance the performance of the project [5]. This research aims to investigate the role of lean management towards achieving sustainability through reducing risks during the architecture design phase through the following objectives:

- Identifying the risks that affect achieving sustainability during the design process.
- Presenting and analyzing previous case studies where lean management was used.
- Highlighting recommendations for the usage of lean management throughout the design process to reduce risks in the architecture design firms.

2. Research methodology

This research methodology consists of; Data collection and, data analysis of previous case studies. Data collection is through the literature review and examples of previous case studies. Data analysis in this research is qualitative. Qualitative analysis is based on the description and classification of collected data followed by categorizing and linking relationships in the literature review and case studies. Finally, highlighting recommendations for lean management usage through the design phase to reduce risks in architecture design firms.

3. Architecture Design Phase

The Architecture design phase is considered as a key phase which plays an important role in decision making, thus delivering sustainable projects. The design phase is such a critical phase that involves several risks impacting the project life cycle [6]. Although the design phase was given a little importance regarding to the factors affecting final construction products, design phase has an important role in influencing project duration and costs. The successful delivery of a project requires coordination between various process parties consisting of architects and engineers of different specialties [7].

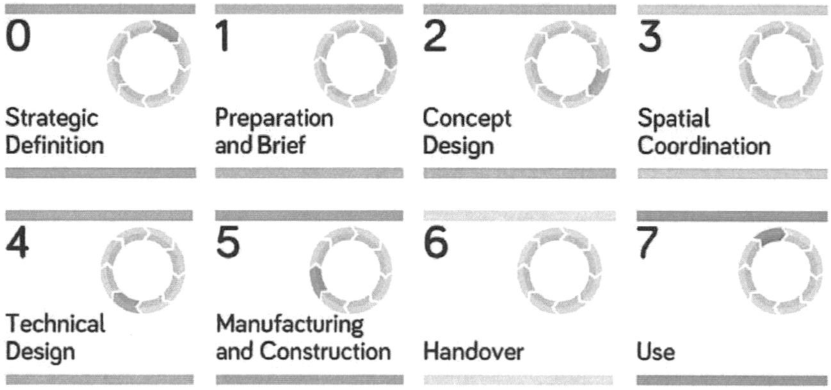

Figure 1. RIBA plan of work stages [8].

The royal institute of British architects presented the design phases in figure 1,and they emphasized the importance of the early stages of the project life cycle where most decisions are made [4].

4. Risk in architecture Design Phase

The Architecture design phase is considered as a key phase that plays an important role that involves several risks impacting the project life cycle [6]. According to Othman & Abdelwahab 2017, there are 18 key risk factors that affect the sustainable delivery of a construction project [9]. These risk factors are related to the concept, developed and technical design phases and involves several parties: architect, client, stakeholders, contractor, project manager.

5. Sustainability

The term "sustainability" has been gaining recognition lately from all countries around the world. It has been one of the biggest challenges of the 21st century to achieve sustainablily with its pillars ans shown in figure 2 [10]. The term sustainability is defined generally as meeting today's need without affecting the needs of the future [11]. However, practicing sustainable construction faces several barriers which

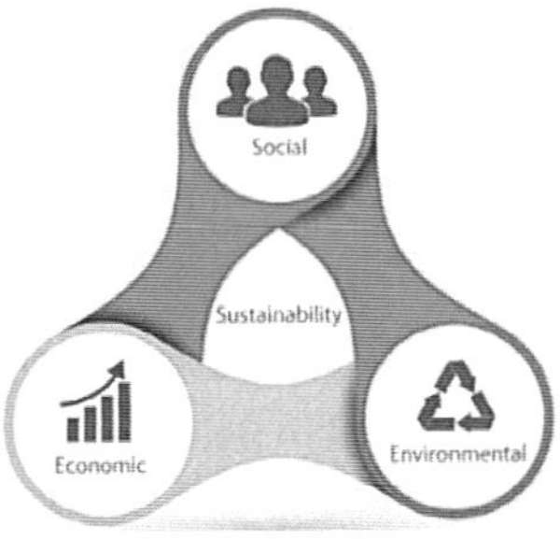

Figure 2. Sustainability pillars.

risk the sustainable delivery of construction projects such as financial, social and technical [12].Lean management aims to improve the project value which will help achieving sustainability [13].

6. Lean management

Lean management is a system that aims at improving the value of a product by satisfying the customer's needs and improving production. By applying lean principles to design process, it will enhance the project in order to reduce risks [13]. The goal of lean is to maximize value and minimize waste; its principles can be utilized to reduce risk factors during the architecture design process [14].

7. Case studies

Examples presented in this paper were chosen from developed countries that used lean thinking to provide an applicable example for lean thinking tools usage to enhance projects delivery. They provide an applicable example for lean thinking tools usage. The case studies presented are limited for focusing on the construction phase rather than design phase where most decisions are made. Thus, applying lean thinking through design phase will achieve sustainable delivery of the project.

7.1. Case study 1

This case example illustrates the concept of using Lean Management techniques in selected construction processes, with the goal of demonstrating how Lean Management may affect the overall cost of a selected process of managing storage area on a building site. The case study Example evaluates expenses and potential savings while incorporating lean methodology into storage area management [15].The office building "Baltic" located in Poznan (Poland) shown in figure 3. The structure has an unusual form that refers to local modernism. It is designed to be about 15 000 m2 [15].Authors chose to study storage area management in depth, taking into consideration cost and savings analyses.

Figure *3.* Baltic Building Poland [15].

7.1.1. *Lean Tools and techniques used*

Lean thinking began with analysis of the existing state of the storage area. The main problem is lack of storage space is divided into 3 main reasons which are materials and machines, site organization, and people and work organization. Some ideas related to lean Management were discussed to enhance the process of storing materials. The building site's management accepted three proposals, which are proposed to enhance the storage area organization in addition to the expected effects and impact on the cost of the project in table . Just in time (JIT) was used to reduction of time in which the materials lie on the construction site. The 5S Method used for sorting, categorizing of materials in terms of usefulness and frequency of use, removal of unnecessary elements, and Visual communication: storing each material at designed place, and training of employees.

7.1.2. *Results of case example 1*

When lean management was introduced, 11 347.50 PLN was saved, and the amount of savings (avoided expenses) for the general contractor is constantly growing. The following comparison of predicted and measured impacts and influence is shown in Table 1.

Table 1. comparison of predicted and measured impacts of lean management [15].

Improvement methods	Expected effects	Expected impact	Measured impact
Just in time (JIT)	-more available space	-high	-high
5S Method	-money savings	-medium	-high
	-quicker access for materials	-high	-high
	-less InSite transportation	-high	-medium

There are activities and processes that are also influenced by Lean Management, however estimating the amount of effect and cost savings using a systematic approach is difficult.

7.2. *Case study 2*

The Norwegian government property developer – Statsbygg – is implementing Lean Construction in the project under consideration by effectively implementing in the tender competition that the prime contractor and designers use Lean Construction principles and a handful of selected methods for delivery system. The research presented in this case is based on the construction phase of a rehab project of a university building, including semi-structured interviews of eight specialists as the client, designer group, and prime contractor, as well as a manuscript study of project documents. [16]. The case is a renovation project for a 115-year-old university building. The structure is part of the Norwegian University of Life Sciences and is located 30 kilometres south of Oslo, Norway shown in figure 4.

Figure 4. the Norwegian University of Life Sciences.

The structure consists of three floors with a basement and attic. The overall gross area is about 8190 m2. The rehab project's goal was to retain the distinctive structure while also adapting it to meet modern general education requirements. The new cost estimate for the project was about 57 million USD. The TFV-theory of production was proposed, which supplements the three production views: transformation, flow, and value (TFV). In pursuit of the TFV-theory, lean project delivery (LPD) systems are introduced.

Traditional project delivery focuses largely on converting resources to products, while ignoring or overlooking flow and value [17]. The goal of Statsbygg's system was to achieve the TFV objectives – a Lean Construction delivery system as shown in table 2.

Table 2. Tools used to initiate lean project delivery [16].

Lean tools used	Phase
Lean six sigma	Announcement of tender
BIM (building information modelling)	Construction
Pull planning, last planner system	Construction
Takt time planning	Construction

The requirement for project managers and foremen to be qualified in Lean Six Sigma. The building was 3D-scanned to create a more realistic BIM model as shown in figure 5. The designers were able to generate better requirements.

Figure 5. BIM Model of the Norwegian University of Life Sciences.

7.3. Results of case example

Pull planning and production control based on the plan hierarchy of the Last Planner System did not perform as expected. At the outset of the construction phase, the pull planning was carried out. The lack of prior experience with the lookahead procedure was identified by the majority of the informants as the primary reason for project control failure. Because the necessary preparation before to meetings was not made. The Takt Time Planning, as well as the pull planning, were ineffective. It was difficult to carry out due to a lack of prior experience. A consultant was hired to establish the Takt Time plans to compensate for the lack of experience. Despite this, after a few months, the Takt Time Planning was abandoned. The bulk of the informants claimed that the Takt zones and sequences were not appropriate for the structure [16].

8. Conclusion

To sum up, lean principals and tools help achieving maximum value and minimum waste. Construction projects are subject to various factors throughout their lifespan, creating uncertainties in the process. Specific actions should be taken to prevent errors from happening at the source in order to eliminate uncertainties. In table 3. a comparative analysis is presented between case example 1 and 2 showing the lean principles and tools used in each example.

Table 3. Comparative analysis of case 1 and 2 (by author).

Lean principals	Case 1	Case 2	Lean tool used
Identify Value		√	Takt time planning
Map value stream	√		5S
Create flow		√	BIM
Establish Pull	√	√	Pull planning, last planner system, and just in time technique
Seek perfection	√	√	Kaizen (six sigma)

Multiple lean tools and techniques were used in the discussed cases. Lean thinking has proven its efficiency in construction risk reduction, therefore the usage of lean thinking in an early stage of design decisions can help reduce design risks.In table 4,there are 18 risk factors presented along with lean principles, tools and techniques that can help reduce risk.

Table 4. lean tools which help reduce design risk (Developed by the author).

Design risk factors	Lean Principles	Lean tools and techniques	References
Design cost overrun.	Create flow, Map the value stream	Value stream mapping ,Just in time,5 S',Takt time planning	[18] [19] [20] [21]
Brief changes by client	Identify value, Customer focus, Establish Pull	Voice of customer,Just in time	[22] [20]
Design variations by architect	Seek perfection	Kaizen, 5 why technique	[23] [18]
Design delay	Establish pull	Last planner system	[20] [22]
Lack of coordination between design parties	Create flow, Culture and people	5 S', Takt time planning	[19] [18] [21]
Incomplete environmental analysis	Identify value, Customer focus	Value stream mapping	[24]
Tight project design schedule	Establish pull	Last planner system	[20] [22]
Design errors and omission	Strive for perfection	Kaizen, Standardization	[23] [18]
Noncompliance to building codes and standards.	Culture and people, Continuous improvement and built-in quality	Standardization	[18] [25]
Skills gap of qualified architects and organizational requirements	Culture and people	Kaizen	[25]
Stakeholders' late changes in the project	Map the value stream	Just in time	[20]
Uncoordinated and incorrect construction documents	Culture and people Create flow	Value stream mapping	[25]
Using outdated construction materials and technology	Continuous improvement and built-in quality	Kaizen	[23]
Lack of consideration of environmental studies	Identify value, Strive fo perfection	Kaizen, 5 S', Takt time planning [21].	[23]
Failure to carry out work in compliance to the contract.	Identify value Customer focus Establish Pull	Voice of customer Just in time	[25] [20] [22]
Lack of considering project life cycle cost.	Create flow Map the value stream Waste elimination	Voice of customer, Valu stream mapping , Just in tim 5 S',Takt time planning	[25] [21]
Lack of communication and coordination between the government and design firms	Culture and people	Kaizen Operation improvement	[25] [23]
Public objections	Identify value Culture and people, improvement	Kaizen, 5 S', Takt time planning [21]	[23]

9. Recommendations

Based on the studied risks in design and lean in addition to the 2 case studies; Recommendations are made to address management, stakeholders and designers and customers to help reduce risks in design.

9.1. Management

- Management training for better flexibility and acceptance of new management changes in addition to enhancing skills and introducing lean benefits towards the projects
- Ensuring suitable firm structure
- Collaboration in taking decisions between top management and other projects parties
- Increase motivation to enhance project value
- Ensure the existence of needed skills while recruiting workers
- Hiring an expert in lean management
- Avoiding several managerial changes that cause interruption in the project

9.2. Stakeholders and customers

- Ensure preparations before meetings
- Ensure activities time and sequence is appropriate,
- Involvement in design decisions
- Realistic estimation of cost and time
- Increase customer awareness about lean benefits and costs

9.3. Designers

- Designers training to enable lean thinking during decision taking
- Increase accuracy of required drawingsand encourage sharing of risks during the design phase
- Communicate with customer to enhance understanding of requirements
- Identifying the level of details required to avoid waste

10. References

[1] Ofori G 2015 "Nature of the Construction Industry, Its Needs and Its Development:A Review of Four Decades of Research " *Journal of Construction in Developing Countries*, pp. 115 - 135.
[2] Oyedele O A 2016 "Impacts of construction industry on socio-economic development of Nigeria .Available:https://www.researchgate.net/publication/308749719_Impacts_of_construction_industry_on_socio-economic_development_of_Nigeria.
[3] Xiahou X, Tang Y, Yuan J, Chang T, Liu P and Li Q 2018 "Evaluating Social Performance of Construction Projects: An Empirical Study" Available: https://www.mdpi.com/2071-1050/10/7/2329?type=check_update&version=1#cite.
[4] Abdellatif M and Othman A 2006 " Improving sustainablity of low income housing projects:The case of residential buildings in Musaffah commercial city,Abudhabi " *Emirates Journal for Engineering Research*, pp. 47-58.
[5] Mazlum S K 2015 Lean design management – An evaluation of waste items for architectural design process *The graduate school of natural and applied sciences of middle east technical university*.
[6] Raftery J 1994 Risk Analysis in Project management, london: routledge.
[7] Aquino J and melhado S B 2002 "*the importance of the design for production in the design process management in building construction*,". [Online]. Available: https://www.researchgate.net/publication/229040939_The_importance_of_the_design_for_production_in_the_design_process_management_in_building_construction.
[8] Royal institute of British Architects, 2020. [Online]. Available: https://www.architecture.com/-/media/GatherContent/Test-resources-page/Additional Documents/2020RIBAPlanofWorktemplatepdf.pdf.
[9] Othman Aand Abdelwahab N 2017 "Achieving sustainability through integrating risk management into the architectural design process" *Journal of Engineering, Design and Technology* pp. 25-43.

[10] Ben-Eli M 2005"*Sustainability The Five Core Principles*,". [Online]. Available: https://www.bfi.org/sites/default/files/attachments/pages/Sustainability-FiveCorePrinciples-BenEli.pdf.
[11] Afzal F, Lim B and Prasad D 2017 "*An investigation of corporate approaches to sustainability in the construction industry*," Kensington,Australia, .
[12] Ametepey O, Aigbavboa C and Ansah K 2015 "Barriers to successful implementation of sustainable construction in the Ghanaian construction industry," Kumasi, Ghana.
[13] Lehman T and Reiser P 2000 "Maximizing value & minimizing waste: value engineering & lean construction". [Online]. Available: http://www.leanconstruction.org.
[14] Saieg P, Sotelino E, Nascimento E and Caiado R 2018 "Interactions of Building Information Modeling, Lean and Sustainability on the Architectural, Engineering and Construction industry: A systematic review" *Journal of Cleaner Production* pp. 788-806.
[15] Nowotarski P, Pasáawski J and Matyja J 2016 "Improving Construction Processes Using Lean Management Methodologies – Cost Case Study" *Procedia Engineering* p. 1037 – 1042.
[16] Haarr K J and Drevland F 2016 "Amandated lean construction delivery system in a rehab project - A case study" in *24th Ann. Conf. of the Int'l. Group for Lean Construction* Boston MA, USA, Boston, MA, USA.
[17] Koskela L J 2000, *An Exploration Towards a Production Theory and its Application to Construction*, Doctoral Thesis, Helsinki University of Technology, Espoo.
[18] Saieg P, Sotelino E, Nascimento D and Caiado R 2018 "Interactions of Building Information Modeling, Lean and Sustainability on the Architectural, Engineering and Construction industry: A systematic review" *Journal of Cleaner Production* pp. 788-806.
[19] Salem O M 2006 "Lean Construction: From Theory to Implementation" *Journal of Management in Engineering*.
[20] Gamal M 2013 *Improving Project performance using lean construction in egypt: A proposed framework* cairo: The American University in Cairo.
[21] Frandson A, Berghede K and Tommelein I D 2013 "Takt time planning for construction of exterior cladding" in *21st Annual Conference of the International Group for Lean Construction 2013* Fortaleza, Brazil.
[22] Swefie M G 2013 *Improving project performance using lean construction in egypt: A proposed framework* cairo: American university in Egypt.
[23] Hargrave M 2021 "Kaizen". Available: https://www.investopedia.com/terms/k/kaizen.asp.
[24] Womack J, Jones D and Roos D 1990 *The machine that changed the world* New york.
[25] Diekmann J , Balonick J, Krewedl M and Troendle L 2016 "Measuring lean conformance". Available: https://www.researchgate.net/publication/264999385_measuring_lean_conformance

Risks of Innovation in the Architectural Design Process in Egypt: An Investigative Study

R Medhat[1,*], A A E Othman[2], and F O Alamoudy[1]

[1] Architectural Engineering Department, Faculty of Engineering, the British University in Egypt, Cairo, Egypt
[2] Head of Architectural EngineeringDepartment, Faculty of Engineering, the British University in Egypt, Cairo, Egypt

*Corresponding Author Email: Rawan174397@bue.edu.eg

Abstract. Achieving sustainability in construction projects is becoming an increasing priority worldwide. Being one of the crucial phases of the construction process, the design process plays a major role in achieving the goals of sustainability. Adopting innovative design solutions is a promising way of enhancing the sustainable performance of buildings. However, the implementation of innovative solutions is risky and needs to be managed carefully to avoid failure. Due to the increasing number of construction projects being developed in Egypt and because of the need to achieve the Sustainable Development Strategy (SDS) 2030, adopting innovative solutions became an essence. Accordingly, this research aims to investigate the risks associated with innovative solutions during the design process. In order to achieve this aim, a research methodology consisted of literature review and case studies was designed to achieve three objectives. First, to build a comprehensive background about the research topic including the concepts of sustainability, innovation and innovation risks, and the design process. Second, depict the relationship between innovation risks, the architectural innovation aspects, architectural design process and sustainability aspects. Final, present and analyze three case studies to explore the innovation management strategies adopted to overcome the risks associated with the adopted innovation solution.

Keywords: Innovation, Risks, Sustainability, Architectural Design Process, Egypt.

1. Introduction

Egypt, as a developing country, faces social, economic, and environmental challenges, and meeting the goals of the country's Sustainable Development Strategy (SDS) by 2030 is critical. The successful implementation of the strategy is essential for addressing crucial issues such as poverty, inequality, and resource depletion [1&2]. In the design and construction industries, there is an increasing need and calls for achieving sustainability. This is duo to the fact that buildings are among the largest consumers of energy and resources. Buildings are responsible for 39% of global carbon emissions, 50% of global material use, and 73% of global electricity consumption [3]. Therefore, there is a need to adopt innovative approaches that enable the design team and construction professionals to achieve the client requirements in a sustainable manner. Egypt is one of many countries that developed building sustainability rating systems, frameworks, codes, and regulations to move towards a sustainable built [1]. However, many challenges block the way such as the complexity and multidisciplinary nature of the concept of sustainability on one hand. On the other hand, the fragmented nature of the design process and ongoing technological advances make it hard to achieve the expected targets [4]. Subsequently, the traditional design approach is no longer sufficient to deal with these challenges. Innovation can be an effective solution to overcome the barriers that arise with developing sustainable built environment. Developing new sustainable building technologies, introducing innovative green and smart materials, applying creative waste reduction processes, and creating novel delivery systems are examples showing how innovation is a crucial ingredient to move towards a sustainable built environment [5]. However, Innovative solutions and techniques, especially in the construction industry, are associated with risks

IOP Conf. Series: Earth and Environmental Science 1056 (2022) 012003 doi:10.1088/1755-1315/1056/1/012003

that stand against their implementation. Addressing these risks and applying innovative techniques at an early stage of the project life cycle increases the probability of success towards achieving sustainability objectives in construction [6]. Therefore, this research aims to investigate the risks associated with innovative solutions and build a relationship between these risks, architectural innovation aspects, architectural design process and sustainability.

2. Research Objectives and Methodology

To accomplish these aims, a research methodology consists of literature review and case studies was designed to achieve three objectives.

- Building comprehensive background through reviewing the concepts of sustainability, innovation and innovation risks, and the design process.
- Depicting the relationship between innovation risks, the architectural innovation aspects, architectural design process and sustainability aspects.
- Presenting and analyzing three case studies to explore the innovation management strategies adopted to overcome the risks associated with the adopted innovation solution.

3. Literature Review

3.1 Sustainability

Sustainability is the development that meets the needs of the present with-out compromising the ability of future generations to meet their own needs. It is the ability to develop and progress with the condition of preserving and utilize natural resource efficiently. The main three pillar of sustainability are environmental, social, and economic pillars. The environmental aspect focuses on protecting the natural resources and physical environment. The social aspect is concerned with providing healthy and safe lives to achieve human prosperity. The economic aspect deals with achieving economic growth without affecting the other two factors [7,8].

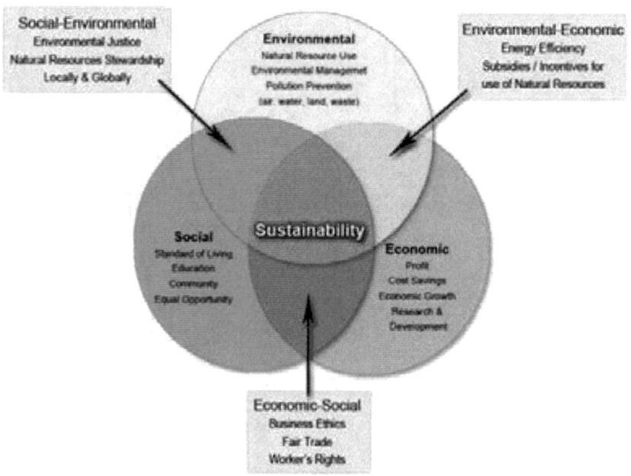

Figure 1. Aspects of Sustainability [9].

The interaction between social and environmental aspects generated a new aspect, which revolves around the right of all individuals to have a fair share of the natural resources of the environment. The economic–social aspect focuses on delivering economic sustainability without compromising society needs through promoting business ethics, ensuring fair trade and preserving workers' rights. Finally, the environmental-economic aspect focuses on achieving environmental objectives of sustainability in an economic way through reducing unnecessary costs, efficient use of energy and natural resources, and offering subsidies and incentives for encouraging research centres and construction organisations to develop creative solutions to achieve economical sustainable environment [4&9], see figure (1).

3.2 Innovation

Innovation is a complex and debatable term that has various definitions. Jacobs and Snijders [10] defined innovation as invention that adds value. The Department of Trade and Industry of UK [11] defined it as the successful implementation of new ideas. A more sophisticated and business-oriented definition by Pillar [12] who defined innovation as the creation, introduction and successful diffusion of products, service, system, processes or even business models, which are new from perspective of the particular organization and/or user. Drucker [13] defined innovation as a tool that embrace and utilize change to improve business. In conclusion, the definitions have two characteristics of innovation in common. First, innovation has to offer a degree of novelty and second, it must add value.

3.2.1 The 4Ps framework of Innovation

For an organization to embrace innovation, it must explore the four main spaces of innovation: person, process, product, and press. The first space focuses on the human factor in achieving innovation and how encouraging new ideas, willingness to take risks, resiliency to uncertainties is essential to achieve innovation [14]. The second space is concerned with the process itself and how to reach desired output most efficiently. The third space is the product and how to create innovative output that is characterized by novelty, elegancy, genesis, and relevancy [6,14,15]. The fourth space is the press, or the paradigm and it is concerned with organizational culture as it focuses on enhancing the working environment and embraces the management support to achieve internal organizational innovation [14,15].

3.2.2 Degrees of Innovation

Throughout the history of innovations, market demand and advance in technology are the main contributors to great innovative ideas. These two factors also define the diverse types of innovation: incremental, disruptive, architectural, and radical. Incremental innovation is the most common type in the organization as it focuses on how existing processes, products and services can be better and more efficient [5,6,15]. Disruptive innovation depends on utilizing modern technologies or solutions to an existing market or problem. It is called disruptive because most often this innovation gets mature with time and overshadows the traditional exiting methods.

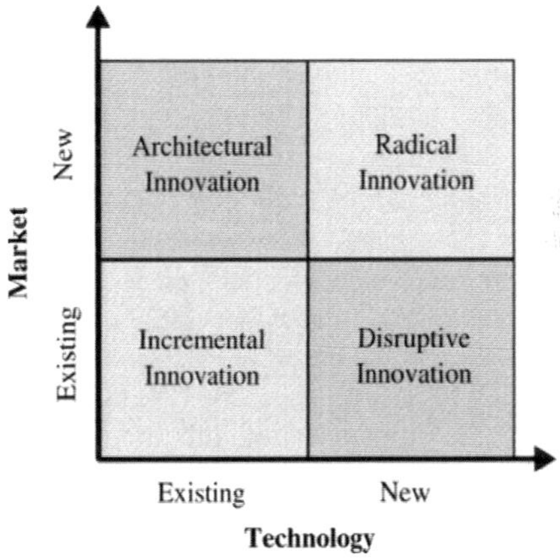

Figure 2. Degrees of Innovation [16].

Architectural innovation is the reconfiguration of existing components in a new and novel way [15]. Finally, radical innovation is the hardest and rarest but has the greatest impact. It utilizes modern technologies to create new market opportunities [5,15], see figure (2).

3.2.3 Innovation in architecture

Contrary to popular belief, the concept of innovation in architecture is not new. It dates back to the Egyptians and Romans, who mastered the use of available resources to create spectacular structures. With the beginning of the industrial revolution, the introduction of new materials such as steel and reinforced concrete, as well as developing inventions such as elevators, aided in shaping radical evolution in architecture [17]. Because of social, economic, and environmental reasons, innovation has gotten a lot of attention over the last two centuries. Buildings are no longer used as social shelters to protect humans. They have a significant impact on their lifestyle, health, and well-being. Economically, rising demand for construction projects causes buildings to cost more today than ever before.

Environmentally, the deterioration of resources around the world necessitates their rationalization [2]. As a result, there is a growing recognition that innovation can address these new challenges. There are various areas of architectural innovation. It could be in the project design, the design thinking process, or the way the project was delivered. The successful implementation of innovation in one area necessitates an integrated approach in which the other areas are addressed to gain the greatest benefit for the firm. Akamija [18] classifies the main factors that influence the incorporation of innovation in architecture as shown in figure 3 and table 1.

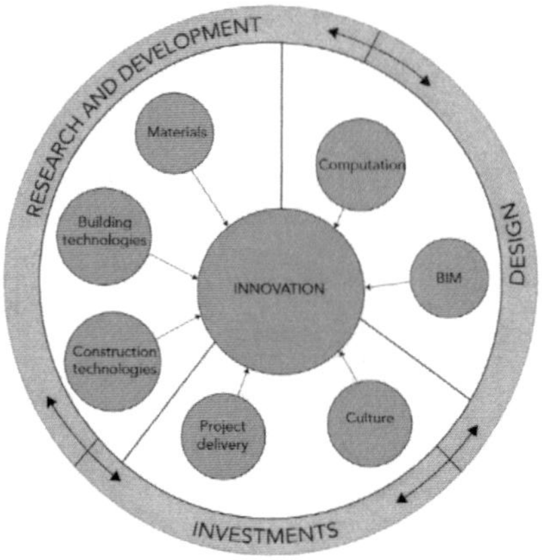

Figure 3. Architectural Innovation Aspects [18].

Table 1. Architectural Innovation Aspects [18].

Advanced material	These are materials with enhanced properties or that are smart and responsive to external stimuli. Materials innovation can be incremental, such as improvements in thermal performance or structure properties of concrete and metals, or radical, such as self-healing materials and nanomaterials.
Building technologies	Building technologies encompass advances in building services such as the façade system, HVAC system, firefighting system, and lighting system. It entails automating all of these systems with sensors and controls to ensure efficiency.
Construction technologies	It is the assembly of building components through advanced techniques such as prefabrication, robotics, and automation.
Project delivery	It is the process that determines how the project is going to be executed. Innovative project delivery such as integrated project delivery (IPD) allows maximum collaboration and quality.
Culture	Innovation can be implemented in the culture of firms and organizations through their strategies, values, the ways of engaging clients, and the way of how they distinguish themselves form other competitors.
Computation	It is the integration of computation in the design process such as using building information modeling (BIM), structural analysis, and environmental analysis.

3.3 Risks of innovation

Innovation is essential ingredient for achieving sustainability [19]. However, it adds additional risks and uncertainties to projects which increase the possibility of their failure [20]. Literature review has identified risks of innovation. Salter & Gann [21] identified 11 barriers of innovation in the design process. On top of the list is lack of time, cost, and lack of interest of project stakeholders. In addition, internal risks related to firms such as the limited understanding of the benefits of innovation, the uncertainty of results and outcome, and the lack of experience and knowledge act as a barrier [22]. Moreover, the lack of interdisciplinary teams and poor communication are contributors to failure of innovation [23]. Table 2 identified risks to innovation in architectural design process. In conclusion, risks of innovation can be categorized into technical risks, financial risks, institutional risks, and schedule risks.

Table 2. Risks of innovation in the design process.

NO.	Risks	Author(s)
R1	Inadequate available design time or failure to deliver on time	[21,24]
R2	Unanticipated Cost of innovation	[21,22,24]
R3	Inadequate top management commitment	[25]
R4	Lack of interest among suppliers within the supply chain	[21,23]
R5	Lack of interest or demand among clients	[21,22]
R6	Too many projects ongoing concurrently	[21]
R7	Lack of organisational skills and expertise	[21, 22]
R8	Difficulty to access information	[21]
R9	Long administrative procedures in organisations	
R10	Lack of co-ordination among different organisational engineering disciplines	
R11	Inability to get feedback from end-users	
R12	Manufacturing technologies or development issues	[22,23]
R13	Uncertain economic return	[22]
R14	Government Bureaucracy	
R15	Lack of belief in value of innovation	[26]
R16	Limited ambition of project participants	
R17	Lack of long-term performance data	[18]
R18	Failing to meet technical criteria	[24]
R19	Lack of client acceptance and marketing risks	[23]
R20	Competitors risks	
R21	Poor organization and project management skills	
R22	Lack of collaboration & communication between multidisciplinary project teams	[27]
R23	Company image and reputation damage	
R24	Lack of access to fund	[28]
R25	Limited organizational resources	
R26	Lack of maintenance strategy	
R27	Organisational culture and resistance to change	[18]
R28	Lack of long-term strategies	
R29	Incomplete codes and standards of innovation	[29]
R30	Lack of quality assessment tools	

3.4 The architectural design process

The design process is a research-based, problem solving and iterative process. It plays a major role towards delivering sustainable projects that meet the client needs and end-user requirements within time, cost and quality standards. Moreover, the decisions made during this process have impacts on the project's performance throughout its life cycle. There are essential factors that affect the design decision making process such as project site, programme, type of client, a few to name. These factors make each project unique [30,31]. According to RIBA work plan, the design phase is divided into 5 stages [32].

- **Strategic Definition:** This is a pre-design phase at which all requirements of the client are defined and fulfilled through initiating construction project. As a result, a business case is developed including a preliminary review of project budget and project risks.
- **Preparatory and briefing:** at this stage, the requirements of the client are translated into a project brief in which the project programme, project outcomes, spatial requirements and feasibility studies are included.
- **Concept design:** The architectural concept is developed in this stage and approved by the client. The concept must be adjusted until it aligns with the project brief and cost plan. Feedback from the project stakeholder is essential at this stage to avoid design changes during further development.

- **Spatial coordination:** This stage introduces different engineering disciplines to the design process. Structure engineer, HVAC engineer, electrical engineer, and specialist work together with the architect to integrate architectural and engineering information.
- **Technical design:** The last phase of the design process that prepares the design for execution. Subcontractors and specialists are integrated in the design process to develop architectural and engineering technical drawings.

Table (3) shows the relationship between risks, architectural innovation aspects, architectural design process and sustainability.

Table 3. Relationship between risks, architectural innovation aspects, architectural design process and sustainability.

Risk	Architectural Innovation Aspects							Architectural design process					Sustainability		
	Culture	Project Delivery	Building Technology	Construction Technologies	Materials	Computation	BIM	Strategic definition	Preparation and briefing	Concept design	Spatial coordination	Technical design	Environmental	Social	Economical
R1		X		X		X	X				X				X
R2	X	X	X	X	X	X	X				X				X
R3	X	X				X	X	X					X	X	X
R4		X	X	X	X			X	X						X
R5	X	X	X	X	X	X		X					X	X	X
R6		X									X				
R7	X			X		X	X				X		X		
R8	X			X		X	X		X				X		X
R9	X														X
R10	X	X		X			X				X	X	X		X
R11		X							X	X			X	X	X
R12			X		X						X	X	X		X
R13	X		X		X			X	X	X					X
R14			X					X	X						
R15	X	X		X	X	X				X	X	X		X	
R16	X													X	
R17								X	X					X	
R18			X		X	X				X	X	X	X	X	X
R19			X			X					X	X	X		X
R20									X	X					
R21	X			X											
R22				X		X					X	X	X		X
R23	X			X		X				X	X	X	X	X	X
R24	X			X		X				X	X	X	X		X
R25	X					X									X
R26			X		X						X	X	X		X
R27			X	X						X	X	X	X	X	X
R28			X									X	X	X	X
R29	X	X	X	X	X					X					X
R30	X	X											X		

4. Case Studies

4.1 Definition and Selection Criteria

A case study is a research approach used to describe and analyse a particular topic, phenomena, event, or project to discover factors, structures, forms, and orders of interaction between the players in the scenario, or to assess work performance or development progress [33]. Three case studies are presented to explore the role of innovation strategies adopted towards overcoming the risk that hinder the implementation of innovation in the architectural design process. The case studies selection criteria were based on the nature of the selected projects, data availability, date of construction, degree of success, and location. All of the case studies featured construction projects that were faced different degree success in overcoming risks of innovation. Because this topic received scant attention in construction literature, few case studies were published, the availability of data was a major factor in selecting the case studies. Finally, the case studies were chosen from France and United Arab Emirates and the learned lessons of the case studies will aid in overcoming the risks of innovation in Egypt. Table (4) explains the research idea through a practical application to the cases studies to illustrate the concept of the idea. These case studies were analysed and compared through investigating (1) the architectural innovation aspect used, (2) the degree of achieving the desired sustainable goals, (3) the degree of innovation, (4) the risks of innovation, (5) learned lessons (recommended innovation management strategies to implement).

4.2 Case Study 1: Arab World Institution, France

Arab World Institution is a well-known building with its unique adaptive façade system. The conceptual design was done by the French architect Jean Novel in 1981 after winning the design competition of the building. What makes the design special is not only the dynamic façade that responds automatically to the external environment allowing a suitable amount of light to the interior of the building, but also, the social aspect behind the design. The façade design is inspired by the traditional Arabian mashrabia. However, only after three years of the opening of the building, serious functional problems started to emerge in the innovative façade system, and within six years the system stopped completely. It is stated that the main reason for the failure of such innovation is related directly to the design process [34, 35].

4.3 Case Study 2: Louvre Museum, UAE

The Louvre Museum in Abu Dhabi is a masterpiece created through the collaboration of the UAE and the French governments in 2007. The main innovative feature of the building is the dome which is considered the largest in the world. Its function is to create a microclimate beneath it and to stand against the burning sun of Abu Dhabi. As a result, the heat gain is ceased to 72%. The museum achieved a Three Pearl Design Rating Certificate, and it is targeting the silver LEED status. Subsequently, this case study managed successfully to overcome risks of innovation and to achieve sustainable values through implementing innovation management strategies [36,37,38].

4.4 Case Study 3: Al-Bahr Towers, UAE

The Al Bahr Towers are twin towers in the financial centre of Abu Dhabi. The towers are listed among the "Innovative 20" tall buildings, in addition to winning the 2012 Council for Tall Buildings & Urban Habitat's Innovation Award. The project deserves these awards as it introduced an innovative adaptive façade system that controls the amount of light and heat entering the building [40]. However, it is stated that this innovation is still far away from adding sustainable values to the project. In conclusion, this case study clearly shows the difficulties and the risks associated with innovations and concludes different innovation management strategies [39] (see table 4).

Table 4. Case Studies Analysis.

	Case Study 1	Case study 2	Case Study 3
Name	Arab world institute	Louvre Museum	Al Bahr Towers
	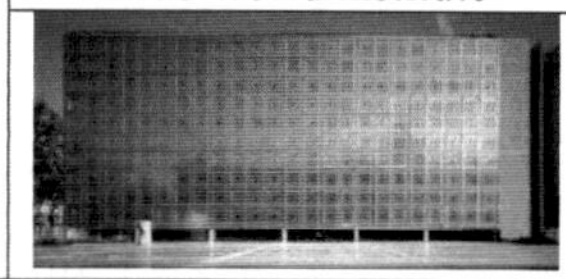		
Location	Paris, France	Abu Dhabi, UAE	Abu Dhabi, UAE
The architectural innovation aspect used	Building technology, Adaptive façade system	Construction technology, Dome structure system	Building technology, Adaptive façade system
Degree of achieving the desired sustainable goals	Failed	Succeeded	Failed
Degree of innovation	Modular	Radical	Modular
Risks of innovation	• High maintenance cost • Fail to meet the technical criteria • Low economic return	• Technical issues • Tight project delivery schedule • Complex manufacturing technologies • Lack of information	• High initial cost • Complex manufacturing technologies • Lack of quality assessment tool • Lack of soft-landing process and post occupancy evaluation
Learned lessons (Recommended innovation management strategies to implement)	This case study failed to overcome the risks of innovation and the following strategies are recommended: • Developing an early-stage, performance-based design approach • Integrating whole life cycle assessments in the design process [34,35]	This case study succeeded to overcome the risks of innovation due to implementing the following strategies: • Developing simulations and computational tests • Creating Large scale mockups • Mitigating risks through partnerships • Using a Design-Build project delivery process [36,37,38]	This case study succeeded to overcome the risks of innovation but failed to achieve its sustainable goal and the following strategies are recommended: • Applying design assessment tests • Developing performance criteria • Using BIM to enhance the team collaboration • Engaging the construction expertise [39,40]

5. Conclusion and Recommendations

Because the adoption of innovative design solutions is a promising way of enhancing the sustainable performance of buildings, innovation in different architectural aspects is gaining increasing attention from the design and construction community. However, risks associated with these aspects, such as higher cost and technological complexity, indicate that risk management strategies need to be developed for encouraging the adoption of innovation in buildings development. During this research, the literature review identified 30 risks that affect the implementation of innovative solution during the design process. In addition, a relationship matrix was developed to relate these risks with the architectural innovation aspects, the design process, and their impact on achieving sustainability objectives. Moreover, 3 case studies were presented and analyzed to explore the innovation management strategies adopted to overcome the risks associated with the adopted innovation solution. Based on that, the research recommends developing an innovation management framework to overcome the risks that hinder the implementation of innovative approaches in the design process.

IOP Conf. Series: Earth and Environmental Science 1056 (2022) 012003 doi:10.1088/1755-1315/1056/1/012003

References

[1] El-Haggar S, Samaha A. Roadmap for Global Sustainability- Rise of the Green Communities [Internet]. Springer; 2019. Available from: http://www.springer.com/series/15883

[2] Chan APC, Darko A, Ameyaw EE. Strategies for promoting green building technologies adoption in the construction industry-An international study. Sustainability (Switzerland). 2017;9(6).

[3] Mavi RK, Gengatharen D, Mavi NK, Hughes R, Campbell A, Yates R. Sustainability in construction projects: A systematic literature review. Vol. 13, Sustainability (Switzerland). MDPI; 2021. p. 1–24.

[4] Othman AAE, Ghaly MA, Zainul Abidin N. Lean Principles: An Innovative Approach for Achieving Sustainability in the Egyptian Construction Industry. Organization, technology and management in construction: An international journal. 2014;6(1).

[5] Nigra M, Dimitrijevic B. Is radical innovation in architecture crucial to sustainability? Lessons from three Scottish contemporary buildings. Architectural Engineering and Design Management. 2018 Jul 4;14(4):272–91.

[6] Tidd J, Bessant J. Managing Innovation: Integrating Technological, Market and Organizational Change. 2018.

[7] Brundtland Commission. 1987.

[8] Aminuddin, A. S. A., & Nawawi, M. K. M. (2013). Investigation of the Philosophy Practiced in Green and Lean Manufacturing Management. *International Journal of Customer Relationship Marketing and Management*, *4*(1), 1–12. https://doi.org/10.4018/jcrmm.2013010101

[9] Rodriguez, S., Roman, M., Sturhahn, S., and Terry, E. (2002). Sustainability Assessment and Reporting for the University of Michigan's Ann Arbor Campus. Master's Thesi. University of Michigan: Ann Arbor: 1-396.

[10] Jacobs D, Snijders H. Innovation routine: how managers can support repeated innovation. Assen. 2008

[11] DTI. Departmental Report. 2005.

[12] Pillar F. Customer-centric innovation. edx; 2021.

[13] Drucker P. Innovation and Entrepreneurship. UK: Pan Business Management. 1985

[14] Cropley, D. H. (2016). Creativity and Culture in Engineering. In *The Palgrave Handbook of Creativity and Culture Research* (pp. 549–571). Palgrave Macmillan UK. https://doi.org/10.1057/978-1-137-46344-9_27

[15] Smith, D. (2015). *Exploring Innovation* (3rd ed.). MCGrowHill.

[16] Rothaermel, F. (2014). *Strategic Management*. McGraw-hill Education.

[17] Ching F, Jarzombek M, Prakash V. A Global History of Architecture. Willey; 2011.

[18] Akšamija A. (2016). Integrating innovation in architecture: design, methods and technology for progressive practice and research.

[19] Seebode D, Jeanrenaud S, Bessant J. Managing innovation for sustainability. 2012.

[20] Bowers J, Khorakian A. Integrating risk management in the innovation project. European Journal of Innovation Management. 2014;17(1):25–40.

[21] Salter A, Gann D. Sources of ideas for innovation in engineering design. Vol. 32, Research Policy. 2003.

[22] Cox G. A Reading Through Cox Review of Creativity in Business: Building on the UK's Strengths. 2005.

[23] Keizer JA, Halman JIM, Keizer JA, Halman JIM. Risks in major innovation projects, a multiple case study within a world's leading company in the fast moving consumer goods. Vol. 48, Int. J. Technology Management. 2009.

[24] Badi SM, Pryke S. Assessing the impact of risk allocation on sustainable energy innovation (SEI): The case of private finance initiative (PFI) school projects. International Journal of Managing Projects in Business. 2016 Apr 4;9(2):259–81.

[25] Chien KF, Wu ZH, Huang SC. Identifying and assessing critical risk factors for BIM projects: Empirical study. Automation in Construction. 2014;45:1–15.

[26] Acklin C. Design-Driven Innovation Process Model. 2010.

[27] Klitsie JB, Price RA, Stefanie C, de Lille H. Overcoming the Valley of Death: A Design Innovation Perspective. 2019.

[28] Yang RJ, Zou PXW. Stakeholder-associated risks and their interactions in complex green building projects: A social network model. Building and Environment. 2014 Mar;73:208–22.

[29] Rahman MM. Barriers of Implementing Modern Methods of Construction. Journal of Management in Engineering. 2014 Jan;30(1):69–77.

[30] Pressman A. Designing Architecture. Routledge; 2012.
[31] Bielefeld B, el Khouli S. Design Ideas. 2013.
[32] RIBA. RIBA Plan of Work [Internet]. 2020. Available from: www.ribaplanofwork.com
[33] Sturman, A. (1997), "Case study methods", in Keeves, J. P (Ed.), Educational Research, Methodology and Measurement: An International Handbook, 2nd ed., Oxford: Pergamon, pp. 61-66.
[34] Luísa Freitas Martinho H, Paula Filipe Tomé António Paulo Teles de Menezes Correia Leitão Manuel de Arriaga Brito Correia Guedes A. Adaptive Façades An Integrated Algorithmic Approach Examination Committee. 2019.
[35] Meagher M. Designing for change: The poetic potential of responsive architecture. Frontiers of Architectural Research. 2015 Jun 1;4(2):159–65.
[36] Hesselgren L, Kilian A, Samar M, Olsson K, Williams C, Sorkine-Hornung Olga. Advances in Architectural Geometry. Chalmers University of Technology; 2018.
[37] Koren BS, Glynn R, Sheil B. Fabricate 2011: Making Digital Architecture [Internet]. Available from: https://www.jstor.org/stable/j.ctt1tp3c6d.43
[38] Ceccato C. Advances in Architectural Geometry 2010. 2010.
[39] Attia S. Evaluation of adaptive facades: The case study of Al Bahr Towers in the UAE. QScience Connect. 2018 Feb 6;2017(2).
[40] Xuereb K. The Al Bahr Towers-Enhancing Sustainability Through Innovation. 2014.

Introducing the Architecture Parametric Design Procedure: From Concept to Execution

Ahmed A. Khamis[1,*], Sherif A. Ibrahim[1], Mahmoud A. Khateb[1], Hesham Abdel-Fatah[2], Mohamed A. Barakat[1]

[1] Architecture Engineering Department, Military Technical College, Cairo, Egypt
[2] Mechanical Design Department, Military Technical College, Cairo, Egypt

[*]Corresponding Author Email: ahmedashraf94@mtc.edu.eg

Abstract The parametric design process is one of the modern architectural trends. It assists in the decision-making process during the early conceptual design stage. It introduces several design alternatives from which the designers can choose to achieve the most suitable design concept. Through this process, the BIM (Building Information Modelling) is considered as the stage of converting the preferred design concept to execution drawings and building information. Furthermore, the computerized digital fabrication stages will assist in building the desired model in different scales for the final decision to be taken. This paper spotlight the overall parametric design process workflow from the “design stage” (using *Grasshopper* which is a plugin in rhino parametric design software) passing through the *Revit* software that is responsible for the working drawings execution to the “manufacturing stage” using *Simplify 3D* software for the preparation of the mock-up digital fabrication model using 3D printing.

Keywords: Parametric design, BIM, Digital Fabrication, Grasshopper, Simplify 3D.

1. Introduction

1.1. Parametric Design

The word "Parametric" is a derivative from the word "parameter,", which is taken from the Greek para, which means "beside" [1]. In architecture, design is a solution to a problem; yet, the nature of the problem is sometimes unclear, therefore the design process also requires learning about the problem. Architectural design, according to Lawson [2], is "a process of finding and solving problems." As a result, numerous alternative solutions may exist, and design, or finding a solution, becomes a selection process. For Simon [3], this should be a sensible decision, yet aesthetic decisions in design may appear irrational to others. "Exploration is exploring what variables might be appropriate" is how Gero defines the design process. as one that “involves exploration, exploring what variables might be appropriate”.

Design is considered as the process of creating a problem description, then generating and searching among options for a solution that meets the problem's requirements. Any quantifiable factor that defines a system or sets its limits is referred to as a "parameter." The term "parametric design" refers to a procedure, in which variables are used to produce a problem description. Further, By altering these factors, a variety of potential solutions can be generated, and then a final solution can be chosen based on a set of criteria, [4].

Jabi [5] describes the parametric design as "a design process based on algorithmic thinking that defines and connects the needed parameter, in which the parameters together reinforce and clarify the connection between architecture design goals and how the design will reply and try to solve problems" in his book "Parametric Design for Architecture.". The notion of parametric design allows the designer to establish relationships between elements or groups of elements, as well as assign values or expressions to organize and govern those definitions, [6]. Dunn went on to say that the process would continue until the intended outcomes were determined based on appropriate "performative" and aesthetic criteria.

Programming or creating code in a certain programming language can be used to construct a parametric model. Various computer-aided design (CAD) applications, on the other hand, maybe modified to allow parametric design. By utilizing their application programming interface (API), they can do a variety of tasks. Recently, CAD systems with parametric functionality have become available, in which the user can manage using a graphical user interface (GUI). This type of software is known as parametric software, and it usually includes the ability to use a scripting three language to further customize the parametric features. Users can use parametric software to create links and associations between geometric objects and items that define variables or functions. Therefore, parametric design is also known as associative design. The development of a digital model based on a set of pre-programmed rules or algorithms known as 'parameters' is known as parametric modeling (or parametric design). That is, rather than being manually edited, the model or components of it are generated automatically by internal logic arguments. Hence, parametric modeling has set the stage for the expression of elements as a set of relations that have variable dimensions [7-9]. To present fully ordered controlled architectural forms, parametric design is defined as a design that is a rule-based producing variant to retain the dependencies and relations between them.

1.2. Building Information Modelling

Building Information Modelling (BIM) has specific international standards that include sharing physical and functional features in a digital representation for a built object that helps in forming a reliable basis for making decisions during stages of execution of buildings" [10]. BIM was created from the fabricated mock-ups [11] which are widely spread in the industry of petrochemical, automotive, or shipbuilding [12]. BIM shows the life cycle of a real building virtually [13, 14]. BIM deals with parametric objects that represent the building components [11, 15]. The designed building objects have geometric or non-geometric features with functional or topologic information [12, 16]. Building information modelling (BIM) has entered a lot of parametric modeling into majority building design. Nowadays these two designing processes, BIM and parametric software have gained credit while sometimes inducing certain forms of confusion.

1.3. Digital Fabrication

The history of fabrication technology began with a mass generation that was presented and created by Henry Ford, founder of Passage Engine Company in early the 19th century [17]. This strategy worked with creating the reproduced show. In this manner, the relationship between amounts, time, and anticipate the quality precisely, and operational costs are cheaper. In Design, as in Lisa Iwamoto's Advanced Creations Structural and Fabric Methods that CAD innovation supplant drawing, advanced creation brings unused understanding and extend the boundaries of engineering shape and development [18], In addition, Seely [19] clarifies that advanced creation impacts the method of engineering plan since of its basic part supporting the making of engineering models. Correspondingly, Dunn [20] clarifies the reasons why understudies and the designer calling make models. Representation of creative thoughts is crucial in design-based disciplines and is exceptionally important in architecture where we frequently cannot see the result, to be specific building that built, until the conclusion of the plan handle. The starting concept was created through a handle that permitted the creator to examine, re-examine, and advance refine thoughts in detail that expanded to the point in such a way that the plan of the extend prepared to construct. Models can be exceptionally flexible objects in this preparation, permitting architects to specific thoughts creatively.

Digital fabrication has filled the crevice between conception and generation. Making a coordinate advanced connect through "file-to-factory" forms of computer numerically controlled (CNC) manufacture [21]. Within the early 2000s, prototyping gear, such as laser cutters and 3D printers, significantly dropped in cost, and Open-Source equipment advance popularized these advances. Advanced manufacturing innovation got to be superior and more available, and the mental exercises empowered by modern innovation got to be more esteemed and critical. Mechanical originators may create prototypes in days rather than months [22]. Furthermore, there may be a significant definition of computerized creation, a preparation that starts with an advanced plan and closes with yield from a manufacturing machine. Computerized manufacturing is an instructive and movement worldview, in which multidisciplinary information consolidates 2D plans, 3D plans, utilize of devices and machines. Advanced creation comes from CAD (Computer-aided plan) at that point exchanged to CAM (Computer-aided fabricating) computer program. The yield from CAM is prepared to fabricate to a particular machine, like a 3D printer or CNC processing machine.

In the late '90s, the first digital fabrication labs are founded in architectural schools, and they were an outcome of the relationship between the architecture and mechanical engineering laboratories [23]. Rapid prototyping and CNC routers were the primary machines for fabricating models in engineering. Within the following years, techniques and methodologies have fast extended. Primarily there are three steps for advanced manufacture work [24], which are:

- Digital design phase. This first step is generating a virtual model using CAD software. The 3d model was exported as triangulated mesh or surfaced.
- The Preparing phase. Step for setting the manufacturing parameters and specific to the fabrication, resulting in a CAM file then send to the machine.
- Fabricate phase. Manufacture parts for fabrication analogy based on the CAM data, with little or no human assistance or interaction. Assembling the fabricated parts might need to achieve their final properties and look before they are ready to use.

Furthermore, the fabrication phase focuses on:1) externalizing ideas and concepts into mock-ups, prototypes, and products; 2) Select the technology needed and digital materials to transform ideas into prototypes and products; 3) Work creatively and try to explore to construct with different fabrication technologies; and 4) Visualize and represent artefacts in various ways while continuously adjusting [25].

Based on the previous introduction and background, this research aims to introduce the parametric design workflow; starting from the architectural design concept to the execution process. This research starts with an introduction illustrating a brief tracking of parametric design evolution, as well as, fabrication processes. Moreover, the methodology of this research will be introduced, followed by workflow description. Finally, ending with the conclusion.

2. Methodology

The workflow of a parametric design starts with using grasshopper to define the early-stage parametric design concept, which gives a lot of design solutions' alternatives. This process can be done through many steps that include:

- Pools of solutions
- Forms Exploration
- Evaluation
- Refinement
- Optimization

3. Case study application

A case study model was designed as a revolved tower comprising a series of constant angles, whereas the plans are increased in a constant ratio whilst revolving. The plans act as shading through the extrusion of plans as an environmental solution for the tower. The scripting is done on canvas of the grasshopper by the creation of a digital model based on a series of pre-programmed rules or algorithms known as 'parameters' as shown in Fig.1.

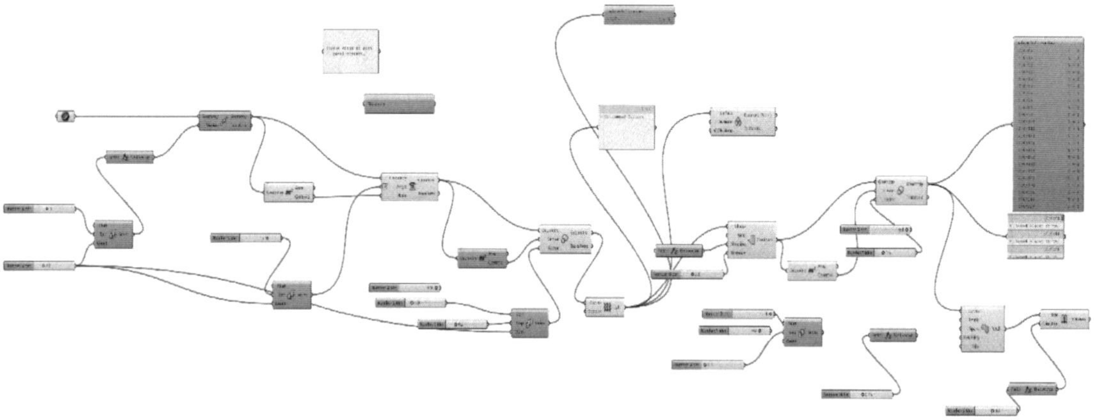

Figure 1. The Scripting of a mathematical equation and algorithmic relation in Grasshopper.

The scripting is being interfaced on the Rhinoceros to be extracted the model on it. The scripting is about algorithmic calculations that have to be done in the correct sequence way to produce the final model. The model that was produced is a twisted tower in Dubai. It was made through mathematical calculations in Grasshopper. Then the final model is interfaced on Rhinoceros to be baked.

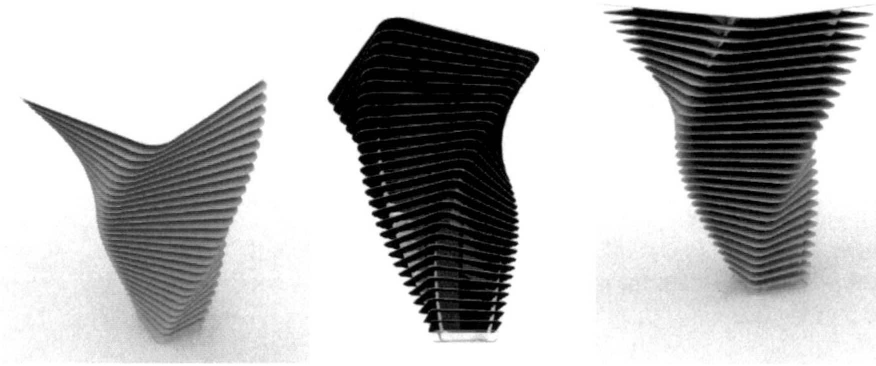

Figure 2. The Scripting of a mathematical equation and algorithmic relation in Grasshopper.

Secondly, the model was exported to DWG file to be imported in Revit (building information modeling program), then each plan is drawn for each floor level according to its concept shape slice in this floor. The walls, floors, windows, and other building information modeling data are entered to produce the final execution workshop drawings.

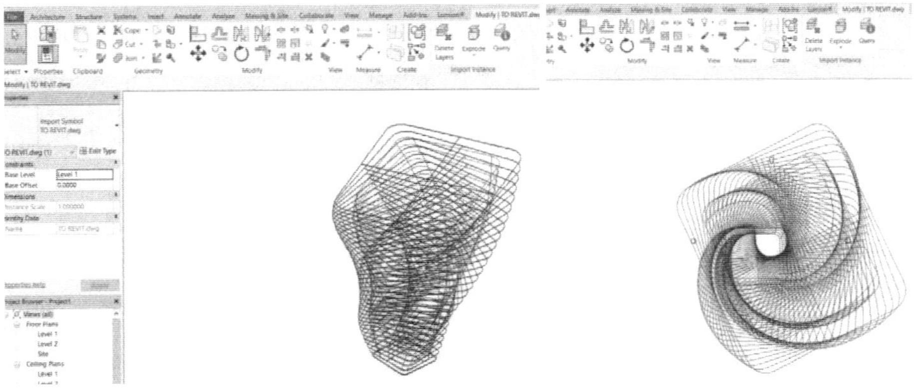

Figure 3. The exported file in Revit BIM to achieve the execution drawings.

Thirdly on finishing the final model concept on grasshopper, the file is being imported to simplify 3D which is a program for preparation for the digital fabrication on 3D printing. The program was used to calculate the amount of plastic that will be used in designing the mock-up phase. It is used also to calculate the time used, cost and give a video simulation for the 3D printer hand working on this mockup. After several fabrications of porotypes of different scales, the Final model is being fabricated with scale 1:1 in which the elevation patterns and cladding are being fabricated.

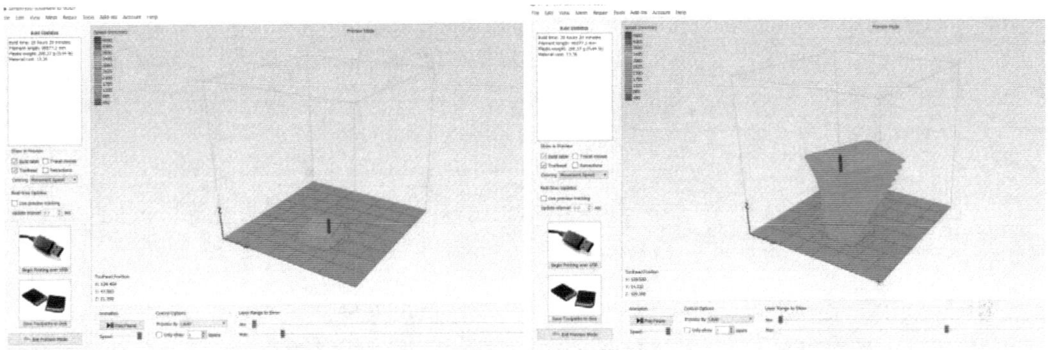

Figure 4. The simulation for the preparation for the fabrication process through simplify 3D and measuring the quantity of material needed and the time for the fabrication of the model.

After the fabrication of the model mock-up, the building is ready to be executed with scale 1:1 on the site after finishing the working drawings and simulation for the civil drawings so the previous stage is considered as a preparation for the fabrication process.

4. Conclusion

This paper is considered as an example for the application on the workflow of the parametric design concept in which the design is formed by Grasshopper then the conceptual form is exported to Revit to achieve the execution of working drawings for the plans where the architectural, structural, and MEP drawings are designed, In parallel, the design is also exported to STL or ObJ format for the preparation of fabrication process and fabricate mock-ups in different scales for plan slicing for laser cutting through Slicer Fusion program or by 3D printing by simplifying 3D. The model is then formed to be ready for real achievement.

Figure 5. The fabricated mock-up which was fabricated by a 3D printing machine.

This flowchart is showing the process for the workflow of the parametric design from concept to fabrication:

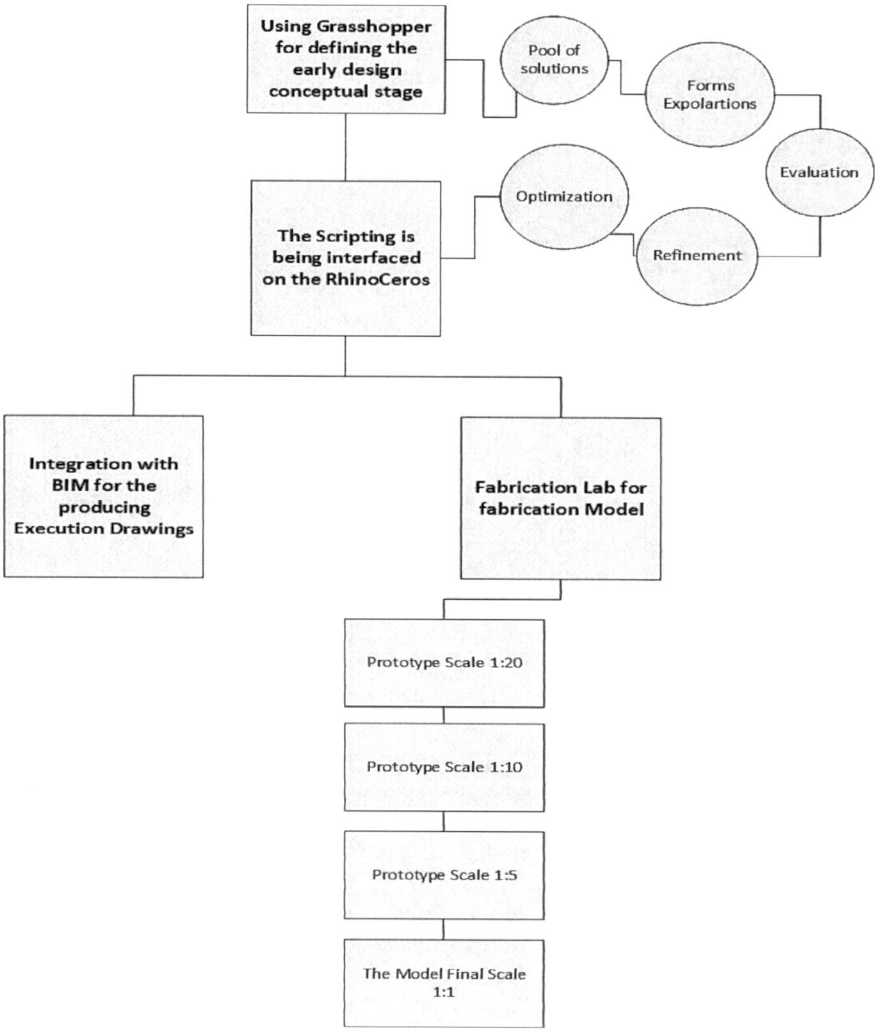

Figure 6. The workflow for a parametric design concept to fabrication.

References

[1] OED, Oxford English dictionary. Oxford University Press, 2002.Another reference

[2] Lawson, B., How designers think: the design process demystified. Elsevier/Architectural, 2006. Oxford; Burlington, MA.

[3] Simon, H., A, The sciences of the artificial. Cambridge Massachusetts: The MIT Press, 1996. Third ed.

[4] Gero, J., *Design prototypes: A knowledge representation schema for design.* Ai magazine, 1990. **11**(4): p. 26-36.

[5] Jabi, W.e., *Parametric Design for Architecture.* Laurence King, London, 2013.

[6] Dunn, N., *Digital Fabrication in Architecture.* London: Laurence King Publishing Ltd., 2012.

[7] Aish, R., *Bentley's GenerativeComponents.* smart geometry conference, 2006. London: Bentley systems institution.

[8] Araya, S., Parametric constructs computational designs for digital fabrication. Massachusettes: MIT institution., 2006.

[9] Kilian, A., Design Exploration through Bidirectional Modeling of Constraints. . Massachusetts: MIT, 2006.

[10] ISO Standard, I.-. Building Information Modeling — Information Delivery Manual — Part 1: Methodology and Format. 2010.

[11] T. Cerovsek, A review and outlook for a "Building Information Model (BIM)": a multi-standpoint framework for technological development, Adv. Eng. Inform. 25 (2011) 224–244. , 2011.

[12] Teicholz Eastman, L.S., BIM Handbook — a guide to building information modeling for owners, managers, designers, engineers, and contractors. All, 2, Wiley, Hoboken, 2011.

[13] Watson, A., Digital buildings — challenges and opportunities. Adv. Eng. Inform. 25 (2011) 573–581. , 2011.

[14] P. Tang, D.H., B. Akinci, R. Lipman, A. Lytle, Automatic reconstruction of as-built building information models from laser-scanned point clouds: a review of related techniques. Autom. Constr. 19 (2010) 829–843, 2010.

[15] C. Nicolle, C.C., Semantic Building Information Model and multimedia for facility management. Web Information Systems and Technologies, Lecture Notes in Business Information Processing2011., 2011.

[16] J. Wong, J.Y., Research and Application of Building Information Modelling (BIM) in the Architecture, Engineering, and Construction (AEC) industry: a review and direction for future research. Proceedings of the 6th International Conference on Innovation in Architecture, Engineering & Construction (AEC), Loughborough University, U.K., Pennsylvania State University, 2010, pp. S. 356–S. 365, 2010.

[17] Hutama, D., Aspek Konteks Pada Wacana Fabrikasi Arsitektur. Https://332lab.Wordpress.Com/2013/12/18/Aspek-Konteks-PadaWacana-Fabrikasi-Arsitektur/, 2013.

[18] Iwamoto, L., Digital Fabrications: Architectural and Material Techniques. Princeton Architectural Press, 2009. p.5.

[19] Seely, J.C., Digital Fabrication In The Architectural Design Process. Massachusetts Institute of Technology, Department of Architecture. Cambridge: Massachusetts Institute of Technology, 2004.

[20] Dunn, N., Architectural Modelmaking Second Edition. London: Laurence King Publishing Ltd, 2014.

[21] Kolarevic, B.e., Architecture in the Digital Age: Design and Manufacturing. New York, NY: Spon Press., 2003.

[22] Blikstein, P., Digital Fabrication and 'Making' in Education: The Democratization of Invention. FabLabs: Of Machines, Makers, and Inventors.Bielefeld: Transcript Publishers., 2013. In J. Walter-Herrmann & C.Büching (Eds.),.

[23] G. Celani, Digital Fabrication Laboratories: Pedagogy and Impacts on Architectural Education, Nexus Netw J –, 2012. Vol.14, No. 3.

[24] Formlabs., DigitalFabrication, https://formlabs.com/blog/digital-fabrication-101/, 2019.

[25] Smit, R.C., Aarhus University, Denmark Towards Digital Smart, Entrepreneurial and Innovative Pupils have been co-funded under the Erasmus+ program KA 2 – Strategic Partnerships |School. Project Number: 2016-1-DK01-KA201-022298., 2016.

Interior Design Guidelines for Reducing the Negative Impacts of Electromagnetic Fields at Residential Workspace

G A Algumbari[1,*], G Nagy[1]

[1] Architecture Engineering department, Faculty of Engineering, the British University in Egypt, Cairo, Egypt

[*]Corresponding Author Email: Gamila156003@bue.edu.eg

Abstract. Pandemic changed human lives, instructions and rules for maintaining social distancing to limit the spread of the virus were imposed. Institutions were forced to adapt working from home mode, the new remote working lifestyle approach emphasizing the strong existence of e-environment facilities (e-learning, e-government, e-commerce) this led to a huge diffusion in the usage of devices that are considered man-made electromagnetic sources in indoor spaces that are poorly designed for long Electro-Magnetic Field (EMF) exposure example of these devices are computers, smartphones, wireless routers, home appliances, and other electronic devices. These EMF sources have several harmful effects on human health, such as stress, headaches, anxiety, and increase risk of cancer. Therefore, this research aims to develop interior design guidelines to reduce the electromagnetic fields negative impacts on the residential workspace in Egypt. First, a qualitative data collecting approach in literature review was conducted then a measurement scale for the protection levels have been defined. Followed by the analytical case studies analysed according to the deduced scale to identify key aspects pertaining to the design for space that reduces (EMF). Research findings provide a guideline for protecting occupants from excessive electrosmog exposure by using light finishing colours, following design standards, right furniture arrangement, ergonomics, relying on natural lighting, ventilation and applying shielding materials.

Keywords: Electromagnetic field radiation, indoor environment, residential workspace, electro-smog, Interior design, exposure to EMF, e-environment.

1. Introduction

At the beginning of 2020, a continuous change has been witnessed in our daily life in many different aspects from performing daily tasks to the mechanism of achieving work requirements due to covid-19 pandemic and the implementation of new restrictions [1]. People are inclined to spend more time at their homes, and as a consequence, homes became the new offices, a place to attend school and even to exercise due to lockdowns [2]. This led to a massive increase in the usage of almost all device categories, especially communication technologies such as mobile communications and wireless data transfer (Wi-Fi, Wi-Max, Bluetooth, etc.). Statistics have shown that in-home data usage in the US during the coronavirus pandemic has increased by 38% upon the daily average resulting in a rise in the devices used on a regular basis [3]. These devices are a source of electromagnetic fields radiation, which can be described as a property of space caused by the motion of an electric charge [4]. Some studies were made to observe the harmful effects of high levels of Electromagnetic fields. Thus, a substantial effect on the wellness of the community caused by emitted radiation from EMF has been noticed these adverse effects are notable if exposed to it in a continuous form. However, Research revealed that undesirable side effects of the high levels of electromagnetic fields exist, never less the experimental case studies have proven their impact on the community health and the adverse effects of long exposure were verified [5], As a matter of fact, The new Pandemic circumstances forced the social distancing leads the workers and students facing the need to move into a more enclosed workspace at their homes [6]. On the other hand, this change proved can be challenging for some since many of the individual arrangements such as: meetings and classes have shifted to a virtually based mode. This has ultimately resulted in increased exposure to screen time and harmful EMF waves [7].

The main aim of this research is to develop interior design guidelines to reduce the effects of Electro-Magnetic Field (EMF) pollution caused by high levels of man-made electronic devices in residential workspaces.

2. Problem

During the covid-19 crisis, the average daily usage of in-home telecommunication, electricity, home appliances, equipment, and many other electronic devices, which are considered EMF radiation sources in homes, has grown remarkably. Consequently, this has created serious health risks and hazards to humans [3]. With this shift to more online approaches, people engage in more screen time than ever before [7]. And spend long durations in indoor areas that are poorly designed for long-term EMF exposure. Furthermore, students and workers are facing the need to set up a home working space [2]. Some studies showed a connection between exposure to electromagnetic radiation and the increased rate of Leukemia, cancer, brain tumors, and other health side effects [8]. Despite the significance of this situation, the subject of developing adequate home workspace design approaches to minimize the effects of these emissions in interior design remains largely understudied [4] EMF can cause a lot of harmful illnesses and with the widespread in the use of technology in our homes and the shift towards homework offices, this could cause a serious problem for inhabitants' health [9].

3. Research methodology

Diverse research methods were used in this study. The key purpose of the research is to develop interior design guidelines to reduce EMF electromagnetic pollution caused by high levels of artificial electromagnetic field (EMF) in the residential workspace area. A mixed-method approach was implemented in this study to fulfill the analysis. First, a qualitative method was conducted through the literature review. The method used archives and documents as a data collection method, for data analysis data reduction and display was used. Moreover, the Quantitative research method was used to fill the gaps found in the literature review, a scale to measure the protection level of each interior design element was concluded, and second a quantitative method was used where an experimental study is implemented to study the protection level of different elements. The scale is used in case studies data analysis to show the outcome of using the proper work space design. Finally, all the data extracted has been used in designing the guidelines.

4. Literature Review

The literature discusses the approaches that can help in reducing EMF by collecting data from different resources. By using archives and documents as a data collection method. The following figure illustrate the methods of EMF reduction in each element of interior design.

4.1. Man-made EMF sources

Man-made sources can include electronic devices such as smartphones, tablets, microwave ovens, radio, and television sets. These devices emit low intensity electromagnetic radiation, which can cause extreme pollution and their EMF levels are far exceeding the natural resources [10].The world's existence depends in great part on this energy which can be found everywhere around us through different EMF sources [11]. These sources are can either be internal in-home sources or external. EMF harmful effects on human health these symptoms are many such as the sleeping disorders, heartbeat irregularity, blood pressure increase [12]. A clear supporting example was in a study by the International Agency for Research on Cancer of the World Health Organization in May 2011 classified the exposure of EMF emitted from wireless devices and cell phones as a possible human carcinogen possible cause.

4.2. Proper Residential workspace design
Designing a workspace plays a significant role in occupants' productivity and wellbeing therefore a workplace design must ensure a healthy indoor environment that is efficient protects the resident from any harmful effects [10]. The interior design of the work area deals with paint colors, furniture orientation, materials used, following design standards, guidelines and using additional elements such as shielding methods to perceive acceptable radiation levels emitted from the devices utilized [13].

4.2.1 Workspace setup design guidelines and standards. When designing a space for study and working activities some architecture guidelines should be followed to ensure the ideal space. The distance between wall and furniture should be 0.05m [10]. The required space to push back the chair from desk and pull out the disk drawer should not be less than 0.9 m and for front of the desk/computer table a min of 1.2 meter space is cleared to permit for walking [11]**.** The reasonable arrangement of the electrical appliances and furniture can at most prevent human beings from absorbing electromagnetic radiation of household appliances [12]. The space between the desktop monitor and human bodey should not be less than arm length.

4.2.2 Indoor environment. Light, air, sound, and thermal comfort are four main aspects that are important when designing a working area [11]. Indoor environmental quality of space affects the wellbeing of human health and their general performance in a significant way [13].

4.2.3 Shielding methods. Different shielding materials such as shielding of doors, windows, ceiling, shielding of walls [14] and flooring [13] can be used to reduce EMF levels inside space which are made of materials that have the properties of blocking these signals some of these shielding methods are mentioned Table 1 in. Another method that can help in reducing EMF is earthing, circuit cut off and EMF detector.

Table 1. Shielding materials (Developed by Auther,2021).

Wall shielding	Floor shielding	Window	Other techniques
Y Shield Paint	Shielding matts	EMF-shielding window film	Earthing
RF-IE50 EMR & RF Shielding Paint	Textile-polymer shieling material	RDF25-window Film/HF shielding	Circuit cut-off switches
Guardia wallpaper		Shield green EMF shielding curtain	EMF detectors

4.3 Literature output
After analyzing and discussing these elements a scale was developed that connects between all the elements and factors affecting the design of reducing EMF to help in choosing the possible component to use in the experiment and prove its ability to reduce EMF. The scale is divided into two main categories. The first is the horizontal column, consists of the workspace interior design, the factors are interior design elements, workplace layout, indoor environment, décor and accessories and architecture elements. The second one, which is the vertical column include level of protection as (inversed relation, high protection, average protection, minimum protection, and no relation) as illustrated in Table 3.

4.4 Residential workspace case studies
In This part the case studies are expected to assess and bring up the cases of residential office spaces, that reduce EMF radiation. Analysts have handled the target of the examination by exploring the design features of home office spaces. The Analysis of cases describes the interior design features and characteristics of those spaces in order to pick up the final checklist to be used in the last part.

4.4.1. Case study selection criteria

Residential building has a workplace: the case studies will be buildings that contains workplace or residential home that has offices of workspaces sections. Interior design components: the case studies will be analyzed through the interior design components and elements of residential workspaces and offices.

Usage of EMF protection method: the case studies will be analyzed through the usage of EMF protection methods as shielding materials, shielding of walls, ceiling and floors, usage of plants, etc.

Table 2. Case Studies (Developed by Authers,2021).

Case study 1	Lighthouse The first case study was designed by Architect Knowhow Shop; it is located in united stated. Moreover, its constructed in 2017 and its area is 154 ft2.
Case study 2	RAM House or the Airplane Mode House The second case study is RAM house designed by Architects Space Caviar and located in USA it constructed in 2015
Case study 3	Timber House The third case study is Timber house designed by Architect: KÜHNLEIN Architecture and it's located in Germany

The case studies act as samples from actual life to evaluate workspace interior design and the low EMF design in residential projects. There are three studies explaining home office design and how the interior design approach helped to monitor EMF exposure. The case studies were analyzed according to developed scale constituent interior design elements where light paint colors were used and analyzed according to the finishing materials, standards and guidelines of furniture spacing and arrangements, indoor environmental factors heating/ventilation, and according to additional components such as shielding materials were used in the case studies.

4.5 Experimental study

An experiment was implemented to study the protection level of certain interior design components. These elements were chosen according to their availability, and the study was performed at a residential building in Egypt. Generating a record of magnetic density to assess if there is an EMF influence on the residential workspace in these building settings. The frequency range is appropriate for noticing radiation from the subsequent sources: computers, laptops, phones, printer, smart phones. An electromagnetic meter providing EMF readings was used) Meter 701A TECEPEL) that include both tesla scale and gauss.

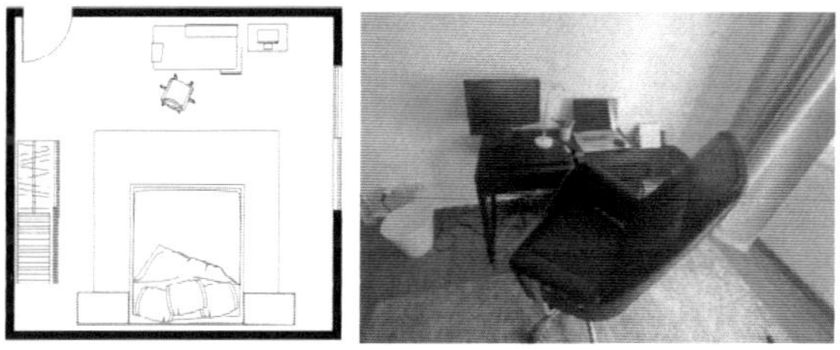

Figure 1. Workstation plan and image (Developed by Authors). Experiment Location El Rehab city, Cairo, Egypt, The aim of the study is to determine the levels of EMF residential work space before and after using methods to reduce EMF. Study Setting: The study was carried in a residential compound (Rehab city) in Cairo. Measurement tool (Electromagnetic Field (EMF) Meter 701A TECEPEL).

4.5.1. Measurement method

- First of all, the measurement distances were chosen and marked inside the selected room, starting with measuring EMF levels when all devices were turned off using an EMF meter. Then, taking the reading of EMF levels when all devices (Laptop, Computer, and printer) turned on.
- Starting with recording readings at 3 cm distance from workstation then 10 cm and 30 cm to estimate the magnitude of emission that human body absorbs when sitting in front of office.
- After that different interior design elements were replaced for 30 minutes then the measurement was recorded again at the same distance points to see the protection level of each. Starting with plants, followed by Wallpaper, then the salt lamp and furniture arrangement.

4.5.2. Experiment findings

The following tables shows the measurements of the experiment in the 7 scenarios:

Table 3a. Different scenarios

Scenarios	Setup	Readings	Min	Max	Protection level
Scenario 1 	Starting with recording the levels inside the room when all devices turned off to determine the initial EMF level and to compare it with other readings.	3 cm	0.02	0.03	The initial case when all devices are turned off.
		10 cm	0.02	0.03	
		30 cm	0.01	0.03	
Scenario 2 	The readings were taken when devices were turned on. They are considered as EMF source in this study.	3 cm	0.09	0.09	This case is considered a base case to compare with other readings.
		10 cm	0.04	0.07	
		30 cm	0.04	0.04	
Scenario 3 	All devices turned on, placing plants (cactus, aloe vera, snake plant) to see its effect in EMF levels.	3 cm	0.09	0.09	No significant variation. Therefore, it has a 0% protection level.
		10 cm	0.04	0.07	
		30 cm	0.04	0.04	
Scenario 4 	All devices turned on, in this case, the furniture arrangement changed so the devices are more spread by following design standards and clear area for each element.	3 cm	0.05	0.07	In this scenario, the levels were decreased rapidly showing maximum protection 35-50%.
		10 cm	0.03	0.05	
		30 cm	0.02	0.02	
Scenario 5 	All devices turned on, in this case, the furniture arrangement changed so the devices are more spread by following design standards and clear area for each element.	3 cm	0.09	0.09	No changes have been noticed when the salt lamp was used. Therefore, it has a 0% protection level.
		10 cm	0.04	0.07	
		30 cm	0.04	0.04	
Scenario 6 	Wallpaper (shielding material for wall) was installed in-wall side that has exterior surface on street side to block signals coming from outdoor sources therefore, resulting in EMF reduction inside the room.	3 cm	0.07	0.08	Installed in-wall side that has exterior surface on street side to block signals coming from outdoor sources.
		10 cm	0.04	0.05	
		30 cm	0.02	0.02	
Scenario 7 	In the last scenario all previous elements were used.	3 cm	0.05	0.07	When all elements were placed a high protection level of 30-50% was promoted.
		10 cm	0.03	0.05	
		30 cm	0.02	0.02	

4.5.3 Experiment results

As seen in Table 3, a reduction in the electromagnetic field had appeared in scenario four when the distance between devices increased by following the design standards, scenario 6 when Guardia wallpaper (shielding material for wall) is used, and scenario 7 when all previous methods were placed. However, readings ensured the fact that measurements of EMF levels decrease with distances. Furthermore, the outcome of the five photos was different, where each had lower or equal measurement compared to the second base scenario. Consequently, that shows that manipulating the suitable design recommendation in the space can affect the EMF radiation, such as changing furniture arrangement by ensuring recommended clearance area between devices. Therefore, increasing distances between these EMF sources resulted in reducing high levels, using shielding material Guardia wallpaper, or by using all of them together to ensure maximum protection in space. About 30-50% decrease in values was shown.

The first and second scenario determines the essential values of EMF inside the room, pointing to EMF level rise when devices are switched on in comparison with reading at the time devices are turned off. After that, check the changes in EMF magnitude at different setups to determine the functional scenario. Although this study is a significant starting point to gather a better understanding of low EMF design in improving residential workspace design, on the other hand, the experimental investigation is considered a pilot experiment due to the limited time different readings may appear concerning longer study duration.

Table 3b. Developed protection level scale, (By Author, 2021)

Workspace interior design / Protection Level	Design elements							Standards and guidelines						Indoor environment				Décor and accessories				Additional techniques								
	Color		Form			Texture/ materials		Furniture arrangements			Ergonomics			Lighting		Ventilation		Plants				Shielding							Other	
																						Wall		Window		Floor		Wire		
	White	Dark	Square	Triangle	Circle	Natural	Artificial	On wall	Parallel	Right angle	Clearance area	Seating ergonomics	Area per person	Natural	Artificial	Natural	Artificial	Snake plant	Aloe vera	Cactus	Salt lamp	Guardia wall paper	Y shield paint	RDF25-Window	Low E-glass	Textile-polymer	Shielding matts	Wire shielding	Circuit cut-off	Earthing
No relation		✓	✓	✓														✓	✓	✓	✓									
Minimum protection 5-15%	✓				✓	✓		✓	✓	✓				✓		✓								✓	✓		✓			
Average protection From 15-30%												✓														✓			✓	✓
High protection 30-50%											✓		✓									✓	✓					✓		
Inversed Relation							✓								✓		✓													

5. Conclusions

After the analysis, studies, and data collection about the residential workspace and EMF radiation, a set of guidelines were established. These guidelines show the steps and elements that could be used while designing residential workspaces to create a certain atmosphere that has low EMF radiation. All the spatial elements affect one another, so working around the materials, light sources, form, textures, furniture, and design elements of the space help in reducing EMF radiation values. These guidelines are the basic steps for creating space with low EMF radiation, however, each element could be manipulated and changed according to the design.

5.1 Interior design guidelines

5.1.1. Materials and texture: Using natural materials for floor finishing, such as wood with an antistatic finish layer. And for paints organic, Zero VOC (clay paint, silicate paint, casein paint) is preferred.
Use fabric and cotton furniture materials granting zero-VOC and antistatic characteristics.
Choosing wallpapers that work as shielding material against electromagnetic waves.
Choose a desk that is made of natural non-magnetic materials and keep away from using furniture that has metal.
Use light colours finishing materials.
Avoid usage of synthetic finishes for walls, floors and furniture led to building up electrostatic charges.

5.1.2. Furniture arrangement: positioning furniture workstations away from walls that have electrical wiring or heavy devices like kitchen equipment.
When choosing a desktop computer, it's recommended to select low certified VDT and CPU desktops.

5.1.3. Clearance area: Leave an arm length about 0.4-0.5m between computer monitor and user position as magnetic fields usually drop off by distance.
Always keep a distance not less than 90 cm between human body and any strong electromagnetic field source.

5.1.4. Lighting: Place slat lamp as a desk lamp since it does not increase EMF levels, therefore, it is the best alternative.
Maximize daylighting as possible to reduce the usage of artificial lighting. Therefore, minimizing EMF from lamps and AC current field's emissions.
Support wired equipment (wired keyboard, mouse, internet connection if possible) is suggested.
Select LED lights because it has low-EMF emissions and high energy efficiency.
Keep away from using lamps that have low-voltage specifications such as fluorescent lamps it emits high EMF radiation.
Ventilation Natural ventilation is always an adequate option.

5.1.5. Shielding materials: Shield power supplies cords to minimize the electric field emissions.
Wire shielding to reduce AC current field emissions.
Grounding (earthing) is preferred.
Shielding mats under office.

It would be valuable to perform an analysis of the impact of EMF radiation within the urban scale and on other spaces within the home. Future work should emphasise the basics which have precedence. Further research should also be shown in other types of buildings, such as large office buildings, retail buildings, etc. Future work on improving e-environment guidelines in the context of interior design should comprise an assessment of the efficiency of the guideline itself on a periodic basis to adapt the guidelines rendering to time-related changes inside the design of the space.

References

[1] D. Carstea, "Electromagnetic field of the large power cables," WSEAS Transactions on. Circuits and Systems, vol. 11, no. 1, pp. 2-5, 2012.
[2] İ. Ayse & Memikoğlu, "). Effects of LED Lighting on Visual Comfort with Respect to the Reading Task.," International Journal of Electrical and Computer Engineering, vol. 11, pp. 930-934, 2017.
[3] National Cancer Institute, "Electromagnetic Fields and Cancer," NIH, USA, 2019.
[4] A.S Farag, "Electromagnetic fields in the home," Electric Power Systems Research, vol. 45, no. 2, pp. 73-89, 1998.
[5] WHO, "world health organization," 4 August 2016. [Online]. Available:https://www.who.int/news-room/q-a-detail/radiation-electromagnetic-fields.
[6] Martin Röösli and Mirjana Moser,, "Symptoms of ill health ascribed to electromagnetic field exposure – a questionnaire survey," International Journal of Hygiene and Environmental Health, vol. Volume 207, no. Issue 2, pp. Pages 141-150, 2004.
[7] Mullen, M. G, & Costa, "Working from Home During the COVID-19 Pandemic: Tips and Strategies to Maintain Productivity & Connectedness," Worcester, MA: University of Massachusetts Medical School, Department of

Psychiatry, Implementation Science and Practice Advances Research Center (iSPARC), Transitions to Adulthood Center for Research., vol. 17, no. 5, 2020.

[8] Kermit George Davis's, "The Home Office: Ergonomic Lessons From the "New Normal"," Ergonomics in Design, 2020.

[9] Gouvernmentpropertymanagement, Workplace Standards and Guidelines for office space, New Zealand: Government Property Management Centre of Expertise, 2014.

[10] A workplace design that reduces employee stress and increases employee productivity using environmentally responsible materials, Eastern Michigan University, 2007.

[11] S. Dyer, Home Office Ideas For Two Person Design Magnificent Decorating Modern, http://www.crismatec.com/, 2016.

[12] Adam, Designing a study room an architect explains, Architectureideas website , 2009.

[13] K. Gustavs, Options to Minimize Non-Ionizing Electromagnetic Radiation Exposures (EMF/RF/Static Fields) in Office Environments, Environmental & Occupational Health Certificate Program University of Victoria, 2008.

[14] OfficespecialistsINC, A GUIDE TO OFFICE SPACE PLANNING & DESIGN, Taskworld, 2016.

[15] S. Hakgudener, EMW SHIELDING CONSIDERATIONS IN BUILDING DESIGN, CALGARY, ALBERTA : UNIVERSITY OF CALGARY , 2018.

[16] W. H. O.-. WHO, "Extremely low frequency fields," Geneva, 2007.

[17] Y. Rashid, A. Rashid and M. Akib, Case Study Method: A Step-by-Step Guide for Business Researchers, International journal of qualitative method, 2019.

[18] pce-instruments, Electromagnetic Field Meter, 2018.

[19] K. Gustavs, "Options to Minimize Non-Ionizing Electromagnetic Radiation Exposures (EMF/RF/Static Fields) in Office Environments," Environmental & Occupational Health Certificate Program University of Victoria, 2008.

[20] V. Gergely Norbert Vizi, "Building materials and electromagnetic radiation," The role of material and shape, vol. 5, no. 2352-7102, pp. 96-103, 2015.

[21] L. Doyle, A.-M. Brady and G. Byrne, "An overview of mixed methods research," Journal of Research in Nursing, 2009.

[22] Statista, "Comparison of average daily in-home data usage in the United States by device type in March 2019 and March 2020".

[23] Office Space Standards and Guidelines, Northwest toerritories canada, 2016.

[24] İ. Ayse & Memikoğlu, "). Effects of LED Lighting on Visual Comfort with Respect to the Reading Task.," International Journal of Electrical and Computer Engineering, vol. 11, pp. 930-934, 2017.

[25] S. R. Shu-Ye Jiang, "Negative Air Ions and Their Effects on Human Health and Air Quality Improvement," International journal of molecular sciences,, 2018.

[26] ILO, "Teleworking during the COVID-19 pandemic and beyond: A Practical Guide," 2020.

[27] WHO, "Virtual press conference on COVID-19 – 11 March 2020," 2020.

[28] S. Kapoor, "Health Implications of Electromagnetic Fields, Mechanisms of," Advances in Biology, p. 24, 2014.

[29] F. Castro, "Timber House / KÜHNLEIN Architektur," Archdaily, 2018.

[30] L. Zimmer, "Timber-louvered house maximizes privacy and light in rural Germany," Inhabitat, 2015.

[31] P. Sisson, "The RAM House Lets You Achieve Digital Detox," curbed.com, 2015.

New Healthcare Design Guidelines for the Diagnostic Radiology Department in the Tertiary Healthcare Facilities

Dalia Eltahlawy[1,*], Mahmoud Zaki[2], Marwa Kamar Eldawla[3]

[1] Department of Architecture Engineering, Faculty of Engineering, Tanta University, Egypt
[2] Professor of Architecture Engineering, President of Tanta University, Egypt
[3] Department of Architecture Engineering, Faculty of Engineering, Tanta University, Egypt

[*]Corresponding Author Email: dalia.eltahlawy@gmail.com

Abstract. The past 10 years have witnessed significant progress in diagnostic imaging through the digital transformation, the evolution of imaging technologies, and future functional requirements. Designers and architects who specialize in healthcare must provide a modern, adaptable built environment that provides the best care for the patient. As a result, the research focused on discussing that problem through three main axes: (the first axis) a presentation of the types of diagnostic radiology methods, (the second axis) studying the relationship between information technology and the programming and planning stage of the diagnostic radiology department, (the third axis) studying the design and technical considerations and their impact on the design of the spaces of the Diagnostic Radiology Department. The results of these axes were applied during the analysis of models for hospitals inside and outside Egypt to derive a general framework to raise the efficiency of hospital design, especially the diagnostic radiology department.

Keywords. Diagnostic Radiology, Hospital Design, Healthcare

1. Introduction - Need for Guidelines

The diagnostic imaging field is constantly progressing. Many factors such as economic concerns, patient-oriented design, and technological developments are now shaping the form of healthcare architecture. A well-designed and planned diagnostic imaging department addresses challenges such as improving the medical experience, integrating flexibility with design, avoiding duplication, and maintaining quality with less expense [1].

Diagnostic radiology is the field of medicine that uses imaging exams to diagnose a patient. The radiology diagnostic techniques are plain radiography, fluoroscopy, R/F, mammography, ultrasound, CT, MRI, and nuclear medicine. The following guidelines are presented through three main considerations: planning, design, and technical considerations.

2. Planning Considerations

The planning phase is the first and most important step to put the basic strategies. Good planning of the radiology department helps to provide a highly efficient design meeting the current and future needs in accordance with ADA design standards. Planning Issues that should be addressed include:

2.1 Information Technology (IT)

IT concept includes digital image acquisition, digital image processing, digital image display, storage, and archiving techniques, digital image communications, and image and information management [2]. Picture Archiving and Communication System (PACS) is the backbone of the digital imaging department. PACS system is subjected to the industry healthcare standards for data and image communications (DICOM and HL-7 standards). For PACS planning, it is necessary to determine where the image archive computer hardware will be located. Planning needs to include not only space but also its support services (HVAC, electrical power, etc.).

2.2 Department location
It can be either on the basement, ground floor, or both for the vertical location. While the horizontal location can be either centralized or decentralized configurations.

2.3 Workflow and circulations
The major drivers of department layout are workflow patterns and the repetitive movement of different users. There are two workflows, interdepartmental and intradepartmental workflow as shown in Figure 2, [3].

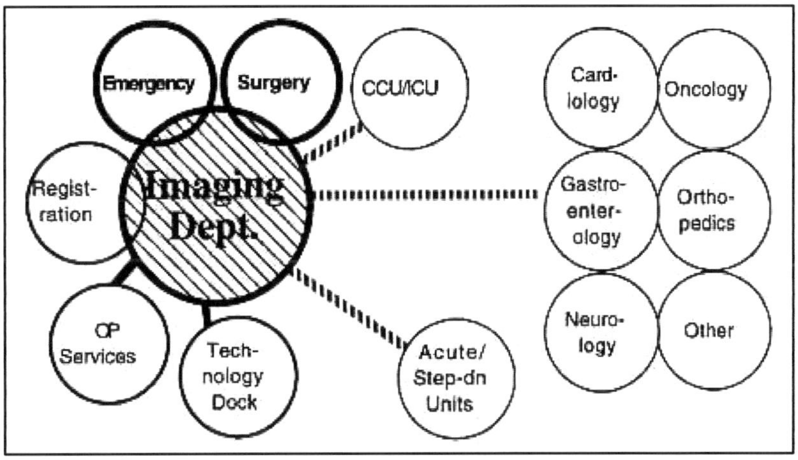

Figure 1. Interdepartmental functional relationships.

2.4 Department configuration
Various systems based on repetitive planning modules have been used to develop the imaging department layout. The concept is based on 2 main points which are [4]:

2.4.1 Functional zones: It includes the following activity zones:

A- Patient Zone
B- Examinassions Zone
C- Central Staff Zone (control corridors, alcoves, and rooms)
D- Personnel Zone
E- Image Archive Zone

2.4.2 Plan typologies: There are 4 typologies as shown in Figure 2 [5]:

A- Single-Loaded Configurations
B- Double-Loaded Configuration
C- Core Configurations
D- Cluster Configurations

2.5 Space requirements
There are two processes for determining the space of an imaging department, functional programming, and space programming.

2.6 Quality of the built environment
The built environment is a key component of the care structure as one of three aspects of quality. There are many strategies to improve the patient experience such as [6]:

A- Sustainability
B- Seismic and Excessive Vibration
D- Disabled Standards (ADA standards)
E- Physical Properties (size, weight, and radiation)

C- Environmental Infection Control (air, water, and surface)

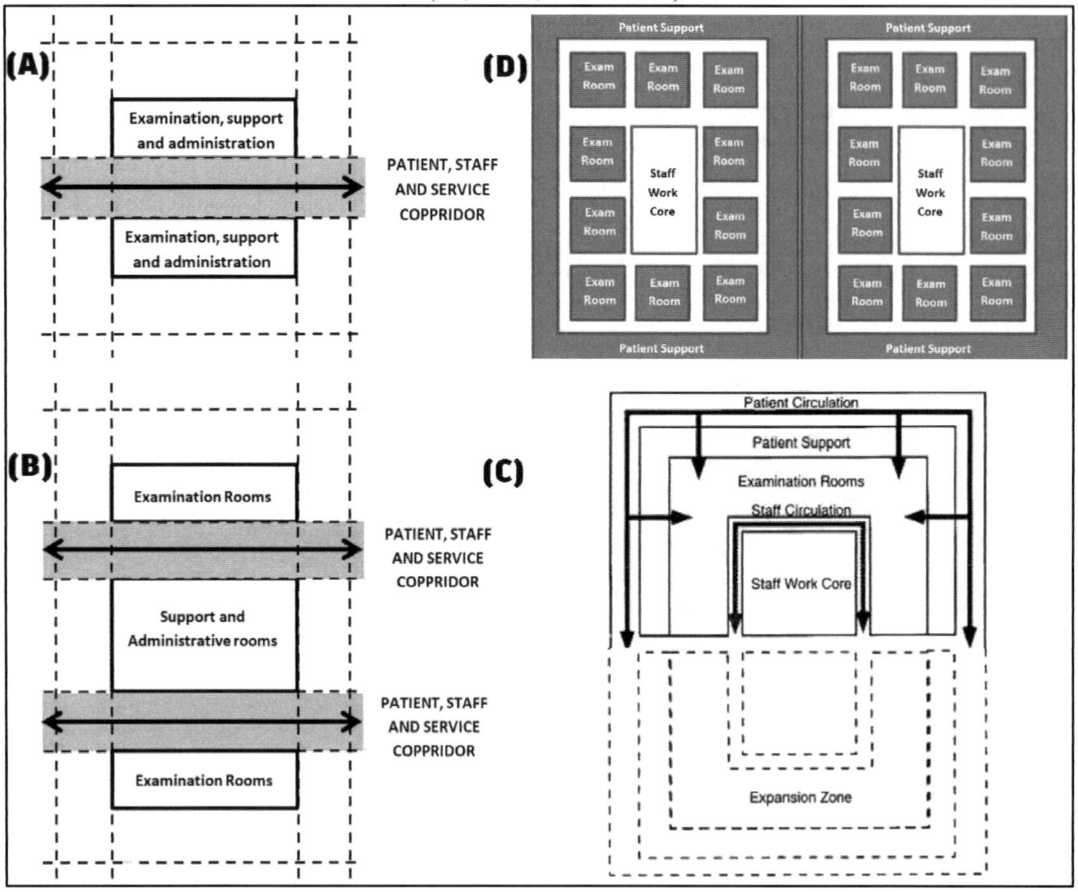

Figure 2. The four layouts of plan typology: (a) single-corridor layout, (b) double-corridor layout, (c) core layout, (d) cluster layout.

2.7 Future expansion and development of new service lines

The department design should be able to expand through two methods either external expansion or internal conversion, or both. Flexibility and adaptability are the main considerations when planning the department to accommodate future expansion.

3. Design Considerations

Good design meets the needs of different users, enhancing interactions between them and technology, and integrating form with function. The design process includes 3 stages, which are:

A- Concept design: A sequence of block diagrams.
B- Schematic design: Block diagrams are evolved into single-line or double-line graphics
C- Design development: Developing the interior and exterior of the schematic design.

The design of the diagnostic radiology department consists of main architectural components that make up each radiology suite within the department such as procedure room, control area, digital image management area, reading room, and support spaces (dressing areas, toilets, and waiting areas). These components can be as a part of the procedure room for a single radiology room. However, they can be shared by more than one room for suite radiology as shown in Figure 3 [7].

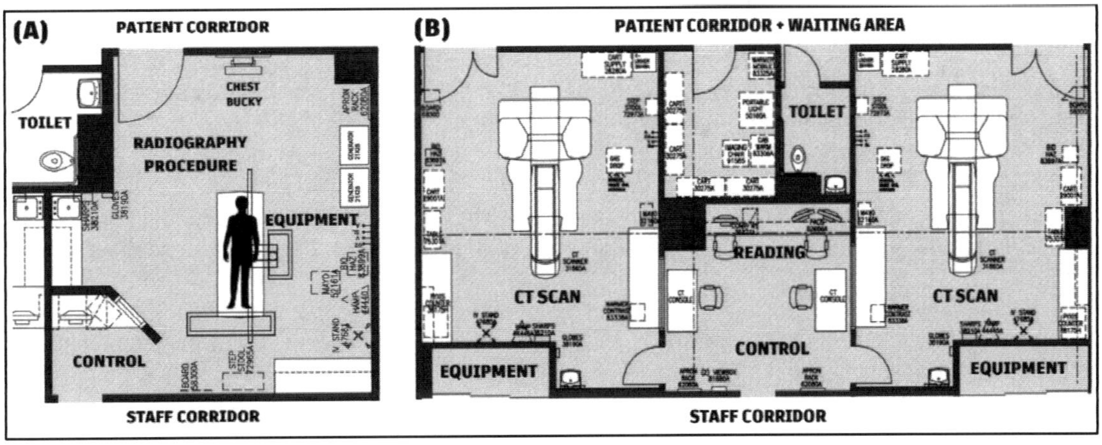

Figure 3. The main component of the radiology department for different rooms: (a) single radiography room, (b) CT suite.

4. Technical Considerations

Many technical considerations must be considered during the planning and design phases such as [8]:

4.1 Structural

Determining the construction system is based on the requirements for the size, weight, and radiology equipment support according to the equipment manufacturer.

4.2 Architectural

All finishing for architectural elements including floors, ceilings, walls, and doors should be coordinated with the equipment manufacturer.

4.3 Heating, Ventilating and Air Conditioning (HVAC)

HVAC installations depend on the number of people and the AC load. The design of the air distribution system shall be in accordance with the HVAC Design Manual criteria of CDC and HICPAC (the Healthcare Infection Control Practices Advisory Committee) [9].

4.4 Plumbing

All plumbing services that include water, waste, and medical gas systems must be planned and designed according to ACR guidelines [10].

4.5 Electrical

Different levels of illumination should be provided. All power installations are subject to the IEEE Recommended Practice for Powering and Grounding Electronic Equipment.

4.6 Life Safety

All design considerations are subject to the latest editions standards of the National Fire Protection Association (NFPA) 101 "Life Safety Code", NFPA 99, Health Care Facilities Code, and the International Building Code.

4.7 Communications

Communications services include telephone, ADP, network, Intercom, nurse-call, public address, and miscellaneous systems such as MATV, CATV.

5. Analytical Study

The analysis is based on all previous considerations that include planning, design, and technical considerations as shown in Figure 4 and Figure 5. The descriptive-analytical study depends on identifying some of the key points, which are:

5.1 Research problem

The design of the diagnostic radiology department does not keep pace with the great development in diagnostic radiology methods, which may lead to evolution or addition or change of some spaces.

5.2 Objectives

Provide healthcare design guidelines for the diagnostic radiology department that accommodate the development of the diagnostic radiology field, through:

- Studying the architectural considerations needed for the diagnostic radiology department in terms of planning, design, and technical considerations.
- Analyzing the gap between what was extracted theoretically and inductively and the local reality through an analysis of hospitals models to provide recommendations support the design process.

5.3 Research Methodology

The research was discussed based on inductive and comparative analytical methods.

- The inductive approach: It studies the different architectural considerations that affect the design of the diagnostic radiology department.
- The comparative analytical approach: Analysis of the diagnostic radiology department models of some hospitals in Egypt (Air force and El Galaa hospitals) to provide recommendations and a suggested vision that support the design process of the diagnostic radiology department.

5.4 Results

After analyzing the hospital models, advantages and disadvantages of each hospital's diagnostic radiology department were determined according to all previous considerations, as shown in table 1:

Table 1. Results of analysis of the two hospitals, Air Force and El Galaa Military hospitals.

	Air Force Specialized Hospital in New Cairo	El Galaa Military Medical Complex
Pros	<u>Planning:</u> - More efficient - Better quality control - Equipment is not duplicated unnecessarily - Future expansion is easy <u>Design:</u> - Digital imaging management is inside the department <u>Technical:</u> - Only power system is up to standard	<u>Planning:</u> - A streamlined workflow - Reduce waiting time - Future expansion is easy <u>Design:</u> - Open MRI is provided - Rooms are new and designed up to standards <u>Technical:</u> - All requirements are up to standards except for AC
Cons	<u>Planning:</u> - Inflexibility - Increase waiting time - Small space compared to hospital size <u>Design:</u> - Too many unnecessary support rooms - Nuclear zone is divided and widely spaced <u>Technical:</u> - Not dilapidated or up to standards	<u>Planning:</u> - Duplication of imaging equipment due to the decentralization of the department - Long patient travel distances <u>Design:</u> - Lake of some imaging equipment - Digital imaging management is outside the department <u>Technical:</u> - Split AC unit should not be used

Comparison	Analysis of Case Studies				
About the Project – Hospital	**Hospital Name**	**Location**	**Design Architects**	**Hospital Type**	**Date**
	Air Force Specialized Hospital	New Cairo, Egypt	Armed Force Engineering Authority	Tertiary Healthcare 400 beds	2014
Planning Considerations – Radiology Depart.	**Department Location**	**Workflow & Circulations**	**Configuration (Functional Zones)**	**Plan Typology**	**Space**
	Ground Floor - Centeral department (without MRI) Basement Floor - Only MRI Unit	Accessible to Both Inpatient & Outpatient Services Accessible to the Emergency Department (ED) Close to the Central Vertical Transportation System	Support Spaces Zone Staff Zone MRI Zone Patient Zone Nuclear Medicine Zone PACS Reading Zone	Cluster Configuration	Approximately 1680 m²
Design Considerations – Radiology Depart.	**Procedure Room**	**Control Areas**	**Reading Room**	**Support Spaces**	**Digital Imaging Management**
	All Imaging Techniques	All Imaging Techniques	Only in Ultrasound, CT MRI & Nuclear Medicine	Distributed to Serve the Entire Department	Inside the Main Radiology Department

Technical Considerations

Procedure Room	Architectural: Ceiling	Architectural: Wall Finish	Architectural: Base	Architectural: Floor Finish	HVAC: Inside Conditions	HVAC: Air Change/hour	HVAC: 100% Exhaust	HVAC: Air Balance	HVAC: Occupancy	HVAC: AC Load-Equipment AC Load-Lighting
Radiography	Acoustical tile ceiling	Paint	Conceret Slab	Static Dissipating Flooring	Temperature: (21- 24)°C Relative humidity: (30- 60)%	4 Air Supply	No	Positive	4 people	Equipment: (1,500 – 4000)W Lighting: 17 W/M²
Fluoroscopy						4 Air Supply	Yes	Negative	4 people	Equipment: 4000 W Lighting: 16 W/M²
Mammography						2 Air Supply	No	Positive	3 people	Equipment: (900 - 1,800) W Lighting: 16 W/M²
CT						6 Air Supply	No	Positive	4 people	Equipment: (5,000 -6,500) W Lighting: 21 W/M²
Ultrasound						2 Air Supply	No	Positive	3 people	Equipment: (1,050 -1,400) W Lighting: 12 W/M²
MRI Closed & Open	No Fulse Ceiling		Resilient Base			6 Air Supply	No	Positive	6 people	Equipment: 3,000 W Lighting: 21 W/M²
Nuclear Medicine	Acoustical tile ceiling					6 Air Supply	Yes	Negative	4 people	Equipment: 2,750 W Lighting: 15 W/M²
PET/CT						6 Air Supply	Yes	Negative	6 people	Equipment: 7,750 W Lighting: 14 W/M²
Bone Densitometry			Concrete			2 Air Supply	No	Positive	3 people	Equipment: 1,600W Lighting: 14 W/M²

Procedure Room	Plumbing & Gases: Cold/Hot Water	Plumbing & Gases: Sanitary Drain	Plumbing & Gases: Medical Gases	Electrical: Lighting	Electrical: Power	Communications/Special Systems: ADP	Communications/Special Systems: Data	Communications/Special Systems: Telephone	Communications/Special Systems: Intercom	Communications/Special Systems: Nurse Call
Radiography	No	No	Yes	General: Fluorescent lights 50 FC Special: Incandescent lights 5 FC	General: Circuit Breaker (230V,16A) Special:Main Switch(400V/145A/160KVA)	Yes	Yes	Yes	Yes	No
Fluoroscopy	Yes	Yes	Yes	General: Fluorescent lights 100FC Special: Incandescent lights 2FC	General: Circuit Breaker (230V,16A) Special:Main Switch(400V/145A/160KVA)			No	Yes	Yes
Mammography	No	No	No	General: Fluorescent lights 40 FC Special: Multi-level illumination	General: Circuit Breaker (230V,16A) Special: Main Switch (400V/10KVA)			No	No	No
CT	Yes	Yes	Yes	General: Fluorescent lights 50 FC Special: Incandescent lights 5 FC	General: Circuit Breaker (230V, 16A) Special:Main Switch(400V/144A/80KVA)			Yes	Yes	Yes
Ultrasound	No	No	No	General: Fluorescent lights 40 FC Special: Incandescent lights 5 FC	General: Circuit breaker (230V,16A)			No	No	No
MRI Closed & Open	Yes	Yes	Yes	General: 30 FC	General: Circuit Breaker(2x230V/16A) Special:Main Switch(400V/150A/88KVA)			Yes	Yes	Yes
Nuclear Medicine	Yes	Yes	Yes	General: Fluorescent lights 50 FC Special: Incandescent lights 5 FC	General: Circuit Breaker (230V, 16A) Special:Main Switch(400V/144A/80KVA)			Yes	No	No
PET/CT	Yes	Yes	Yes	General: Fluorescent lights 50 FC Special: Incandescent lights 5 FC	General: Circuit Breaker (230V, 16A) Special:Main Switch(400V/144A/80KVA)			Yes	Yes	Yes
Bone Densitometry	No	No	No	General: Fluorescent lights 30 FC Special: Multi-level illumination	General: Circuit Breaker (230V,16A)			No	No	No

Figure 4. Analysis of the Air Force specialized Hospital.

IOP Conf. Series: Earth and Environmental Science 1056 (2022) 012006 doi:10.1088/1755-1315/1056/1/012006

Comparison	Analysis of Case Studies				
About the Project	Hospital Name	Location	Design Architects	Hospital Type	Date
Hospital	El Galaa Medical Complex For Armed Forces	El Nozha, Cairo Government, Egypt	Armed Force Engineering Authority	Tertiary Healthcare 922 beds	2014
Planning Considerations	Department Location	Workflow & Circulations	Configuration (Functional Zones)	Plan Typology	Space
Radiology Depart.	Decentralized (each hospital has it's depart.) Ground Floor: All departments Basement Floor: Only Military Emergency Building's Department	Accessible to Both Inpatient & Outpatient Services Accessible to the Emergency Department (ED) Close to the Central Vertical Transportation System	Patient Zone Staff Zone Support Spaces Zone	Mixed between Single-Loaded & Double-Loaded Corridor Configuration Plan	According to the Hospital Space Area
Design Considerations	Procedure Room	Control Areas	Reading Room	Support Spaces	Digital Imaging Management
Radiology Depart.	All Imaging Techniques	All Imaging Techniques	Only in CT, MRI & Nuclear Medicine	Distributed to Serve Each Department	Outside the Diagnostic Depart. Central Management of the Hospital

Technical Considerations

Procedure Room	Architectural: Ceiling	Architectural: Wall Finish	Architectural: Base	Architectural: Floor Finish	HVAC: Inside Conditions	HVAC: Air Change/hour	HVAC: 100% Exhaust	HVAC: Air Balance	HVAC: Occupancy	HVAC: AC Load-Equipment AC Load-Lighting	Plumbing & Gases: Cold/Hot Water	Plumbing & Gases: Sanitary Drain	Plumbing & Gases: Medical Gases	Electrical: Lighting	Electrical: Power	Communications/Special Systems: ADP	Communications/Special Systems: Data	Communications/Special Systems: Telephone	Communications/Special Systems: Intercom	Communications/Special Systems: Nurse Call
Radiography	Acoustical tile ceiling	Paint	Concreet Slab	Vinyl Tile	Temperature: (20- 24)°C Relative humidity: (30- 60)%	4 Air Supply	No	Positive	4 people	Equipment: (1,500 – 4000)W Lighting: 17 W/M²	Yes	Yes	No	General: Fluorescent lights 50 FC Special: Incandescent lights 5 FC	General: Circuit Breaker (230V,16A) Special:Main Switch(400V/145A/160KVA)	Yes	Yes	No	Yes	Yes
Fluoroscopy						4 Air Supply	Yes	Negative	4 people	Equipment: 4000 W Lighting: 16 W/M²			Yes	General: Fluorescent lights 100FC Special: Incandescent lights 2FC	General: Circuit Breaker (230V,16A) Special:Main Switch(400V/145A/160KVA)			No	Yes	Yes
Mammography						2 Air Supply	No	Positive	3 people	Equipment: (900 - 1,800) W Lighting: 16 W/M²			No	General: Fluorescent lights 40 FC Special: Multi-level illumination	General: Circuit Breaker (230V,16A) Special: Main Switch (400V/10KVA)			Yes	No	No
CT						3 Air Supply + 1 Split AC	No	Positive	4 people	Equipment: (5,000 -6,500) W Lighting: 21 W/M²			Yes	General: Fluorescent lights 50 FC Special: Incandescent lights 5 FC	General: Circuit Breaker (230V, 16A) Special:Main Switch(400V/144A/80KVA)			Yes	Yes	Yes
Ultrasound						1 Air Supply	No	Positive	3 people	Equipment: (1,050 -1,400) W Lighting: 12 W/M²			No	General: Fluorescent lights 40 FC Special: Incandescent lights 5 FC	General: Circuit breaker (230V,16A)			No	No	No
MRI			Resilient Base			9 Air Supply	No	Positive	3 people	Equipment: 3,000 W Lighting: 21 W/M²			Yes	General: 30 FC	General: Circuit Breaker(2x230V/16A) Special:Main Switch(400V/150A/88KVA)			Yes	Yes	Yes
Nuclear Medicine						2 Split AC	Yes	Negative	4people	Equipment: 2,750 W Lighting: 15 W/M²			Yes	General: Fluorescent lights 50 FC Special: Incandescent lights 5 FC	General: Circuit Breaker (230V, 16A) Special:Main Switch(400V/144A/80KVA)			Yes	Yes	No
PET/CT						2 Split AC	Yes	Negative	6 people	Equipment: 7,750 W Lighting: 14 W/M²			Yes	General: Fluorescent lights 50 FC Special: Incandescent lights 5 FC	General: Circuit Breaker (230V, 16A) Special:Main Switch(400V/144A/80KVA)			Yes	Yes	Yes
Bone Densitometry			Concrete			2 Air Supply	No	Positive	3 people	Equipment: 1,600W Lighting: 14 W/M²	No	No	No	General: Fluorescent lights 30 FC Special: Multi-level illumination	General: Circuit Breaker (230V,16A)			No	No	No

Figure 5. Analysis of El Galaa Military Medical Complex.

6. Conclusions

There are several solutions proposed to improve the design of the Diagnostic Radiology Department at Air Force Specialty Hospital such as transforming the functionality of some non-essential support spaces and using them to provide space for digital image imaging management. The separate nuclear medicine zones can be combined into one zone to avoid duplication. In addition to upgrading the architectural finishes and technical systems according to the standards, as well as developing a plan for future expansion.

On the other hand, the design of the Diagnostic Radiology Department at El Galaa Military Medical Complex can be developed through some planning and design solutions. First, design a central, complex department either in a separate, independent central building or located within a central hospital building. Second, providing space for digital image management within the diagnostic department. Third, adding DEXA in the women's unit within the radiology department of the Military Emergency building. Fourth, the integration of the MRI unit with the radiology department at the Children's Hospital to reduce the travel distance for patients and have direct access to the outpatient zone. Fifth, adding fluoroscopy to the radiology department at the Children's Hospital, and not only to the operations department in the civil emergency building. Finally, replacing the split AC units with a central AC system subjected to the standards (CDC and HICPAC). In addition, considering the continuous development of radiology services in line with continuing technological progress.

References

[1] Rostenberg B, FAIA and FACHA 2006 *The Architecture of Medical Imaging: Designing Healthcare Facilities for advanced Radiological Diagnostic and Therapeutic Techniques* (New Jersey: Wiley) pp 3–8
[2] Auer J, Stick J, Kümmerle J and Prange T 2019 *Equine Surgery-E-Book (Fifth Edition)* (U.S.: Elsevier Health Sciences) pp 1836 -1882
[3] Rostenberg B, FAIA and FACHA 2006 *The Architecture of Medical Imaging: Designing Healthcare Facilities for advanced Radiological Diagnostic and Therapeutic Techniques* (New Jersey: Wiley) pp 184–201
[4] Fischer H W 1982 *Radiology Department: Planning, Operation, and Management* (U.S.: Edward Brothers, Inc.) pp 45–90.
[5] Rostenberg B, FAIA and FACHA 2006 *The Architecture of Medical Imaging: Designing Healthcare Facilities for advanced Radiological Diagnostic and Therapeutic Techniques* (New Jersey: Wiley) pp 214–217.
[6] Fowler E, MacRae S, Stern A, Harrison T, Gerteis M, Walker J, Edgman-Levitan S and Ruga W 1999 The built environment as a component of quality care: understanding and including the patient's perspective *The Joint Commission Journal of Quality Improvement* **25** (7) pp 352–362.
[7] Rostenberg B, FAIA and FACHA 2006 *The Architecture of Medical Imaging: Designing Healthcare Facilities for advanced Radiological Diagnostic and Therapeutic Techniques* (New Jersey: Wiley) pp 234–277
[8] Sykes D, Tocci G and Cavanaugh W J 2012 *Sound & Vibration 2.0: Design Guidelines for Health Care Facilities (Handbook 2010)* ed ARC (New York: Springer Science) pp 33–53.
[9] Sehulster L *et al* 2004 (updated 2019) *Guidelines for Environmental Infection Control in Health-Care Facilities; Recommendations from CDC and the Healthcare Infection Control Practices Advisory Committee (HICPAC)* (Chicago IL: American Society for Healthcare Engineering/American Hospital Association) pp 27–35.
[10] American College of Radiology (ACR) 2020 *ACR Manual on MR Safety (Version 1.0)* (U.S.: ACR) Appendix B: Magnetic Resonance Facility Safety Design Guidelines pp 48–51.

Architectural design criteria for infection control in hospitals during construction and development

Emad Marawan[1, *], Hosam Rezk[2, 3], and Hesham Sameh[4]

[1] PhD Applicant, Faculty of Engineering, Cairo University, Cairo, Egypt.
[2]Chairman of General Authority for Construction and Housing Cooperative
[3]Architecture Engineering Department, Military Technical, Cairo, Egypt
[4]Architecture Engineering Department, Faculty of Engineering, Cairo University, Cairo, Egypt

*Corresponding Author Email: emad.marawan67@gmail.com

Abstract: Hospitals are the type of buildings which need development and expansion continuously so it can achieve all the demands of the progress in the medical field, medical devices, and the updates in the infrastructure of the hospitals. Due to the spread of many bacteria that are related to the construction and development process, many countries had to take into considerations many strict precautions to reduce the danger of the infection spread. These precautions are important either inside the building of the hospital or outside, so the maximum reduction of infection spread should be reached as there are many patients suffer from lack of immunity. This research aims to reach a rational design criterion for hospital designers which gains the required demands to combat the infection spread during construction and development process meanwhile continuing to present the medical services. It is strongly recommended to introduce flexible structural module that can be reused in different activities. Also, isolation of patient's spaces during construction becomes a mandatory manner. Moreover, it is advised to design an adaptable ventilation system which facilitates different sectors separation, either patient, medical stuff, or workers paths.

Keywords: Hospitals Design, Hospitals Infection Control, Architecture of Hospitals Buildings, and Hospitals Constructing and Developing.

1. Introduction

Buildings of hospitals have the features of growth and expansion to keep up with the progress in all the specializations and therapeutic methods. This growth leads to many architectural adjustment and development for all different systems whether in the infrastructure of the building or what consequents due to the oldness of the buildings and their infrastructure. To achieve such a goal, many studies and researches should be performed to investigate the possibility of the development conducting in different hospital parts while continuing to present medical services without being interrupted by the expansion or development works.

Development works could be confronted by some challenges which had to be conquered during the execution stage such as noises and vibrations caused by the construction and finishing equipments which are not desirable in the health care area, in addition to its bad effect which may be caused to the diagnostic and therapeutic devices due to vibration [1]. Also, ventilation paths conflict between medical services and construction areas could expose patients to polluted air. Moreover, some medical services could be stopped until finishing construction works due to lack of places to be replaced. One extra challenge is existence of many materials and labor which make conflict with the user's paths in the hospital either patients, doctors, or administrators. Bad psychological effects and spread of infection because of the pollution due to renovation work should also be taken into account [1].

Hospitals keep presenting services 24 hours every single day of the week, so the works must keep up with it without stopping. So, Stopping the infection spread is one of the biggest issues that faces the designers and constructors due to the developments in the health care area. The construction and renovation process in hospitals requires digging and demolition, what require coordinating with the responsible of presenting the medical services to reach the minimum level of danger of infection spread.

Many kinds of infection spread which are caused by bacteria related to the construction process had been monitored all over the world. It was found that most people who are susceptible for infection are those with immunodeficiency due to underlying disease or their associated therapies such as chemotherapy. The construction process in any site might be dangerous for this population category, but it becomes much more serious in health care buildings because of the high number of people who are in danger in those sites [2]. For example, the death rate due to infection in Canada hospitals among 250000 patients per year, is approximately 8000 to 12000 cases in hospitals, and it was found that around 7-8% of them due to the constructions and maintenance in the hospitals, what means around 500 -1000 case per year [3].

So, it could be concluded that the construction process comprises the danger of getting infected for patients, and conflicts with all the aspects of patients care all around the world. Nowadays, many health institutions in some countries put precautions, and enforce the designers and constructors to apply them so the pollution could be decreased to guarantee that the execution of the construction works in the health care site are safely done in a way that decreases the danger of infection [2].

This research aims to reach reasonable design criteria for hospitals which achieves infection control in hospitals during the construction and renovation process to make sure that it's safely done in a way that reduces the danger of infection spread for the patients.

2. Methodology

2.1 A brief history of infection control - past and present

To appreciate the current development in the healthcare epidemiology field, it's important to understand the history of infection control in the hospitals. The available historic resources have been reviewed for four historic periods of time, middle, modern, advanced ages and after the Second World War by focusing on the cases that are related to the hospital's infection and its effect on the hospitals design. The infection control doctors, committees and nurses, in addition to confining the changes, problems and development in infection control nowadays are highlighted as well.

Investigating the hospitals infection has been started since the first half of the 18th century to the beginning of the bacteriology era. All the contributions began in Scotland such as what was done by the pioneers (Pasteur, Koch, Lister and Doctor Sir John Pringle) who strongly believed that overcrowding and poor ventilation are the main reasons for the hospital's infection spread problem. Besides after 100 years in 1858 what Florence Nightingale had done after her experiments in the military hospitals during the Crimean war, that she worked on reforming the hospitals after discovering that the patients zones should be designed in a way that allows a good ventilation to reduce the infection spread. She also became a pioneer in designing the patient's wards in the pavilion type (figure 1), and had a ward called by her name. Then at the end of the 9th century it was found out that the air itself could be a source of infection, [4].

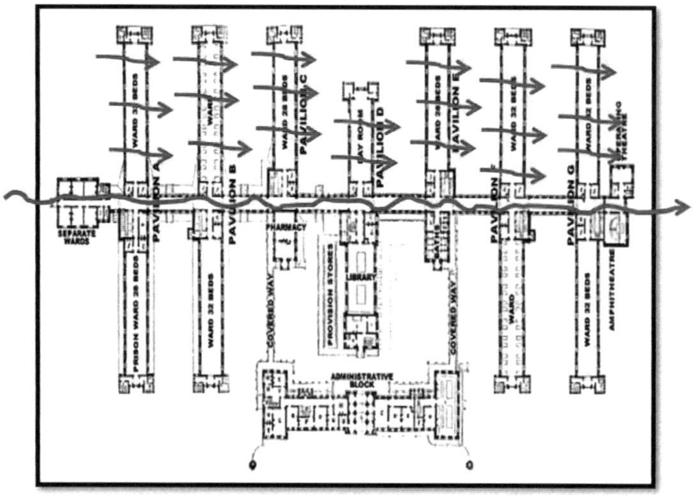

Figure 1. Pavilion type suggested by Florence Nightingale 1858 [4].

It was observed that the real challenge over the 60 past years in hospitals are the high coasted and complicated problems to control infection, what was a motivation to develop the control system. The first official step in the control system was taken in the United Kingdom in 1941 by the British Medical Research Council (MRC). It was recommended that to avoid surgical poisoning, specialized employees with full-time working should be hired to supervise the infection control, and pursuant to the recommendation there was a job created and named by "infection controller", but it was a part-time job. then the name has been changed to the part-time control of infection officer (CIO) in 1988, and the doctor was called the infection control doctor (ICD).

Then in 1944, the Mekong river committee recommended that every hospital should have a commit for infection control that consists of doctors, nurses and administrators. Then in the middle of 1950s, Brindand Mor has proposed the idea of the infection control nurse (ICN) and the first ICN was appointed in the United Kingdom in 1959. Since then, the infection control committee has taken place in most of the big hospitals all around the world [5].

2.2 Danger evaluation standards for patients

There are many countries that had put precautions to control infection during the construction and renovation process in the hospitals. These precautions aim to achieve the highest level of isolation between the spaces where there are construction works and the ones where the patients stay, to control infection as much as possible especially the bacteria and virus leak.

To control infection of those bacteria which are related to construction works, there are some important precautions to be applied whether inside the hospital building or outside for expansion works and renovation in the main site. These precautions which were applied in the USA, Australia, Alaska and Canada are all almost the same.

The danger for patients who are in the same location or nearby the construction site should be evaluated before initiating any works, especially when those works include demolition inside or outside the healthcare area which has the highest level of danger for patients who suffer from lack of immunity (table 1).

All the doctors should be aware of all the possible dangers which are related to executing the project to know all the effects and examination ways to deal with it medically. There are 4 main aspects to deal with that evaluation and its related danger. Those aspects have different steps of every stage in the construction and renovation projects:

- First step: specifying the type of the construction activity.

- Second step: conducting a layout of the patients' wards and making them groups based on the danger category (table 3)
- Third step: developing a matrix to categorize the precautions of censorship based on the type of construction activity and the groups, based on danger category for patients (table 4).
- Fourth step: implementation and quality control for the precautions
  All the projects should be evaluated individually as not all precautions work in all the categories in the matrix. Table 2 shows a sample of types of construction, renovation, installation and maintenance activities. [6-9]

Table 1. Classification of dangerous levels (groups) based on hospital spaces users.

Dangerous Level	People Category
First group No evidence on danger	• Employees, services presenters and contractors. • All patients who aren't in the second to fourth groups.
Second group Increased danger	• Patients with long cycles of steroids with high doses, especially the ones which continue for a long time in the hospital. • Lack of immunity/ AIDS patients • Patients who are treated with chemotherapy • Patients who are subjected to artificial respiration • Blood disease • Dialysis patients
Third group High danger	• Blood disease which needs less than 14 days after the chemotherapy treatment • Newborn in the (ICU) • Solid organ transplant
Fourth group Very high danger	• Marrow transplant within 12 months if it's more than 12 months, the blood disease specialist should be consulted • Autologous peripheral blood stem cell transplantation, during recovery with long time, more than 14 days after chemotherapy treatment or immunosuppressive therapy such as: 1. Acute leukaemia (AML) 2. Acute lymphoblastic leukaemia (ALL) 3. Lymphoma 4. Primary myeloid nervous system lymphoma 5. Aplastic anemia 6. Children who suffer from: 7. Severe combined immunodeficiency syndrome (SCIDS) 8. Chronic granulomatous disease in childhood (CGDC)

Table 2. Types of construction, renovation, installation and maintenance activities

Types of Constructions	Work Activity
Type (A) Inspection and general follow up activities.	Contains for example: • Removing the false ceiling for visual inspection (for 5 m^2) • Walls coating. • Electrical works. • Plumping works that interrupt the water supply for patients caring area (like one room) for less than 15 minutes. • Any activities that don't generate dust or cause cuts in the walls or the ceiling.
Type (B) Small and short period activities which generate the minimum level of dust.	Contains for example: • Installation of telephone and computer cables. • Reaching areas above the false ceiling. • Plumbing works that interrupt the water supply for more than one patient's caring area (like 2 wards) for less than 30 minutes. • Cutting walls or ceilings to control the dust diffusion.
Type (C) Any activity that generates a medium to high level of dust.	Contains for example: • Demolition of building components • Sand that remains from painting or covering walls • Removing floors/wallpapers and the false ceiling • Building a new wall • Any electrical works or any works above the false ceiling • Any works in the main cables • Plumbing works that interrupt the water supply for more than one patient's caring area (like 2 wards) for more than 30 minutes but less than an hour
Type (D) Demolition and construction projects	Contains for example: • Heavy demolition works. • Removing a whole false ceiling system. • Main paths or electrical works above the ceiling. • Plumping works that interrupt the water supply for more than one patient's caring area (like 2 wards) for more than one hour. • New building.

IOP Conf. Series: Earth and Environmental Science 1056 (2022) 012007 doi:10.1088/1755-1315/1056/1/012007

Table 3. Grouping of hospital spaces based on level of danger exposure.

Group (1) Low Danger	**Group (2) Medium Danger**	**Group (3) Medium-high Danger**	**Group (4) Highest Danger**
• Areas where there are no patients. • Offices.	• Workers and visitor's cafeteria. • Food and drinks area. • Labs which are not included in group (3). • Loading and discharging units for various supplies. • Laundry services for food and patients. • Material management. • Patients care and other areas which are not included in group (3) or (4) • Physiotherapy, rehabilitation, and speech therapy. • General paths for patients	• Respiratory bronchoscopy. • Obstetrics rooms. • Hemodialysis. • Heartbeat. • Emergency department. • Laparoscopic units. • Higher accreditation Unit. • Sub-acute units for long stay • Medical imaging: public MRI CT scan Ultrasound • Medical units. • Microbiology laboratories. • Newborn nurseries. • Pediatrics. • Pharmacy. • Recovery Rooms (PACU). • Kidney units. • Surgical units. • Virology laboratories.	• Anesthesia rooms and respiratory pumps. • Angiography rooms. • Cardiac catheterization units. • Chemotherapy. • Food Services. • Large scale food preparation. • Hematology unit - inpatient and outpatient. • Intensive care units. • Oncology Unit - Inpatient and outpatient. • Operating rooms. • Outpatient invasive operating rooms, including imaging-guided safes. • Mixture of pharmacy/clean rooms. • Radiotherapy Unit. • Sterile supply units. • Transplant surgery units.

Table 4. Matrix of danger groups and construction activities.

			Construction Activity						
			Type (A)	**Type (B)**		**Type (C)**		**Type (D)**	
Groups of danger	Low Danger	**Group 1**	First	Second		Second		Third	Fourth
	Medium Danger	**Group 2**	First	Second		Third		Fourth	
	Medium-High Danger	**Group 3**	First	Third		Third	Fourth	Fourth	
	Highest Danger	**Group 4**	Third	Third	Fourth	Third	Fourth	Fourth	

2.3 Infection control polices during construction and renovation

Before the beginning of any construction or renovation project there are some issues that should be handled as the infrastructure employees and the engineers must cooperate with the workers who work in the infection control field to:

- Design the function of the new space.
- Evaluate the infection danger from virus and bacteria for the patients, workers, and the visitors of the hospital.
- Putting strategy to reduce the danger of infection spread which is related to the construction works like dust combat and the polluted water…etc.
- Monitoring the requirements that are mentioned during construction works.

There is a main role for infection prevention unit (IPU) to control and prevent infection in the construction and renovation projects in hospitals, which can be summarized as follow:

First: *Maintenance and installation works:*

The infection prevention unit (IPU) should be consulted in all the maintenance and installation works which been categorized as class 3 or 4 of danger (Table 4) in the site, or any construction works in the nearby area or the surrounding paths (Table 3).

Second: *Mega construction projects and the private ones:*

- All the construction and renovation works should be clearly specified because it may affect the spaces where the patients who are in highly danger stay. This could need strategies which make them safe, and continuous monitoring and applying these strategies is mandatory.
- The reduction of danger standards should be subjected to the instructions of the nation health policies.
- To reduce the danger, engineering rules should be added to the contracts between the health services and the contractors who are going to execute the project.
- The danger for the patients who are in the construction site should be evaluated and to put strategies to reduce this danger such as: moving them to safe places, affording the personal protection equipment and the antifungal prophylactic treatments.
- There should be an official evaluation for the danger of infection control in the construction, renovation, and installation activities as well as the strategies which is there to reduce danger before starting any works.
- The strategies should be monitored and signed that they correspond with the IPU requirements.
- There should be an IPU representor in the construction and renovation projects team.

Figure 2 shows a flowchart for cooperation process between infection protection unit and construction workers.

3. Discussion

By studying the precautions that some countries had put to be applied during the construction and renovation works in the built hospitals with continuing to present the medical services to the patients while the patients and visitors of the hospitals are nearby the site, and as it was presented in the research, and to control the infection danger, it was found out that the majority of the countries are the same in those polices and precautions which are:

- Specializing the type of the construction activity
- Putting a layout of the patients' wards and making them groups based on the danger category (table 3)
- Putting a matrix to categorize the precautions of censorship based on the type of construction activity and the groups which is based on danger category for patients (table 4)
- Implementation and quality control for the precautions

Figure 3 shows a flowchart that explains the precaution policy which followed by many countries.

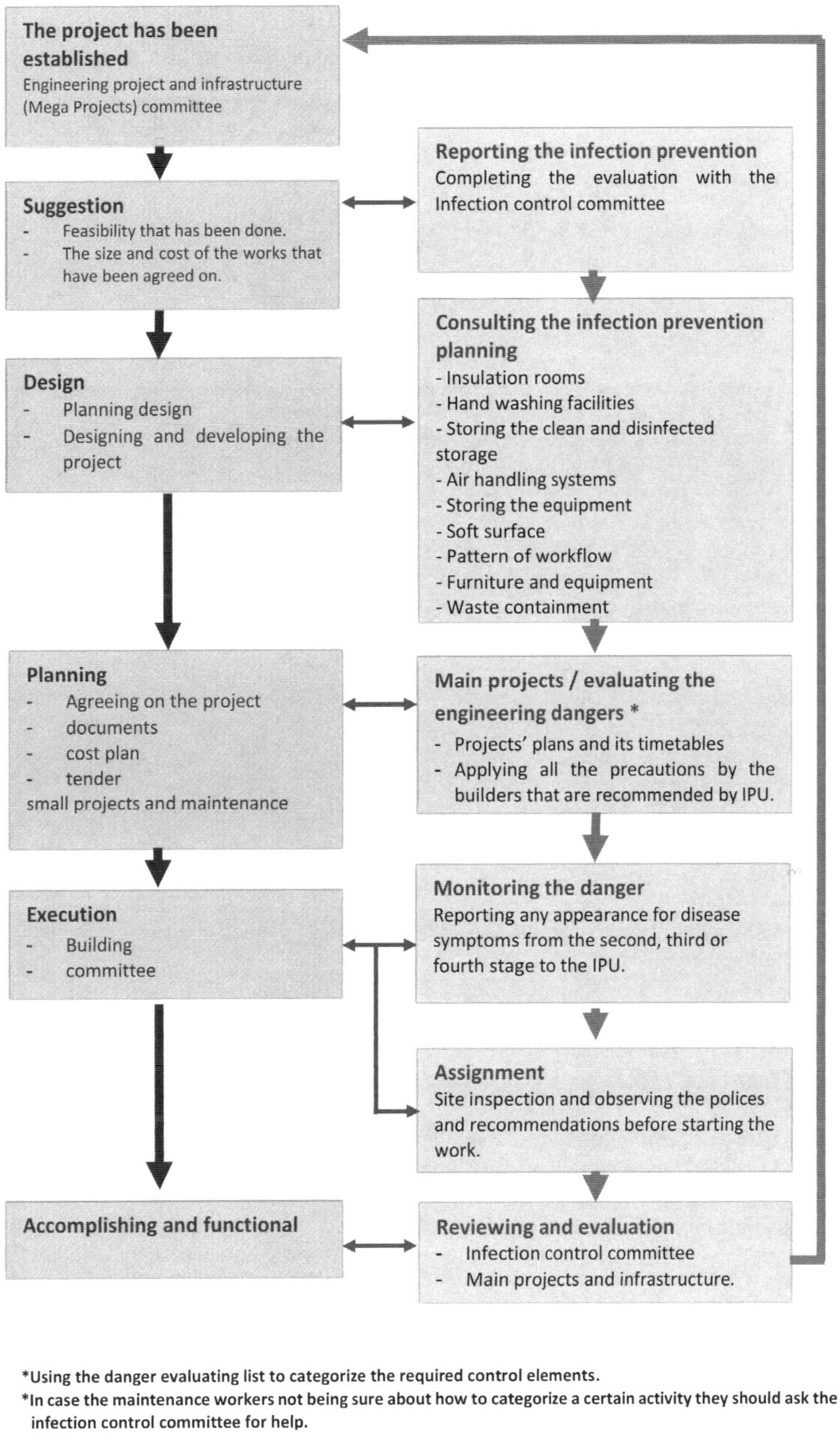

Figure 2. Flowchart for cooperation process between infection protection unit and construction workers [6].

IOP Conf. Series: Earth and Environmental Science 1056 (2022) 012007 doi:10.1088/1755-1315/1056/1/012007

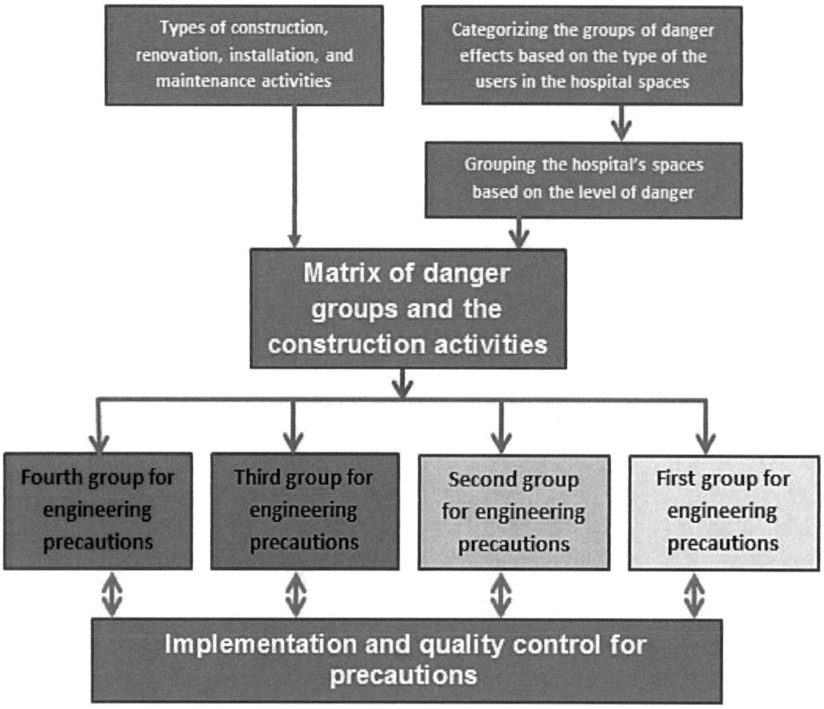

Figure 3. Flowchart that explains the precaution policy which followed by many countries like USA, Australia, Alaska, and Canada.

This matrix has released five groups of engineering precautions that should be applied during the construction and renovation works in the hospital with continuing to present the medical services. These precautions are to control the infection in the hospitals during the construction and renovation process, and it was found out these precautions participated in lowering the danger from the fourth to the first group, and they are:

a) Dust controlling:
- Executing by using systems that decrease the dust during the construction activities.
- Using dust control methods.
- Shutting the doors and the windows which aren't used with anti-permeability tapes.
- Covering the air and oxygen sources in the construction area.
- Before establishing the project, impermeable dust barriers should be put (made of plaster boards, pressed cardboard, or plywood) for the room from the actual ceiling (including the false ceiling) to the floor.
- Making sure that the windows, doors, plumping and electrical openings and the exhaust outlets is covered well in the construction and renovation area, also the ventilation paths and above the ceiling.

b) Labors, equipment, and construction waste:
- During the demolition, and every work that causes dust or ceiling works.
- Controlling that all the employees and workers who get in the construction site should put on a shoe cover then take it off in the waiting area when they leave the site.
- When the holes in the barriers are defined, they should be covered immediately or within 60 minutes then it must be fixed permanently within 4 hours.

c) Ventilation:
- Insulating the air condition system in the working areas.
- Monitoring the neediness for changing or cleaning the filters of the construction area.
- Keeping the negative pressure in the construction site.

- Making sure that the air has passed directly outside the openings, or it has been filtered by HEPA filters to reuse it.
- Keeping the negative pressure inside the waiting room and the construction area.

d) Completing the project:

- The dust barrier shouldn't be removed until the project is completed, and the site is completely cleaned.
- Removing the dust barrier carefully to reduce the dust diffusion.
- Wiping the horizontal and vertical work surfaces with wet and clean fabrics by water and detergents.

e) The medical stuff and controlling danger during the construction process:

- Making sure to wear the protective cloths which is took off before going to the patients care areas to control the infection, bacteria and virous, also the medical staff should be away from the construction site.
- If possible, moving the patients who are in wards nearby the site of construction.
- Making sure that the patients are away from the construction area.
- The patients should wear masks if it was necessary to cross the construction area.
- The resident patients who are highly in danger should come first in reaching it (as the positive pressure rooms are provided which its air is filtered by the HEPA filters) more than the places which don't have this category of patient staying areas.

Figure (4) shows a sample of pictures which clarify previous precautions applying on site.

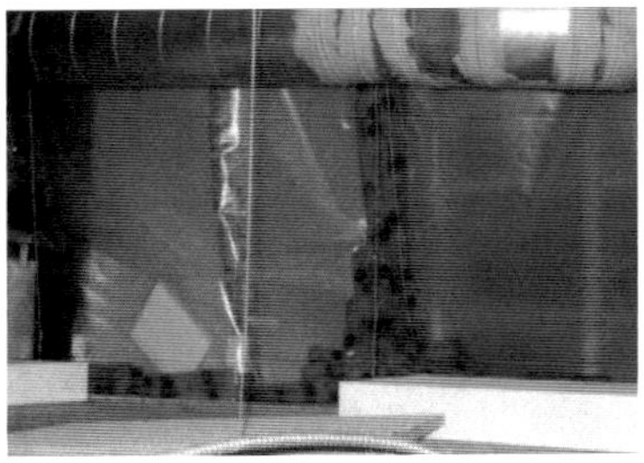

a- Insulation system between spaces above the false ceiling.

b- Exterior renovation works and what it causes due to the demolition works.

c- Insulation system for the maintenance works area and intaking the polluted air to prevent it from reaching the health care areas.

d- System of intaking the polluted air to prevent it from reaching the health care areas.

Figure 4. Sample of pictures clarifying onsite applied precautions, [3-5].

4. Conclusions

Setting the design criteria for the hospitals helps in decreasing the spread of infection during the renovation and the new construction process on stages with presenting the medical services in the completed parts of the hospitals, by:

- Separating the patients' spaces who are in danger as they could be completely separated from the whole hospital parts with secured entrances in case of the construction or renovation works inside the hospital.
- Designing the ventilation system as it could be easy to separate every part of the hospital and completely isolating it from the main ventilation system in the hospital during the construction and renovation for this part.
- Designing a flexible ventilation system to be able to separate between the patients and medical stuff paths and the workers and construction material paths during the construction and renovation for this part.
- Developing a structural module which achieves the flexibility as the same spaces in the hospitals could be reused in new or different activities.

References

[1] Larry Lee, and Chi, "Infection prevention and construction", Alaska Infection Prevention Conference, October 8, 2019.

[2] Canadian Standards Association Z317.13-17. Infection Control During Construction, Renovation, and Maintenance of Health Care Facilities, 2017.

[3] Fundamentals of Infection Control in HCF Issue 2 Rev. 3, Guidelines for the Canadian Construction Industry, 2004.

[4] Emad Marawan, "Technology effect on hospital buildings", *MSc., Military Technical College, Cairo, Egypt, 2001.*

[5] Association of professionals in infection control and epidemiology, *Mosby Inc., 2012.*

[6] Australian government, Northern territory government, NT hospital policy, 2021.

[7] A.F. Talento, M. Fitzgerald , B. Redington, N. O'Sullivan, L. Fenelon g , and T.R. Rogers, "Prevention of healthcare-associated invasive aspergillosis during hospital construction/renovation works", *Journal of Hospital Infection, 103, 1-12, 2019.*

[8] Hajime Kanamori, William A. Rutala, Emily E. Sickbert-Bennett, and David J. Weber, "Review of Fungal Outbreaks and Infection Prevention in Healthcare Settings During Construction and Renovation", *Healthcare Epidemiology, 61, August 2015, 433-444.*

[9] Francine Kidd, Clark Buttner, and Amy B. Kressel, "Construction: A model program for infection control compliance", *Association for Professionals in Infection Control and Epidemiology, Inc. Vol. 35 No. 5, 347-350, 2007.*

Towards a Model for the Assessment of Low-income Housing: A case study in Egypt

Hady Gamal Mabrouk[1,*], Sherif Elfiki[2], Amgad Fahmy[2]

[1] Misr International University, Faculty of Engineering, Department of Architecture
[2] Architectural Engineering and Environmental Design Department, College of Engineering and Technology, Arab Academy for Science, Technology and Maritime Transport, Cairo

[*]Corresponding Author Email: Hady.Mabrouk@miuegypt.edu.eg

Abstract. Egypt is currently witnessing an unprecedented development in the sector of low-income housing. So far, little research has investigated the extent to which such projects have contributed to the genuine satisfaction of the residents. Residential satisfaction (RS) describes the sense of fulfillment when someone has what he needs in a home. RS is a well-known criterion for planners, architects, experts, and decision-makers. Due to its remarkable significance, many models have been developed to assess RS in different places of the world. Residence is a highly complex concept with physical, psychological, and social implications. That is why it is believed that the particularity of every community shall prioritize the elements of RS in different ways. The present research is meant to develop an appropriate model for assessing residential satisfaction in low-income housing projects in Egypt. Towards this objective, the study undertakes an analytical review of previous RS models and examines them against the Egyptian context. The sought-after model shall address the characteristics of three primary domains: the residential unit, the neighborhood facilities, and the social environment. Accordingly, the outcome of this study provides an effective framework that can help developers and decision-makers to enhance their future housing schemes based on a more profound understanding of the individuality of the local context. Hence, contributing to the improvement of the built environment in Egypt.

Keywords: Low-income Housing in Egypt, Residential Satisfaction, Social Environment, Neighborhood Facilities, Residential Unit.

1. Introduction

Residential satisfaction is difficult to quantify since it is highly subjective and is dependent on the location, time, participation of a varied range of individuals, and the research objectives [1]. Numerous research on residential satisfaction have produced inconsistent findings. The term "Residential Satisfaction" has been used in various contexts, [2]. Understanding its main idea is crucial for obtaining reliable statistics, [3].

The term "housing satisfaction" or "residential satisfaction" should be defined and separated accordingly into_"housing" and "satisfaction". Firstly, Housing is a composite term that refers to the system's overall physical and social components. It is a complex subject that considers location, political jurisdiction, tenure, and structural type [4]. Secondly, satisfaction is the process of evaluating what was obtained or received compared to what was expected. Alternatively, it might be described more accurately as the perceived gap goal and achievement. Satisfaction is described not just in physical/materialistic terms but also in terms of the ability to establish relationships between residents. Moreover, it is described as a personalized response to an external environment [5]. As previously stated, residential satisfaction is the sense of fulfillment associated with having or achieving what one wants or wishes in a home [6]. It may also be defined as a reflection of homeowners' assessments of the overall quality of their life or as the extent to which an individual's housing expectations are met [7]. Eventually, residential satisfaction also refers to the subjective assessment of the residents' quality of life in a particular residential setting and serves as a trigger element impacting residential mobility [8].

2. Literature Review

Several experts and specialists have conducted empirical research on residential satisfaction to meet the objectives of their studies, which have used various techniques that are mostly articulated in the form of questionnaires to evaluate residential satisfaction [8]. The general qualities of these questionnaires are classified into numerous areas to derive perceived environmental quality indexes; in addition to assessing the participants' satisfaction with their home environment, several questionnaires' sections collect data on people's social, personal, demographic, and behavioral characteristics in their residence. In comparison, other sections emphasize features of the built environment such as housing conditions, neighborhood, physical and mental health. Besides, socio-economic factors such as financial results and social interaction [9]. On the other hand, several researchers focused on measuring residential satisfaction using criteria such as "security and walkability", "destination accessibility", "travel network" and "social network" [10]. A review of empirical data about cross-cultural residential satisfaction [4] was also studied [8], divided into four sections: sociodemographic features, housing features, neighborhood, and social environment. Furthermore, another research study discovered that the building's efficiency, the owners' maintenance culture, the social, neighborhood, management and housing unit's characteristics all have a notable influence on the inhabitants' satisfaction. [4]. Thus, according to this review, further studies are still needed to evaluate residential satisfaction in Egypt's low-income housing areas.

From Onibokun in 1974 and Amerigo and Aragones in 1997 to Mohit in 2010, researchers suggested combining relevant factors into an interaction system to investigate how those variables influence the residents' degree of residential satisfaction [11]. They advanced their theories to create models that could measure residential satisfaction in a holistic approach. Accordingly, in 1974, Onibokun first defined the integrated residential satisfaction index as the housing habitability System, which included four subsystems: tenant, home, environment, and management [12]. Then in 1990, Amerigo and Aragones created a more elaborate conceptual framework that explored and analyzed the active interplay between individuals and their living situations and defined it as a systemic model measuring residential satisfaction based on Onibokun's 1974 habitability [8] and focusing on analyzing the cognitive, emotional, and behavioral processes in the interaction process. Finally, in 2010, Mohit proposed a conceptual model of residential satisfaction that focuses on the interplay of objective and subjective factors of the home environment, influencing the level of residential satisfaction [13]–[16]. Residential satisfaction was shown to be determined by subjective and objective factors. In addition, the socio-economic qualities of the occupant and his or her household are influenced by his/her evaluation [13]. According to recent research done by Mohit 2010 and confirmed by Ibem et al. 2013 residential satisfaction is classified into four domains: Housing Unit that includes both Housing Unit Characteristics (HUC) and Housing Unit Supporting Services (HUSS). Neighborhood that includes both Neighborhood Characteristics (NC) and Housing Estate Supporting Facilities (HESF). Lastly, Social environment and the Socio-Demographics characteristics of the Inhabitants [13] and [14].

3. Methodology

Literature from prior studies was collected and compared. This literature review focuses on studies aimed at quantifying and evaluating residential satisfaction in low-income areas built by both the private and public sectors. The analysis aims to examine previously investigated domains and attributes and identify unexplored gaps that require more studying [17] and [18]. Additionally, to create an appropriate model for evaluating residential satisfaction in Egypt's low-income housing areas.

Although the contents were very similar, the titles and objectives differed slightly. Numerous studies have been undertaken to determine the resident satisfaction in areas such as housing, social environment, and neighborhood like Mohit et al. 2010 [13], Ibem et al. 2013[19], Alnsour et al. 2016[20], Tan 2016 [7], Cho 2020 [21], Wenjia 2018 [11] and others.

doi:10.1088/1755-1315/1056/1/012008

Some studies examined only one domain, such as the housing quality only like Xue et al. 2016 [22], Barmark 2015 [23] and Afacan et al. 2016 [24] or neighborhood satisfaction only like Bonaiuto et al. 2004 [25], Berkoz et al. 2009 [26], Afacan 2015 [27] and Hadavi et al. 2016 [28]. In contrast, others assessed resident satisfaction regarding a specific component or quality, such as the performance of housing, like Hashim et al. 2012[29], or performance in terms of safety like Ciorici et al. 2019 [30] These studies used a Likert scale to measure inhabitants' satisfaction with the specified characteristics. However, in some instances, additional qualitative work was carried out to verify, validate, and explain the findings, such as field trips or interviews that are semi-structured [11, 29, 31-32]. In 1976 Campbell et al. proposed three distinct domains for assessing resident satisfaction: "Housing", "Neighborhood" and "Social Environment" [4, 11, 14, 17, 32-34]. While not all reviewed research used the same terminologies, some researchers classified them into other categories.

Nonetheless, all the analyzed characteristics fall into one of three categories: housing, neighborhood, or social environment. Accordingly, these characteristics are gathered, categorized, and assessed with referencing the sources from which they were generated in the upcoming tables. Figure 1 describes the filtration process of the relevant studies in a flow diagram to reach the accurate number of studies used in this review paper.

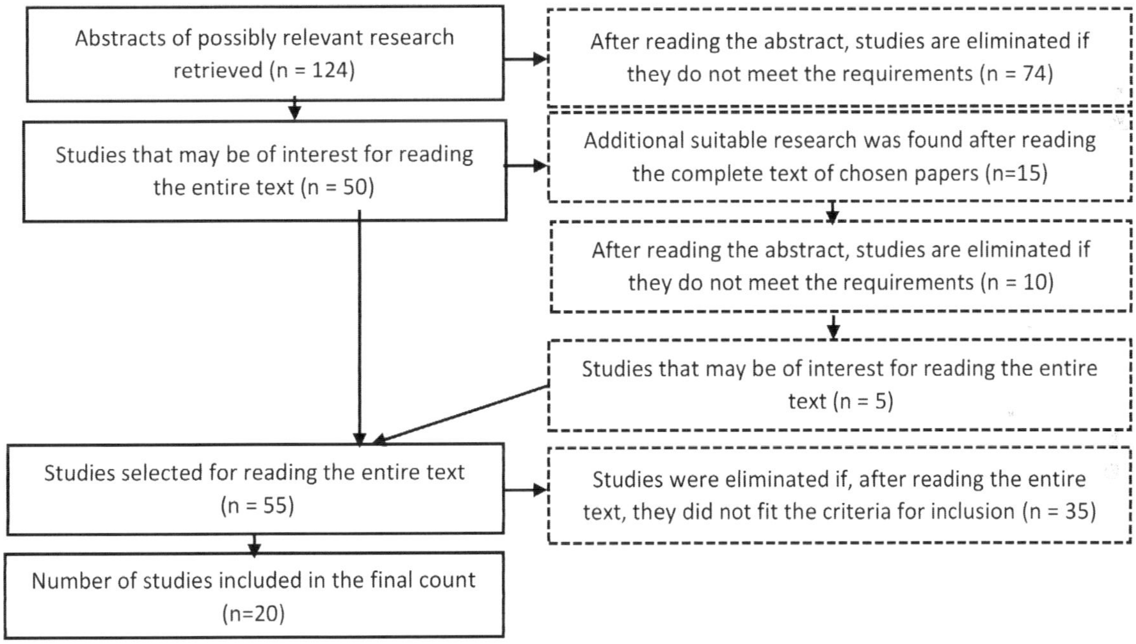

Figure 1. A flow diagram demonstrating the process of selecting research to be included in the review.

3.1 Housing Unit

Housing satisfaction is a recurrent theme in studying residential satisfaction. That is why various studies, such as Xue et al. 2016 [22], Barmark 2015 [23], and Afacan et al. 2016 [24], have concentrated exclusively on it. As illustrated here, housing satisfaction attributes were categorized into two broad categories: Housing Unit Characteristics (HUC) and Housing Unit Services (HUS), as shown in Table 1. According to a previous study, residents' satisfaction with their dwelling is significantly influenced by the structural indicator of housing. [11]. There are many different housing features, including physical qualities like laundries and washrooms, dining and living rooms' sizes, the level and number of outlets and the number of bathrooms and bedrooms. Other aspects of housing include housing affordability, privacy, and quality.

While developer-provided housing services include security and maintenance. All these aspects and services are mentioned in the indicators' list of Notably, comfort, building quality, layout, and dwelling size are all essential factors in determining residential satisfaction [35]. Meanwhile, another study stated that , a tiny room that has a shared kitchen and toilet with neighbors contributes to the unit's low rating, as does the unit's insufficient air circulation and ventilation, which results mainly in the apartment having low temperature in the winter and high temperature in the summer [36]. After gathering and classifying the attributes examined, the traits are categorized and listed in Table 1 in order of their frequency of occurrence and the literature from which they were taken.

Table 1: Housing Unit Domain; Attributes, Number of assessments in literature and its references.

Domain	Attributes and their frequency of occurrence	References
Housing Unit Supporting Services (HUSS)	Corridor (10), Staircase (7), Garbage collection (16), drain "Sanitary Services" (12), overall Cleanliness (7), Electrical repair services (18), Water supply (14), Gas supply (1), Television transmission (1), Management and maintenance practices (11), Rules and regulations within the housing estate (9), Handling of residents' complaints (3), Security/ safety (7), Overall support services (2), Plumbing (5), Street lighting (7), Lift (5), Swiftness in emergency repairs (3), Common area - maintenance (1), broadband network (1), Firefighting (6).	Alnsour and Hyasat, 2016a [20]; Cho, 2020a [21]; Dinç et al., 2014 [35]; Ibem et al., 2019a [37]; Ibem and Aduwo, 2013a [33].
Housing Unit Characteristics (HUC)	Living area (19), Dining area (18), Kitchen space (20), Master Bedroom-1 (19), Bedroom-2 (18), Bedroom-3 (18), No Of Bedrooms (6), toilet (9), Bathroom (12), Ventilation of House (17), Natural lighting (15), Orientation of flat (1), building materials used (14), Windows (8), doors (5), Walls (2), Floors (3), Façade appearance (3), Privacy in the residence (11), building features (3), structure types (1), size of residences (5), Overall physical condition (3), natural way of life (5), Terrace / balcony (6), Acoustic quality "Noise" (6), Washing & Drying Area (6), Car porch (3), Family hall (3), Store room (8), type of residence (5), Color of the building (1), Entrance lobby design (3), Community space (1), interior design (1), Front Yard (3), Back yard (3), Guest room (3), Maid room (3), Study room (3), Utility room (3), No of Toilets/Bathroom (4), Multipurpose room (3), Counseling room (1), Children's play room (1), Children's afterschool care room (1), Design of fence or boundary (1), Floor to ceiling clear height (1), Proportion of windows: walls (1), Water tightness from rain (1), Sanitary fittings (1), Unit temperature in winter (2), electrical sockets provided (4), Space utilization of flat layout (3), location of residence in the estate (4).	Ibem and Amole, 2013a, 2013b, 2014c [19], [38], [39]; Li & Song, 2009 [40]; Liu, 1999 [41]; M. A. Mohit et al., 2010a [13]; M. A. Mohit and Adel Mahfoud, 2015a [16]; M. A. Mohit & Azim, 2012b [42]; M. A. Mohit and Nazyddah, 2011a [43]; Mridha, 2015 [44]; Potter & Cantarero, 2006 [45]; Ukoha and Beamish, 1996 [46]; WENJIA, 2018a[23].

3.2 Neighborhood

After the dwelling domain, the neighborhood is the second most assessed domain. According to the literature, its features were classified into two broad categories: Neighborhood Characteristics (NC) and Housing Estate Supporting Facilities (HESF), as shown in Table 2, along with the sources cited. Certain surveys were conducted exclusively only to assess neighborhood satisfaction Bonaiuto et al. 2004 [25], Berkoz et al. 2009 [26] and Hadavi et al. 2016 [28]. Most neighborhood dissatisfaction occurs due to the increased distances to school, jobs, shopping, and medical centers. Additionally, community factors such as accident and crime prevention relate to increased satisfaction with one's home. Access to public transit, physical environment, communal and commercial attractions all contribute to neighborhood satisfaction.[47]. Another study affirms this, indicating that crime, lack of services, industrial expansion, and employment location are all potential contributors to neighborhood dissatisfaction [48].

Table 2: Neighborhood Domain: Attributes, Number of assessments in literature and its references.

Domain	Attributes and their frequency of occurrence	References
Neighborhood Characteristics (NC)	Distance to the workplace (11), Distance to school (11), Distance to Hospital (11), Distance to Health clinics (9), Distance to Shopping centers (11), Distance to Market (10), Distance to Bus station / Taxi Stand (10), Distance to train station / Metro (3), Public transportation facilities (5), Distance to Ferry services (1), location of housing estates (3), Distance to the town center (6), Nearness urban services (2), Distance to Children's play area (1), Distance to sporting facilities (4), Overall Distance "General" (2), Overall neighborhood facilities (1), I would recommend my immediate neighborhood to a friend (1), Closeness to friends and relatives (1), Distance to recreation facilities (4), Distance to Religious Building (5), Distance to the public library (3).	Alnsour and Hyasat, 2016a [20]; Cho, 2020a[21]; Dinç et al., 2014[35]; Ibem et al., 2019a[37]; Ibem and Aduwo, 2013a [33]; Ibem and Amole, 2013a, 2013b, 2014c [19], [38], [39]; Li & Song, 2009 [40]; Liu, 1999 [41]; M. A. Mohit et al., 2010a [13]; M. A. Mohit and Adel Mahfoud, 2015a [16]; M. A. Mohit & Azim, 2012b [42]; M. A. Mohit and Nazyddah, 2011a [43]; Mridha, 2015 [44]; Potter & Cantarero, 2006 [45]; Ukoha and Beamish, 1996 [46]; WENJIA, 2018a[11].
Housing Estate Supporting Facilities (HESF)	Parking (11), Perimeter road (7), Pedestrian walkways (5), Children's play area (9), Recreational area /Park (11), the landscape of the neighborhood (5), Cleanliness of public areas (6), Maintenance of residential block (6), Overall satisfaction with neighborhood facilities (2), Overall public facilities (6), Public infrastructure and urban services (4), Local shops (5), Shopping facilities (4), Food stalls (2), Kindergarten (3), Clinics (Medical/healthcare facilities) (4), School (Education Facilities) (4), Prayer hall (4), The accessibility for handicapped (1), the lighting level of public areas (2), Fitness equipment (3), Postage service (1), Multi-purpose hall (2), Public phone (2), Affordability of day care services (2), Counseling service (1), Psychiatrist service (1), Parenting education service (1).	

3.3 Social Environment

Less research, such as Addo 2016 [32], Mohit et al. 2010 [13], Wenjia 2018[11] and others, concentrated on assessing and quantifying the social environment. Occasionally, it has been shown that the social environment has a high correlation with overall life satisfaction [3], [15], [30] and [32]. Several types of research that evaluated housing units and/or neighborhood satisfaction included certain aspects of the social environment as neighborhood characteristics. Eventually, they were separated and included in Table 3 for the social environment. Because the social environment has received the least attention, the features evaluated are limited and diverse. The most highly rated variables were social ties and community safety, and they were shown to be related to one another. Nevertheless, communities where there are apparent relationships among the people, and they participate in a shared activity can aid in the reduction of crime levels in their area and improve their overall quality of life [30].

3.4 Socio-Demographics

In the area of socio-demographics some studies have found that the education level of the residents has an impact on their degree of residential satisfaction [33] and [49]. Another study confirmed this argument, which discovered a good association between home satisfaction and age group, income rate, educational level, and employment status simultaneously [6]. According to other studies, age has a beneficial effect on residential satisfaction, with older people reporting a higher sense of satisfaction regarding their homes than younger ones [50]. Furthermore, higher incomes allow households to upgrade their properties in more desirable areas, perhaps leading to higher satisfaction levels, based on a recent study [51]. In addition, the employment sector and household size significantly affect the level of residential satisfaction [33].

The findings in this part focus mostly on the inhabitants; including their age, income, period of residency, home ownership and household type, all of which affect the residents' satisfaction, either positively or negatively [4]. However, that is not all. Other studies indicate that sociodemographic characteristics such as the household size, residence duration, and tenure type are positively connected with total residential satisfaction [5]. The socio-demographics factors that affect residential satisfaction are summarized in Table 4.

Table 3: Social Environment Domain: Attributes, Number of assessments in literature and its references.

Domain	Attributes and their frequency of occurrence	References
Social Environment	Level of security (18), Level of crime (13), Accident situation (4), Noise level / Quietness (12), Anti-social activities (6), Police protection (2), Relationship with neighbored (9), Relationship with the community (7), Communal activities (10), The social differences of people in the community (1), Friendliness of neighborhood (1), I have good relationships with neighbors (1), I get emotional support from neighbors (1), I have many friends among my neighbors (1), Overall social environment (2), Privacy from neighbors (1), Special requirements for disabled (1), Elderly centers (1), Have cultural relations gotten better or worse? (1), Cross-cultural understanding (1), life in "......." is very stressful (1), Living in "......" is beneficial for my family (1), Racial discrimination (1), Educational status of the neighbors' (1), I have a good relationship with staff (1), I get emotional support from staff (1), Attitude of other children and youth in the apartment building (1).	Cho, 2020a [21]; Dinç et al., 2014 [35]; Ibem et al., 2019a [37]; Ibem and Aduwo, 2013a [33]; Ibem and Amole, 2013a, 2013b, 2014c [19], [38], [39]; Li & Song, 2009 [40]; Liu, 1999 [41]; M. A. Mohit et al., 2010a [13]; M. A. Mohit and Adel Mahfoud, 2015a [16]; M. A. Mohit & Azim, 2012b [42]; M. A. Mohit and Nazyddah, 2011a [43]; Mridha, 2015 [44]; Potter & Cantarero, 2006 [45]; Ukoha and Beamish, 1996 [46]; WENJIA, 2018a [11].

Table 4: Socio-Demographics Domain; Attributes, Number of assessments in literature and its references.

Domain	Attributes and their frequency of occurrence	References
Individual and Household's Socio-Economic Characteristics	Gender (13), Marital Status (11), Age (14), Family size/persons per bedroom (Household Size) (14), Educational level (14), Monthly family income (Monthly net income) (12), Employment sector (13), Length of residency (11), Vehicle owned "Main means of transportation" (5), Tenure Status (7), Occupation sector (2), Floor level (1), Ethnicity (1), Occupation "Housing" type (3), Unit type (1), Number of children (1), Youngest child's age (1), Number Of Bedrooms In the Dwelling Unit (4), Working wives (3).	Alnsour and Hyasat, 2016a [20]; Cho, 2020a[21]; Ibem et al., 2019a[37]; Ibem and Aduwo, 2013a [33]; Ibem and Amole, 2013a, 2013b, 2014c [19], [38], [39]; Li & Song, 2009 [40]; M. A. Mohit et al., 2010a [13]; M. A. Mohit and Adel Mahfoud, 2015a [16]; M. A. Mohit & Azim, 2012b [42]; M. A. Mohit and Nazyddah, 2011a [43]; Potter & Cantarero, 2006 [45]; Ukoha and Beamish, 1996 [46]; WENJIA, 2018a[11].

4. Discussion

After showing each of the characteristics by their occurrence in prior case studies and literature, it is vital to grasp these studies' relevance and conclusions. As they can aid the government and the commercial sector in interpreting residents' perspectives in the residential environment where they inhabit, allowing them to design future housing developments that will improve residents' Residential Satisfaction and encourage social inclusion [52]. It is also important to highlight some key points that should be explored in future research. **First:** The researchers assert that questionnaires are preferred when working with low-income individuals, as their results can be more generalized and highly applicable to more comprehensive ranges. They can also forecast future outcomes and study causal relations [36]. However, relying solely on a Likert scale to assess resident satisfaction prevents residents from explaining their responses, limiting the research's capacity to address the problem effectively [17].

Individuals may be dissatisfied with the same quality for various reasons, and their lack of an adequate method of expression may prohibit researchers from discovering the proper repair [53] and [54]. Qualitative data may be required to additionally describe and identify quantitative conclusions. **Second:** RS should be viewed as a dynamic phenomenon rather than a static outcome since humans are ever-wanting creatures, and once one need is met, another takes its place [8] and [36]. Thus, in response to the proposed research question, residential (dis)satisfaction is the positive or negative psychological outcome of the evaluative process between perceived and expected features of the living environment at various scales, in which socio-physical characteristics of the living environment are evaluated at various scales using the individual's mental reference. This subjective reaction to the objective surroundings may result in a sense of fulfillment or deprivation [55]. However, RS is not always generated by total congruence between the present and ideal residential environments. Indeed, inhabitants can be satisfied under less-than-optimal conditions [12] via psychological processes such as need revision, cognitive dissonance reduction, and cognitive adaptation [2] and [56]. **Third:** A brief scan of some earlier attempts reveals their flaws. For example, several RS models and classifications exclude distinctions between physical and social components of the home environment [50] and [57]. However, canter and Rees (1982), in a similar manner to the stated classifications, stressed the relevance of this division in every RS study. Additionally, RS researchers have recently underlined the importance of diverse experts and the division of responsibilities in this field [25] and [36]. Thus, specialization should occur, and experts should address each factor.

5. Conclusion

Residential satisfaction is a broad research that may be used in various contexts. Residential satisfaction studies vary according to location, individuals, culture, and, most importantly, the house itself. Therefore, this research will result in varying degrees of satisfaction, depending on the study's purpose. Nonetheless, residential satisfaction is notoriously difficult to be defined accurately for three reasons [6]. First, it has two terms: residential and satisfaction. Secondly, measurement mostly relies on the subject's subjective judgment, inherently unstable. Thirdly, because the topic of RS involves several variables, a diverse group of specialists is required to develop a comprehensive model. Finally, this study implies that more research on residential satisfaction may be conducted on each project according to its specific case to help in guiding housing design and regulation.

References

[1] J. D. Lord and G. S. Rent, 1987 Residential Satisfaction in Scattered-Site Public Housing Projects, The Social Science Journal, Vol. 24, No. 3, pp. 287-302.
[2] G. Galster, 1985 Evaluating Indicators for Housing Policy: Residential Satisfaction Vs. Marginal Improvement Priorities, Social Indicators Research, Vol. 16, pp. 415-448.
[3] A. Emami and S. Sadeghlou, 2021 Residential Satisfaction: A Narrative Literature Review Towards Identification of Core Determinants and Indicators, Housing, Theory and Society, Vol. 38, No. 4. Routledge, pp. 512-540.
[4] M. A. Mohit, A. Mahfoud, M. Al-Khanbashiraja, and M. Al-Khanbashi Raja, 2014 Residential Satisfaction-Concept, Theories and Empirical Studies, Malaysia.
[5] M. A. Mohit and M. Azim, 2012 Assessment of Residential Satisfaction with Public Housing in Hulhumale', Maldives, Procedia - Social and Behavioral Sciences, Vol. 50, pp. 756-770.
[6] S. Weidemann and J. Anderson, 1985 Home Environments: A Conceptual Framework for Residential Satisfaction, New York.
[7] T. H. Tan, 2016 Residential satisfaction in gated communities: A case study of Desa Park City, Kuala Lumpur, Malaysia, Property Management, Vol. 34, No. 2, pp. 84-99.
[8] M. Amerigo and J. Aragones, 1997 A Theoretical and Methodological Approach to The Study of Residential Satisfaction, Journal of Environmental Psychology, Vol. 17, pp. 47-57.
[9] R. C. Fauth, T. Leventhal, and J. Brooks-Gunn, 2004 Short-term effects of moving from public housing in poor to middle-class neighborhoods on low-income, minority adults' outcomes, Social Science and Medicine, Vol. 59, No. 11, pp. 2271-2284.
[10] E. Leslie and E. Cerin, 2008 Are perceptions of the local environment related to neighborhood satisfaction and mental health in adults, Preventive Medicine, Vol. 47, No. 3, pp. 273-278.

[11] X. Wenjia, 2018 The Residential Satisfaction of the Low-Cost Housing in The New Second-Tier City of Jiangsu Province, China Xi Wenjia Faculty of Built Environment University of Malaya Kuala Lumpur 2018, Malaysia.
[12] M. Amerigo and J. Aragones, 1990 Residential Satisfaction in Council Housing, Madrid.
[13] M. A. Mohit, M. Ibrahim, and Y. R. Rashid, 2010 Assessment of residential satisfaction in newly designed public low-cost housing in Kuala Lumpur, Malaysia, Habitat International, Vol. 34, No. 1, pp. 18-27.
[14] M. A. Mohit and N. Nazyddah, 2011 Social housing program of Selangor Zakat Board of Malaysia and housing satisfaction, Journal of Housing and the Built Environment, Vol. 26, No. 2, pp. 143-164.
[15] M. Mohit and Z. Rahim, 2012 Summary of Research Report Assessment of Residential Satisfaction in High-Rise Condominium and Terrace Housing: Case Studies from Kuala Lumpur.
[16] M. A. Mohit and A. K. Adel Mahfoud, 2015 Appraisal of residential satisfaction in double-storey terrace housing in Kuala Lumpur, Malaysia, Habitat International, Vol. 49, pp. 286-293.
[17] A. Ahmad, Y. Mansour, M. Gabr, and M. H. Khalil, 2020 Quality of Life and the Architecture of Low-income Neighborhoods, in Architecture and Urbanism: A Smart Outlook, Springer International Publishing, pp. 341-355.
[18] B. van Wee and D. Banister, 2016 How to Write a Literature Review Paper, Transport Reviews, Vol. 36, No. 2, pp. 278-288.
[19] E. O. Ibem and D. Amole, 2013 Subjective life satisfaction in public housing in urban areas of Ogun State, Nigeria, Cities, Vol. 35, pp. 51-61.
[20] J. A. Alnsour and A. S. Hyasat, 2016 Residential Satisfaction with Low Income Housing in Jordan: Salt City as a Case Study, Jordan Journal of Economic Sciences, Vol. 3, No. 1, pp. 31-44.
[21] M. Cho, 2020 Residential satisfaction among low-income single-mother households: the case of residential welfare facilities in South Korea, Archnet-IJAR, Vol. 14, No. 3, pp. 359-378.
[22] P. Xue, C. M. Mak, and Z. T. Ai, 2016 A structured approach to overall environmental satisfaction in high-rise residential buildings, Energy and Buildings, Vol. 116, pp. 181-189.
[23] M. Barmark, 2015 Social determinants of the sick building syndrome: Exploring the interrelated effects of social position and psychosocial situation, International Journal of Environmental Health Research, Vol. 25, No. 5, pp. 490-507.
[24] Y. Afacan and H. Demirkan, 2016 The influence of sustainable design features on indoor environmental quality satisfaction in Turkish dwellings, Architectural Science Review, Vol. 59, No. 3, pp. 229-238.
[25] M. Bonaiuto, M. Bonnes, and M. Continisio, 2004 Neighborhood evaluation within a multiplace perspective on urban activities, environment, and Behavior, Vol. 36, No. 1, pp. 41-69.
[26] L. Berkoz, S. S. Turk, and Ö. L. Kellekci, 2009 Environmental quality and user satisfaction in mass housing areas: The case of Istanbul, European Planning Studies, Vol. 17, No. 1, pp. 161-174.
[27] Y. Afacan, 2015 Resident satisfaction for sustainable urban regeneration, Proceedings of the Institution of Civil Engineers: Municipal Engineer, Vol. 168, No. 4, pp. 220-234.
[28] S. Hadavi and R. Kaplan, 2016 Neighborhood satisfaction and use patterns in urban public outdoor spaces: Multidimensionality and two-way relationships, Urban Forestry and Urban Greening, Vol. 19, pp. 110-122.
[29] A. E. Hashim, S. A. Samikon, N. M. Nasir, and N. Ismail, 2012 Assessing Factors Influencing Performance of Malaysian Low-Cost Public Housing in Sustainable Environment, Procedia - Social and Behavioral Sciences, Vol. 50, pp. 920-927.
[30] P. Ciorici and P. Dantzler, 2019 Neighborhood Satisfaction: A Study of a Low-Income Urban Community, Urban Affairs Review, Vol. 55, No. 6, pp. 1702-1730.
[31] H. N. Husin, A. H. Nawawi, F. Ismail, and N. Khalil, 2015 Correlation Analysis of Occupants' Satisfaction and Safety Performance Level in Low-Cost Housing, Procedia - Social and Behavioral Sciences, Vol. 168, pp. 238-248.
[32] I. A. Addo, 2016 Assessing residential satisfaction among low-income households in multi-habited dwellings in selected low-income communities in Accra, Urban Studies, Vol. 53, No. 4. SAGE Publications Ltd, pp. 631-650.
[33] E. O. Ibem and E. B. Aduwo, 2013 Assessment of residential satisfaction in public housing in Ogun State, Nigeria, Habitat International, Vol. 40, pp. 163-175.
[34] J. A. Alnsour and A. S. Hyasat, 2016 Residential Satisfaction with Low Income Housing in Jordan: Salt City as a Case Study, Jordan Journal of Economic Sciences, Vol. 3, No. 1, pp. 31-44.
[35] P. Dinç, E. Özbilen, and M. B. Bilir, 2014 A multi-dimensional scale for measuring residential satisfaction in mass housing projects, Indoor and Built Environment, Vol. 23, No. 6, pp. 864-880.
[36] C. O. Aigbavboa and W. D. Thwala, 2012 An appraisal of housing satisfaction in South Africa low-income housing scheme, International Journal of Construction Management, Vol. 12, No. 1. Chinese Research Institute of Construction Management, pp. 1-21.

[37] E. O. Ibem, E. A. Ayo-Vaughan, A. O. Oluwunmi, and O. A. Alagbe, 2019 Residential Satisfaction Among Low-Income Earners in Government-Subsidized Housing Estates in Ogun State, Nigeria, Urban Forum, Vol. 30, No. 1, pp. 75-96.
[38] E. O. Ibem and D. Amole, 2013 Residential Satisfaction in Public Core Housing in Abeokuta, Ogun State, Nigeria, Social Indicators Research, Vol. 113, No. 1, pp. 563-581.
[39] E. O. Ibem and D. Amole, 2014 Satisfaction with Life in Public Housing in Ogun State, Nigeria: A Research Note, Journal of Happiness Studies, Vol. 15, No. 3, pp. 495-501.
[40] S. M. Li and Y. L. Song, 2009 Redevelopment, displacement, housing conditions, and residential satisfaction: A study of Shanghai, Environment and Planning A, Vol. 41, No. 5, pp. 1090-1108.
[41] A. M. M. Liu, 1999 Residential satisfaction in housing estates: A Hong Kong perspective.
[42] M. A. Mohit and M. Azim, 2012 Assessment of Residential Satisfaction with Public Housing in Hulhumale', Maldives, Procedia - Social and Behavioral Sciences, Vol. 50, pp. 756-770.
[43] M. A. Mohit and N. Nazyddah, 2011 Social housing program of Selangor Zakat Board of Malaysia and housing satisfaction, Journal of Housing and the Built Environment, Vol. 26, No. 2, pp. 143-164.
[44] M. Mridha, 2015 Living in an apartment, Journal of Environmental Psychology, Vol. 43, pp. 42-54.
[45] J. Potter and R. Cantarero, 2006 How does increasing population and diversity affect resident satisfaction? A small community case study, environment, and Behavior, Vol. 38, No. 5, pp. 605-625.
[46] O. M. Ukoha and J. O. Beamish, 1996 Predictors of Housing Satisfaction in Abuja, Nigeria, Housing and Society, Vol. 23, No. 3, pp. 26-46.
[47] A. OZO, 1990 Low-cost Urban Housing in Nigeria Strategies, Nigeria.
[48] G. Mulligan, J. Carruthers, and M. Cahill, 2005 Urban Quality of Life and Public Policy: A Survey, in Urban Dynamics and Growth: Advances in Urban Economics, pp. 730-802.
[49] E. Vera-Toscano and V. Ateca-Amestoy, 2008 The relevance of social interactions on housing satisfaction, Social Indicators Research, Vol. 86, No. 2, pp. 257-274.
[50] M. Lu, 1999 Determinants of Residential Satisfaction: Ordered Logit vs. Regression Models, United States.
[51] B. Frank and T. Enkawa, 2009 Economic drivers of dwelling satisfaction: Evidence from Germany, International Journal of Housing Markets and Analysis, Vol. 2, No. 1, pp. 6-20.
[52] Z. Huang and X. Du, 2015 Assessment and determinants of residential satisfaction with public housing in Hangzhou, China, Habitat International, Vol. 47, pp. 218-230.
[53] S. Dawadi, S. Shrestha, and R. A. Giri, 2021 Mixed-Methods Research: A Discussion on its Types, Challenges, and Criticisms, Journal of Practical Studies in Education, Vol. 2, No. 2, pp. 25-36.
[54] Linda N. Groat and David Wang, 2013 Architectural Research Methods, John Wiley & Sons, Inc., Hoboken, New Jersey.
[55] J. R. Hipp, 1992 Specifying the Determinants of Neighborhood Satisfaction: A Robust Assessment in 24 Metropolitan Areas, North Carolina, [Online]. Available: http://sf.oxfordjournals.org/
[56] S. J. T. Jansen, 2014 Why is Housing Always Satisfactory? A Study into the Impact of Cognitive Restructuring and Future Perspectives on Housing Appreciation, Social Indicators Research, Vol. 116, No. 2, pp. 353-371.
[57] A. G. Onibokun, 1976 Social System Correlates of Residential Satisfaction, Environment and Behavior, Vol. 8, No. 3, pp. 323-344.

Application of COVID-19 guidelines in workplace design

Salma M. Elskalakany[1, *], Osama Tolba[2], Amgad Fahmy[3]

[1] Department of Architecture Engineering and Environmental Design, Arab academy for science, technology, and maritime transport, Department of Architectural engineering and Environmental design, Cairo, Egypt.

[*]Corresponding Author Email: Salma.alshalkany@hotmail.com

Abstract. The spread of COVID-19 has been recorded as one of the most severe worldwide pandemics. Workplaces are a germicidal environment for spreading viruses due to their high occupation. As a response, international health, safety, and architectural organizations have published pandemic-related guidelines to create a safe working environment for the employees. Covid-19 has impacted reshaping the internal architectural workspace design. The research shall review the published international organizations' guidelines on workspace design and its architectural interpretations established by international design companies, which seek to create safer designs for workplaces post-pandemic. This study is a concise review of the workplace-related guidelines and precautions published by health and safety and architectural organizations such as WHO, CDC, OSHA, ASHARE, AIA, AHIA, and ILO. Secondly, aggregate data has been developed to frame and compile the design interpretations of the published pandemic guidelines recognized by several international architecture and furniture companies which has been interpreted into architectural plans of the common spaces of the office buildings. As a result, a matrix table was created which includes the common spaces in office buildings and their post-COVID architectural solutions. This matrix shall be used as a tool to be employed as a guide for design consultants in designing healthy workplaces that are resilient to airborne pandemics, which shall benefit in revisiting the architectural standards according to the published workplaces Pandemic guidelines.

Keywords: Design workplaces, COVID-19, Health, Architectural standards, Office spaces.

1. Introduction

The outbreak of COVID-19 was declared as a pandemic by WHO on 11 March 2020. Evidence highlighted that social mobilization plays a vital role in spreading infectious diseases [1]. Chang, et al., stated that close contact and exposure can play a major role in SARS-CoV-2 transmission, the prolonged short-range aerosol transmission of the virus is possible in crowded and poorly ventilated enclosures [2]. As COVID-19 proved to be a highly portable and deadly infectious disease, we can prevent its spread by preventing overcrowding and good ventilation in buildings.

After the spread of Covid-19, several official health and safety organizations published new precautions and guidelines related to workplace designs to overcome this pandemic (airborne viruses). The recommendations/guidelines of the workplaces have been collected from the World Health Organization (WHO), the Occupational Safety and Health Administration (OSHA), the Centers for Disease Control and Prevention (CDC), the American Society of Heating, Refrigerating and Air-Conditioning Engineers (ASHARE), the American Institute of Architects (AIA), the Association of Healthcare Internal Auditors (AHIA) and International Labour Organization (ILO).

WHO issued a standard to maintain a distance from one person to another with at least a 1-meter distance [4]. Accordingly, Anggrain stated that physical distancing has had an impact during the design process. According to Work precedents, physical distancing was a significant challenge to develop new design solutions [5]. ILO organization have used the new work hybrid model of the future post-Covid-19 office. This model promotes a relaxed working atmosphere that allows workers to easily reconfigure their workplace as often as they want to satisfy their needs [6]. Brodie, et al., indicated that companies raised the awareness that instead of retrofitting the existing office; they can rethink the footprint needed by establishing several new work models such as co-working flex space and remote work. These workplace concepts will affect the design of the basic workplaces, improve how work is done, and lead to savings in many different aspects like the office capital cost, rents and operations and maintenance and management, and other expenses [7].

Nediari, et al., tried to improve traditional working standards by highlighting how office design evolves in response to the necessity of worker health and safety in the workplace [8]. All the published measures and architects' applications have been focused on many aspects like promoting physical distance and reducing people flow

1.1. Research problem
The research problem according to Mortice that post pandemic, new social norms affect how people use the workplaces subsequently it affected the design of the workplaces, which led to designing spaces that avoid close face-to-face contact, reduce blind corners [3] and minimizing the danger of spreading airborne viruses.

1.2. Research aim and objective
This research focused on workplace design and how it responds to the new normal of the Covid-19 pandemic. The research aim is to review the published precautions and guidelines related to workplace design post-COVID-19, in addition to collecting the commonly developed architectural and furniture solutions and compiling them in one matrix table. This table includes the following:

1- Horizontal rows include the common spaces of the office buildings.
2- Vertical columns includes the commonly architectural solutions interpreted by the international and architectural furniture companies

This table shall be considered as a guide for design consultants in designing healthy workplaces that are resilient to airborne pandemics and guide architectural standards writers in identifying the newly required design guidelines for pandemics like the COVID-19 pandemic.

1.3. World health, safety and Architectural organizations guidelines post COVID-19
ASHARE uniquely published guidelines specified the design, operation, and maintenance of heating, ventilating Air-conditioning systems post-covid-19 to help reduce the virus transmitted through the air and the guidance of using the UV technology with shorter wavelength 254-nm UVC [9]. Where Columbia University reported that UV technology kills viruses before we breathe them. Although it can penetrate and kill viruses floating in the air, as viruses are incredibly small [10].

CDC also recommended using short wave UVC energy to kill viruses like COVID-19. ASHARE highlighted that it could be used in the mechanical ventilation paths to treat the air indirectly before entering the space. And also, it can be used as a light spectrum to inactivate the virus [11]. Providing employer's safe and healthy workspace was one of the OSHA organizations' missions post-covid-19.

AIA recommended measures are programmatically translated for the publications of the CDC and world health organizations. It focuses on how spaces will be changed, and the buildings will be more than different usage post-COVID-19. All the international organizations' guidance will recap that architecture building standards will be recapped to maintain post-COVID-19. AIA's guides recommend the usage of outdoor spaces as a tactic for healing after covid-19. AIA's recommendations for curing offices after covid-19 are similar to school guidance, highlighting the main aspects of social distancing and people flow and circulation patterns. The below table concluded the most common guidelines published by Health and safety and architectural organizations post-covid-19. (See Table 1) Health, safety and architectural guidelines post-covid-19.

Table 1. Health and safety and architectural guidelines post-covid-19.

Guidelines	References
Request a physical distance of 6 feet (1.83 meters) and apply social isolation	[4, 12]
Reduce the number of people in the building during working hours;	[13, 14] [4] [15]
Work-from-home policies (telework or absenteeism)	[14]
Floorplans to be revisited, if appropriate, remove or rearrange office furniture to keep the recommended physical distance.	[4, 12, 16]
Employees must decrease from face-to-face contact	[12, 14] [17]
Sharing desks are forbidden	[15, 18] [12]
Limit the no of staff and make flexible working hours. And meeting limits of 10 individuals or less and encourage virtual meetings.	[14, 19] [12]
Occupation rate of 107.64 ft^2 (10 m^2) per person	[12]
Think for more office usage for outdoor spaces	[16]
Increase plants to purify the air	[20]
Practicing hand hygiene is a simple yet effective way to prevent infections	[21]

2. Methodology

First, this paper is a concise review of the published precautions and guidelines by the official health, safety, and architectural organizations related to workplace design. In addition to reviewing the most common solutions and design interpretations of post-COVID-19 guidelines applied by several international architectural and furniture companies. The criteria for choosing the international companies are:

- Their worldwide ranking.
- The significant number of workplace design and executed projects.
- Considering their designed and executed architectural solutions as a response to COVID-19.

Those international companies have interpreted the published guidelines into architectural plans and physical architectural solutions, either in their constructed workplaces or in their future workplace projects.

Secondly, the development of an aggregate model for those proposed solutions. Aggregate methods are produced by combining information from multiple sources instead of information from an individual. The Aggregate data of this paper are collected from the developed solution by the different companies for each space in the office buildings. These data have been compiled into a single matrix table, (see table 2) in order to be considered as a guide for design consultants in designing healthy workplaces that are resilient to airborne pandemics and to acquire the most applicable solution for architectural standardization.

3. Architectural interpretations of COVID-19 guidelines

Nediari, et al., added that designers have begun to plan a new sustainable design approach in workplace design as a result of COVID-19 pandemic [8]. Several international architectural and furniture companies have developed a guidebook includes the post-pandemic architecture applications to prevent the spread of COVID-19 in office buildings. This guidebook has been developed based on the interpretations of the published guidelines by the safety and health organizations. Which includes safety instructions, new configurations for their existing office buildings, new design solutions for upcoming new workplaces and the innovative furniture products used post COVID-19. However, the most common architectural solutions for different zones in the office building are mentioned below and represented in architectural plans (see Table 2).

3.1. Entrance

Whole building design guide (WBDG) specified that workplace entrances are the most fertile environment for passing viruses from people, products, and equipment and have continuous high densities [22]. There are four applied Architectural applications for the entrance. The first application is the addition of a gate with a particular type of UV light (far-UVC light with a wavelength of around 220 nm). The second application is to create a sanitization space, as a space must be added to the workplace layout between the main entrance and the main lobby. The third application declare that aqueous and specific gaseous treatments were proven to be effective against microbial viruses in building entrances. Many international and local companies used it to disinfect people, several international companies applied the fourth application during COVID-19 to disinfect the shoes due to their high occupancy workplaces. Notrax company introduced the disinfecting floor mat as a shoe sanitizer system to prevent workplace virus transmission through shoes [23].

3.2. Open office space

There are four applied architectural applications for the open office space. The first application highlighted the ability to change furniture arrangements to be seated back-to-back instead of the traditional face-to-face setting, emphasizing the zigzag pattern arrangement. The second application represents Wakabayashi's solution, applied by Google, which is the concept of using team pods rather than rows of desks, each pod is a flexible wheeled piece arranged together to create a work environment [24].

The third application represents Fehlbaum's solution, which offered a workstation that comply with COVID-19 regulations; it used separated partitions to protect workers [25]. Where panels are the main elements that must be installed between desks to protect them. A six-foot office, the distance between employees, is a new addressed element in the workplace that affects the office layout. Where the fourth application represents Cushman & Wakefield Company's solution that applied the 6 feet rule. The floor around the desks was designed to take the circle of the 6 feet as a visual barrier for employees [26].

3.3. Meeting room
There are four applied architectural applications for the meeting room. Gensler company application is presented in the first architectural application, that introduced the outdoor meeting room to access fresh air and encourage people to collaborate [27]. Meanwhile, the second application represents Office snapshot magazine introduction of the triangular meeting table concept, obtaining the three-sided table to achieve the physical distance rule [28].

The third application represent Perkins & Will solution, which introduces the single occupant and co-op styles concept in which the meeting room model has been articulated to the concept of a lounge [29]. Brodie, et al., application is presented in the fourth architectural application, which is the introduction of conference meeting rooms rather than large meeting tables assigned too many employees at the same time, where the concept of blending between remote and office workers, between virtual and physical presence [7, 22, 30].

3.4. Vertical circulation (stairs)
Architects are now featuring stairs as the main vertical circulation for the major projects to encourage movement and healthier workplaces. Meanwhile, there are four applied architectural applications. Weiss proposed solution is presented in the first application, which is using the double stairwell, it is one of the optimized stairs shall fit the social distance guidelines. It distributes the people flow into two stairs in the same area. It can also be designed as up and down stairwells and fits the small footprint areas [31]. Active metal company solution presented in the second application, which is the concept of designating open/atrium stairwells [32].

The third and fourth applications are presented from Gensler company published solutions. The third solution describe the company's recommendation to its employees to use the fire stairs. This shall reduce the employee traffic from the main stairs and designate up stairwell and the other to be down stairwell for one-way vertical traffic. The fourth application is the outdoor stairs, which give occupants more space and fresh air to move [33].

3.5. Shared office
There are four applied architectural post-pandemic applications for the shared office. Profirm's solution is presented in the first application, which reconfigures the desks in a crisscross pattern [34]. The second application presents Bordie, et al., solution of redesigning the workplace to support company priorities and blend between private offices and cubical [7].

BOS company solution is presented in the third application was the innovation of a new shape representing the 6-foot rule [35]. Meanwhile, the fourth application describes Perkins & Will company solution which focuses on the desks arrangements to be back to back [36].

3.6. Cafeteria
There are four applied architectural post-pandemic solutions for the cafeteria. Fehlbaum's solution is presented through the first application, which introduced the dancing wall; it is a movable wheeled mobile wall partition that can be used as a plant wall, whiteboards, pinboards, and can be moved by the user as a separator [25]. Meanwhile, the second solution describes Gensler proposed solution where the cafeteria layout is designed to maintain the physical distance rule and reduce the number of people (27). Perkins & Will company proposed solution for the cafeteria is presented in the third application, where the space program is exposing the plan to relocate the cafeteria to an outside covered heated/cooled facility [29]. The fourth application is designing the cafeteria tables with a main transparent partition to prevent virus transmission.

3.7. Reception desk
There are three applied architectural post-pandemic solutions for the reception desk. The first application was applied either by maximizing its size to fit the minimum physical distance rule or designing an element in front of the reception desk to maintain the recommended physical distance and reduce the people flow around the reception desk. The second application represents National Institutes of Health (NIH) solution, where it proposed the traditional solution, applied by companies worldwide, of installing screens to protect visitors and the receptionist from virus transmission [37]. The third application is proposed by Gensler, which used the digital design, which is a digital customer experience that helps the customer talk to digital screens and the virtual receptionist is existed on screens to help customers [27].

3.8. Waiting area and lounge

There are four applied solutions in the waiting area and lounge. The first application presenting the application of Vitra company which introduced the individual seating alcove high back, as a safe setting [38]. BOS, a furniture design firm, solution is presented through the second application which introduced the concept of separated modular pieces and configuring the layout to consider small groups in the waiting area [39]. Snapshots magazine published solution for the waiting and lounge area is introduced in third application, which proposed the concept of space segregation, the visitors are more secure by avoiding people's crowded circulation [40]. The fourth application removed the waiting area and replaced it with reappointments and digital reservations.

3.9. Horizontal circulation (corridors and pathways)

There are three applied applications for the corridors and pathways. Gensler application for this area is presented through the first application, which changed the horizontal circulation to be prepared on one-way foot traffic circulation [27, 41]. Perkins & Will solution for this area is presented through the second application which consider narrowing pathways to force employees to move in a one-way direction [36]. The third application represents Profirm solution, which recommended using ceiling barriers in public areas, it can be installed with different elements like plants, screens, plexiglass panels [34]. In the same context, UCONN proposed installing the plexiglass panels from the ceiling in the two-way traffic corridor to avoid virus transmission between people [42].

3.10. Public areas (toilets)

There are three applied applications in the toilets post-pandemic. The first application represents Snapshots magazine solution for this area, which highlighted the importance of using individual scattered toilets all over the architectural workplace layout to serve more employees [40]. The second application is minimizing the number of people using the toilets to minimize infection where the architectural layout will be subtracted to fill another space gap. CDC declared that hand hygiene and washing practice is the most effective way to prevent the transmission of virus and prevent infection [21]. In the same context, the third application represents Snapshots magazine's another solution for this area, which stated that an international design firm designed sanitary facilities to encourage the employees to wash their hands frequently with easy access to sinks without opening doors or human contact [40].

4. Conclusion and future work

Focus on health in designing the workplaces is the new approach of designers post-pandemic, the conclusion of this research raised awareness of how workplaces and office buildings are changing in response to the importance of employee health and safety.

The developed matrix table that describes the common solutions for each space in the office buildings post-pandemic, these solutions shall be used as a guide for design consultants in designing healthy workplaces that are resilient to airborne pandemics.

Additionally, the architecture standards must conform to the newly published workplace health and safety guidelines and be revised to occupy the required changes. Thus the table shall be used as guidance in revisiting the current architectural standards along with the published health and safety guidelines.

A detailed survey template shall be developed in future research based on the above matrix table that sheds light on the different architectural solutions that have been interpreted from the published pandemic guidelines by the international architectural and furniture companies. This survey template shall be used in future research to quantify the Egyptian design experts' opinions to evaluate the most applicable interpretation/architectural design alternatives for each space that has been mentioned in the literature review.

Table 2. Matrix table includes the Architectural applications interpretations of COVID-19 Guidelines.

	Application 1	**Application 2**	**Application 3**	**Application 4**
1-Entrance	Ultraviolet gate 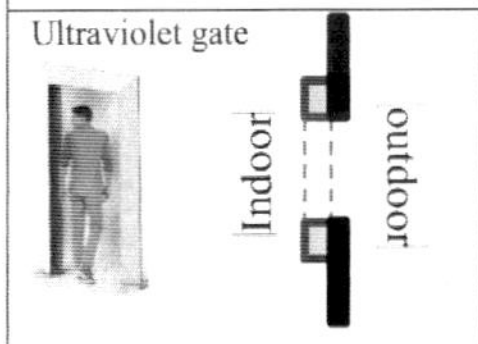	sanitizing space 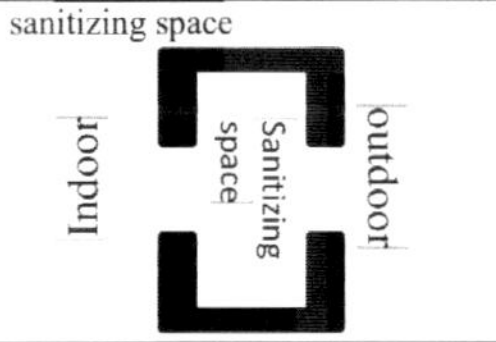	Air/water sanitizing gate 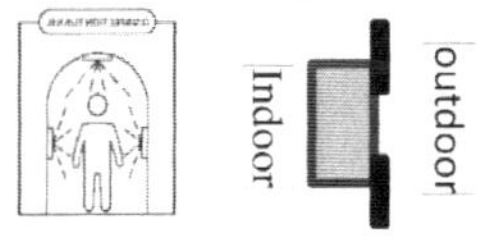	Embedded self-disinfecting Floor material 
2-Open office space	Cubicles 	zigzag pattern Workspace 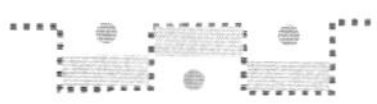	back to back Workspace 	six-foot black circles tiles 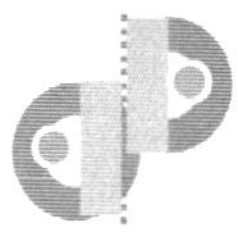
3-Meeting room	reduce density –partition-one way circulation, outdoor meeting 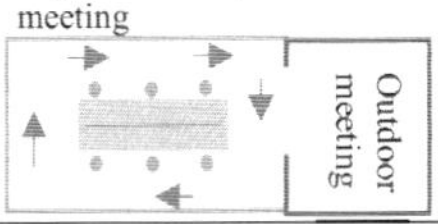	Triangular table creates physical barriers. 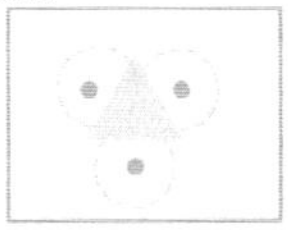	single occupant and co-op style 	small conference rooms 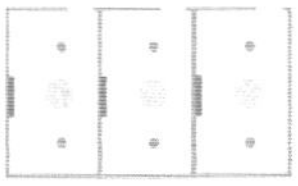
4-Vertical circulation (stairs)	Double stairwells 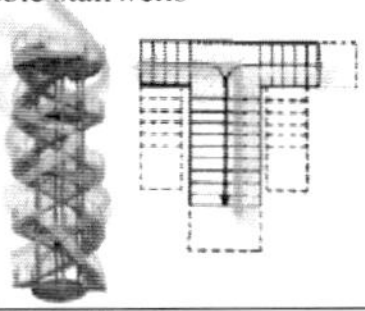	fire stairs-specify each up and down 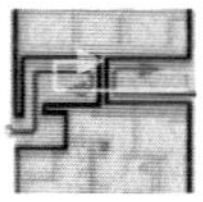	Open well stairs/atrium stairs 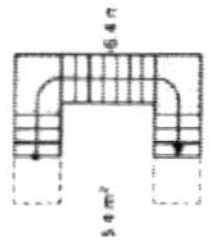	Outdoor stairs Direct flight 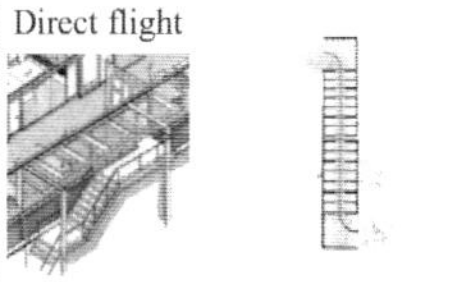
5-Shared offices	Shifting offices 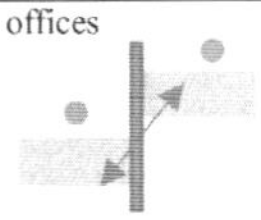	Individual offices 	New shape with a partition (physical distance) 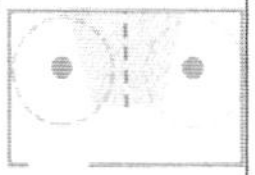	back-to-back 
6-cafeteria	Dancing plant wall 	minimize no. of people 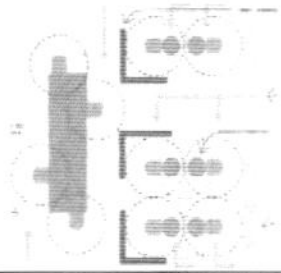	half indoor / half outdoor 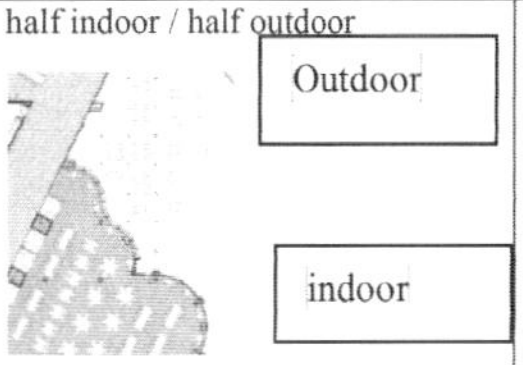	partition 
7-Reception desk	wider counter 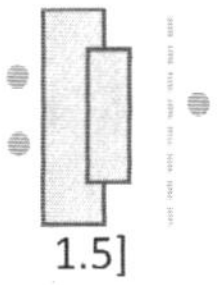	transparent barrier face to face interaction 	self serve visitor screenings 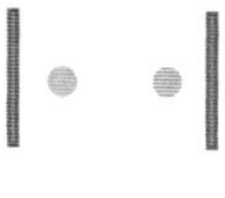	
8-waiting area/ lounge	single setting 	reduce density 	Space Segregation/ back to back 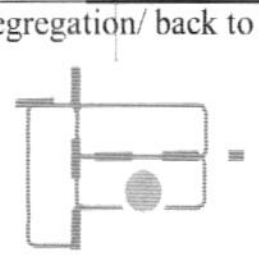	remove waiting area / wait outside 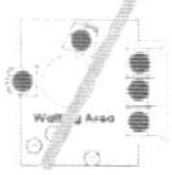
9- Horizontal circulation (corridors)	One way direction with wider corridor (signage) 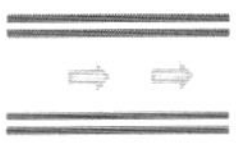	Narrow corridor 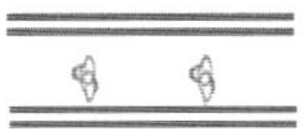	installed partition(one-way vs. two-way) 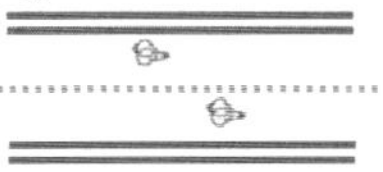	
10-Public spaces-toilets	Individual toilets 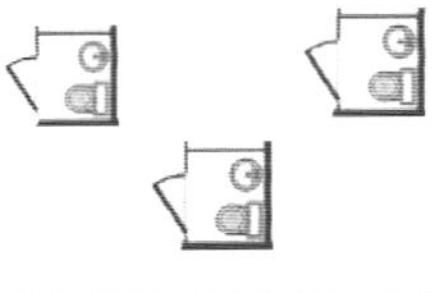	sinks in cabinets change, reduce capacity toilet layout 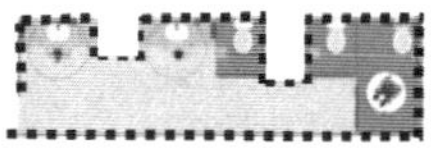	design sinks the open office space or individual outside the toilet 	

References

[1] WHO. *Coronavirus outbreak* 2022 [updated january 19. Available from: https://covid19.who.int/.
[2] Chang L, Zhao L, Gong H, Wang L, Wang L. *Covid-19 outbreak associated with air conditioning in restaurant, guangzhou, China*. 2020;Vol. 26, No. 7,:4.
[3] Mortice Z. *What will architecture design look like after covid-19? Flexible and resilient* 2020 [updated 8 20. Available from: https://redshift.autodesk.com/architecture-design-covid-19/.
[4] WHO. *Covid-19: Physical distancing* 2021 [updated December 8. Available from: https://www.who.int/westernpacific/emergencies/covid-19/information/physical-distancing.
[5] Anggrain SP, editor *Physical distancing concept of public space placemaking in post covid-19: A study precedent*. 6th International Conference on Sustainable Built Environment (ICSBE) IOP Publishing; 2021; Yogyakarta: IOP Conf. Series: Earth and Environmental Science 933 (2021) 012036.
[6] ILO. *Teleworking During The Covid-19 Pandemic And Beyond (A Practical Guide)*. Switzerland: International Labour Organization 2020.
[7] Brodie B, Aoron DS, Rob P, Aditya S. *Reimagining The Office And Work Life After Covid-19*. Toronto: Mckinsy and company; 2020.
[8] Nediari A, Roesli C, Simanjuntak P, editors. *Preparing post covid-19 pandemic office design as the new concept of sustainability design*2021; Jakarta, Indonesia: IOP Conference Series: Earth and Environmental Science.
[9] ASHARE. *In-Room Air Cleaner Guidance For Reducing Covid 19 Air In Your Space/Room*. The American Society of Heating, Refrigerating and Air-Conditioning Engineers; 2020.
[10] University C. *Using the power of light: Preventing the airborne spread of coronavirus and influenza virus* 2021 [Available from: https://www.crr.columbia.edu/research/using-power-light-preventing-airborne-spread-coronavirus-and-influenza-virus.
[11] AHSARE. *Ashrae Epidemic Task Force Filtration & Disinfection (Ashare Guidance Document)*. The American Society of Heating, Refrigerating and Air-Conditioning Engineers (ASHARE); October 2021.
[12] WHO. *Getting Your Workplace Ready For* COVID-19. World Health organization; 2020.
[13] AIHA. *Reopening: Guidance For General Office Settings*. American Industrial Hygiene Association; 2020.
[14] OSHA. *Guidance On Preparing Workplaces For COVID-19*. Occupational Safety and Health Administration; 2020.
[15] CDC. *Interim guidance for implementing safety practices for critical infrastructure workers who may have had exposure to a person with suspected or confirmed Covid-19* 2020 [updated December 3. Available from: https://www.cdc.gov/coronavirus/2019-ncov/community/critical-infrastructure-sectors.html.
[16] AIA. *Re-Occupancy Assessment Workplace Controls*. New york, Washington: The American institute of Architects 2020.
[17] CDC. *Strategies for optimizing the supply of N95 respirators* 2021 [updated September 16. Available from: https://www.cdc.gov/coronavirus/2019-ncov/hcp/respirators-strategy/index.html.
[18] CDC. *If You Are An Essential Critical Worker Who Has Been Exposed To Covid-19 If You Are An Employer Of Workers Exposed To Covid-19*. Atlanta: Centers for disease and prevention; 2020.
[19] ASHARE. *ASHRAE Position Document on Indoor Air Quality*. USA: The American Society of Heating, Refrigerating and Air-Conditioning Engineers; 2020.
[20] El-Tanbouly R, Hassan Z, El-Messeiry S. *The role of indoor plants in air purification and human health in the context of covid-19 pandemic: A proposal for a novel line of inquiry*. Frontiers in Molecular Biosciences. 2021:13.
[21] CDC. *Handwashing: Clean hands save lives* 2020 [Available from: https://www.cdc.gov/handwashing/index.html.
[22] WBDG. *Workplace reimagined: Covid-19 reshapes the modern office experience* 2020 [updated December 3. Available from: https://www.wbdg.org/industry-news/workplace-reimagined-covid-19-reshapes-modern-office-experience.
[23] Notrax. *A safer workplace floor in a post-covid-19 world* 2020 [updated june 26. Available from: https://www.notrax.com/blog/disinfecting-shoes-workplace-floor-mats/.
[24] Wakabayashi D. *Google redesigns the office for gen z and a post-pandemic era* 2021 [updated May 1. Available from: https://www.sanjoseinside.com/silicon-valley/google-redesigns-the-office-for-gen-z-and-a-post-pandemic-era/.
[25] Fehlbaum N. *Paving The Road Back*. VITRA; 2020.
[26] Cushman&Wakefield. *6 feet office* 2020 [Available from: https://www.cushmanwakefield.com/en/netherlands/six-feet-office.
[27] Gensler. *Reconnect Design Strategies For A Pots Covid World*. GENSLER RESEARCH INSTITUTE; 2021.
[28] Architecture F. *26 return-to-workplace solutions for the post-covid office* 2020 [Available from: https://officesnapshots.com/articles/26-return-to-workplace-solutions-for-the-post-covid-office/.
[29] Perkins&Will. *Let's take it outside* 2020 [updated September 23. Available from: https://perkinswill.com/news/lets-take-it-outside/.

[30] George Athens A. *Working outside the box post-covid-19,if social distancing and remote work are to remain, it will have a profound impact on how we occupy space and how we work together moving forward* 2021 [Available from: https://www.workdesign.com/2020/05/working-outside-the-box-remote-work-post-covid-19/.
[31] Weiss S. *Stairs on the rise* 2021 [Available from: https://www.naiop.org/en/Research-and-Publications/Magazine/2021/Summer-2021/Development-Ownership/Stairs-on-the-Rise.
[32] Metal A. *Stairs – the answer to post-covid building design* 2020 [updated September 10. Available from: https://www.activemetal.com.au/2020/09/10/covid-benefits-stairs/.
[33] GENSLER. *Back To The Office Return Strategies For The Workplace And Office Buildings*. Gensler; 2020.
[34] profirm. New reality make it a benefit.: profirm; 2020.
[35] BOS. *Let's return to work together* 2020 [Available from: https://go.bos.com/rtw/p/1.
[36] Perkins&Will. *Road Map For Return Guidance For A Return To Office During Covid-19*. perkins&will; 2020.
[37] NIH. *Nih Safety Guidance & Covid-19 Safety Plan For Working Onsite During The Coronavirus Pandemic*. National institutes of health 2021.
[38] Vitra. *The Future Of Shared Spaces*. Vitra; 2020.
[39] BOS. *Experience a productive return to the workplace*: BOS Inspiring workspace furniture 2020 [Available from: https://go.bos.com/rtw/p/1.
[40] Snapshots. *26 return-to-workplace solutions for the post-covid office* 2020 [Available from: https://officesnapshots.com/articles/26-return-to-workplace-solutions-for-the-post-covid-office/.
[41] Van R, D.stone W, Romero V, dyke ford J. *Covid-19 indoor exposure levels: An analysis of foot traffic scenarios within an academic building*. science direct. 2020;volume 7(100185):3.
[42] UCONN. *Guidance For Plexiglass Barriers In Support Of Covid-19 Prevention Efforts*. Environmental health and safety: Univeristy CONNECTICUT; 2020.

Impact of Urban Compactness, Architectural Form and Climatic Considerations: an Analysis in Traditional Islamic Buildings

Marwa F. Abdelalim[1, *]

[1] Architecture Engineering Department, Prince Sultan University, Riyadh, KSA

[*]Corresponding Auhtor Email: mabdelalim@psu.edu.sa

Abstract. The special metropolitan texture of Islamic urban areas mirrored its social, political construction which was reflected in the architectural forms organized along the central avenue with the city's mosques, Madrasahs, Sabils, and Bimaristans, worked as the public substance of the city and had their own particular arrangement of rules and guidelines. This was strongly echoed in the trademark design of the age, set inside a well-weave metropolitan texture of human extents a cityscape that served to enhance the central core of the Islamic cities. This paper reviews literature on the traditional Islamic city's urban surroundings, architectural form, socio-cultural and climatic factors that influenced the formation of such unique Islamic Buildings, where these factors helped to shape the engineering styles and utilization of spaces inside traditional Muslim public buildings, especially the application of microclimatic modifiers such as open courtyards in Islamic houses, where in the middle age metropolitan places of Islamic urban areas were envisioned to make positive microclimatic conditions in their inside such as water fountains, Planted trees, Malqafs, Mashrabiyyahs, Semi open spaces, and Iwans. Consciousness of the multifactorial nature of the impacts on the Muslim view of houses and the utilization of spaces is essential for building planners, architects, and developers to be appropriately prepared to address the issues of customers. By offering alternatives for whatever is lowering people's quality of life or efficiency. This research should contribute to Saudi Arabia's 2030 Vision, which aims to implement more successful Sustainability ideas as those previously used in traditional Islamic cities.

Keywords: Islamic Buildings, Cairo, Compactness, Climate, Urban, Architecture

Nomenclature
Bimarstan; Hospital
Iwan; Vaulted portal opening onto a courtyard
Hammam; Bath
Sabil; Regular Path
Mashrabiyyah; Projecting oriel window enclosed with carved wood lattice work
Malqaf; Wind Catcher
Madrasah; School
Qaysariyya; Mall
Khan; Store
Suq; Commercial path
Qasaba; Main path

1. Introduction

At the point when Europe was enduring in obscurity ages, this was an age that supported the recognized and generally surprising practices of information, generosity, and morals, [7]. In Medieval Islam, religion, law, instruction, and government were coordinated to the point that a customary Muslim would barely attempt to recognize them as discrete elements, in this way, the exemplary mosque was to work as a strict substance, an official courtroom, and a scholarly and instructive focus, [8]. The medieval Islamic concept of the city was unlike the Hellenistic concept; in fact, it was more closely related to the concept underlying the ancient cities of Mesopotamia. The Muslim conquerors gradually obligatory their method of life on the Hellenistic cities, and though they maintained the fortification walls of town for defensive reasons, they transformed the urban material inside, [8]. The research intended to address through giving common topographies represented by architectural forms of some religious-civil typologies; Mosques, and traditional housing, as well as traditional applications of microclimatic modifiers in Heritage Buildings' typologies in Arab regions, such as Old Cairo they are all representing traditional heritage areas that can lead implementing new sustainable cities, applying the same traditional applications to reach the goal of an integrated sustainable buildings.

2. The urban fabric transformation concept

Islamic culture provided coherence by applying a flexible urban planning structure, [6], that balanced opportunity and Privacy, modesty, hospitality, [5]. These are the key qualities of thriving cities nowadays, Figure.1.

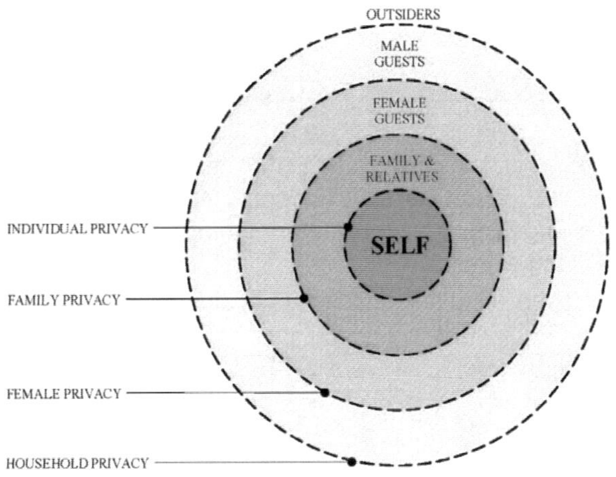

Figure 1. Coherence, the ability to integrate systems, and concepts of applied privacy in Islamic traditional cities [5].

The Muslim conquerors gradually imposed their way of life on the Hellenistic cities, and although they retained the fortification walls of the city for defensive reasons, they changed the urban fabric within, [1], Figure. 2.

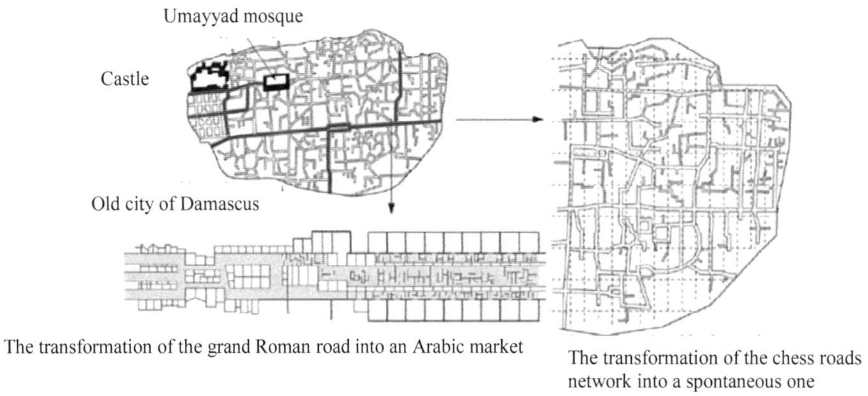

Figure 2. Impact of changed urban fabric on Public Islamic buildings [10].

The important features of Islamic cities shape show up fairly consistent and then exhibit a sure rationality; which represents a ‘coherent’ urban system. As a result of uniform legislative guidelines, an almost same socio cultural framework, created via Islam, similar climatic prerequisites and development techniques within most of the Islamic world; the production of incredible similarities in the approach to the metropolis building process. This resulted in the beehive city pattern at most of the Islamic cities, though differences did occur. The agora gave way to the jami 'mosque (the religious and political center of the city), the colonnaded avenues became the suq, [1], or bazaar streets; the basilica, the Qaysariyya (market hall) or khan (inn or lodging for travelers as well as a storehouse). The baths naturally remained as public baths, or hammams.

These adaptive changes were relatively minor compared to changes that took place in the street network; in time, the gridiron street pattern was gradually eliminated as new buildings replaced the old in a more oriental and less rigid urban pattern. The central part of the mosaic picture of the medieval Islamic city was and still is the mosque, Figure.3.

The conventional suq was a labyrinth of restricted shopping paths frequently canvassed and for the most part nearby the jami' mosque. Figure.4. It contained stores as well as various workshops. In an average organization like style, related exercises were assembled and the area of these different exercises comparative with the mosque followed a specific foreordained example, [8].

All public institutions followed a design concept based on the utilization of central open courts. Thus, the mosques, Madrasahs, Hammams, Bimaristans, Khans (markets), and Qaysariyyahs (Malls) were all inward-looking buildings facing at least one courtyard which is considered as one of the most important microclimate modifiers Figure 5-6-7, that can effectively improve energy efficiency in contemporary residential buildings, [9].

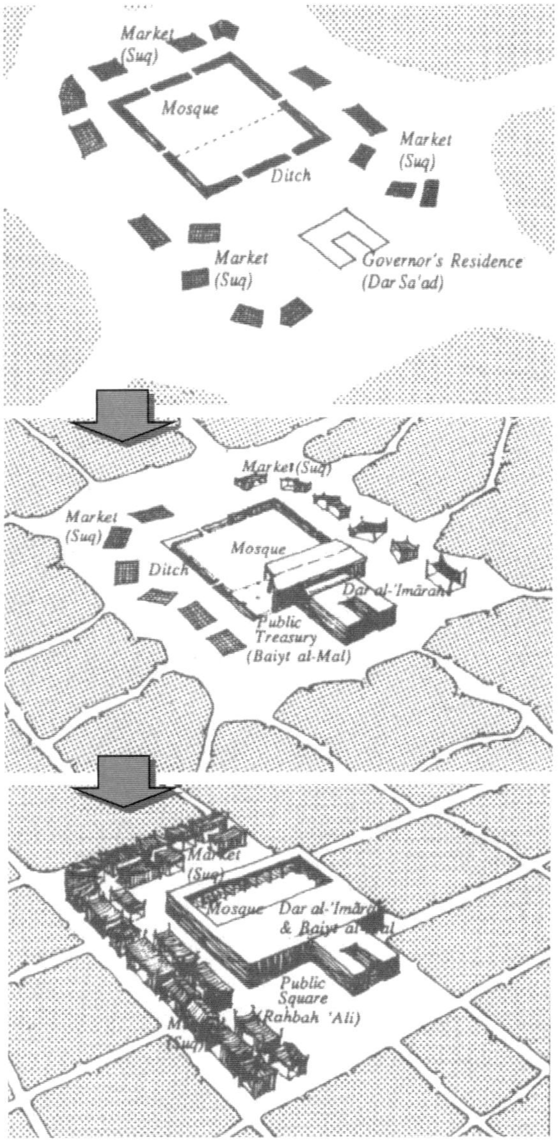

Figure 3. The mosque, as a political religious nuclear core of Islamic city structure, [8].

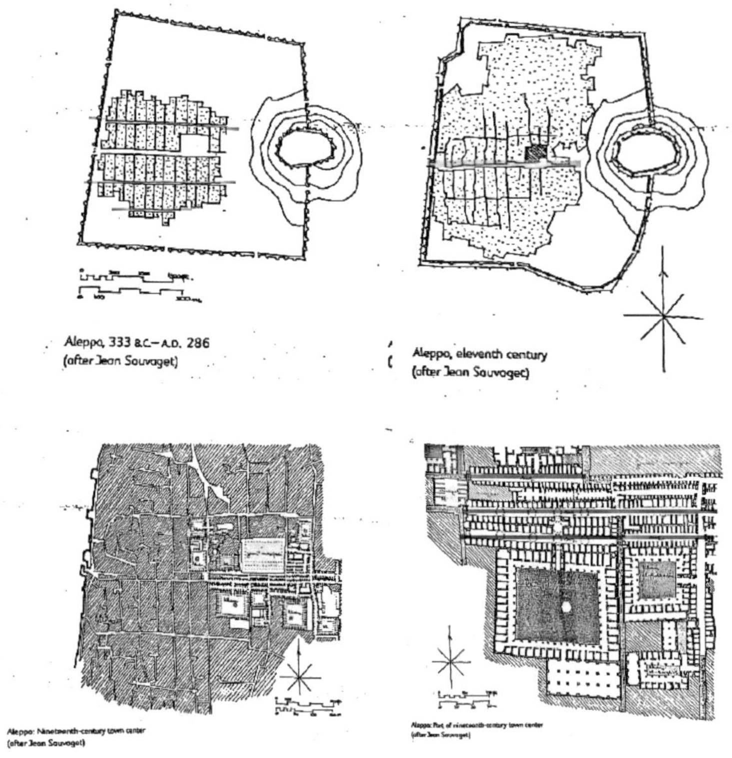

Figure 4. Jamie (Mosque) as a nuclear core of urban Islamic fabric in Aleppo, [8].

The architectural style of the prophet's Mosque in Medina had the great impact on the architectural style of the Mosque throughout the entire history of Islam where open courts were adopted in mosques typologies. Education was always closely connected with worship, and from the beginning mosques could be used for prayer as well as for teaching. The result was the tabernacle mosque or madrasa. Its layout seems to go back to the Khurasan houses and resembles Iranian Mosque Layout: a rectangular courtyard with an Iwan in the middle of each side. Classes are held in the Iwans and students live in cells arranged along the partitions.

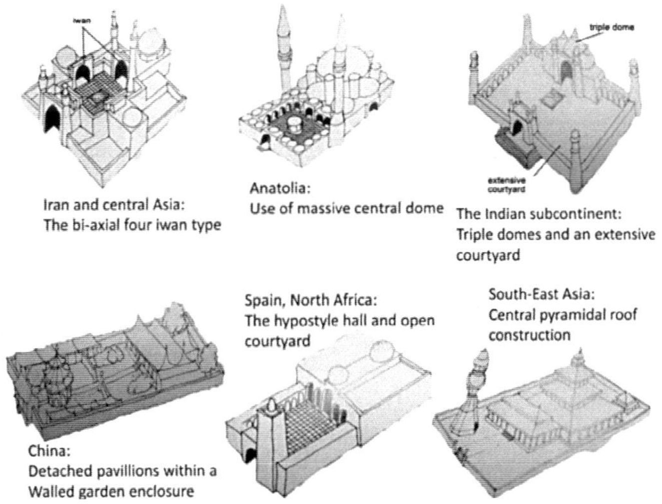

Figure 5. Application of open courtyards in Mosques' typologies all over Iran& central Asia, Anatolia, Indian, China, South east Asia and Spain, North Africa, [11].

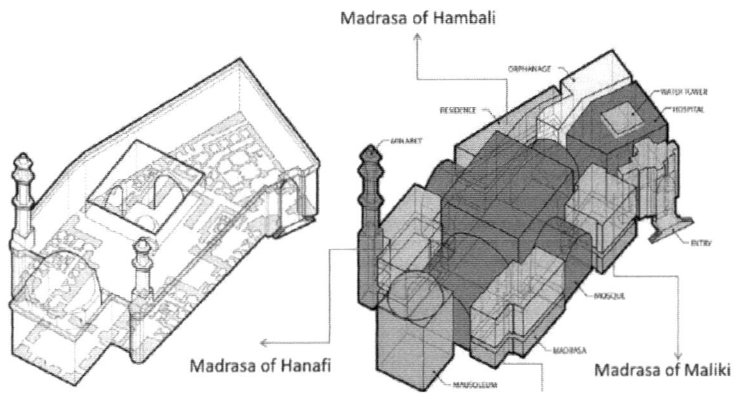

Figure 6. Applied Courtyard& Iwans in Madarasa Al Sultan Hasan- Mamluk era, Egypt, [7].

3. Methodology:

3.1. Socio-Cultural factors affecting Islamic buildings' typologies:

The Islamic means was the religious center. Therefore, the mosque was placed in the center of the city's fabric. Right at the intersection of two main arteries, similar to its place in the heart. The social and educational aspects of society were centered in the mosque of the early Islamic city. The mosque was a religious authority, a court of justice, and an intellectual and educational center. Arabs believe in maximum privacy, where houses were divided into two areas:

A) Salamlik: public part of the house where male visitors and friends were received.
B) Haramlik: private and secluded sanctuary reserved for the family. The exterior windows in the haramlik were protected by a barred bay window or Mashrabiyyahs to satisfy curiosity about the outside world without being seen. Figure 7. As well as the Broken entrance: the design of the entrance is indirectly protected by a screen wall to inhabit a direct view from the street.

Figure 7. Mashrabiyah in Alshehemi House, Old Cairo, [7].

Muslims were very well known for their hospitality; they provide a place to meet visitors and friends. They even provide bedrooms to accommodate family or friends from another city or province. This was introduced, as Caravansarai-like warehouses with accommodations, which was the beginning of hospitality buildings in history such as hotels and motels. Also in North Africa the Khan is called a Funduq, but structure and purpose remain the same throughout Islam, where economy based on different trades along with the Qasaba, Figure 8.

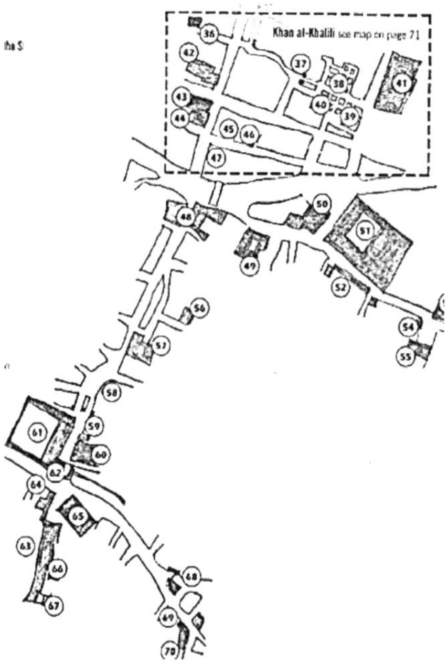

Figure 8. The Qasaba, from Khan Al- Khalili, Old Cairo, [7].

3.2 Microclimatic modifiers as a model example in urban Islamic buildings

The records of architecture characterize a delightful correlation between the urban surroundings and public buildings whether it is religious, civil, or military Islamic typologies, which have been designed with careful attention to climatic necessities and sociocultural contexts, [3]. Traditional courtyard houses in the hot and dry regions of Egypt can additionally be regarded successful climate-representative structure that responds to many continual environmental challenges, [2] These houses use features of microclimatic considerations such as open courtyards, Iwans, T-shaped spaces, water basins to provide thermal comfort for their occupants and irregular walls for privacy, Figure. 9, Table 1.

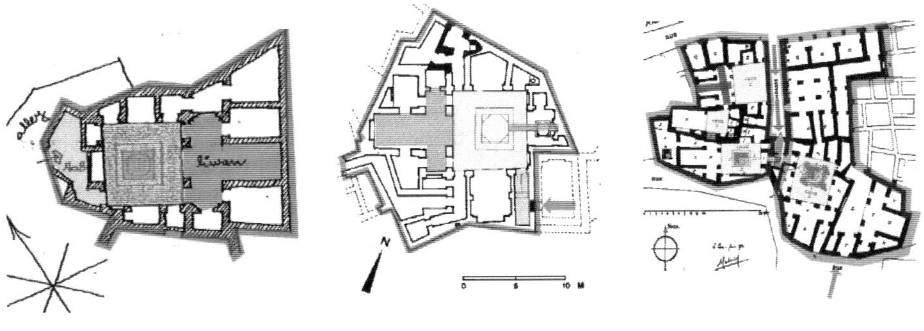

a b c

Figure 9; Graphical analysis of three traditional Islamic courtyard houses in Egypt; (a) Fustat House no.8. (b) Fustat House no.3- (c) Fustat House no.2. All featured with the T shaped spaces, open courtyards, water basins as microclimatic modifiers. [8] Edited by the researcher

Courtyards	Iwan	Entrances	Plot boundaries

Table 1. Analysis for architectural forms for the three traditional Islamic Courtyard houses.

House No.	Architectural Form	Court-yards	Entrances	Iwans	Plot Boundaries
Fustat House no.8	Small dwelling had only one courtyard.	1	Privacy wall blocked the view from the street into the interior, corridor linked the entrance hall to the court.	The principal suite of living spaces, consists of a T-shaped space made of a large and deep central hall, flanked by two smaller spaces, and triple-arched Iwan opened to the court.	The plot of this house was irregular, the interior, which has an axial layout, appeared symmetrical and rectangular.
Fustat House no.3	Larger single court garden dwelling.	1	The street entrance door faced a spirit wall and a short corridor linked with the court.	West side contained the typical T-shaped Iwan group, and south of this suit was a bent corridor which linked led to a side entrance.	The plot of this house was irregular, the interior, which has an axial layout, appeared symmetrical and rectangular.
Fustat House no.2	Still larger, consisted of two separate houses joined by a narrow passageway.	4	South the main entrance to the formal section of the house had no privacy wall.	Harem section had a smaller court garden with triple-arched Iwan group.	Neighbouring house in north of house No.2 was a medium sized home with entrance from the street. Perhaps it was shopkeepers. It had also two courts and had both formal and family living areas.

As the Jami" Mosque" was the focal point of the Islamic Urban community, so the sakan or maskan, the Arab words for house, was the center of the basic social unit, the family. The word sakan is "related to the word sakinah meaning peaceful and tranquility. The inward-looking sakan, open to the calm of the sky, made cool by the element of water, self- contained and peaceful, the deliberate antithesis of the harsh public world of work, warfare and commerce, is the place where the early Arab family found its sakinah", [5].

4. Discussion:

Three traditional courtyards houses were analysed, in terms of the applied courtyards, Iwans, Entrances and the plot boundaries. It was noticed that for the Iwans they were common in adopting the T shaped space, all of which are having bent entrances that enhance the concept of privacy and the emphasize the meaning of compactness urban tissue which helps in supporting the applications of climatic modifiers that are adopted in the traditional houses.

5. Conclusion:

The research aimed to address while presenting common features represented by architectural forms of some religious-civil typologies; Mosques, and traditional housing, as well as traditional applications of microclimatic modifiers in Heritage Buildings' typologies in Arab regions, such as Old Cairo they are all representing traditional heritage areas that can lead implementing new sustainable cities, applying the same traditional applications to reach the goal of an integrated sustainable buildings, adopting such courtyarded architectural design will help definitely in enhancing and improving the concepts of sustainable buildings.

References

[1] Bianca, S. Urban Form in the Arab World. Past and Present, 2000.
[2] Fahmy, M., Mahmoud, S., Abdelalim, M., & Mahdy, M. (2019). Generic Energy Efficiency Assessment for heritage buildings; Wekalat El-Ghouri as a case study, Cairo, Egypt. *Energy Procedia, 156*, 166-171.
[3] Gramlich, A. (2013). A concise history of the use of the rammed earth building technique including information on methods of preservation, repair, and maintenance.
[4] Custodian of the Two Holy Mosques King Salman Bin Abdelalaziz Al Saud, (2016) Vision 2030 Document, [online] Available from https://www.vision2030.gov.sa/en, [Accessed 25 April 2016].
[5] Othman, Z., Aird, R., & Buys, L. (2015). Privacy, modesty, hospitality, and the design of Muslim homes: A literature review. *Frontiers of Architectural Research, 4*(1), 12-23.
[6] Rose, J. F. P. (2016). The Well-Tempered City: What Modern Science, Ancient Civilizations, and Human Nature Teach Us About the Future of Urban Life Hardcover
[7] SCA. (2001). *Historic Cairo,Al Muizz Ldin Allah Street; The Great street*. Cairo, Egypt: Supreme Council of Antiquities.
[8] Schoenauer, N. 6,000 Years of Housing.(2003)
[9] Soflaei, F., Shokouhian, M., & Shemirani, S. M. M. (2016). Traditional Iranian courtyards as microclimate modifiers by considering orientation, dimensions, and proportions. *Frontiers of Architectural Research, 5*(2), 225-238.
[10] Hasan Mansour. (2015). The Lost Identity of the city; The case of Damascus [online]. Available from https://www.researchgate.net/figure/the-Islamic-Plan-of-Damascus_fig4_286252741 [Accessed September 2015].
[11] Shahril Khairi. (2011). Presentation[online], Available from https://www.slideshare.net/shahrilkhairi/history-of-islamic-architecture, [Accessed 15 September 2011].

Towards a Taxonomy of The 21st century Architectural Practices in the age of Sustainability and Technology

M.T.Saleh[1,*], Y. Mansour[2], S. Kamel[1], K. Dewidar[3], A.A.Farid[1]

[1] Department of Architecture Engineering, Faculty of Engineering, Ain Shams University, Cairo, Egypt
[2] Department of Architecture Engineering, Faculty of Engineering, Ain Shams University, Cairo, Egypt
[3] Department of Architecture Engineering, Faculty of Engineering, the British University in Egypt, Cairo, Egypt

[*]Corresponding Author Email: Eng_Mennatarek@yahoo.com

Abstract. The architectural practices witness drastic transformations in the 21st century globally, which put the architectural community in a situation of crisis; facing challenges to identify different trajectories and their characteristics for a reliable evaluation of the approaches and strategies. Conferences and research laboratories started to seek a philosophy interpreting new practices, mostly known is the 2003 Pompidou exhibition in France discussing "Non-standard Architectures". This research investigates the interrelation between architecture, and the most-two dialectic terms of the age "sustainability and technology", linking influencing beliefs and philosophies with the design approaches and strategies through a comprehensive design theory. It emphasizes the paradigm shift from modernism architecture that tied the design to craft to a richer knowledge paradigm ties design, technology and craft in a complex-system. This has more impressive and complex drivers corresponding to the reflects of "The Knowledge Power" coined by Foucault that shapes the whole cultural paradigm and its sub-systems of any society across history. Such a philosophy can reach a meta-theory about current practices of architecture seen as isolated islands or fragmented practices through revealing the radical drivers of current iconic projects and identifying hypothetical boundaries that interprets these typologies and their relevancy to each other promoting a new holistic paradigm of architecture.

Keywords: Sustainability, Digital technology, Philosophies, radical influences, Paradigm shift, I-Praxis, criticism, iconicity.

1. Introduction

It is recognized in the 21st century the upheavals of new beliefs, values and ideologies that influence the architectural practices and building characteristics. Previous writings and contributions can't identify current practices reaching a level of failure of criticism due to the successive transformation waves, [1]. A new phenomenon evolves with new malleability of high-rise towers, smart-complex projects, archi-urban typologies that merge the logic of building and city in a single product. Besides, the wide practices oriented towards sustainability being green, environmental or ecological. New efficiency measurements also appeared with formative awards that buildings thrive like: LEED, BREAAM, and living building challenge. The current meaning of iconicity has been drastically changed to a vague complex of dimensions that need precise definitions for assessment methods.

However, the wide spectrum of contemporary practices calls for a novel approach to reach a philosophical meta-theory that interprets these diverse practices. Despite the fragmented knowledge and multi-versions of reality existing in the essence of technology. It is important to stand on the critical drivers that shapes the contemporary architectural paradigm after the "Death of Modernism" [2].

The research is a qualitative study to critically investigates the current state of knowledge standing on exploring the relation between architecture and the two-dominant terminologies of the age "sustainability" and "technology", which leads to realizing the transformations of philosophies, concepts, and frameworks of action from previous architectural paradigm to a new one. This is obtained by studying the diverse practices to reach a collective form of sub-divisions, through gathering projects with similarities not essentially identical but with symbolic generalizations to reach a meta-analysis.

Hence, the research extracts a supreme paradigm of architecture linking current knowledge power with architecture, while emphasizing the true nature of the different architectural languages of the age. Besides, linking the visual dimension of architecture with rational dimension through a new theoretical frame work. This explains the interplay between existing sub-ordinates or styles that aims to richer a higher knowledge, as well as the radical transformations in the contemporary design theories. Also, it suggests possible correlations in the future of architecture and its smart-sustainable impact.

2. Sustainability as a belief

Sustainability is an idea initially exists since50's, threats to the finite global resources started to rise and call for the efficient use of these resources to save life on earth. Kenneth Boulding published his book in 1965 "The Economics of the Coming Spaceship Earth" and argued that "man should treat earth's resources the same way as people in spaceship carefully manage their resources in order to survive" [3]. Also, among the significant publications of the 20th century was "Limits to Growth"; which based on a computer simulation highlighted the severe illness of the consumption-based model of modernism and rise the need to an efficiency-based model. Later, critical threats expanded to new aspects to find solution to the current regressions with real challenges of Ecological crisis, limited resources, Poverty, Population growth, and the economic crises. Especially after the situation of the 70s Great Depression.

2.1. Definition of Sustainability

The conceptual meaning of sustainability defined by the World Commission on Environment and Development (WCED) or the Brundtland Report also known as Our Common Future as "Meeting the needs of the present generation without compromising the ability of future generations to meet their needs." [4]. A conflict upraises later in the concept of present and future needs, that promoted rephrasing statement identifying a new terminology a: "Sustainable development is the development that does not compromise the ability of future generations to meet their own needs" [5].

2.2. The Pillars of Sustainability

The sustainable development as a comprehensive framework embraces three main pillars "figure1":

- The Economic Pillar (Profit): "Economic sustainability" implies a system or a technology of production that satisfies present consumption levels without compromising future needs. It seeks the 'sustainability' of the economic system that emphasizes fairness and equity, [6].
- The Social Pillar (People): The theory of 'social sustainability' calls for economic growth constrained by the requirements of social equity. In order to link these, an enabling environment must be created that regards the human needs and well-being, optimizes resource use, prioritizes resource allocation, and regards equitable resource distribution despite the different classes, [7].
- The Environmental Pillar (Planet): Which involves ecosystem integrity, carrying capacity and biodiversity. Resources must be harvested no faster than they can be regenerated. Wastes must be emitted no faster than they can be assimilated by the environment [7].

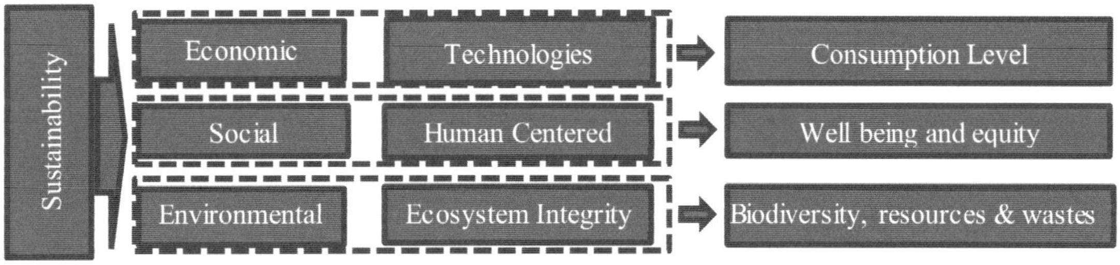

Figure 1. Sustainability as a complex belief of three pillars or (sub-oriented systems).

2.3. The Evolution of Sustainability as a belief of the paradigm of architectural practices

The architecture of the 20th century is envisioned through a concept that deals with forms, spaces, and materials, often discussed as a style with a mechanized philosophy and industrial design approaches. Meanwhile, the 21st architectural practices are characterized by seeking for efficiency, economy, and saving resources in order to create a better world. This idea is now synthesized by the belief of "Sustainability" which focuses on "the absolute primacy of the process over the style, looking for quality of life"[8]. Hence, the paradigm of the contemporary practices of architecture announces a shift from its preceding, and give attention to the matter of human centered design, environmental impact and efficient use of resources as main design conceptions. This is accomplished by new terminologies in architecture like: green, environmental, ecological, smart-resilient and regenerative buildings "Table1", which are all considered sustainably oriented practices that seek to eliminate any damage that affect the future of the coming generations, despite their different design trajectories leading to diverse types of practices as discussed briefly below [9-12]:

Table 1. The evolution of the sustainable oriented architectural practices in the 21st century.

Green practices	Which is the practices concerned only by the environmental. It seeks to eliminate the negative impact of the built environment on the natural environment despite the quality of the process inside.
Environmental practices	Which is the practices concerned by one two pillars of sustainability; the social and environmental. It works to enhance the relation between man and nature and seeks proper approaches for this reconciliation.
Ecological practices	Which is also a practice concerned by the eco-system and biodiversity. It integrates human, nature with its biotic and a biotic habitants aiming to reach a balanced system that embrace diversity and continuity of species.
Smart-resilient practices	Which is a practice that is concerned mainly by the economic and the environmental pillars to make the best use of the current technologies and systems to adapt and respond to the changes in the natural environment seeking the best efficiency of the renewable resources of energy.
Regenerative practices	Which emphasizes the environmental and economic pillars. It seeks to maximize the use of renewable resources reaching a higher level of zero, neutral or self –sufficient practices. Aims at generating a positive impact that compensate the previous damage subjected to nature and eco-system.

3. Digital technologies and the philosophies of architecture

Clearly, the beginning of the 21st century announces technological breakthrough; that information and digitalization are found on the base of scientific, economic, social, cultural processes. Any specialists works in a certain field face complex problems, they only require the use of computer and software products to solve them. Computing will forcibly reshape the possibilities of whatever ideologies arise for the future with the growing sciences of computer, information and communication technologies, precisely in the domains of BIG Data, artificial intelligence (AI) and internet of things (IoT). The age of the technological revolution announces drastic changes, where the early perception of these changes is perceived as:"The industrialized modern systems of cities based around wealth created from commodity production for a mass market place is replaced by a new system based around the generation of wealth from information services which are globally rather than nationally organized"[13].

Architecture as a cultural discipline is highly influenced by the new technologies and the capabilities of digitalization and computerization since modernism and the work of Frank Ghery, Luigi Moretti and Peter Eisenmen communicated with the world through the new language producing new complex forms like the influencing "Bilbao museum" in Spain. Later this evolved from being only computer aided drafting systems to a new phase of digital simulation, fabrication, materialization and industrialization as in the famous "Swiss-re tower" in London. A paradigm shift is clearly recognized from the previous industrial modern philosophy that ties art and industry to the new one bridging ties between art, industry and technology.

The current reality of architecture re-identifies the main actors of the architectural paradigm by the indulgence of the machine logic with the human logic as the moderators of the design practice. By turn, the new paradigm expresses its logic digitally through scripting on the new software systems that can generate multiple forms and compositions with optimum solutions and performance., Consequently, this generate new architectural languages that either reflect new or traditional philosophies, and identified as a reflection of a revolutionary complex-technologies. This announces a new discourse of architecture in the 21st century that introduce an intelligent pattern of practices with a radical philosophy behind that can be called "I-Praxis" presented in "figure 2".

Figure 2. I-Praxis influencing the new pattern of architectural practice in the 21st century.

3.1. The changing nature of criticism in the 21st century

During the last decade, digital design has become central in the domain of architectural practices. An evolution is obviously observed on the logic and legitimizations of concepts, conceptions and approaches, all developing the new "Digital Design Thinking" [14]. A new philosophy that aims to organize the interactional relation between with the designer, information and image, announcing the new "Non-standard Architecture" as described in the Pompidou Centre in 2004. These new philosophies challenge with respect the old philosophies of architectural practices that were discussed through the free slabs, abstract shapes, inclined roofs or innovative materials and curtain wall systems, to open up to new discussions with different semeiotic meanings and principles.

The new "typology of criticism" relates to the new knowledge of software systems, scripting languages, performance measurements, interpretations of elements and sub-systems. A different data models that regard set of parameters like: form, site, environment and human behaviour analysis, all gather in a cross-disciplinary holistic system generated by new conditions and seen through a different perspective originated from a more complex-reality. Meanwhile, criticism has a real challenge to cope with the rapid changes of the digital design practices. A new approach to discuss the principles and approaches of the design thinking with the new-formal characteristics of the generated buildings as a part of the whole. This methodology has to identify the sub-stages and sub-tasks involved in the practices to crystalize certain styles that have similarities in their characteristics and design ideologies inside the boundaries of the digital architecture. These efforts focus on the different styles or schools of practices like: BIM, Parametricism, Sustainable Architecture and Smart-Buildings, with different models of interrelation between the design components. Although, it also has to expand to a meta-level of discussion to obtain a holistic theory about the contemporary practices influenced by the complex-web of scientific knowledge and the wide comprehensive social knowledge of sustainability. The subordinate styles seen as isolated islands can then be related to diverse digital design conceptions that result-up distinguish morphologies standing upon these complex theoretical foundations.

4. Observing The Iconic Architectural Practices of The Early 21st century

The critical analysis of three iconic projects aims at extracting their radical beliefs and philosophies as real drivers of the design practices of iconic buildings in the early 21st century

4.1. CCTV Tower – A New Iconic Idiom of Skyscrapers

Designed by Rem Koolhaas in China in 2012, koolhas describes a revolutionary phenomenon as seen in "figure3", saying: "The building we created for CCTV could never have been conceived by the Chinese and could never have been built by Europeans. It is a hybrid by definition." [16].

4.1.1. Building Philosophy. CCTV reveals a set of new philosophies summarized as: The integral design approach changing the traditional horizontal spatial configuration of studios and locations to vertical ones in spatial introvert quality. The 3-Dimensional complex experience through the distorted changing 3D planes of the building along its sides. The fragile equilibrium of the urban void that has a dimension of city containment through an overhead plane. The Impressive structural technology and its ability to demonstrate strong rational visual quality. The power of the parametric design with digital revolution of Beijing representing a gate to the city by its iconic loop morphology also implemented on circulation.

4.1.2. Design process. Recognizing a set of design strategies dealing with parameters like: Technology: Maximizing the smart-applications, the use of structural technology and the capabilities of parametric design in the use of the urban void and the flying bulk head cantilever (structural expressionism). Human: emphasizing quality of indoor spaces and human cantered approach. Site: Demonstrating its urban context. Environment: Neutral approach with 70% sun shading glass linking the inside to the outside. Economy: Moderate concerns. Materials & Resources: selecting the advanced and impressive.

4.1.3. Building Morphology. Discussed through a set of attributes: Form generation: Malleable composition emphasizing the innovative diversity of form which change from side to side with its innovative urban void and the flying cantilever using the computer aided design. Using basic shapes and subtraction compositions. Profile: The building introduces a new dynamic changing closed-Loop morphology through its huge mass and innovative structure. Critical Expressive vocabularies: the dynamic urban void and the flying bulk-head cantilever. Facades: Exotic sharp irregular diagrid. Steel-Glass facades. Skyline: Emphasizing verticality. Interiors: Modern style resilient open spaces.

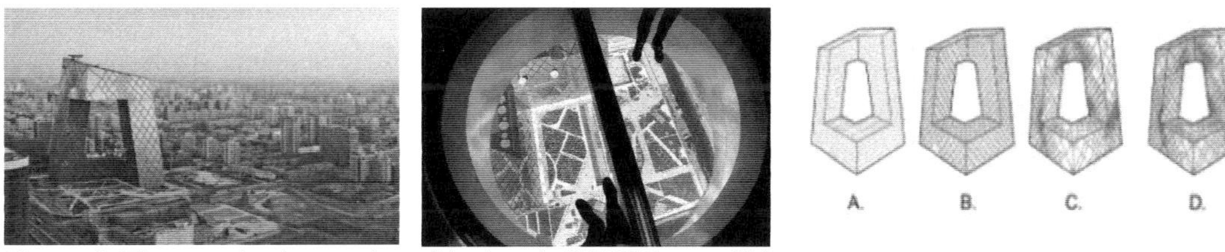

Figure 3. The complex-innovation of the iconic CCTV Tower, China.

4.2. Louvre Abu Dhabi – A Cultural Floating Oasis

Understanding the culture of the site and the creative interpretation of its natural and built environment was the base for the iconic dome design. It is mainly influenced by the Arabian desert sun rays that broke down in between the fronds of swaying palm trees in the local oasis and the traditional homes roofed by the interwoven palm leaves[18]. This seen as an architecture rises from its place "figure 4".

4.2.1. Building Philosophy. Revealing new philosophies expressed as: The Archi-Urban Fabric with contextualism approach regarding the Arabian traditional urban context with its districts and narrow street grids evolving to a floating Arabian oasis transforming the traditional spatial configuration of (building in the city) to (city in the building). The floating museum city as a futuristic city expansion approach that transforms the traditional (water inside the building) to (building inside the water) "C-stands". Redefining the spatial layout planning transforming the (Indoor-Indoor circulation) of adjacent functional spaces to (Indoor-Outdoor-Indoor circulation). A 3D vertical plane digital Sky articulation extending the traditional building roof to a super scale shading system covered up by complex perforated metal sky. The new complexity of the sky-Dome by The use of the computer aided design to create different parametric models to attain the best lattice morphology with economic structure.

4.2.2. Design Process. Realizing the strategies with: Technology: Maximizing the smart-applications, the use of structural technology and the capabilities of parametric design and adopting the new material technology using 3,900 panels of ultra- high performance fiber concrete (UHPC) that serves the structure. Human: Intimate-human scale with multiple indoor qualities. Site: strongly influenced by its site and regional context extracting a tripod between UAE original citizens, the water element and the built urban environment. Environment: with climate control approach and some passive design solutions in the energy gain, humidity range, optimized consumption targeting LEED Silver Rating. Economy: Moderate concerns. Materials & Resources: selecting the most healthy, durable and clean materials.

4.2.3. Building Morphology. Discussed through its visible attributes: Form generation: a matrix of interlocking-stepped cubical composition generated by subtraction that has a hypothetical rectangular base looks like and extends to the metal perforated metal sky. Profile: A simple rectangular short building silhouette that has an iconic dome above. Critical Expressive vocabularies: The water carrying the building, and the dome. Facades: white plain facades with some Arabic writings and few narrow openings covered steel claddings in the used skin motifs. Skyline: Emphasizing horizontality. Structure: steel frames and skeleton system. Interiors: ultra-modern indoor space with light colours of paint and wide spans using the latest technologies in showing and serving the artefacts.

Figure 4. The genius locai in Dubai Louvre in the age of Parametricism, UAE.

4.3. Apple Business Park - The Futuristic Spaceship.

Designed by Norman Foster in California in 2019. The Apple campus is a smart- sustainable city, an innovation that rise from its natural environment with a futuristic theme using the advanced materials and embedded technologies associated by the highest possible performability "figure 5".

4.3.1. Building Philosophy. Exploring its innovative philosophies expressed as: Creating a new typology of head-quarter through emphasizing smart-sustainable iconicity emphasizing the domains of technology, sustainability and innovation to create a new typology. The recreation of urban landscape towards a countryside in the city. Introducing the business park that strongly regard the large landscape and open spaces around the building and the different functions exist to serve 80% of land area with the food harvesting areas, meadows and artificial ponds in a sustainable smart complex city. A Futuristic-experience through the innovative circular form that looks like a space agent with pure techno-functional inspiration offering smart panels, virtual dashboards, wireless communication systems, energy cars and more advances. Human-centred design: enhancing socio-cultural dimension and productivity through introducing a new smart-radial prototype that allows more communication between large number of users and influence the team's cooperation inside, providing natural daylight and ventilation, direct out door views, standards and circulation that regard physical and spiritual wellbeing. A sustainable holistic approach embraces the optimum performance, social parameter and the natural environment.

4.3.2. Design process. Emphasizing the strategies concern: Technology: the use of the smart technologies and advanced applications inside the spaces of building and in the construction of its structural elements. Human: human scale with perfect and smart user experience inside. Site: aiming to enhance the landscape around and increase the green open spaces. Environment: high regard to the environmental parameters, consumption of resources, the building emissions, CO2 footprint emphasizing sustainable design approach. Economy: high concerns. Materials & Resources: selecting the most clean and efficient materials, besides their life cycle.

4.3.3. Building Morphology. Observing its main visible attributes: Form: A ringular donut shape with atrium extracted from its center emphasizing its light weight, it is characterized by fully exploited functional potentials. Profile: thin soft organic circular profile that emphasizes integration with natural around. Critical Expressive vocabularies: the folded complete glass facades, the full-rise glass doors, shade of theatre. Facades: all in glass with small ratio of flying concrete floor slabs. Skyline: Low-rise smooth roof. Structure: steel structure with sloped concrete slabs and prefabricated inclined glass panels with specific specs. Interiors: smart-futuristic style with resilient spaces associated by all amenities and encouraging social interaction with glass facades that open to the outside landscape building extension.

Figure 5. The Apple City as a new smart-sustainable typology of headquarters, California.

5. Conclusions

The 21st century architectural practices represent undoubtedly a revolutionary shift in the paradigm of architecture from modernism to a new complex-knowledge paradigm. It is identified by its knowledge of sustainability as radical belief monitoring the whole practices inside the paradigm but with relevant approaches embracing different strategies and priorities towards its three pillar: social, economic and environmental. Also, ontological philosophies are all influenced by the technological revolution distinguishes the age; seen through main three developing trajectories: first Computational Design Capabilities; represented in the new design softwares, building information modeling, the algorithm designs shifted the traditional architectural forms to more innovative and complex forms that satisfy the strive of creativity of contemporary architects like Rhino, Grasshopper, and visual LISP. Second: The Building Material Technologies; that develop the physical properties of materials to enhance the performance of the structural systems and the used materials internally and externally in building construction. Third: The Building management systems; depending on the new technologies of the Internet of Things (IoT) and the Artificial intelligence (AI) through the smart sensors that monitor whatever it is programed to detect, they can detect motion, occupant's density, natural light intensity, air quality, indoor temperature and other parameters, based on its smartness they can take decisions to close or open complete sectors in buildings, adjust the light intensity inside, recycle the water to reuse in landscaping and so on.

The case studies show that the sustainability as a belief, and the new philosophy of digital intelligence regulates contemporary practices globally influencing the early design conceptions, design process and materialization, fabrication and construction all contributes to generate a new pattern of architecture that may embrace smaller styles or practices seen as individual languages of creativity and innovation. The holistic experience here can drive to a meta-theory that promotes a paradigm shift in the architectural discourse with the new pattern of architecture "I-Praxis". This pattern integrates the human and machine logic in formulating the new architecture, aiming to fulfill the desired quality of the process through using technology to serve the environment and biodiversity, human needs and well-being and economic efficiency. Achieving a clear understanding to the different existing styles has ultimate essential to evaluate the design key-concepts and approaches "Subjective dimension" behind the iconic visible morphologies of buildings "Objective dimension". This help to stand on the true nature of the practices and the legitimization behind their strategies formulating the various subsystems of concepts like we see: "Parametricism" of CCTV Tower. "contextualism" as in Dubai Louvre, "smart-adaptive architecture" as Apple City and etc. Moreover, some passive practices that extend from previous paradigms like the famous Bosco Towers by Stefano Boeri as an evolving ecological practice.

By these means, we can reach a comprehensive understanding to the contemporary practices and identify a generic-boundary of a sustainable complex "Architectonic" practices. Besides, identifying the new pattern of philosophical architecture "I-Praxis"; Within which smaller attempts vary in their design ideologies and conceptions but also consistent through a holistic digital design theory thriving to reach sustainability and make best use of the technological advances to achieve creativity and innovation of architectural idioms as reflection of the contemporary power of complex-knowledge.

References

[1] Charles Jencks, *"The Architecture of Jumping Universe "*. 1997.
[2] C. Browne, "Postmodernism, ideology and rationality," *Rev. Int. Philos.*, vol. 251, no. 1, pp. 79–99, 2010.
[3] K. E. Boulding, "The economics of the coming spaceship earth," in *Environmental Quality in A Growing Economy: Essays from the Sixth RFF Forum*, vol. 3, 2013.
[4] G. H. Brundtland, "Our common future - Call for action," *Environ. Conserv.*, vol. 14, no. 4, 1987, doi: 10.1017/s0376892900016805.
[5] A. A. Bartlett, "Reflections on sustainability, population growth, and the environment-2006," in *The Future of Sustainability*, 2006.
[6] K. Murphy, "The social pillar of sustainable development: A literature review and framework for policy analysis," *Sustain. Sci. Pract. Policy*, vol. 8, no. 1, 2012, doi: 10.1080/15487733.2012.11908081.
[7] A. D. Basiago, "Economic, social, and environmental sustainability in development theory and urban planning practice," *Environmentalist*, vol. 19, no. 2, 1998, doi: 10.1023/A:1006697118620.
[8] A. Tostões, "Modern and Sustainable," *Mod. Sustain.*, no. 44, 2011, doi: 10.52200/44.a.y5jilqmj.
[9] B. Rudofsky, *Architecture without architects, an introduction to nonpedigreed architecture*. The Museum of Modern Art: Distributed by Doubleday, Garden City, N.Y., 1964.
[10] S. S. Vuckovic and K. Kujundzic, "Architecture of the 20th century as a Cultural Heritage : Architectural Communication – Memorial Hall in Kolasin, Montenegro," *Archit. Urban Plan.*, vol. 9, no. November, p. 47, 2015, doi: 10.7250/aup.2014.007.
[11] C. Aigbavboa and W. Thwala, *The Construction Industry in the Fourth Industrial Revolution*. 2020.
[12] S. Attia, *Regenerative and Positive Impact Architecture Learning from Case Studies Introduction*. 2018.
[13] D. C. Thorns, "The Transformation of Cities," *Transform. Cities*, 2002, doi: 10.1007/978-1-4039-9031-0.
[14] R. Oxman and T. P. Technion, "Theory and design in the first digital age," vol. 27, pp. 229–265, 2006, doi: 10.1016/j.destud.2005.11.002.
[15] avier S. Merina, "EURAU18 Alicante: Retroactive Research: Congress Proceedings," *EURAU18 Alicant. Retroactive Res. Congr. Proc.*, 2018, doi: 10.14198/eurau18alicante.
[16] P. Fraioli, ""The Invention and Reinvention of the City: An Interview with Rem Koolhaas.,"" *J. Int. Aff. 65.2*, vol. 113, no. 19, 2012.
[17] ""Kool Enough for Beijing?,"" *China Daily.*, 2004.
[18] M. Mudhafter, "Cultural artifice: the Louvre Abu Dhabi, United Arab Emirates, by Ateliers Jean Nouvel," *The Architectural Review*, 2019. [Online]. Available: https://www.architectural-review.com/buildings/cultural-artifice-the-louvre-abu-dhabi-united-arab-emirates-by-ateliers-jean-nouvel.

Causes Influencing Construction Waste Generation During the Design Process: An Analytical Study

A S Saad[1,*], A A E OTHMAN[2], F O Alamoudy[1]

[1] Architectural Engineering Department, Faculty of Engineering, the British University in Egypt, Cairo-Suez Desert Road, Al-Shorouk, Cairo, Egypt.
[2] Head of Architectural Engineering Departement, Faculty of Engineering, the British University in Egypt, Cairo-Suez Desert Road, Al-Shorouk, Cairo, Egypt.

[*]Corresponding Author Email: aya165939@bue.edu.eg

Abstract. Despite the social and economic contributions of the construction industry towards achieving sustainable development objectives, it has an adverse effect on the environment such as the generation of waste, water, and energy consumption. Construction waste (CW) is the loss of materials that are produced during or after the construction process and has no remaining value. Several countries promote waste elimination plans at early stages to reduce CW at its source. The Egyptian government is expanding its investments in new mega projects and infrastructure facilities like new administrative capital, bridges, roads, energy projects which resulted in increasing the rate and amount of generated CW. The literature review stated that 33% of CW generated in the primary design stage is the result of design flaws such as design errors and lack of waste management strategies used during the design process. Accordingly, this research aims to investigate the factors influencing CW generation during the design process. To achieve this aim, the development of proper research methodology took place, where it consisted of qualitative analysis of previously published literature and examining case studies to accomplish two objectives. Firstly, an investigation of previous literature focused on studying the design process, CW, and the global and national waste management strategies. Secondly, three case studies were introduced and analysed to study the role of design out waste strategies towards reducing CW during the design process.

Keywords: Design Strategy, Construction Waste, Strategies Evaluation, Waste Reduction.

1. The Design Process

1.1 Nature of the Design Process

The design phase is known for its significant impact on the construction industry. It makes a substantial contribution to the delivery of sustainable projects that transform client requirements into technical drawings and specifications. Furthermore, decisions taken during this phase have a long-term impact on the project's performance across its lifecycle. Iteration, innovation, and complexity are some of the characteristics of the design process as it offers alternative solutions to meet existing and developing client expectations, meet user requirements, cope with regulatory changes, capitalize on business opportunities, adapt to technological improvements, provide value, and handle related risks [1].

1.2 Design Process Stages

The RIBA Plan of Work divides the pre-construction design phase into the following stages.

- **Strategic Definition (SD):** This stage aims at defining client requirements to develop a business plan for alternative provision, including an analysis of project risks and budget, choosing the option that best meets customer requirements, examining previous project feedback, and conducting site inspections [2].

- **Preparation and Briefing (PB):** This stage includes various tasks related to the identification of project aims and objectives through the project brief that includes the program, site information, and analysis. Furthermore, the expected outcome of the project is set as quality standards, sustainability outcome, spatial relations, and requirements as well as feasibility studies, budgeting, and execution plan. [2].

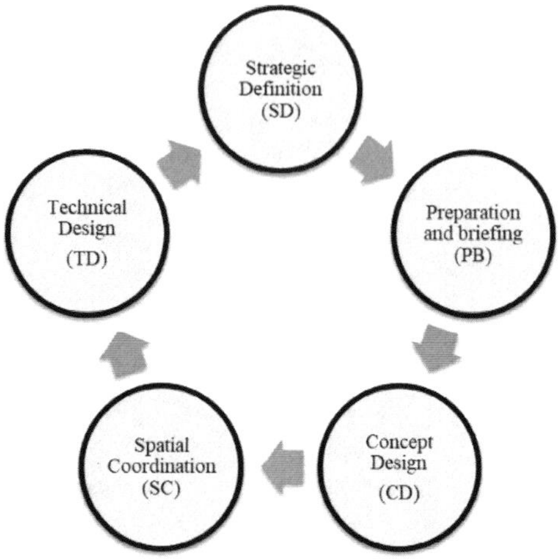

Figure 1. Design Process stages.

- **Concept Design (CD):** The third stage includes innovation and creativity in developing a design concept and philosophy and relating them with engineering strategies to fulfil the cost plan and project specifications. Then, the concept is reviewed with project stakeholders to further develop the project program [2].

- **Spatial Coordination (SC):** This stage focuses on design studies and analysis and confirms the project alliance with the required budget through cost exercises. Furthermore, the testing of the project concept with spatial requirements and relations take place, and further confirmation of the project alliance with the updated cost plan and project specifications [2].

- **Technical Design (TD):** Technical documents are prepared, developed, and coordinated in this stage by various design team building systems [2], see figure (1).

2. Construction Waste

2.1. Definition and Nature

Construction waste is defined as anything discarded or with an intention to discard [3]. This definition applies to all waste types whether they are intended for recycling or disposal operations. Compared to other activities worldwide, this industry is one of the most regarded producers of waste, for example, the construction waste generated in the European Union (EU) represents 35% of its total waste generation which equals to 4 times more than the total household waste production [4]. Construction waste is any product that is made by human or industrial activity that does not have remaining value. It is hard to estimate the amount of construction waste generated in the average site precisely. However, studies estimated that from 4% to 30% of the delivered building materials turned into wastes due to several reasons like loss and negligence [4]. Getting rid of construction waste affect natural resources as it occupies large landfill areas, it also negatively affects the surrounding environment and human health. Accordingly, the best way to avoid the impact of the waste is to avoid its production by implementing waste minimization strategies during the design phase [3].

2.2. Types of Construction Waste

Construction waste is divided into two types. First, physical construction waste is known as solid waste which includes sand, bricks, steel, concrete ruins, other organic materials, produced due to renovation activities and demolition process which includes formation, land excavation, civil building construction, roadwork, site clearance, building renovation, and demolition activities. Second, Non-physical waste takes place during the process of construction and describes as non-value-adding activity as the overrun of time and cost for a construction project [3], see figure (2).

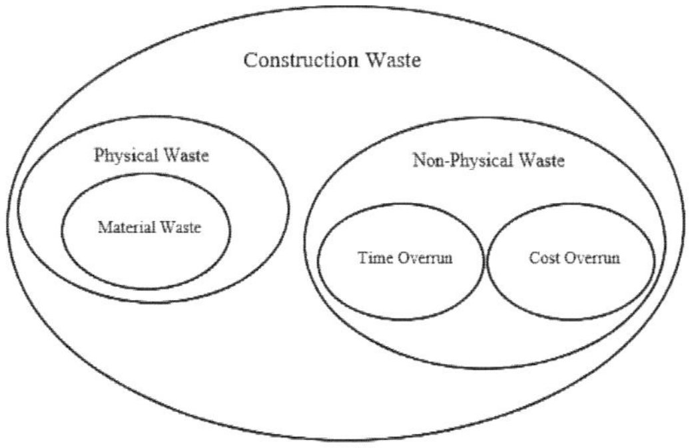

Figure 2. Types of Construction Waste [4].

3. Causes, Classifications and Impacts of Construction Waste Generated During Design Process

Construction waste is a kind of waste generated at the construction site which is directly or indirectly related to the design process. Previously published literature was investigated to pin down the causes of CW generated during the process and indicate the design stages in which construction waste is generated. In addition, these causes were categorized under 15 categories, see table (1). Furthermore, table (2) correlates the causes of CW and waste output of these causes.

Table 1. Construction waste causes, categories & design process stages (Developed by the Authors).

Category of Waste Causes	Causes of Construction Waste	Design Process Stags				
		SD	PB	CD	SC	TD
Communication, Clarity & Understanding Related Causes	Project Brief is Unclear and Incomplete [5].	✓	✓	✓	✓	✓
	Inappropriate Communication Between Client and Designer [5].	✓	✓	✓		
	Designers' Ignorance to Client Requirements and Behave Unilaterally [5].	✓	✓	✓		
	Late Information Flow Among Parties [5].	✓	✓	✓	✓	✓
	Unclear Client Requirement [5].	✓	✓	✓		
	Complicated Design [5].			✓	✓	✓
	Absence of Design Visualisation [5].			✓	✓	
	Interaction Between Various Specialists [5].			✓	✓	✓
	Frequent Design Changes by the Architect [5].			✓	✓	✓
Feasibility Study	Inappropriate Feasibility Studies [6].	✓	✓			
Value	Initiating Value Engineering Changes [7].	✓	✓	✓	✓	✓
Project User	Lack of Involvement of Users in Briefing Process [8].	✓	✓			
	Disregard of Users' Culture and Traditions [5].	✓	✓			
Coordination and Accuracy	Errors in Construction Drawings and Documents [9].				✓	✓
	Design Errors and Omissions [10].			✓	✓	✓
	Lack of Coordination Among Parties [10].	✓	✓	✓	✓	✓
	Poor Design Quality [10].			✓		✓
	Poor Quality Control [10].	✓	✓	✓	✓	✓
	Inexperience Designer [10].	✓	✓	✓	✓	✓
	Lack of Integrating of Waste Reduction Strategies During Design [10].	✓	✓	✓	✓	✓
Inadequate Provision of Information	Late Provision of Information in late stages of Design & Construction [5].	✓	✓	✓	✓	✓
	Lack of Information Provision [5].	✓	✓			
	Lack of consideration of Environmental Requirements [11].	✓	✓	✓	✓	✓

Regulations	Poor Coordination and Communication between Firms and Authorities regarding Approvals and Planning [12].	✓	✓	✓	✓	✓
Tech. Advancement	Response to Technological Advancement [12].				✓	✓
Quality and Sustainability	Lack of Requirements and Construct Ability [12].				✓	✓
	Inconsideration of the Complete Project Lifecycle [12].	✓	✓	✓	✓	✓
	Upgrading Project Facilities [12].	✓	✓	✓	✓	✓
	Poor Quality of Specified Materials [10].				✓	✓
	Eliminate Proven Low-quality Materials and Equipment [12].				✓	✓
Cost	Restricted Design Fees [13].	✓	✓			
Time	Insufficient Design Timeframe [13].	✓	✓	✓		
Unforeseen Conditions	Unexpected Conditions [14].	✓	✓	✓	✓	✓
Market Condition	Responding to Market Demand [15].	✓	✓	✓	✓	✓
User Demands	Users Exaggerate their Needs [15].	✓	✓			
	Late Involvement of Users Across Design Stages [15].				✓	✓
	Late Adjustments of Project Requirements by Stakeholders [15].				✓	✓
	Frequent Design Changes by the Client [5].	✓	✓	✓	✓	✓
Design Expertise	Absence of Design Experience [16].	✓	✓	✓	✓	✓
	Availability of Materials in Market [16].				✓	✓

Table 2. Cause of construction waste and waste output (Developed by the Authors).

Waste output / Causes of construction waste	Materials	Defects	Waiting	Cost Overrun	Rework	Delay	Client Dissatisfaction	Extra Processing	Unused Talents
Project Brief is Unclear and Incomplete		✓	✓	✓	✓	✓		✓	
Inappropriate Communication Between Client and Designer		✓		✓	✓	✓	✓	✓	
Designers' Ignorance to Client Requirements and Behave Unilaterally		✓		✓	✓	✓	✓	✓	
Late Information Flow Among Parties		✓	✓	✓	✓	✓		✓	
Unclear Client Requirement		✓	✓	✓	✓	✓	✓	✓	
Complicated Design	✓	✓		✓	✓	✓		✓	
Absence of Design Visualisation	✓			✓	✓	✓	✓	✓	✓
Interaction Between Various Specialists			✓	✓	✓	✓			
Frequent Design Changes by the Architect			✓	✓	✓	✓		✓	
Inappropriate Feasibility Studies	✓	✓		✓					
Initiating Value Engineering Changes		✓					✓		
Lack of Involvement of Users in Briefing Process				✓	✓	✓	✓		
Disregard of Users' Culture and Traditions	✓	✓	✓	✓	✓	✓	✓	✓	
Errors in Construction Drawings and Documents	✓	✓		✓	✓	✓	✓	✓	
Design Errors and Omissions	✓	✓		✓	✓	✓	✓	✓	
Lack of Coordination Among Parties	✓	✓	✓	✓	✓	✓	✓	✓	
Poor Design Quality	✓	✓	✓	✓	✓	✓	✓	✓	
Poor Quality Control	✓	✓	✓	✓	✓	✓	✓	✓	
Inexperience Designer	✓	✓	✓	✓	✓	✓	✓	✓	✓
Lack of Integrating of Waste Reduction Strategies During Design	✓	✓	✓	✓	✓	✓	✓	✓	✓

Late Provision of Information in Late Stages of Design & Construction	✓	✓	✓	✓	✓	✓		✓	
Lack of Information Provision	✓	✓	✓	✓	✓	✓		✓	
Lack of Consideration of Environmental Requirements	✓	✓	✓	✓	✓	✓		✓	
Poor Coordination and Communication between Firms and Authorities regarding Approvals and Planning	✓	✓	✓	✓	✓	✓		✓	
Response to Technological Advancement	✓		✓	✓	✓	✓		✓	
Lack of Requirements and Construct Ability	✓	✓	✓	✓	✓	✓	✓	✓	
Inconsideration of the Complete Project Lifecycle	✓	✓	✓	✓	✓	✓	✓	✓	✓
Upgrading Project Facilities			✓	✓	✓	✓			
Poor Quality of Specified Materials	✓	✓	✓	✓	✓	✓	✓	✓	
Eliminate Proven Low-Quality Materials and Equipment	✓		✓	✓	✓	✓		✓	
Restricted Design Fees		✓							
Insufficient Design Timeframe		✓					✓		
Unexpected Conditions	✓	✓	✓	✓	✓	✓	✓	✓	
Responding to Market Demand	✓		✓	✓	✓	✓			
Users Exaggerate their Needs	✓	✓	✓	✓	✓	✓		✓	
Late Involvement of Users Across Design Stages	✓	✓	✓	✓	✓	✓	✓	✓	
Late Adjustments of Project Requirements by Stakeholders	✓	✓	✓	✓	✓	✓	✓	✓	
Frequent Design Changes by the Client	✓	✓	✓	✓	✓	✓		✓	
Absence of Design Experience	✓	✓	✓	✓	✓	✓	✓	✓	✓
Availability of Materials in Market		✓	✓	✓	✓	✓	✓	✓	

4. Global Design Out Waste Strategies

The design out waste (DOW) strategy is known as a set of rules for utilizing assets efficiently across various stages of the design process in any construction project to reduce waste generation. This is done through the assessment of waste generation and identification of waste elimination methods so that project constructability is improved as well as project quality saving cost and time. This strategy was developed by the Waste and Resources Action Programme (WRAP) in the United Kingdom to assist architects in planning waste reduction from the early design phase using five main principles for designing [4]: (1) Waste Efficient Procurement, (2) Materials Optimization, (3) Off-Site Construction, (4) Reuse and Recycling, (5) Deconstruction and Flexibility. Furthermore, Early supplier involvement (ESI) and early contractor involvement (ECI) during the design phase are two strategies adopted to utilize the technical experience of suppliers and contractors to enhance the building performance, reduce life cycle costs, compress delivery duration, incorporate cutting-edge construction methods and reduce construction waste during the design process [17-19].

5. Waste Management Strategies in Egypt

According to the Egyptian Environmental Affairs agency [20], the waste generated in Egypt increased by 3.4% per year. The National Solid Waste Management Program (NSWMP) was developed to address the air pollution crises which is called the "Black Cloud" in 1999-2000. The program was not implemented due to the lack of institutional, legal, and waste management framework leading to the absence of guidance during the implementation process. In Egypt, there are currently two laws that address solid waste collection, treatment, and disposal. First, Law 38/1967 relates to general cleaning, and secondly law 4/1994 addresses environment protection. Both laws deal with the construction wastes after their occurrence. No rules or framework are to be followed to prevent the generation of construction wastes from their source which is the design process. In addition, the Green Pyramid Rating System (GPRS) which was developed to increase resource efficiency, protect the environment, and improve people quality of life established credit points to projects that use recycled materials and prefabricated elements and obtain credits by the provision of where, and how they will get rid of demolished work [21]. According to [22] GPRS has no rules or procedures for diverting landfills' waste using the 3Rs techniques (Reuse Reduce / Recover). However, GPRS demand an introduction of project materials' tables. Furthermore, Recycling is severely disregarded as an industry in Egypt. Therefore, it is not recommended as a CDW management strategy. Furthermore, The National Centre for Housing and Building Research (HBRC) in Egypt has no code that concerns the prevention of construction waste generation during the design process. The existing codes are concerned only with how to deal with construction and demolition waste after its existence [23]. Moreover, Egypt vision 2030 and Cairo Vision 2050 have not mentioned how to eliminate the construction waste generation from the design process. They only focused on enhancing the solid waste management system by increasing the percentage of solid waste collection and management [24, 25].

6. CASE STUDIES

6.1 Definition, Objective and Selection Criteria

Cases study is a research method used across various research disciplines to describe and analyse a single subject, event, and occurrence with the goal of identifying different variables, interactions, and relations between different factors or evaluating performance or development progress [26]. Within this research, three case studies were illustrated to explain the effect of applying design-out waste strategies during the design process to reduce CW. The case studies selection criteria were based on a number of factors including the nature and size of the selected projects, data availability, types of waste, waste-out-strategy adopted, location, and degree of success. All of the case studies featured construction projects that were faced different types of CW and adopted different strategies. Because this topic received scant consideration in construction literature, few case studies were published, the availability of data was a major element in selecting the case studies. Finally, the case studies were chosen from Saudi Arabia, China, and Thailand with different cultures and traditions and the learned lessons of the case studies will aid in reducing CW during the design process in Egypt. Table (3) discusses the research idea through a practical application to case studies to show the concept of the idea. It summarized the case studies through comparing the design out waste strategy used, and the amount and type of construction waste reduced.

6.2 Case Study (1): Sciences and Computer College, Dammam University, Saudi Arabia

The Sciences and Computer College in Dammam, Saudi Arabia, has eight laboratories, forty-eight classrooms, and offices. The project design took 4 months to complete using 2D drawings and cost 36 million Saudi Riyal (SAR) in total. The project was developed by researchers utilising Revit to transform the 2D designs into a 3D model; the modelling procedure took around 2 months to complete. BIM contributes to reducing design time by nearly 50%, resulting in a 200,000 SAR reduction in project cost, eliminating construction waste by creating an accurate model free of design errors, resulting in fewer unexpected changes during the construction phase, and improving communication between design parties, which improves decision-making during construction [27].

6.3 Case Study (2):26-storey High-rise Residential Building, China

A 26-story residential building made of concrete brick located in China was used to examine the effect of using prefabricated components instead of caste-in-situ elements on the amount of construction waste generated by using simulation software called Oracle Crystal Ball. The building design finished in February 2017, the simulation limits only to waste produced by timber concrete formwork and mortar and brick of the masonry work. The results show that in-situ generated waste for timber concrete formwork is 10.52% kg/m2, and for the masonry work is 4.77 kg/m2 on the other hand by using prefabricated units the construction waste will be reduced by 36.04% and by 25.53% independently [28].

6.4 Case Study (3): A Façade Architectural Design of a 3-Storey Residential Building, Thailand

An architectural design for a façade of a residential building composed of three stories located in Bangkok, Thailand subjected to design for construction waste minimization guidelines (DFCWM) guidelines during the design phase for reducing construction waste, cost, and used materials. The façade is made of galvanized steel structure and perforated aluminium sheet cladding. The façade design undergoes four stages which are preparation and briefing, concept design, spatial coordination, and technical design. The guidelines are proposed to each stage as a new way to be applied through the design practices to reduce the construction waste. The DFCWM guidelines application during the design phase of the façade design decreased the construction cost, made the production and assembly process easier, and avoided a significant amount of waste [29].

Table 3. Case Studies Analysis.

	Case 1	**Case 2**	**Case 3**
Name	Sciences and Computer College, Dammam University	26-storey High-rise Residential Building	A Façade Architectural Design of a 3-Storey Residential Building
		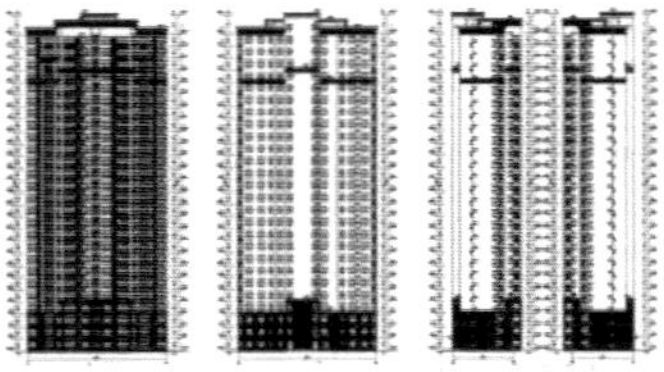	
Location	Saudi Arabia	China	Bangkok, Thailand.
Applied Design Out Waste Strategy	Building information modelling (BIM)	Prefabricated Components	Design for Construction Waste Minimization Guidelines. (DFCWM Guidelines)
Outcomes	• Wastes generation sources were eliminated as ○ Errors in Design errors ○ Poor communication ○ Unexpected design changes in the execution phase This resulted in reducing the waste, waiting time, and reducing the total project's cost [27].	• The rate of waste generation from masonry structures decreased by 53.6 % by using prefabricated walls. • The rate of waste generation from timber formwork decreased by 30% by using prefabricated components. • Waste reduction through using of prefabrication components can help to reduce landfill congestion, lower the risk of unlawful dumping, and improve environmental and social issues [28].	• Process Simplification • Resource Reduction • Waste Minimization • Cut costs by approximately 46% • Value Added to the façade design • Assisted the design development process to meet all stakeholders' requirements without sacrificing the design concept [29].

7. Conclusion and Recommendations

Despite the efforts done by experts in the construction industry to move forward towards sustainable development objectives, the adverse impact on the environment is highly regarded such as CW generation. CW is the loss of materials that are produced during or after the construction process and has no remaining value. 33% of CW generated is a result of deficiencies in the design process. Several countries are promoting waste elimination plans at early stages to reduce CW at its source. The expansion of investment in construction projects in Egypt resulted in increasing the rate of CW. In this study, Qualitative analysis of previously published literature was carried out and identified 40 causes of CW generated during the design process where they were categorized into 15 groups and their waste output was highlighted. Three case studies were discussed to emphasize the role of designing out waste strategies towards reducing CW. The research explained the national and international waste management strategies. It has been realized that despite the several strategies developed in Egypt, there is a lack of strategies or codes that promote CW reduction during the design phase.

References

[1] Othman AAE, Hassan TM, Pasquire CL. Drivers for dynamic brief development in construction. Vol. 11, Engineering, Construction and Architectural Management. 2004. p. 248–58.
[2] Osmani M, Villoria-Sáez P. Current and Emerging Construction Waste Management Status, Trends and Approaches. In: Waste. Elsevier; 2019. p. 365–80.
[3] Nagapan S, Abdul Rahman I, Asmi A. Factors Contributing to Physical and Non-Physical Waste Generation in Construction Industry. International Journal of Advances in Applied Sciences. 2012 Jun 3;1(1).
[4] Osmani M, Villoria-Sáez P. Current and Emerging Construction Waste Management Status, Trends and Approaches. In: Waste. Elsevier; 2019. p. 365–80.
[5] Barrett and Stanley. Better Construction Briefing by Peter Barrett, Catherine A. Stanley (z-lib.org), 1999.
[6] de Valence. Design and Construction: Building in Value. 1999.
[7] Stocks, S. and Singh, A. (1999), "Studies on the impact of functional analysis concept design on reduction in change orders", Constr. Mgmt Econ., Vol. 17 No. 3, pp. 251-67.
[8] Kernohan, D., Gray, J., Daish and Joiner, D. (1992), User Participation in Building Design and Management, 1st ed., Butterworth-Heinemann Ltd, Oxford.
[9] O'Leary, A. (1992), Construction Administration in Architectural Practice, 1st ed., McGraw-Hill, New York, NY.
[10] Nagapan S, Rahman IA. A Review of Construction Waste Cause Factors Critical Impact Energy for Local Impact Effects (penetration, Scabbing, and Perforation) of hard Projectile on Concrete Slabs View project SWID Web (Solid Waste Illegal Dumping Web) View project [Internet]. 2011. Available from: http://www.uthm.edu.my
[11] Best, R. and Valence, G. (1999), "Getting it right at the start", in Rick, B. and Valence, G. (Eds), Building in Value: Pre-Design Issues, Wiley, New York, NY.
[12] CIB (1997), Briefing the Team, (Working Group 1), Construction Industry Board, Tomas Telford, London.
[13] ICE (1996), Creating Value in Engineering: Design and Practice Guide, The Institute of Civil Engineers, Thomas Telford, London.
[14] O'Brien, J. (1998), Construction Change Orders, 1st ed., McGraw-Hill, New York, NY.
[15] Smith, J. and Wyatt, R. (1998), "Criteria for strategic decision making at the pre-briefing stage", in Hughes, W. (Ed.), Association of Research in Construction Management (ARCOM), Proc. of the 14th Annu. Conf., 9-11 September, Reading University, UK, pp. 300-9.
[16] Tenah, K. (1985), The Construction Management Process, 1st ed., Reston Publishing Company Inc., VA.
[17] Ahmed, A., Othman, E., Hosny, A., Elgendawy, S., & El-Gendawy, S. (2017). Improving Construction Sustainability through Integrating Suppliers and Manufacturers in the Design Process PhD Research-Materials Procurement Conceptual Framework for Minimizing Waste in the Egyptian Construction Industry View project (Bachelor dissertation) Talent Management: A Novel Lean-Integrated Approach to Maximize Creative Heritage Communities Development in Egypt. View project Improving Construction Sustainability through Integrating Suppliers and Manufacturers in the Design Process. https://www.researchgate.net/publication/271270305.
[18] Laryea, S., & Watermeyer, R. (2016). Early contractor involvement in framework contracts. Proceedings of Institution of Civil Engineers: Management, Procurement and Law, 169(1), 4–16. https://doi.org/10.1680/jmapl.15.00012.
[19] Bakar, A., Hamid, A., & Krishnapillai, G. (2006). The impact of purchasing and Early Supplier Involvement (ESI) in a manufacturing firm Enhancing Brand Equity through Sustainability Tourism Marketing: A study on homestays in Malaysia View project Sustainable Entrepreneurship View project. https://www.researchgate.net/publication/279463127.
[20] Egyptian Environmental Affairs Agency. (2011). National Solid Waste Management Programme (NSWMP) Egypt Main Report.

[21] GPRS. (2011). The Arab Republic of Egypt Ministry of Housing, Utilities and Urban Development Through The Housing and Building National Research Center In conjunction with The Egyptian Green Building Council.
[22] Daoud, A. O., Othman, A. A. E., Ebohon, O. J., & Bayyati, A. (2020). Overcoming the limitations of the green pyramid rating system in the Egyptian construction industry: a critical analysis. Architectural Engineering and Design Management. https://doi.org/10.1080/17452007.2020.1802218
[23] Abdelhamid, M. S. (2014) Assessment of different construction and demolition waste management approaches, HBRC Journal, 10 (3), pp. 317–326. DOI:10.1016/j.hbrcj.2014.01.003.
[24] Egypt Vision 2030. (2016). sds_egypt_vision_2030.
[25] Cairo Vision 2050. (2008). Cairo 2050 Within a national vision of Egypt Video.
[26] Sturman, A. (1997), "Case study methods", in Keeves, J. P (Ed.), Educational Research, Methodology and Measurement: An International Handbook, 2nd ed., Oxford: Pergamon, pp. 61-66.
[27] Eldeep AM, Farag MAM, Abd El-hafez LM. Using BIM as a lean management tool in construction processes – A case study: Using BIM as a lean management tool. Ain Shams Engineering Journal. 2022 Mar 1;13(2).
[28] Hao J, Chen Z, Zhang Z, Loehlein G. Quantifying construction waste reduction through the application of prefabrication: a case study in Anhui, China. Available from: https://doi.org/10.1007/s11356-020-09026-2.
[29] Laovisutthichai V., Lu W, Bao Z. Design for construction waste minimization: guidelines and practice. Architectural Engineering and Design Management. 2020; DOI:10.1080/17452007.2020.1862043.

Improving the Efficiency of Hospital Nursing Unit Design: an analytical study

E Ibrahim[1,*], A Sherif[2], H Serag Eldin[3]

[1] Cairo University, Giza, Egypt
[2] Architecture Engineering Department, The American University in Cairo, New Cairo, Egypt
[3] Cairo University, Giza, Egypt

[*]Corresponding Author Email: elshima.arc.81@gmail.com

Abstract: The inpatient nursing unit is one of the most important departments in a hospital building. It is where patient treatment and healing occur. Improving the efficiency of the nursing unit design may improve the length of patient stay, reduce medical errors and increase patient satisfaction, thus improving the performance of the medical services provided in the hospital. This paper addresses the architectural design factors which affect performance in a typical inpatient nursing unit. A two-step approach was used in this paper. First, the factors which influence the nursing unit performance were identified from literature. Then, typical nursing unit floor plan shapes were identified. Four floor plan shapes were selected from these for analysis. Analysis focused on the factors of flexibility of operation, ease of patient monitoring, provision of patient privacy, reduction of nurse travel time, and provision of a supportive healing environment. The racetrack floor plan has shown a higher level of performance in comparison with the other floor plan shapes. It has shown superior qualities in regard to flexibility, patient monitoring, and nurse travel distances. However, for this shape to be successfully used, sound and noise control measures should be adopted to improve performance regarding patient privacy and noise disturbance. The courtyard and radial floor plan shapes provide a medium level of performance. The radial floor plan is superior in regard to patient monitoring capabilities. Although the single corridor plan has shown higher levels of sound privacy and lower levels of noise disturbance, it did not provide good flexibility, patient monitoring, and travel distances.

Keywords: hospital, nursing unit, architectural design, design efficiency, flexibility, patient privacy.

1. Introduction:

The inpatient nursing unit is one of the most important departments in a hospital building. It is where patient treatment and healing occur. Improving the efficiency of the nursing unit design may reduce the length of patient stay, reduce medical errors and improve patient satisfaction, thus improving the efficiency of the medical services provided in the hospital. There is a lack of knowledge on the design factors that may affect the functional performance of nursing units in hospital buildings.

This paper raises the research question: "How can the design efficiency of a nursing unit be improved so that it achieves its role in terms of flexibility and efficiency?" Typical formations of the inpatient nursing unit floor plans are studied in regard to flexibility of operation, ease of patient monitoring, provision of patient privacy, reduction of nurse travel time, and provision of a supportive healing environment.

A two-step approach was used in this paper (figure 1). First, the factors which influence the nursing unit performance were identified from literature. Then, typical nursing unit floor plan shapes were identified. Four floor plan shapes were selected from these for analysis.

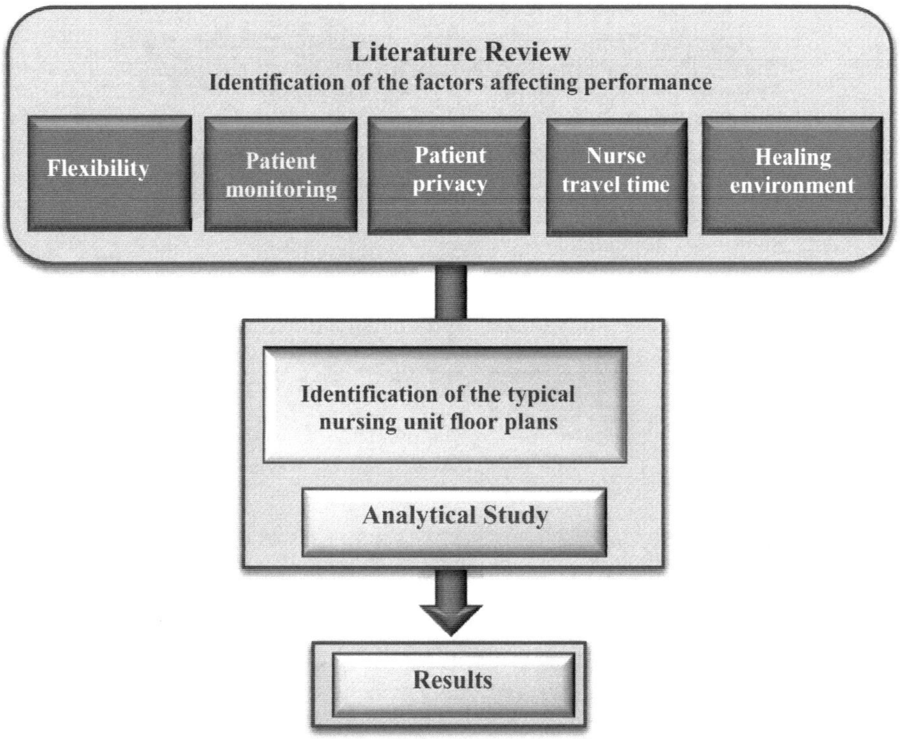

Figure 1. The two-step research approach.

2. The factors affecting performance in the nursing unit

The nursing unit floor plan shape has an impact on several factors influencing the healthcare performance. This paper focuses on four main aspects [1]. These are: flexibility, patient monitoring, patient privacy, nurse travel time and provision of a supportive healing environment.

2.1 Flexibility

It is one of the most important design aspects that should be taken into account when designing hospitals. It helps meet the ever-changing requirements resulting from the introduction of new methods of treatment and the emergence of new state-of-the-art medical equipment. It also helps address the operational efficiency needed with the uncertain changing operational environment of healthcare settings [2]. There are many aspects that may be addressed when considering the nursing unit design flexibility. These include the possibility of dividing the unit into separate parts, the provision of multiple patient visibility points, the provision peer lines of sight for nurses, the centrality of support, and the possibility of future expansion.

The possibility of dividing the nursing unit into separate parts (multiple division/zoning options) can be achieved when the location of the division lines is carefully planned early in the design stage. The unit should be able to be divided into a number of functional parts, allowing for the possibility of using each division as a semi-independent subunit resulting in an integrated flexible design. An example is the floor plan of the medical-surgical nursing unit of the St. Francis Hospital, Charleston, South Carolina. The nursing unit can be subdivided into four semi-independent sections (figure 2). Each sub-section consists of 10 patient rooms, a nursing station with common support services [3].

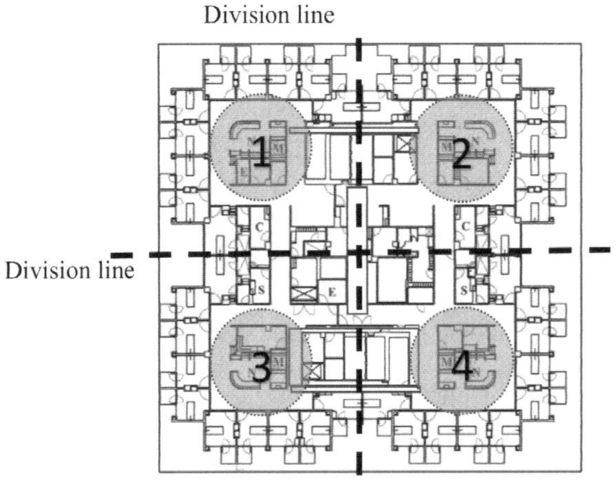

Figure 2. The possibility of subdividing the nursing unit at St Francis, Charleston, South Carolina [3].

In order to allow of the above flexibility, the nursing station should be split into a number of substations allowing each part of the nursing unit to be split into a functional entity. The above example illustrates the use of four nursing stations, allowing for the planned subdivision of the unit [3]. The use of decentralized nursing stations also helped improve operational efficiency by bringing the nurse bases closer to their patient rooms. In the above example of the St. Francis Hospital, Charleston, South Carolina, each nursing unit was responsible for supervising a limited number of closely associated patient rooms. This arrangement allowed for good lines of sight between the nurse stations and the patient rooms (Figure 3).

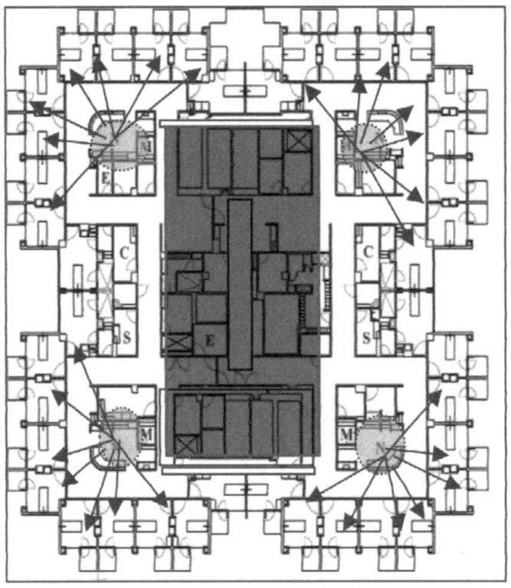

Figure 3. The decentralized nursing stations and the central support services in the St Francis, Charleston, South Carolina [3].

By contrast, the design of the nursing unit of the Clarian West Medical Center, Avon, Indiana, relied on only one main nurse station. Although the floor plan of this nursing unit is divisible, the use of a single nurse station did not allow for the needed design flexibility. It also resulted in reduce lines of sight between the nurse station and the patient rooms as illustrated in figure (4) [4].

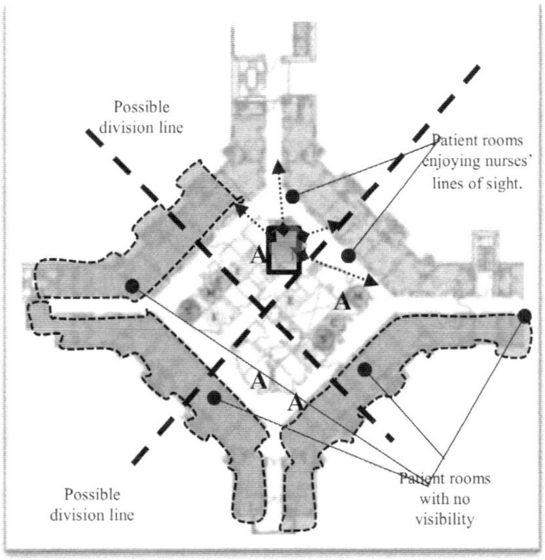

Figure 4. A single nursing station in Clarian West Medical Center, Avon, Indiana [4].

Adopting a teamwork approach by the healthcare providers is one of the important factors affecting operational flexibility and efficiency. This depends on allowing the healthcare team members to help one another, exchange their places and provide more services to different groups of patient rooms. This also allows for social communication between the healthcare providers. This approach has a positive impact on staff job satisfaction, thus helping raise the efficiency of patient care and speed of recovery. Good communication between the nurse stations is essential to the success of this approach. A direct line of sight between nursing stations is necessary to achieve this collaborative approach. Figure (5) illustrates the lines of sight that were planned in the design of the St. Francis Hospital, Charleston, South Carolina mentioned above [3].

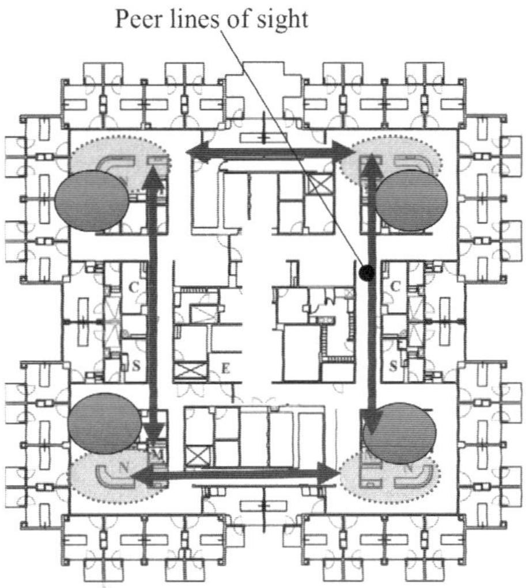

Figure 5. Peer lines of sight at St Francis, Charleston, South Carolina [3].

Support services are essential for the functioning of the nursing unit. These include the clean and soiled utility rooms, linen and food services, the medication station, etc. They should be placed in locations as close as possible to the healthcare providers in order to reduce the time wasted in traffic [3]. Placing the support spaces at the heart of the nursing unit allows for the needed flexibility. The central location also allows for dividing the unit, while accessing the centrally placed support services. This can be illustrated in the design of nursing unit of the St. Francis Hospital, Charleston, South Carolina as shown in figure (3) above.

The possibility of future expansion of the nursing unit is one of the aspects that should be considered. The design should allow for the possibilities of these expansions [5]. This may be needed to accommodate additional patient, family and/or support spaces. Design of the Clarian West Medical Center, Avon, Indiana illustrated above allows for this possibility. The open-ended corridors allow for the addition of future spaces.

2.2 Patient Monitoring

The ease of accessing patient rooms is one of key the factors contributing to the efficiency of operation of the nursing unit. The healthcare team should be able to efficiently supervise their patients. This principle was established as early as the Florence nightingale's open ward, where the word ward was derived from the concept of "guarding" or "surveillance" [3]. Allocating the nurse stations near the patient rooms allows for their visibility and accessibility. This doe not only help the nurses in performing their tasks, but also facilitates better interaction with patients. In addition, it helps improve the sense of safety among patients, thus helping in the creation of a healthy environment [6]. The use of decentralized nurse stations can help in this regard. Also, the use of glass windows between the patient rooms and the nursing unit corridor can provide better visual accessibility to the patients, especially when the patient room bathroom is placed outboard [7].

2.3 Patient Privacy

The nursing unit design can have an influence on patient privacy. This includes both visual privacy and also the sound privacy of the communication between patients and the family members [8]. Privacy can be achieved through two main elements [9]. These are as follows:

Visual privacy can be achieved by carefully designing the patient room layout. The inboard patient room design where the patient toilet is placed at the corridor side of the room improves patient's visual privacy [10]. This can be especially effective when the nursing unit is formed of single patient rooms [11]. Patient privacy can also be achieved by using other simple means such as the curtains. It should be noted that visual privacy should be carefully considered as it counter acts the efficiency of operation and the patient safety feel discussed above.

Sound privacy entails preventing patient room sounds from being heard in the outside. It also relates to preventing the healthcare team discussions from being heard by the patients, thus maintaining the confidentiality of patient information. The nurse station is one the main sources of this sound, especially to and from the patient rooms located nearby [3]. Patients in their rooms should not be allowed to hear the healthcare team conversations about their patients. Providing a suitable distance between the nurse station and the adjacent patient rooms can help improve sound privacy. Separating the corridor and workspaces of staff from those of patient rooms can help isolate the sound of work discussions from patients and also prevent overhearing patient room sounds. This paper will focus on sound privacy as it relates to the formation of the nursing unit floor plan.

2.4 Reduction of Nurse Travel Time.
The healthcare team spends a substantial amount time travelling between the nurse station, patient rooms, the medication and clean utility rooms, and the soiled utility room. This represents the greatest daily work burden on nurses. A study [12] was conducted to calculate the walking distances of nurses during day and night work. It was found that during the day, nurses walk 3.30 to 3,500 km on average for 12 hours. This causes staff fatigue, affecting the efficiency of operation and the safety of patients.

Careful design of the nursing unit can help reduce the time and effort wasted in these frequent trips. This reduces staff stress and provides a better service for patient care, and thus support the overall efficiency of the hospital operation [13]. The tasks conducted by nurses can be summarized as follows: patient care, which includes room arrangement, whitewashing, medical and health care for patients, assessing their condition, and passing with doctors to follow patients; processing and disbursement of medicine, which includes processing and delivery of medication according to a doctor's consultation; and administrative and service activities, which includes equipment processing, transportation of patients between different sections, preparation and writing reports within the nursing station and other administrative work. The travel distance of each nurse is the resultant of multiplying the distance travelled between the different parts mentioned above and the frequency of trips conducted for each task.

Location of the nursing station in relation to patient and the service rooms has a great influence on nurse travel distances. Also, the floor plan shape can have a great effect, where a round-form nursing unit design, with a central nurse station, results in shorter walking distance. The rounded form, where nurse station would be surrounded by patient rooms, allows for direct access to and better visual observation of all patient rooms.

2.5 Provision of a supportive healing environment
Design of the nursing unit must take into account the psychological as well as the functional needs of patients, families and staff [14]. It should provide an environment supportive to patient comfort and self-healing, as well as patient safety and the reduction of the incidence of disease [15]. The elements of a supportive environment include design of the patient room, control of noise, and the provision of supportive spaces such as visitor, patient and staff lounges and children play areas. The patient room design should provide an environment which stimulates healing, by creating interior design modes that enhance the harmony between mind, body and soul. In addition, patient safety should be observed, where the patient bathroom should be easily reached from the patient bed without crossing the room, thus reducing the chances of patient falling on the floor [16]. This paper will focus on the control of noise since it has a strong influence on patient comfort.

3. Analytical Study
The analytical part of this study starts with the identification of the typical nursing unit floor plan shapes. These were derived from literature sources. Then, four prototypical floor plans were analyzed in regard to the factors identified above.

3.1 The typical nursing unit floor plan shapes
The nursing unit floor plan shape has evolved in response to technological and medical developments. The most commonly floor plan shapes can be categorized as follows [17-18]: i) open (Nightingale) floor plan, ii) single corridor floor plan, iii) duplex (Nuffield) floor plan iv) double corridor (racetrack) floor plan, v) courtyard floor plan, vi) cruciform (cluster) floor plan, vii) radial floor plan, and viii) triangular floor plan. Their floor plans can be either simple, complex or other shapes. These are illustrated in figure (6).

3.2 The tested nursing unit floor plan shapes
Four prototypical nursing unit floor plan shapes were developed for analysis. They were based on the typical shapes identified above. These were: single corridor, racetrack, courtyard and radial nursing unit. All design

IOP Conf. Series: Earth and Environmental Science 1056 (2022) 012013 doi:10.1088/1755-1315/1056/1/012013

elements were normalized in these floor plans. They were shaped to have an equal number of beds (20 beds). They all had standardized patient room sizes and shapes.

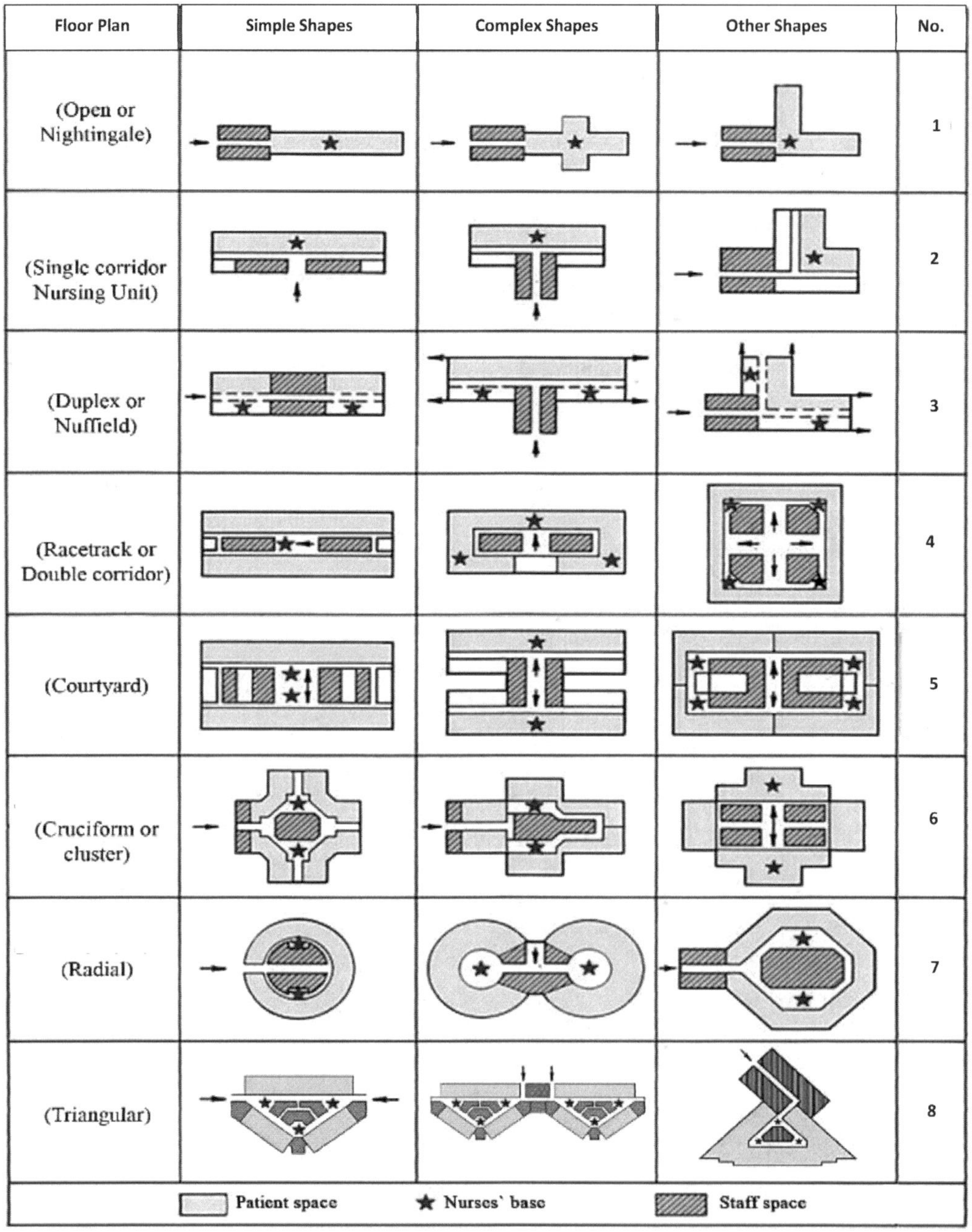

Figure 6. Typical nursing unit floor plan shapes [17-18].

The tested shapes have the following general characteristics. The single corridor nursing unit floor plan (shape A) utilizes one long corridor linking all nursing unit components. The support services are placed in two parts alongside the nursing unit entrance. The racetrack nursing unit floor plan (shape B) divides the nursing unit into four sections or small units, with decentralized independent nursing stations serving five patient rooms each. The four parts share common services. The courtyard nursing unit floor plan (shape C) takes the form of a rectangle having a central nursing station. This plan provides a central corridor for the healthcare staff, separate from those of the patient rooms. The radial nursing unit floor plan (shape D) takes the form of a two-circle composition with patient rooms placed at the perimeter and a central nursing station. In this shape, service areas were placed in a separate section away from the patient rooms, yet close to the nursing station. Figure (7) illustrates the four tested floor plan shapes.

Single corridor nursing unit floor plan

Racetrack nursing unit floor plan

Courtyard nursing unit floor plan

Radial nursing unit floor plan

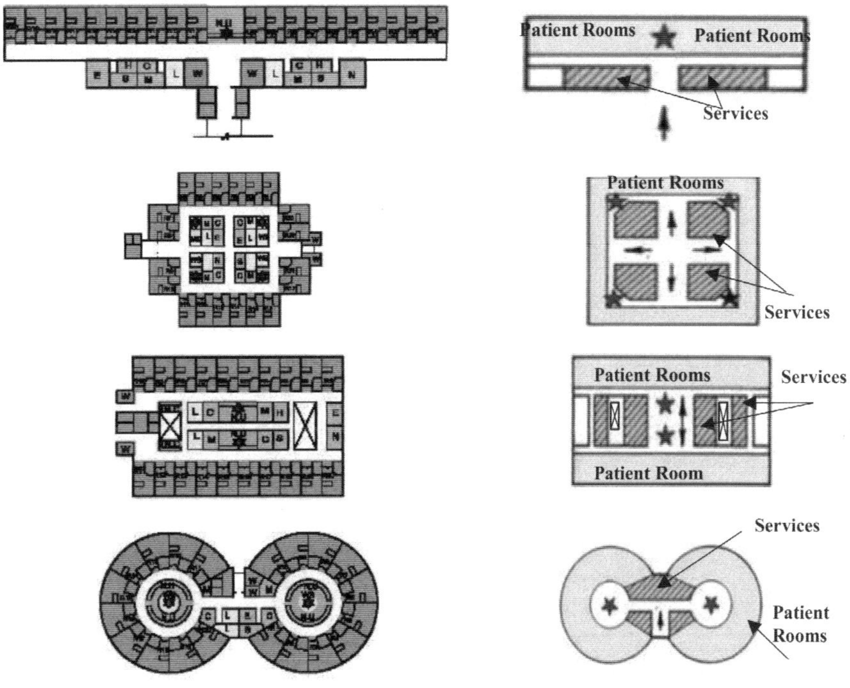

Figure 7. The tested nursing unit floor plan shapes.

3.3 Analysis of the tested floor plan shapes

The prototypical floor plans were examined in relation to the five factors identified above. These were: flexibility, patient monitoring, patient privacy, nurse travel time, and provision of a supportive healing environment.

3.3.1 Flexibility. Table (1) illustrates that all floor plan types are divisible into two or more parts. The racetrack and courtyard shapes were more successful in this regard since their central shared facilities can easily serve the divided segments. Regarding patient room access and visibility from the nurse station, the courtyard and radial shapes were superior in comparison with the two other shapes. The racetrack shape came second, while the single corridor allowed for limited accessibility and vision. The provision of peer lines of sight was variable between the four shapes. The racetrack and radial floor plans provided more clear lines of sight, while the courtyard shape was moderate in this regard.

Accordingly, it could be deducted that the racetrack shape is the most successful shape in regard to the flexibility factors. Second comes the courtyard and radial shapes. The single corridor shape seems to provide the lowest performance regarding flexibility.

Table 1. Flexibility aspects of the nursing unit.

A-Single corridor nursing unit	**B** –Racetrack nursing unit	**C**-Courtyard nursing unit	**D**-Radial nursing unit
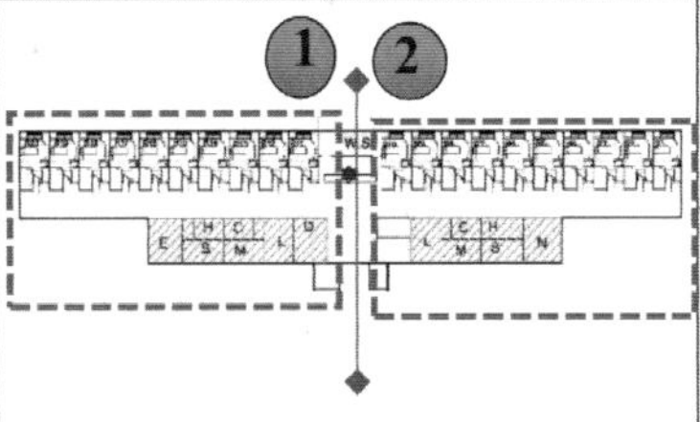	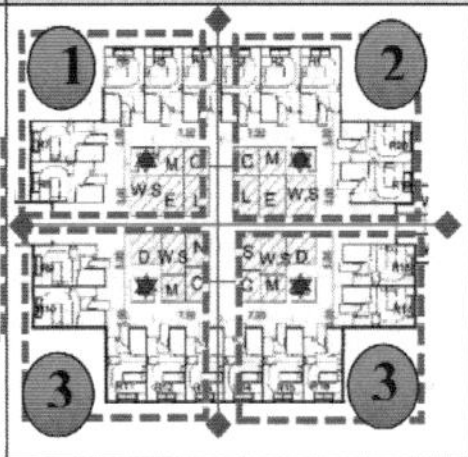	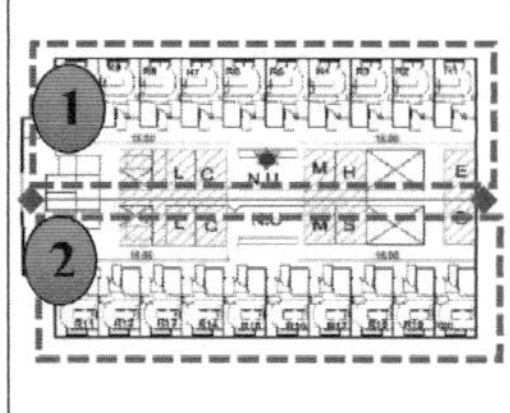	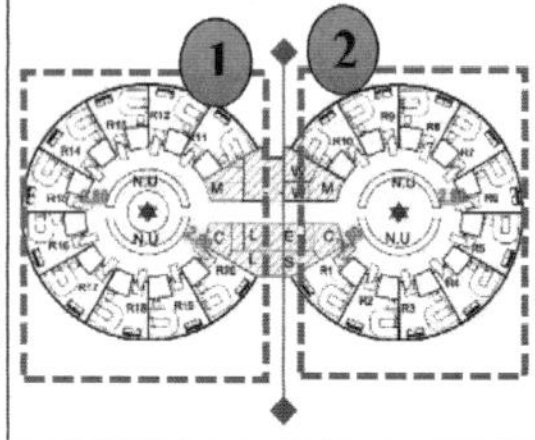
Nursing unit plan can be divided into two parts with limited shared support – **Low**	Nursing unit plan can be divided into four parts with central shared support - **High**	Nursing unit plan can be divided into two parts with central shared support – **High**	Nursing unit plan can be divided into two parts with limited shared central support - **Medium**
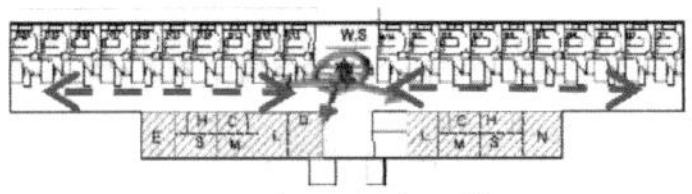 An arrow shows the distance between nursing stations and patients' rooms	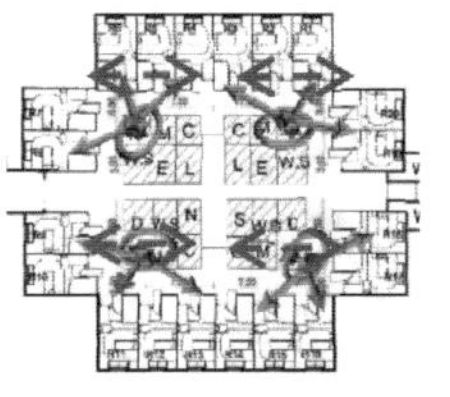	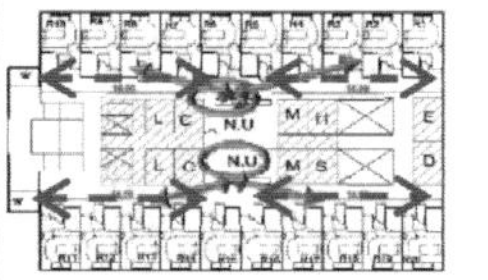	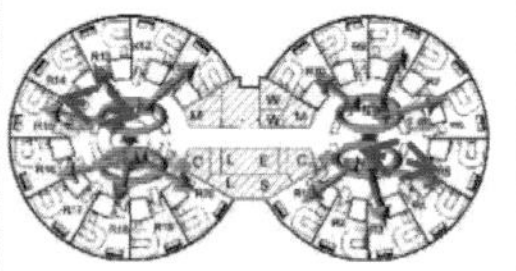
Availability of a nursing station for every **10 rooms** – **Low**	Availability of a nursing station for every **5 rooms - High**	Availability of a nursing station for every **10 rooms** - **Medium**	Availability of a nursing station for every **5 rooms** – **High**
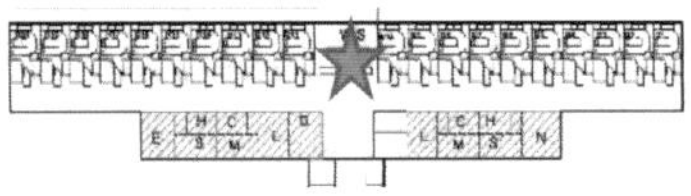	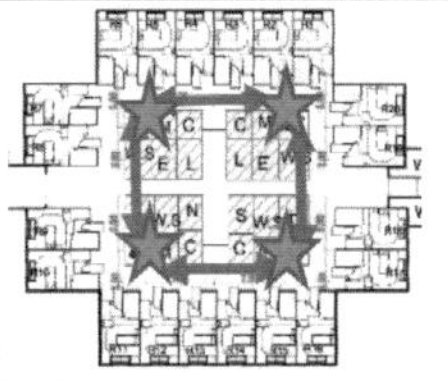	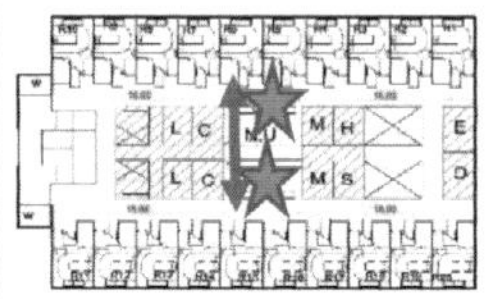	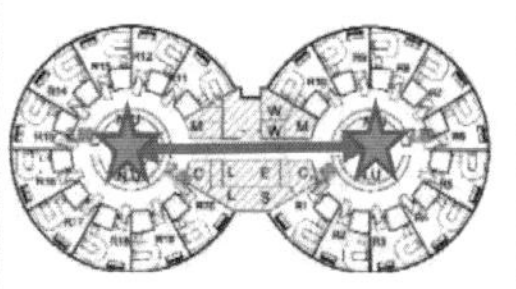
Peer lines of sight **Low**	Peer lines of sight **High**	Peer lines of sight **Medium**	Peer lines of sight **High**
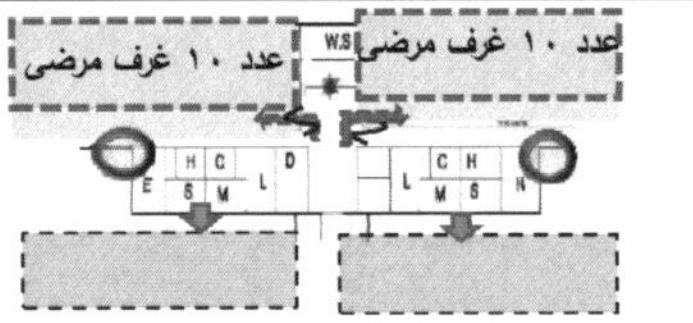	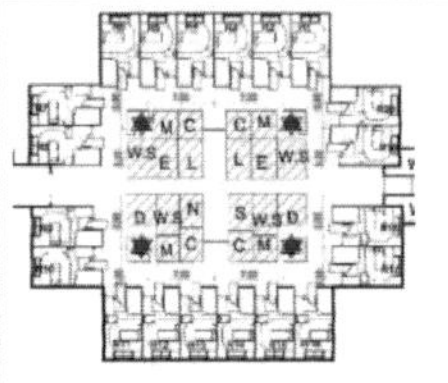	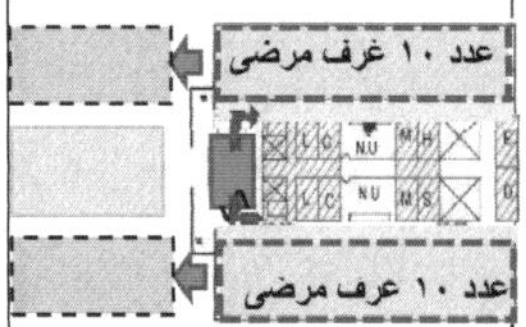	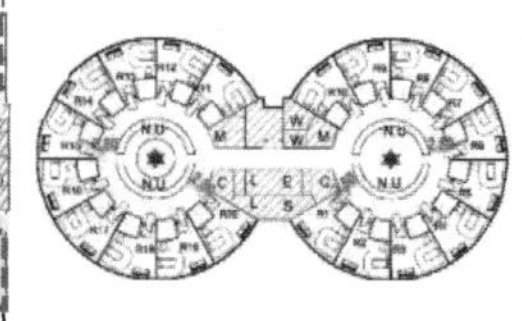
Easy future expansion **High**	Limited future expansion **Low**	Easy future expansion **High**	No future expansion **Low**
Overall Low	**Overall High**	**Overall Medium**	**Overall Medium**

3.3.2 Patient Monitoring. Table (2) demonstrates that patient monitoring is directly related to the positioning of the nursing unit. The racetrack and radial forms provide high levels of patient room visibility since their shapes allow for placing the nurse stations close to the patient rooms. The courtyard shape provides a medium level of visibility, while the single corridor shape has a limited level of visibility.

Table 2. Ease of patient monitoring.

***A**-Single corridor nursing unit*	***B** –racetrack nursing unit*	***C**-courtyard nursing unit*	***D**-radial nursing unit*
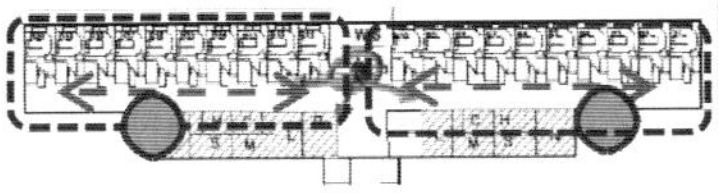	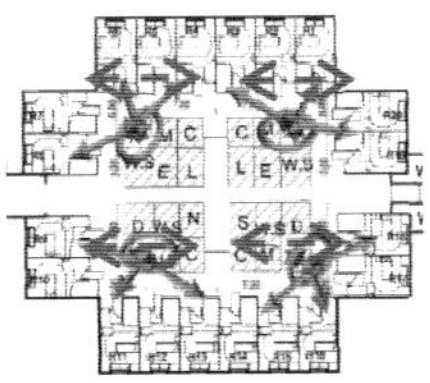	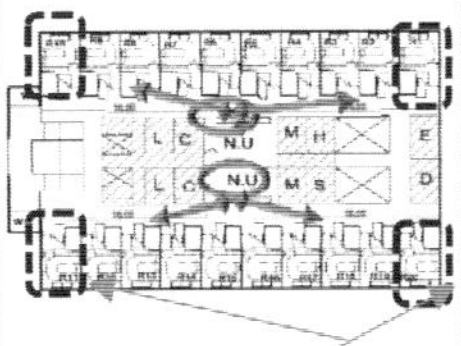	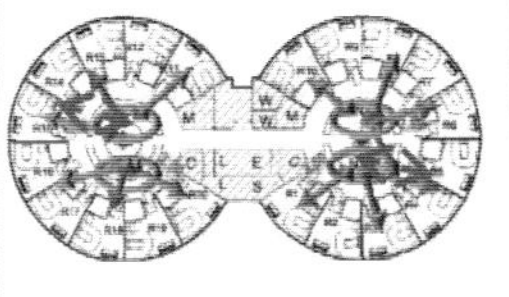
visibility - **low** • Not enough is achieved to supervise and follow patients, given the length of the corridor. • Room relationship of nursing station - **low**	visibility – **high** • It's got control of the decentralized nursing station, which has four patient rooms. • Room relationship of nursing station - **high**	visibility – **medium** • Achieve a better level of visibility for 16 patient rooms • The nursing station's relationship to the rooms is **medium**	visibility – **high** • It's got control of the centralized nursing station, which has four patient rooms. • Room relationship of nursing station - high visibility -**high**

3.3.3 Patient privacy. The proximity of patient rooms to the nurse station reduces sound privacy where the nurses can hear the speech in the room, and the patients can also hear staff discussion in the nurse station. Proximity of the nurse station to patient room doors was calculated for each floor plan shape. The sound privacy levels were classified as low, medium and high. Table (3) illustrates that the highest levels of sound privacy was achieved in the single corridor floor plan shape, while a medium level was achieved in the courtyard shape. The racetrack and radial shapes were low since the nurse station was in close proximity to a large number of patient rooms.

Table 3. Sound privacy.

***A**-Single corridor nursing unit*	***B** –racetrack nursing unit*	***C**-courtyard nursing unit*	***D**-radial nursing unit*
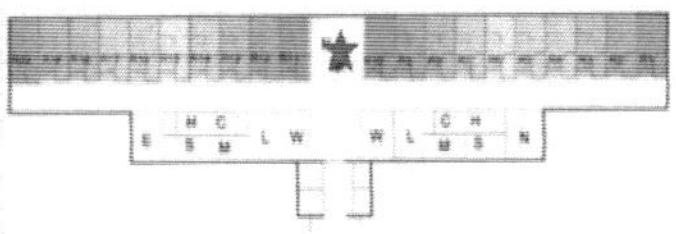	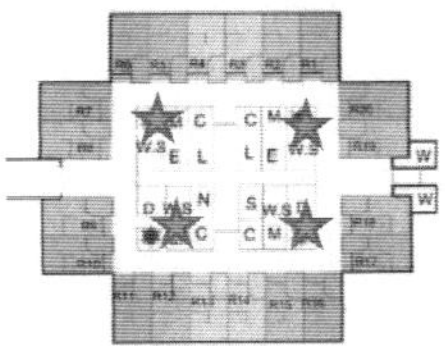	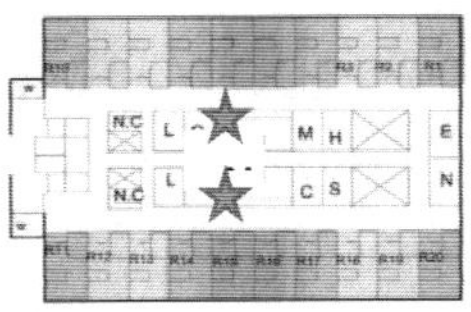	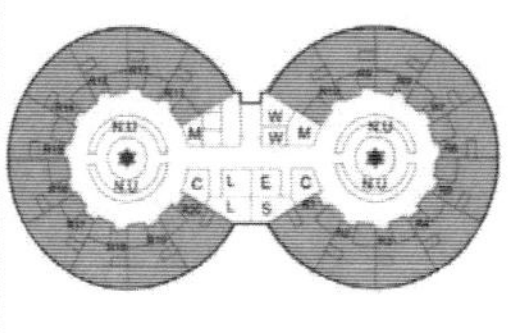
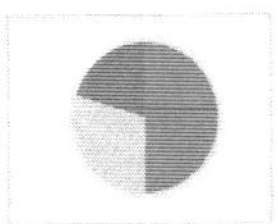 **Low Medium high** **20% 30% 50%**	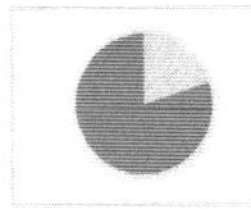 **Low Medium high** **80% 20% 0%**	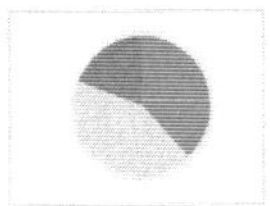 **Low Medium high** **20% 40% 40%**	 **Low Medium high** **100% 0% 0%**

3.3.4 Nurse Travel Distances. The nurse travel distances were calculated for each of the four floor plan shapes. Four nurses were assumed to be available in each shift. Each nurse was assumed to be responsible for five rooms within the nursing unit. Figure (8) illustrates the method used for calculating travel distances.

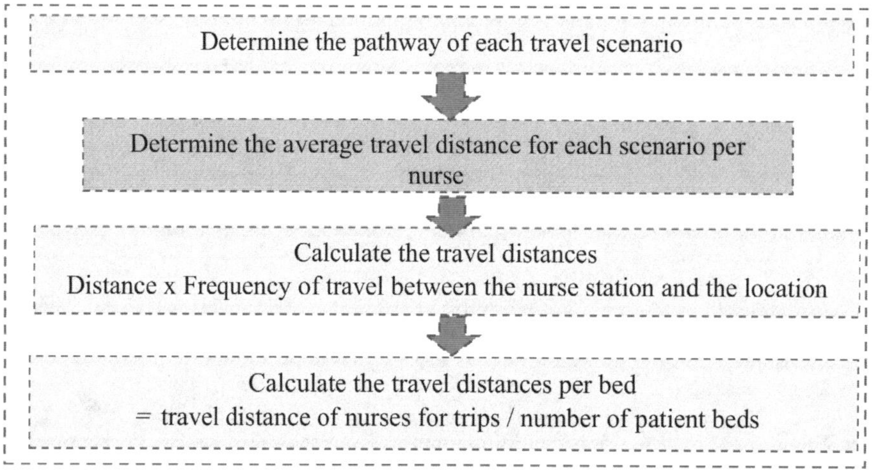

Figure 8. Calculation of nurse travel distances.

Table (4) illustrates the movement paths for each nurse. A number of possible travel scenarios were assumed, and the average movement distances were taken for each scenario. The travel scenarios were based on previous studies [19]. For example, in one scenario, it was assumed that the nurse administers medication 3 times/shift for each patient. This number was confirmed by the author's empirical research. In this scenario, it was assumed that the nurse visits the medication room before administering the medication to each patient in the patient room. This aligns with current patient safety standards. In another scenario, it was assumed that the nurse visits the patient rooms in a sequence for measuring patient vital signs, such as blood pressure and blood sugar. In a third scenario the nurse was assumed to visit each patient room 4 times to take care of the patient, such as receiving the patient, taking care of the patient, assessing the patient's condition, passing with doctors. In this scenario, the nurse was assumed to move from the nurse station to patient room 1, then back to the nursing station before going to patient number 2, etc.

The table demonstrates that the shortest travel distance was achieved when using the racetrack floor plan shape. The second low distance was related to the radial floor plan shape. The highest travel distance was associated with the single corridor floor plan shape. The travel distance of the courtyard shape was medium.

Table 4. The nurse travel distances.

Single corridor nursing unit

- Average total distance per patient care trip = 2014
- Average total distance for trips to process and deliver medicine to patients = 3392 m
- Average total distance for flights 5406 m
- The Conclusion of the average distance per bed can be from 5406/20 bed = **270 m.**

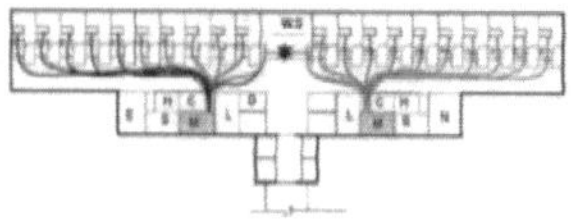

IOP Conf. Series: Earth and Environmental Science 1056 (2022) 012013 doi:10.1088/1755-1315/1056/1/012013

Racetrack nursing unit • Average total distance per patient care trip = 950 m • Average total distance for trips to process and deliver medicine to patients = 1584 m • Average total distance for trips = 2534 m • The Conclusion of the average distance per bed can be from 2534/20 beds **= 126 m.**	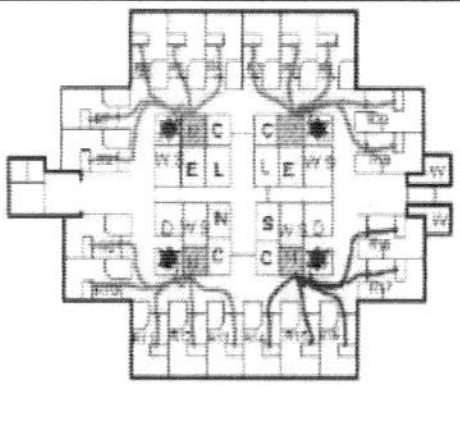
Courtyard nursing unit • Average total distance per patient care trip = 1186 m • Average total distance for trips to process and deliver medicine to patients = 2431 m • Average total distance for flights = 3617 m • The average distance per bed can be inferred from 3617/20 bed = **180 m.**	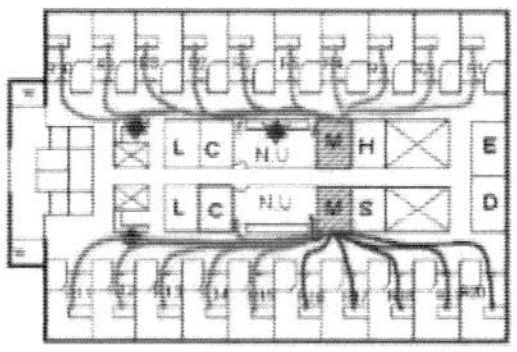
Radial nursing unit • Average total distance per patient care trip = 1186 m • Average total distance for trips to process and deliver medicine to patients = 2203 m • Average total distance for flights = 3419 m • The Conclusion of the average distance per bed can be from 3419/20 beds **= 161 m**	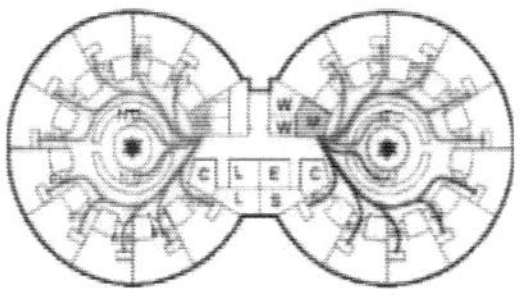

3.3.5 Provision of a Supportive Healing Environment. The nursing unit shape has an influence on the noise levels affecting patient comfort. The nurse stations and the associated workspaces are considered the main sources of noise in the patient unit. Table (5) provides an illustration of the noise levels expected along the corridors in front of patient rooms. The lowest levels of noise were achieved in the single corridor floor plan where a large number of rooms are located away from the work areas. Second best is the courtyard floor plan shape since it has a separate corridor for service. The lowest performance in this category was associated with the racetrack and radial floor plan shapes.

Table 5. Nurse unit noise levels.

***A**-Single corridor nursing unit*	***B** –racetrack nursing unit*	***C**-courtyard nursing unit*	***D**-radial nursing unit*
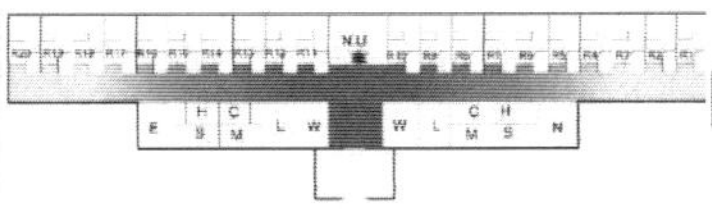	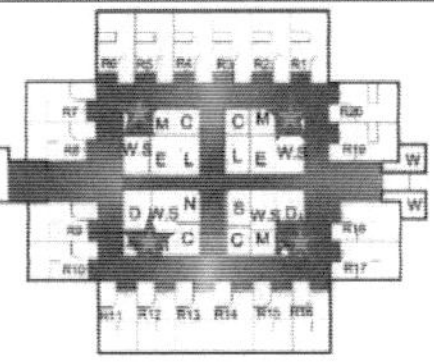	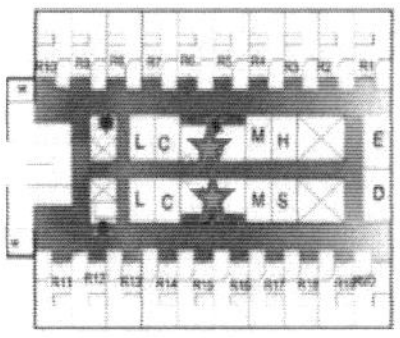	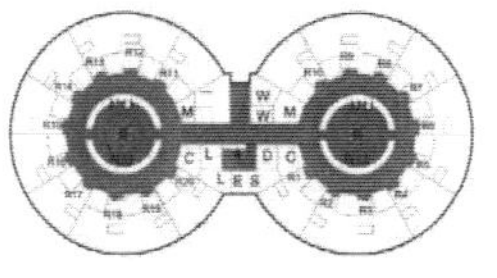
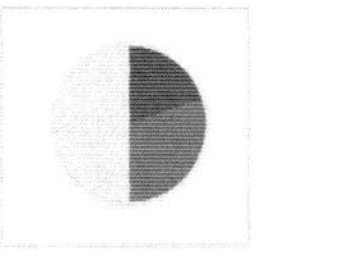 **Low Medium high** **50% 30% 20%**	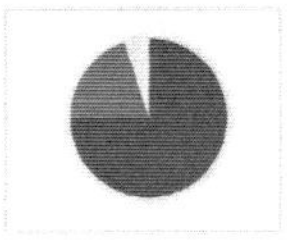 **Low Medium *high*** **5% 20% 75%**	 **Low Medium high** **20% 45% 35%**	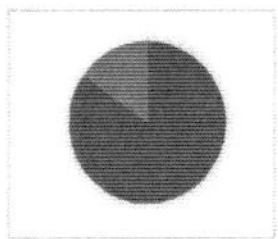 **Low Medium high** **0% 15% 85%**

4. Conclusion

Each of the floor plan shapes offered its own unique qualities. The following table (table 6) provides an overall evaluation of the four tested floor plan shapes in regard to the set criteria.

Table 6. Overall evaluation of the tested floor plans.

Factors affecting performance in the nursing unit	Floor plan shapes			
	A Single corridor	B Racetrack	C Courtyard	D Radial
	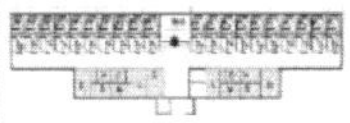	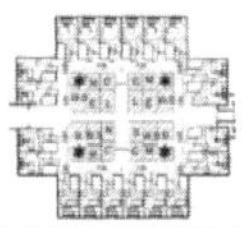	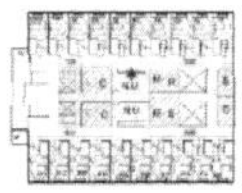	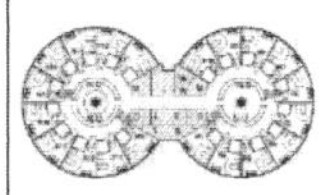
1. Flexibility	● (light grey)	● (black)	● (dark grey)	● (dark grey)
2. Patient monitoring	● (light grey)	● (black)	● (dark grey)	● (black)
3. Patient privacy	● (black)	● (light grey)	● (dark grey)	● (light grey)
4. Nurse travel distance	● (light grey)	● (black)	● (dark grey)	● (dark grey)
5. The healing environment	● (black)	● (light grey)	● (dark grey)	● (light grey)
Overall	**Low**	**High**	**Medium**	**Medium**

The above table demonstrates that the racetrack floor plan shape provides the highest level of performance in comparison with the three other shapes. It has shown superior qualities in regard to flexibility, patient monitoring, and nurse travel distances. However, for this shape to be successfully used, sound and noise control measures should be adopted to improve performance regarding patient privacy and noise disturbance. The courtyard and radial floor plan shapes provide a medium level of performance. The radial floor plan is superior in regard to patient monitoring capabilities. Although the single corridor plan has shown higher levels of sound privacy and lower levels of noise disturbance, it did not provide good flexibility, patient monitoring, and travel distances.

References

[1] Pati, D., Harvey, T., & Cason, C. (2008). Inpatient unit flexibility: Design characteristics of a successful flexible unit. Environment and Behavior, 40(2), 205-232.

[2] Friwan, Ahmed Abdullah Yusuf. (1999)." Flexibility in architectural design: Case Study of Hospital Buildings in the Hashemite Kingdom of Jordan, "Master's thesis, Faculty of Engineering, University of Jordan, p.19.

[3] Harvey Jr., Thomas E. & Pati, Debajyoti & Evans, Jennie & Waggener, Laurie T. and L. Cason, Carolyn (2008), " Inpatient Unit Design: Defining the Design Characteristics of a Successful Adaptable Unit", AIA Report on University Research, Vol. 2, American Institute of Architects, p16

[4] Mercy Heart Hospital Springfield Opens Phase. (2018) https://christnerinc.com/uploads/wysiwyg/ASHE_Mercy_Heart_POE_presentation_Final.pdf

[5] Babbu, Abdul Halim, "Flexibility": A key concept in Hospital Design, International Journal of Application or Innovation in Engineering & Management, Volume 5, Issue 5, May 2016, pp24-28.

[6] Hendrich, A. (2003). Optimizing Physical Space for Improved Outcomes: Satisfaction and the Bottom Line.Minicourse sponsored by the Institute for Healthcare Improvement & The Center for Healthcare Design, Atlanta, GA Ulrich, R. S. (2006). Essay: Evidence-based health-care architecture. The Lancet, 368, S38-S39.

[7] EDRA.(2008). Impacts of different patient room configurations on patient care activities in adult medical-surgical units, Retrieved from https://www.slideshare.net/CADREResearch/edra2008inboard-outboard-study

[8] Abdl Kadir, Walid Abdul Munim. (2000). "Modern Concept of Psychiatric Health Care Buildings," doctoral thesis, Faculty of Engineering, Cairo University, p. 203.

[9] Becker, F. D., Carthey, JS. (2008). Evidence-Based Design, Managing Risk, and a Healing Environment. Retrieved from http://.unsw.edu.au/fapi/datastream/unsworks:9010/source01?view

[10] Heba Mohamed Abdou, (2016), "The Impact of Human Needs on Internal Design Standards at Oncology Treatment Centres," Ph.D. thesis, Faculty of Engineering, Department of Architecture, University of Munaseb 71.

[11] Chiang, Y. (2010). Nurses' stations. Health Facility Design Dilemma, Cornell University, 2. Retrieved fromhttps://cpb-us-e1.wpmucdn.com/blogs.cornell.edu/dist/a/3723/files/2013/09/Nurses-stations-qr3u4x.pdf

[12] Measuring How Far Nurses Walk .(2015). the world of design at Herman Miller Retrieved from https://www.hermanmiller.com/content/dam/hermanmiller/documents/research_topics/NurseWalking.pdf

[13] Pati, D., Harvey Jr, T. E., & Thurston, T. (2012). Estimating design impact on waste reduction: Examining decentralized nursing. JONA: The Journal of Nursing Administration, 42(11), 513-518

[14] Healing Environments: How to Design Wellbeing.(2016). Retrieved from https://www.bolidt.com/en/healing-environments-how-to-design-wellbeing.

[15] Andrade, C. C., & Devlin, A. S. (2015). Stress reduction in the hospital room: Applying Ulrich's theory of supportive design. Journal of Environmental Psychology, 41, 125-134.

[16] Aripin, S. (2006, November). Healing architecture: A study on the physical aspects of healing environment in hospital design. In 40th Annual Conference of the Architectural Science Association (ANZAScA), Adelaide, Australia (pp. 22-25).

[17] James, W. P., & Tatton-Brown, W. (1986). Hospitals: design and development. Architectural press.

[18] Cai, Hui. Making" invisible architecture" visible: A comparative study of nursing unit typologies in the United States and China. Georgia Institute of Technology, 2013.

[19] Hendrich, A., Chow, M. P., Bafna, S., Choudhary, R., Heo, Y., & Skierczynski, B. A. (2009). Unit-related factors that affect nursing time with patients: Spatial analysis of the time and motion study. HERD: Health Environments Research & Design Journal, 2(2), 5-20.

Effect of Bio-material in Thermal Insulation Case-Study: Energy Saving in Residential Building in Aswan City

R E Abd El-Hady[1,*], A A Zayan[1], A F Mohamed[1]

[1] Architectural Engineering & Environmental Design Department, Arab Academy for Science, Technology & Maritime Transport, South Valley, Egypt.

[*] Corresponding Author Email: raniaemad@aast.edu

Abstract: The buildings industry represents about 34% of the total greenhouse gas emission percentage of the whole planet. Therefore, global warming is the main problem that countries are looking forward to finding a solution for by using a sustainable strategy to create new communities with zero carbon effects through decreased energy consumption and sustainable materials and design. One of the main reasons for the climate problem is that 64% more energy is used. The building material industry has a role in the rising energy and climate differences. Although increasing energy consumption depends on many reasons, one of the characteristics of building material used in the construction phase as fired clay brick is one of the significant construction industries. It produces a massive amount of greenhouse gases. On the other side, it will make these materials expensive and scarce in the long run, leading to substantial uncontrolled degradation of the natural environment. So, some different materials and forms of construction have developed in different parts of the world due to climatic differences to be commensurate with the surrounding climate environment.

Keywords: Bio-Materials, Energy Consumption, Thermal Insulation, Hot-Arid Climate, Aswan City.

1. Introduction

According to Fitch Ratings ' report, the construction industry in Egypt will grow in the next decade, with an average annual growth rate of 9% from 2020 to 2024. So, it will help Egypt move on in this field to be ranked first instead of fourth by 2029, worth more than 89 billion dollars [4]. According to BUR's report in 2018 in Egypt, energy use is one of the highest categories of environmental pollution. In addition, one of the energy use sectors in the construction industry is to produce building materials used in new constructions and operate buildings. Thus, these activities exacerbate the Greenhouse gases problems [12]. Energy use is one of the highest categories of environmental pollution and greenhouse gases. The percentage of pollution in this category represents about 64.5% of the total emissions for the year 2015, according to BUR's report in 2018 [8]. According to the Mobilization and Statistics Authority, the Construction Industry represents about 2.5% of the total energy use emissions for the year 2015. So, conventional materials produce massive greenhouse gas emissions and environmental pollution [6]. Bio-based materials are considered a good resource for buildings in the twenty-first century due to their sustainability and versatility than conventional materials [1].

1.1. Problem statement

In this century, one of the biggest problems is greenhouse gases emission. Buildings are responsible for more than 1/3 of this emission through increasing energy consumption to achieve thermal comfort to residents. One of these causes is the building material used in the construction phase. It has the value of thermal conductivity, which allows heat to transfer through them, which raises the temperature inside the building more than the default.

1.2. Research aim

Because of the effect of energy consumption in producing building materials and the effect of the buildings on the environment, the main objective is to enhance building-material characteristics by using bio-materials or agriculture wastes to increase its resistivity. Moreover, to prevent a large amount of heat from transferring through building materials like walls, roofs, etc. The use of bio-materials blocks in a building will decrease its energy consumption. Also, it will decrease the heat transfer into the buildings to improve human health and thermal comfort and decrease the use of air conditioning.

1.3. Methodology

The following steps adopt the methodology of this research to achieve its aim and objective:

- Evaluate the performance of the bio-material example in India.
- Determine the amount of energy consumption for a residential building in Aswan City using conventional materials.
- Determine the energy saving resulting from replacing fired conventional building techniques with Biomaterial (Brick).

2. Construction Industry in India

Fire clay brick is considered one of the primary materials in the construction industry in India. However, it produces a massive amount of greenhouse gases [12]. So, finding new sustainable materials and technologies can reduce environmental pressure (directly and indirectly).

2.1. Hempcrete bricks

Research understudying suggested that new manufacturing technology (hempcrete) is considered the best way to manage agricultural waste generated in India.

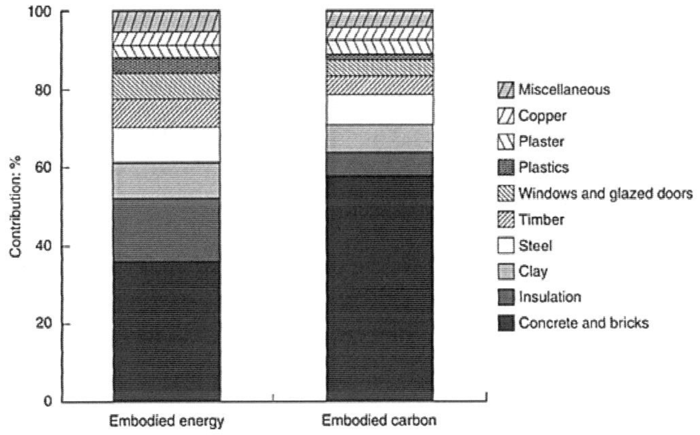

Figure 1: Constituent raw materials for building.

This new building material works a critical role in heat insulation and sound absorption. The effect of hempcrete brick acts as a carbon skin through production phases. Firstly, the plant absorbs carbon dioxide from the atmosphere in the photosynthesis process. Secondly, by carbonation, calcium hydroxide is generated at bricks after agricultural waste is produced, absorbing carbon dioxide to form more stable calcium bicarbonate as shown in figures 2, 3, 4.

Figure 2: Constituent raw materials.

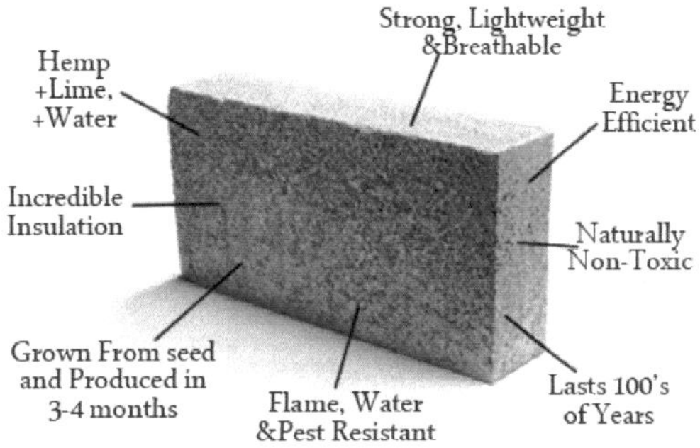

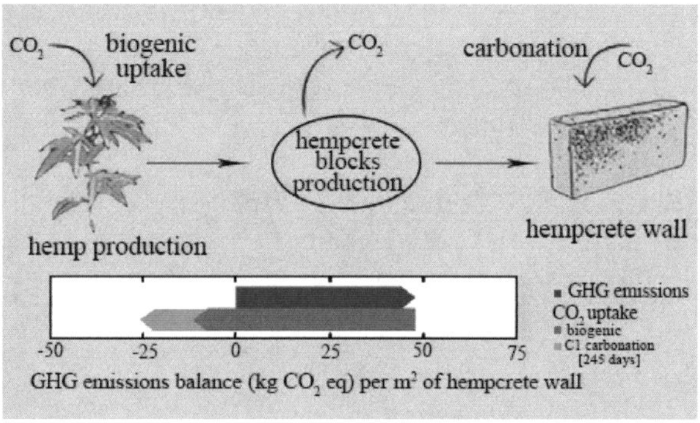

Figure 3: Benefits of hempcrete bricks. **Figure 4:** Negative carbon footprint of hempcrete.

3. Construction Industry in Egypt

According to Egypt's mobilization and statistics authority, the construction industry represents about 2.5% of the total energy use emissions for 2015. This percentage represents the energy used in producing common building materials [6]. So, this percentage and the growth rate in the industry field show that they are alarming, which will need more energy to produce common building materials and operate existing buildings.

4. Case Study: Residential Building in Aswan City

Aswan city is considered a hot-arid climate region. So, conventional construction materials with high Thermal transmittance (U-value) cannot provide indoor thermal comfort (naturally: in passive means) in such a harsh environment.

Case study District covers (0.12 km^2) area with built up about 40% buildings from the total area and the building heights are approximately 18 m. All Buildings were built with conventional materials. The characteristics climate of the study area is calculated on the hottest day in Aswan on June 05 in the year 2006. The maximum and minimum annual temperatures of 47°C -29.5°C [2].

Figure 5: Case Study area from Google Earth.

Using Design-Builder energy simulation 5.5 software helps compare how much heat gain or losses indoor building using conventional brick (default Scenario) and hempcrete bricks. Note that using EPW file for Aswan city in two scenarios and installing all factors like HVAC system (Fan Coil Unit(4-pipe), Air cooled Chiller), opening size, floor and roof layers.

4.1. Simulate default scenario

The first step in this research is to prepare the test model of one block from the case study district. To determine the amount of cooling unit needed to achieve thermal comfort indoor blocks and show the role of bricks heat transfer. According to climate consultant program software, the Comfort zone of Aswan citizens is between (19.5- 26) °c according to climate consultant program software. So, buildings will need to be cooled by air condition to achieve thermal comfort and human health, which will be caused increasing in CO_2 emission. One of the causes that affect increasing CO_2 emission is construction materials used in the construction process. Bricks have an essential role in using air conditions, as shown in figure 7. The wall significantly affects heat transfer, embodied carbon, and equivalent CO_2. As shown in Table 2.

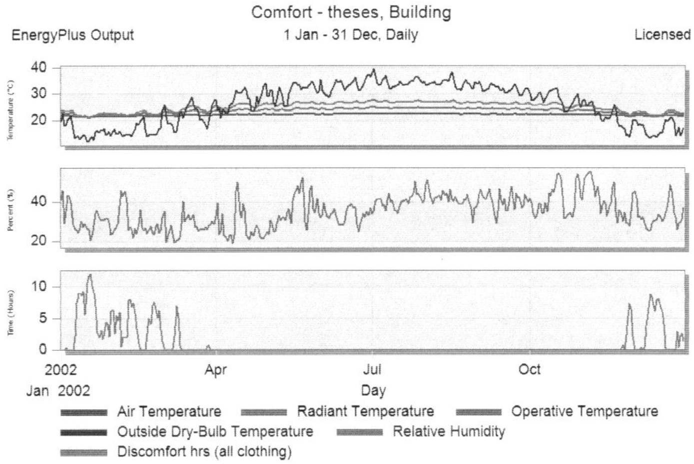

Figure 6: Climate Conditions in Aswan city.

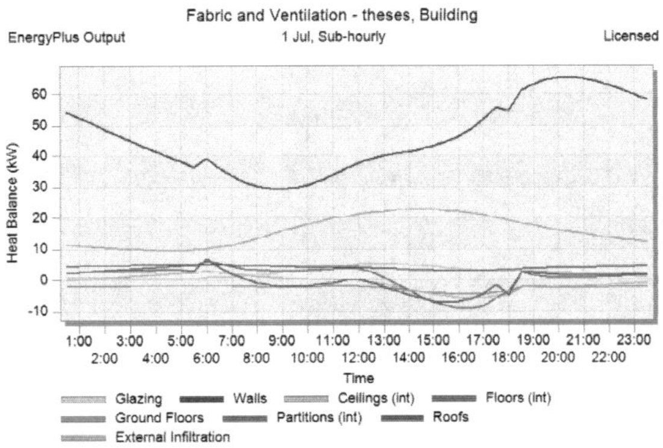

Figure 7: Heat balance in construction elements.

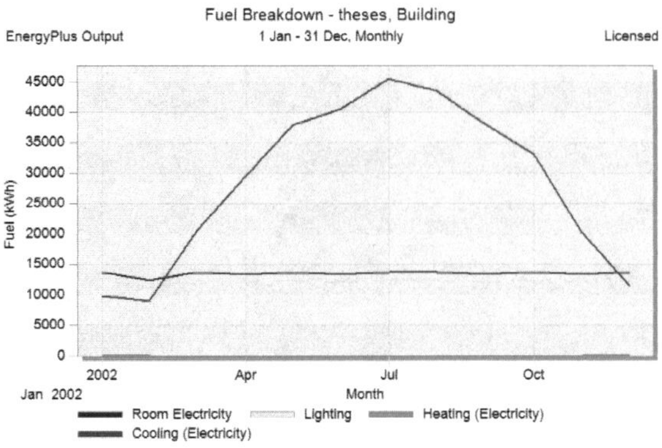

Figure 8: Electricity Distribution in residential building.

Table 1: Construction embodied carbon and equivalent co$_2$ in default case

Constructions Embodied Carbon and Inventory	Area (m^2)	Embodied Carbon ($kgCO_2$)	Equivalent CO_2 ($kgCO_2$)
Roof	299	45172.4	47415
25cm Wall	1673.9	131765.6	137162
Wooden door	104.4	0.0	0.0
12cm Wall	873.2	31164.5	32670.0
Floor	1495.1	82295.3	83670.8
G. Floor	299	24212.1	24550
Sub Total	4744.6	314609.87	325467.91

According to graphs and tables, conventional bricks have adverse effects in the environmental context and increase carbon emission through produced and long-run operating processes.

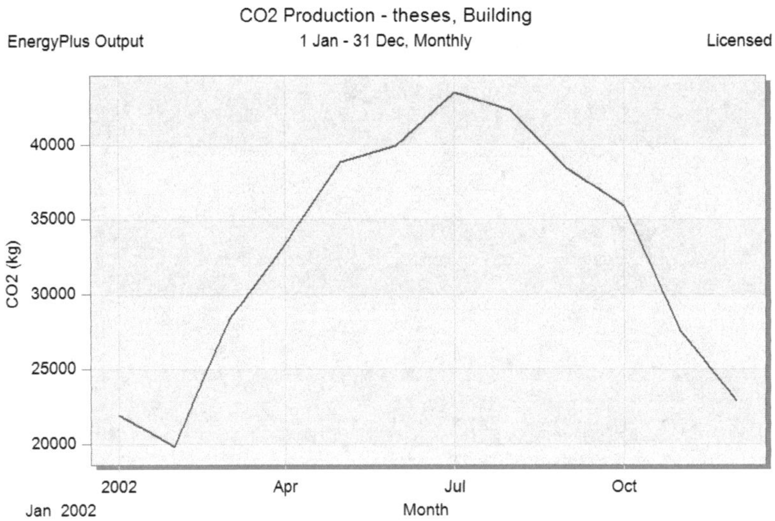

Figure 9: CO_2 production in residential buildings

Table 2: Materials embodied carbon and equivalent CO_2 in traditional materials

Materials embodied Carbon and Inventory	Area (m^2)	Embodied Carbon ($kgCO_2$)	Equivalent CO_2 ($kgCO_2$)	Mass (kg)
Cement /plaster /mortar – gypsum plaster	3420.3	6846.6	7417.2	57055
Cement /plaster /mortar –cement plaster	3468	28791.2	28791.2	151532.8
Cement /plaster /mortar –cement mortar	2093.2	19686.1	19686.1	103611.1
Brick - burned	2547.1	139291.4	145622.8	633142.7
Earth common	299	4365.7	4365.7	218286.1
Painted Oak	104.4	0.0	0.0	2557.5
Ceramic /porcelain	1794.1	26822.3	28472.9	41265
Concrete Tiles (roofing)	299	2763	2763	12558.9
Sandstone	2093.2	0.0	0.0	269717.8
MW Glass Wool (rolls)	299	274.5	301.4	179.4
Concrete Reinforced (with 2% steel)	299	33370.9	35523.8	107647.9
Cast Concrete	2392.2	51192.6	51192.6	639907.1
Bitumen pure	598	1205.7	1331.2	2511.8
Sub Total		314609.8	325467.9	2239973.2

4.2. Simulate hempcrete bricks scenario

In this step, enhancing exterior skin of building (Bricks) by using agricultural waste of hempcrete plant to create hempcrete bricks which their properties are conductivity 0.05 (W/m-k), Specific Heat 1530 (J/kg-k), and Density 480 (Kg/m^3). After using hempcrete bricks, the amount of heat transfer through walls has been reduced, as shown in figure 11, and the heat balance graph of walls becomes smoother than Scenario one; it is shown to minimize the difference between values.

So, bio-materials will enhance the human health of spaces and thermal comfort inside buildings, decreasing the need for cooling, as shown in figure 10, which shows the amount of energy that passes through walls and other elements. CO_2 emission and GHGs effect will be decreased due to enhancing thermal transmittance (U value) of material and do not need for cooling indoor buildings as shown in figure 12. However, from another perspective, this type of bio-bricks (hempcrete bricks) will decrease energy consumption indoor building, as shown in figure 13.

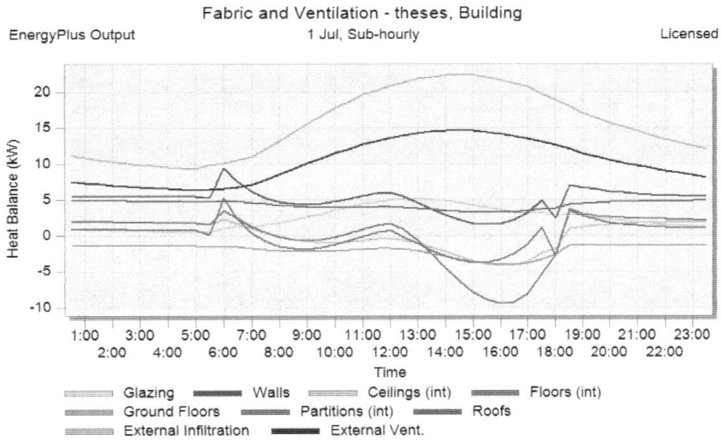

Figure 10: Heat balance in construction elements after using hempcrete bricks.

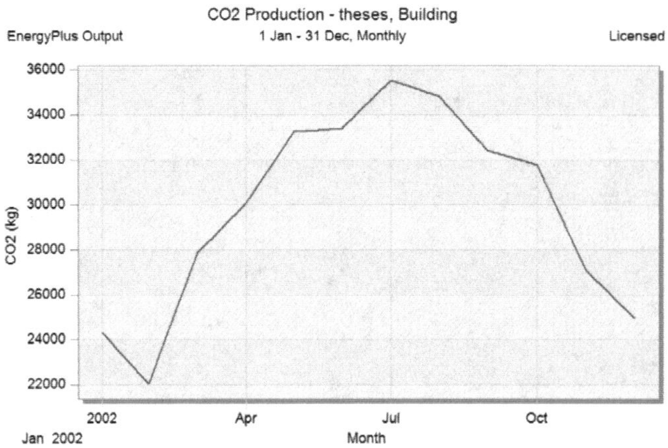

Figure 11: CO_2 production in residential building in second scenario.

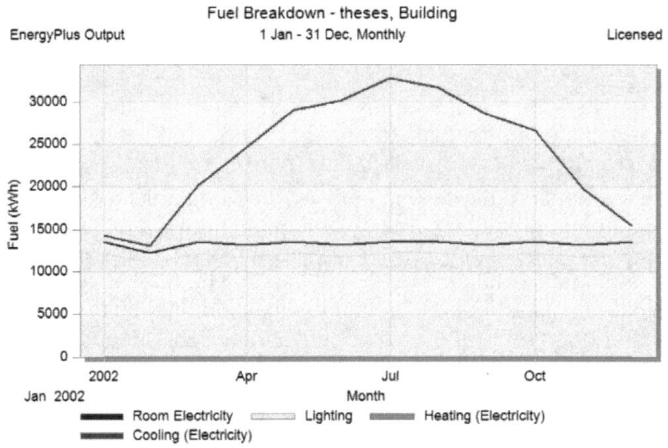

Figure 12: Electricity distribution in the residential building after enhancing process.

Table 3: Comparison of thermal characteristics between two scenarios

Item	Traditional Bricks	Hempcrete Bricks
Conductivity (W/m-k)	0.85	0.05
Specific Heat (J/kg-k)	840	1530
Density (Kg/m3)	1500	480
Total Cooling loads (Wh/m2)	45472.32	32788
Embodied Carbon (kg CO2)	139291.4	53413

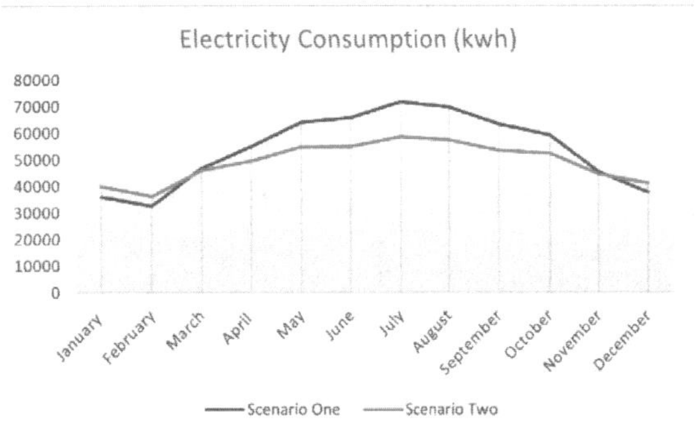

Figure 13: Comparison between two Scenarios in electricity Consumption.

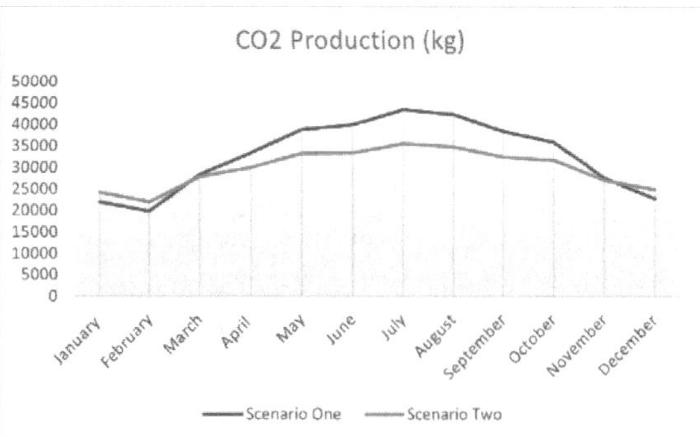

Figure 14: Comparison between two Scenarios in CO_2 production

IOP Conf. Series: Earth and Environmental Science 1056 (2022) 012014 doi:10.1088/1755-1315/1056/1/012014

5. Conclusions

Buildings are responsible for about 1/3 of greenhouse gas emissions. This is because it represents building materials like (bricks, concrete, tiles, etc.). Since the results of simulation software use traditional materials to calculate total cooling loads, then the amount of CO_2 emission will be reduced by using biomaterial.

- Using manufacturing has a more negligible effect than traditional ways in the surrounding environment.
- Making strategy to manage the waste of agriculture without adverse effects on the environment.
- Making guidelines for selecting alternative building materials for the more sustainable building to enhance the indoor environment and human health.
- Enhancing the thermal transmittance of materials will help in reducing electricity consumption in residential buildings.

References

[1] A. Sandak, e. (2019). Biomaterials for Building Skins. In J. S. Anna Sandak, Bio-based Building Skin (pp. 27-64). Hong Kong: Springer Open.

[2] Africa-Region 1. (n.d.). Retrieved from Climate.OneBuilding: https://climate.onebuilding.org

[3] Braticius, J. (2020, July 09). Cutting greenhouse gas emissions through circular economy actions in the buildings sector. European Environment Agency.

[4] Economy news. (2020, 6 4). Retrieved from state information service web site: https://sis.gov.eg/Story/147456/Fitch-forecasts-Egypt%E2%80%99s-construction-sector-to-outperform-peers-in-Middle-East%2C-North-Africa-by-2029

[5] El Menshawy, A. M. (2022). A comparative study on green wall construction systems, case study: South valley campus of AASTMT. Case Studies in Construction Materials, e00808.

[6] ElGendy, A.-B. (2017). Environmental Statistics Report. Cairo: Central Agency for Public Mobilization and Statistics.

[7] EM-DAT: Disasters Of The Week. (2021, Octobar 11). Retrieved from EM-DAT: https://www.emdat.be

[8] Fouad, Y. (2018). Egupt's First Biennial Update Report. Cairo: Ministry of Environment, Egyptian Environmental Affairs Agency.

[9] Mohamed, A. (2019). Economic Evaluation of Renewable Energy Systems, Case Study: An Eco-House Powered by Nano-Crystal PV in New Aswan, Egypt. IOP Conference Series: Earth and Environmental Science.

[10] Mohamed, A. F. (2020). Comparative study of traditional and modern building techniques in Siwa Oasis, Egypt: Case study: Affordable residential building using appropriate building technique. Case Studies in Construction Materials.

[11] Mohamed, A. M. (2019). Comparative study of bio-materials versus conventional materials in building construction methods; economical evaluation of different building materials. SN Applied Sciences, 1179.

[12] Rautray, Priyabrata , Roy, Avik , Mathew, Deepak John , Eisenbart, Boris. (2019). Bio-Brick - Development of Sustainable and Cost. international Conference on Engineering Design (pp. 3171-3180). India: ICED19.

A Guideline for Developing Resilient Office Buildings using Nanotechnology Applications

Gehan Nagy[1, *], Haidy Adnan[2]

[1] Associate professor of architecture, the British university in Egypt, Cairo
[2] College of architectural Engineering, the British University in Egypt, Cairo

[*] Corresponding Author Email: gehan.nagy@bue.edu.eg

Abstract. Resilient buildings have become a widely researched topic in the engineering discipline. A building has enabled connectivity between building systems, including air conditioning, ventilation, safety and security system. They are even more important with growing pandemic risks and the spread of airborne viruses such as Covid-19. The literature indicates that nanotechnology applications can have a great role in the designing resilient office buildings. Nanotechnologies play a major role in architectural design; building materials combined with nanotechnology became smaller, lighter, and more efficient than conventional materials. Thus, this research aims to develop design guidelines for resilient office buildings in Egypt using nanotechnology applications. The methodology applied includes a review on nanotechnology and resilient office buildings key design features, following a simulation using design builder program where conducted to modify the nanomaterials in resilient office buildings recorded through quantified data. After applying the nanotechnology concept on walls, the roof and glazing types by inserting nanomaterials. The research concluded that the use of vacuum insulation panel, Nano-gel glass and Nano-painting had the highest impact on decreasing the total energy consumption by 14.5% and carbon dioxide emissions by 42%.

Keywords. Nanotechnology, Resilient office buildings, simulation, energy consumption.

1. Introduction

Office buildings are considered one of the most important models in which the importance of the various structures and their work are connected in an organized manner and with the mechanism of high efficiency and performance [1]. On the construction of administrative buildings. Administrative buildings with smart architecture are very useful [2]. Researchers found that nanotechnology can be a useful approach when designing resilient buildings. Resilient buildings involve all the systems integrated together such as lighting, ventilation, cooling and heating, fire system that are adapted by utilization of new technologies and addition of advanced technological systems to support the building [3]. In order to meet the needs of building users, they enable connectivity between building systems that are flexible enough to adapt easily [16]. Moreover, resilient buildings determine what happens within and outside it, providing a comfortable and healthy environment. It can be effective and increase employee productivity through UV protection, reduce carbon dioxide emissions, air purification and many other systems [6]. This paper seeks to tackle an essential topic which is the use of nanotechnology applications in resilient office buildings design. Nanotechnology was used for the first time in 1974 by Norio Anguish to refer "the precise and accurate tolerances needed machining and finishing materials" Nanotechnology improves Community Resilience through better building materials, and applications.

Nanotechnology "refers to the creation, investigation and application of structures, molecular materials. Internal interfaces or surfaces with at least one critical dimension or manufacturing tolerances of typically less than 100 nanometers" are defined as the use of structures between 1 nanometer and 100 nanometers in size. Nanotechnology applications consist of two main categories: Nano-devices and nanomaterials [8].

1.1. Problem formulation

As a result of Covid-19 that caused huge crisis such as environmental and biological changes [1], clearly show that societies need to strengthen their resilience. The lack of a clear criteria in designing for resilient buildings in Egypt led to a lot of smart technologies. But till now, one of the best technologies, i.e. nanotechnology application, is not implemented thoroughly [2]. Moreover, implemented nanotechnology application will aid in adapting with these changes, and help provide human comfort and create suitable indoor environment [5]. Taking into consideration that guidelines are not followed and familiar to many offices in Egypt [4]. Consequently, office buildings become unhospitable leading employees to be less creative and innovative in their workplace [1].

2. Research methodology

The key purpose of the research is to develop guidelines for designing resilient office buildings in Egypt using nanotechnology applications. Qualitative and quantitative methods, a review of literature where nanotechnology and resilient office buildings are first defined, and their key design features are identified. Then a simulation is done using design builder program to modify the nanomaterials in the building recorded through quantifies data [8], to ensure accurate measurements and findings in order to specify the precise range of total energy consumption and carbon dioxide emissions in relation to each nanotechnology material. Finally, all the data extracted has been used in designing the guidelines [18].

3. Literature review

3.1. Nanotechnology applications

Nanotechnology is an important role in our life, it is a scientific field of growth, as shown in table [1].

Table 1. Nanomaterials (developed by author, 2021).

Nanomaterials			
Structural materials	Non- structural materials	Insulation materials	Nano-painting materials

3.1.1. Nanomaterials

- Structural materials: The components of structural materials are steel, concrete, and Nano-carbon tubes [7]. However, the steel structural material can resist corrosion, characterized by its binding and ductility, where the concrete can reduce the sound absorption by 240%. Furthermore, the Nano-carbon tubes are 250 stronger than steel [9].
- Non-structural materials: The non-structural materials are glass, dry-wall and Nano-wood [5]. However, glass that is treated with nanotechnology is characterized by the addition of titanium dioxide as it contains and control air pollution, where dry wall can resist water and consume less energy consumption [18]. Furthermore, the Nano-wood simulates the water and oil repellence [20].
- Insulation materials: The insulation materials consist of Nano-gel, as it can transmit light 91% per cm^2, reduces heat gain, where it also consists of thin film insulation as it can block the sunlight and solar absorbing windows that can transmit 70% of net light [12].
- Nano-painting materials: Nano coating materials appear in this regard with powerful impact because these coatings have additional properties and advantages that render them outperform

the traditional coating materials [19]. The Nano coating materials vary according to their different functions as well as characteristics and features [1], as shown below the Nano-painting materials in table [2].

Table 2. Nano-paintings (developed by the author, 2021).

Nano-paintings materials		
Anti-fingerprint	Anti-reflection	Anti-scratching
Anti-graffiti	Self-cleaning lout's effect	Anti-bacterial
Easy to clean coating	Self-cleaning photocatalytic	

3.1.2. Nano-devices. The components of Nano-devices are lighting, filtration, water filtration, solar energy and storage energy [17]. However, Nano-led saves energy from 82% to 93%, where the filtration decomposing harmful substances and pollutants [13]. Furthermore, water filtration eliminates the pesticides. Solar energy can be recycled, and it is characterized by its higher efficiency of plastic cells 30% greater. Lastly, the stored energy can improve battery efficiency and decrease carbon emission [6].

3.2. Resilient buildings

A resilient building is a building that achieves a collection of important aspects such as the social aspects that is represented in the improvement of the internal environment of the building [16].

3.2.1. Characteristics of resilient buildings. The buildings determine what happens within and outside it [19], where it can provide a comfortable condition and a healthy environment through air purification. Furthermore, it can be energy efficient by reducing electricity costs. However, it also can increase employee productivity and enhancing life safety and security [15] as shown below in figure [1].

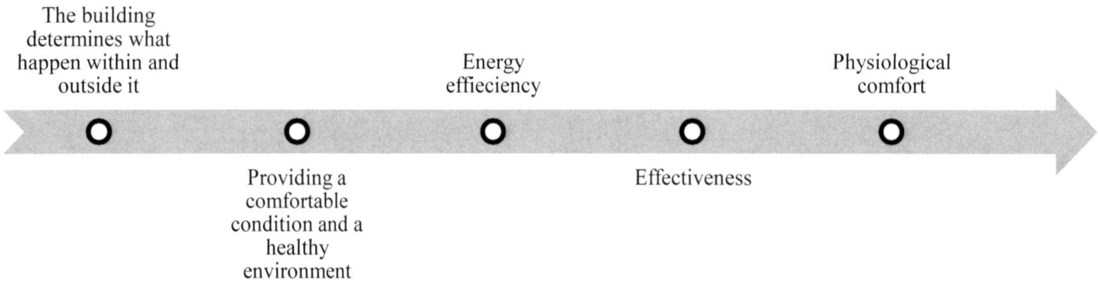

Figure 1. Characteristics of resilient buildings (developed by the author, 2021).

3.2.2. Benefits of resilient buildings

Figure [2] below shows the benefits of resilient buildings [4].

Figure 2. Benefits of resilient buildings (developed by the author, 2021).

3.3. Local case study simulation

3.3.1. Case study selection criteria. The selection of this office building was based on:

(1) National/ local case,
(2) Availability of online data and resources,
(3) Suitable for the research content (office building),
(4) Availability of resilience characteristics.

The aim was to develop guidelines for resilient buildings using nanotechnology applications. The study was carried in an office building. Simulation program used was design builder software v5.03. 007.

Figure 3. Galleria 40, office building in Sheikh Zaid.

Table 3. Building features description, (Forbes & Fikretoglu, 2018).

Construction phase	New construction	**External wall**	Building materials with high recycled content sourced from local vendors
Building type	Non-residential building	**Roof**	Flat roof
Detailed building type	Mixed-use, office and retail	**Glazing type**	Double glazed
Net floor area	41.687 m^2	**g-value glass**	0.48
Land plot size	17.000 m^2	**Overall U-Value windows**	2.7w/(m^2*k)
No. of stories	5 stories	**Space cooling system**	Centralized chiller, usually a central system located in the building
Concentration type	Concrete	**Construction year**	2016

3.3.2. Simulation process. A simulation model will be formed through importing dimensions and specifications of buildings to the program, the model will simulate the real building energy consumption. The input layers in walls, roof, and glass in the building features are going to be changed in each simulation scenario. However, the variables factors will be the roof layers, walls layer, and glass layers. for the case study will be analysed through this program. Then the building plan was drawn using 2D AutoCAD and exported to the program to carry out the simulation test as shown below in figure [4].

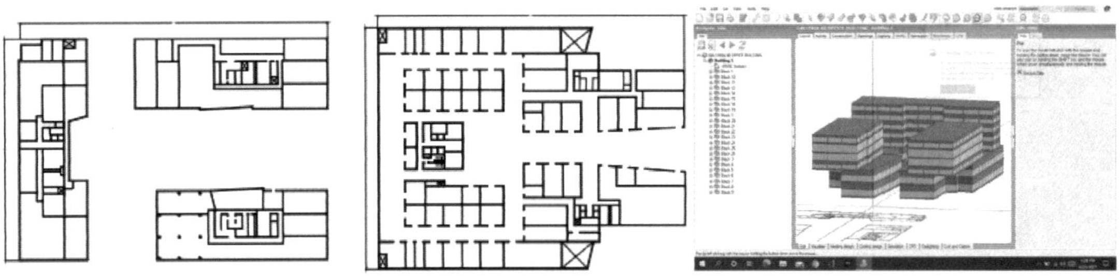

Figure 4. Simulation model (developed by the author, 2021).

3.3.3. Simulation methods. First of all, gathering all the data and thermal properties of each constructed material from the Egyptian energy code and research papers [1]. Then, putting all the input data in the activity and construction menu into the program [18]. After that, insert Nanotechnology material in three alternatives simulation after gathering thermal properties of each material [15]. Repeating the process of each type in the three alternatives, measuring energy consumption of each case, measuring carbon emissions for each case Gathering information into a detailed comparison to examine the differences [1].

3.3.4. Simulation results. The following graphs show the impact of nanomaterials on total energy consumption and carbon dioxide emissions of three alternatives and base case. Base case result simulation results show that the carbon dioxide emission shows its annual production. Carbon emission production is about 1.4 kg*10^6 per year. however, the total annual energy consumption is about 2.80 m kwh in cooling set point temperature = 24 while in cooling set point temperature =26 is about 2.61 m. The results show that using of vacuum insulation panel provides a less carbon dioxide emission production and total energy consumption than the base case study. When all materials were inserted and placed with the vacuum insulation panel, Nano-gel and Nano-painting, a fast increase occurred in simulation readings than all previous simulation, this led to reach the least carbon dioxide emissions and the least total energy consumption of the alternatives.

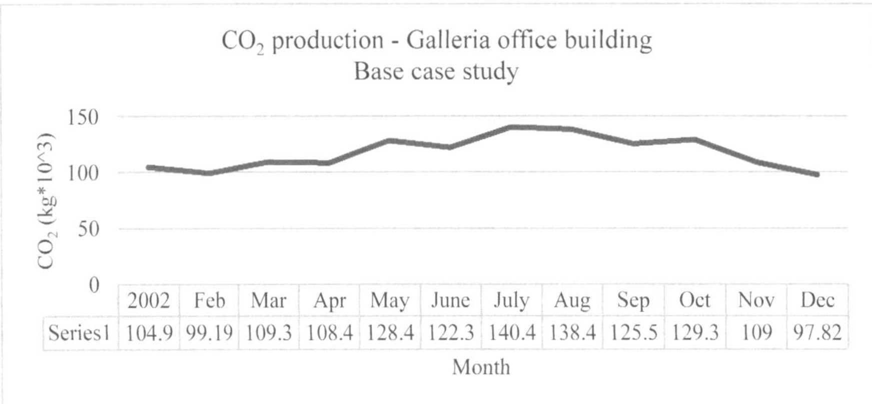

Figure 5a. Carbon dioxide production results of the base case study.

IOP Conf. Series: Earth and Environmental Science 1056 (2022) 012015 doi:10.1088/1755-1315/1056/1/012015

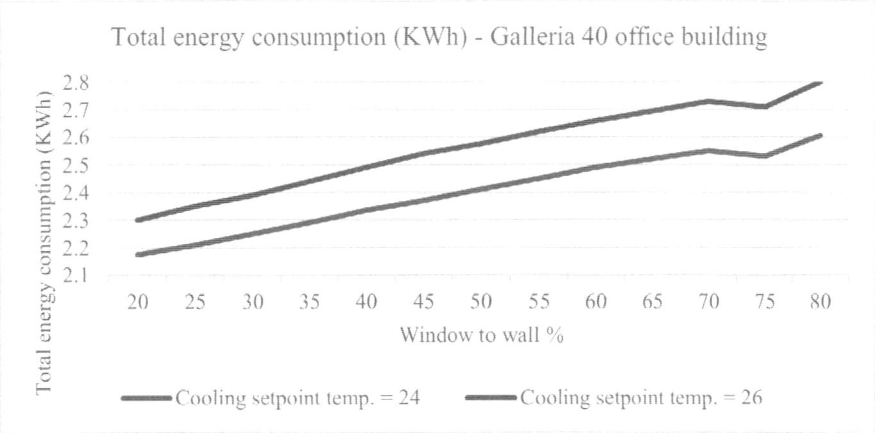

Figure 5b. Annual total energy consumption results of the base case study.

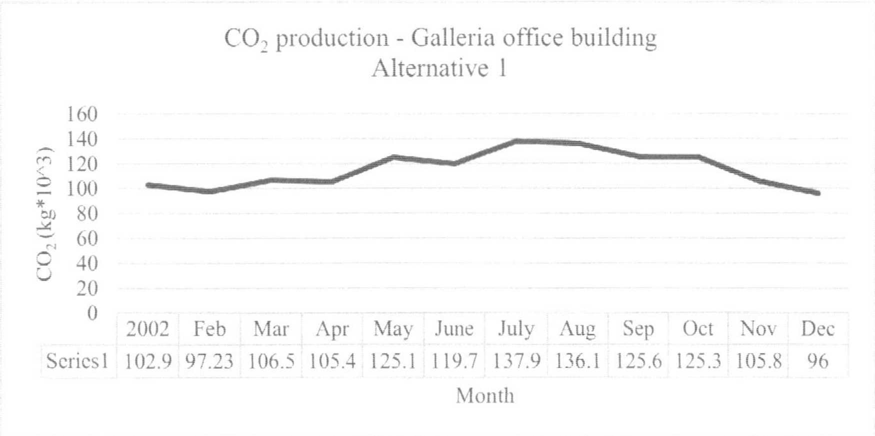

Figure 6a. Carbon dioxide production results of alternative 1 case study.

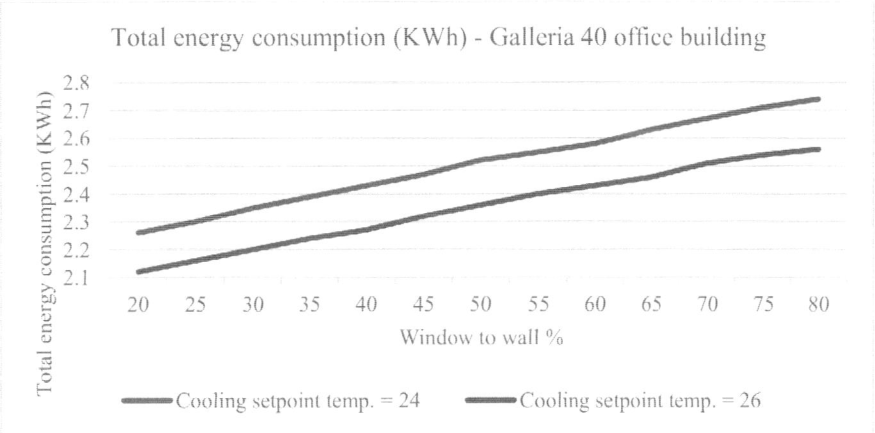

Figure 6b. Annual total energy consumption results of alternative 1 case study.

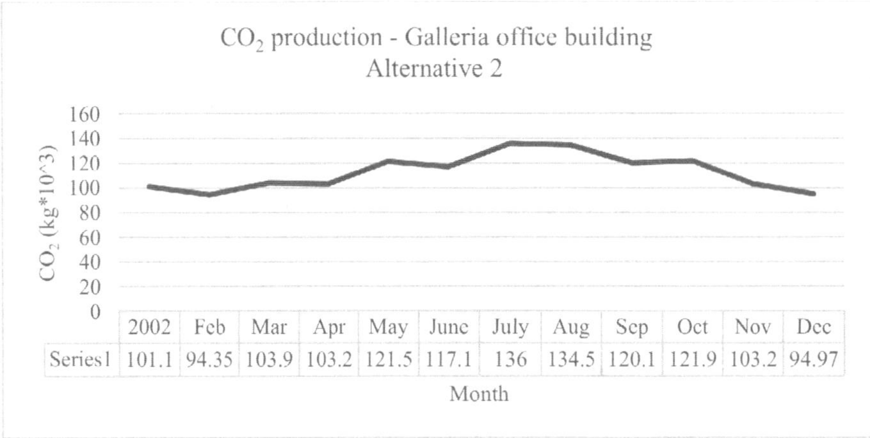

Figure 7a. Carbon dioxide production results of alternative 2 case study.

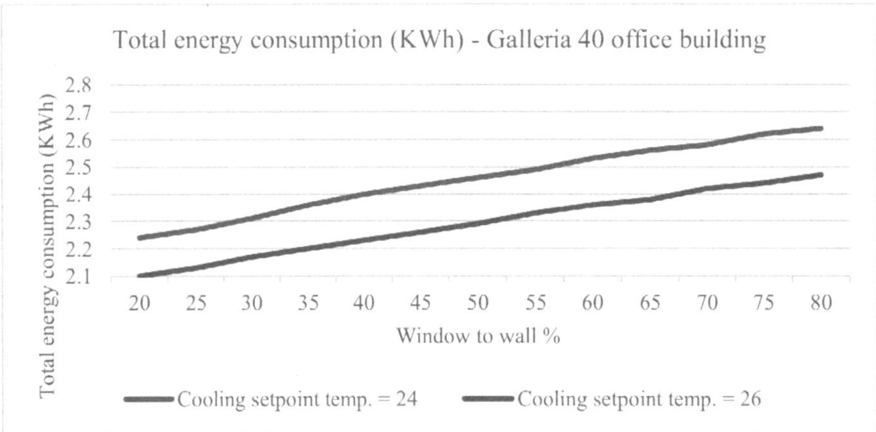

Figure 7b. Annual total energy consumption results of alternative 2 case study.

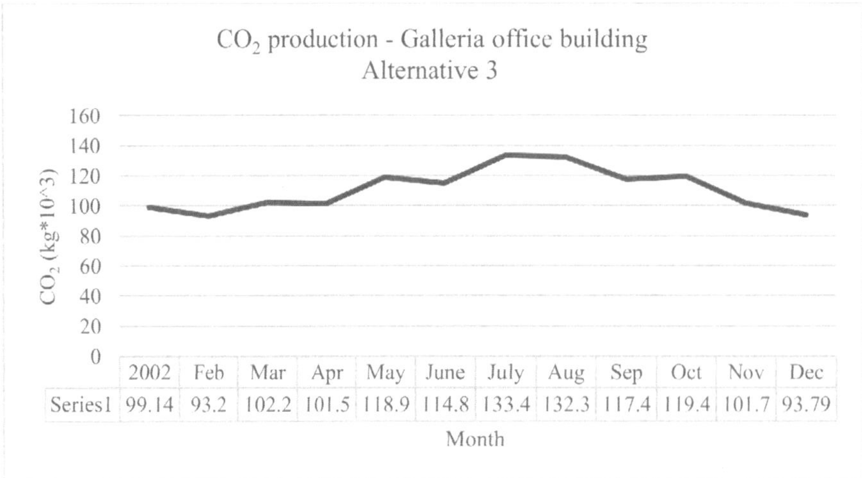

Figure 8a. Carbon dioxide production results of alternative 2 case study.

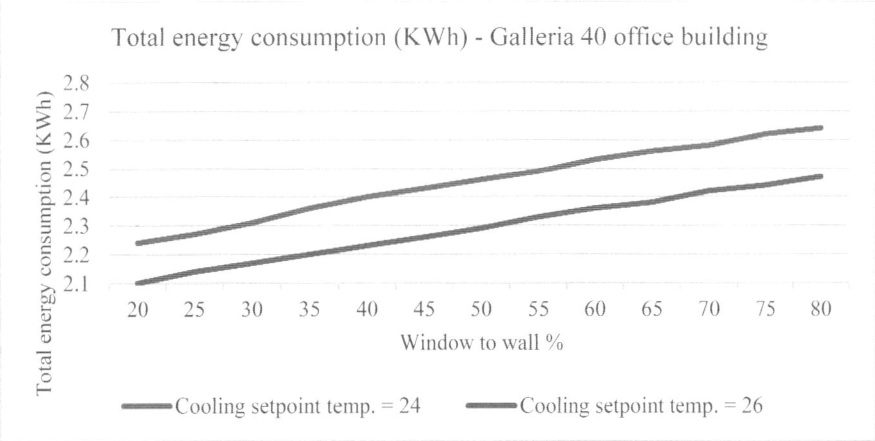

Figure 8b. Annual total energy consumption results of alternative 2 case study.

4. Conclusion

A comparison of the simulations results was done between the base case and the three alternatives simulation, illustrating the calculations of the carbon dioxide emission and total energy consumption [10]. Furthermore, the carbon dioxide emission and the total energy consumption of the third alternative which use the vacuum insulation panel, Nano-painting and Nano-gel glass was the best alternative and better than the base case. Furthermore, it reduced the total energy consumption about 14.5% and 42% in carbon emission dioxide than the base case and the other two alternatives as shown below in figure [9].

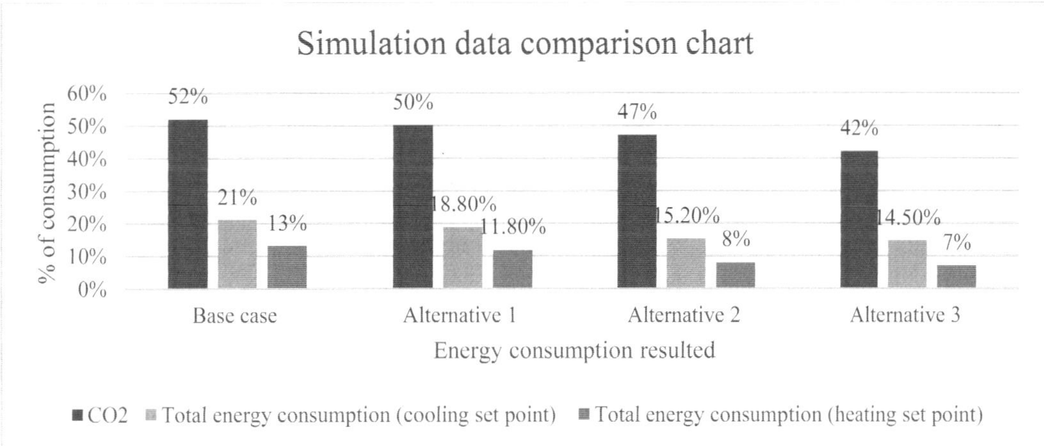

Figure 9. Comparison chart (developed by the author, 2021).

4.1. Guidelines for achieving the resilience characteristics through nanotechnology applications

Guidelines to present the different nanotechnology applications that can be implemented in resilient office buildings in Egypt after the analysis studies and data collection to enhance the total energy consumption and carbon dioxide emission which are deduced in the output matrix of the literature review based on what was applied in our real life.

Table 4. Guidelines for achieving the resilience characteristics.

Characteristic To Be Achieved	Nanotechnology Applications	Effect on Characteristics
Comfortable condition and healthy environment	Using structural materials, non-structural materials, insulation materials, Nano-painting, filtration, water filtration, and store energy.	Reduces carbon dioxide emissions and control air pollution.
Energy efficiency	Using non-structural materials, insulation materials, Nano-painting materials, lighting, solar energy, and store energy.	Reduces maintenance costs, less energy consumption, and reduce electricity costs.
Increasing employee productivity	Using Filtration and insulation materials can attain this characteristic.	Provide thermal comfort and thermal conductivity
Effectiveness	Using Structural materials and non-structural materials provides the building with effectiveness.	Provide safety, security, and fire protection.
Physiological comfort	Using insulation materials and Nano-painting	Provide thermal comfort and controlling noise.
Safety and security	Using structural materials	For more safety and security
Environmentally friendly	Using insulation materials and store energy	Less energy consumption
External walls, roof and floors construction	Using insulation material such as the Vacuum insulation panel (VIP) technique. Provides high-performance thermal insulation in buildings because of its ultra-low thickness (5–50 mm) and its very low thermal conductivity (0.002–0.004 W/(m.°K)) compared to the conventional insulation materials.	These values are ten times less than the conventional insulation materials used in office buildings today.
Windows, facades, walls construction	Using Nano-gel	Used in energy efficiency building and daylighting systems.

4.2. Recommendations

- More investigation on the relationship between different types of nanomaterials and characteristics of resilience's
- More detailed experiments on the nanotechnology applications and how each of the applications affect energy consumption and carbon dioxide emission.
- Research into how employees interact with different resilient characteristics.
- Research into more different nanotechnology applications and how to be applied.

References

[1] T. N. I. o. B. S. (NIBS), "Good Practices In ResilienceBased Architectural Design," the 12th Edition of Architectural Graphic Standards, 2018.
[2] A. R. Abdin, A. R. El Bakery and M. A. Mohamed, "The role of nanotechnology in improving the efficiency of energy use with a special reference to glass treated with nanotechnology in office buildings," Ain Shams Engineering Journal, vol. 9, no. 4, p. 2671–2682, 2018.
[3] M. G. Al Marri, M. A. Al-Ghouti, S. Vasanth C. and N. Zouari, "Date pits based nanomaterials for thermal insulation applications—Towards energy efficient buildings in Qatar. ," PLOS ONE, vol. 16, no. 3, 2021.
[4] S. Burroughs, "Development of a Tool for Assessing Commercial Building Resilience. ," Procedia Engineering, vol. 180, p. 1034–1043, 2017.
[5] J. K. Calautit and B. R. Hughes, "Sustainable Buildings: opportunities, challenges, aims and vision. ," Sustainable Buildings, vol. 1, 2016.
[6] A. S. Dahlan, "Smart and Functional Materials Based Nanomaterials in Construction Styles in Nano-Architecture. ," Silicon, vol. 11, no. 01, p. 1949–1953, 2019.
[7] J. Dohaney, R. Mairéadde , A. Rhian and K. Sutherland, "Benefits, barriers, and incentives for improved

resilience to disruption in university teaching," International Journal of Disaster Risk Reduction, vol. 50, 2020.
[8] S. Hernández and . M. Cristina Solache de , "Nanotechnological products in architecture and construction," Hernández-moreno & torre, vol. 2, no. 34, 2017.
[9] B. J. Jennings, E. D. Vugrin and D. K. Belasich , "). Resilience certification for commercial buildings: a study of stakeholder perspectives. ," Environment Systems and Decisions, vol. 33, no. 2, p. 184–194, 2013.
[10] M. T. Kelly, J. Soto-Muñoz and L. Marín-Restrepo, "Adaptive thermal comfort for resilient office buildings," Journal of Physics: Conference Series, vol. 1343, no. 012148, 2019.
[11] M. D. Le Hong, "The application of traditional "tube house" in water revitalization," in Vernacular Architecture: Towards a Sustainable Future, 2014, pp. 245-250.
[12] F. M. Linares, "Nanotechnology: An Urgent Need," Soft Nanoscience Letters, vol. 03, no. 03, pp. 45-46, 2013.
[13] D. Marchese, E. Reynolds, M. E.Bates, H. Morgan and I. Linkov, "Resilience and sustainability: Similarities and differences in environmental management applications," Science of The Total Environment, pp. 613-614, 1275–1283, 2018.
[14] P. Mathew, S. Lino and S. Hoon Lee, "Assessing the Energy Resilience of Office Buildings: Development and Testing of a Simplified Metric for Real Estate Stakeholders," Buildings , vol. 11, no. 3, 2021.
[15] P. Mathew, L. Sanchez, S. H. Lee and T. Walter, "Assessing the Energy Resilience of Office Buildings: Development and Testing of a Simplified Metric for Real Estate Stakeholders," Buildings, vol. 11, no. 3, 2021.
[16] F. Márquez and M. Carmen , "Nanomaterials for Sensor Applications," Soft Nanoscience Letters, vol. 05, no. 01, 2015.
[17] R. Phillips, L. Troup, D. Fannon and J. Matthew , "Do resilient and sustainable design strategies conflict in commercial buildings? A critical analysis of existing resilient building frameworks and their sustainability implications," Energy and Building, vol. 146, p. 295–311, 2017.
[18] I. Tien, "Resilient by design: the case for increasing resilience of buildings and their linked food-energywater systems.," Elementa: Science of the Anthropocene, vol. 7, 2018.
[19] A. de Waal and J. Linthorst, "Future-Proofing the High-Performance Organization.," Sustainability, vol. 12, no. 20, p. 8507, 2020.
[20] S. Forbes and D. Fikretoglu, "Building Resilience: The Conceptual Basis and Research Evidence for Resilience Training Programs," Review of General Psychology, vol. 22, no. 2, p. 452–468, 2018.

Self-Bricks in the Field of Rapid Construction

Mai Mohamed Matareek[1,*], Mohammed El-Sayed Hassanien[2]

[1] El-Shrouk Academy, Shrouk City, Cairo, Egypt
[2] Free researcher, Cairo, Egypt

[*]Corresponding Author Email: mai.m.matareek@gmail.com

Abstract. Orientation to the use of modern technology in the field of rapid construction through an innovative engineering design for a unit represented by a brick which enjoys the capacity of self in jaw and installation by using mechanical assembly in the manner of vertical and horizontal interlock it also has the possibility of manufacturing from a variety of raw materials and different sources to allow the building of walls in multiple sizes, and can be formed in different shapes to facilitate the construction of the building, especially the implementation of temporary installations in times of crises and disasters and rapid buildings with the aim of achieving sustainable green construction. The research problem is focused on The construction and equipment operations of the units that can be disassembled, installed, assembled and transported lack the principle of alignment with psychological comfort, The stages of implementing the buildings in the targeted manner require long periods, The inability to transfer the facilities "units" established by similar methods to the targeted places, The need to find a quick way to suit the requirements of successive events locally and globally, which require rapid construction, Using building units once with no possibility of recycling again, either through reusing the same units or recycling and re-manufacturing them. The Targets are Summarized in Presenting a new idea of building with the aim of preserving the environment by making a new design for building units that reduce the use of materials that pollute, whether in the stages of manufacture or transfer operations, Contribute to solving problems related to displacement by quickly implementing temporary facilities in times of crises and disasters, and psychological adaptation of the user's needs within a vacuum of environmental comfort using the newly developed building unit, Flexibility and speed of disassembly, installation and assembly with the possibility of transportation and storage in the narrowest spaces for the elements, which works to reuse more than once for a variety of functions, Ease of building by the unit in locations that are difficult to reach with the needs of traditional construction methods and the use of modern means of manufacturing, Achieving sustainable green construction by recycling waste and converting it into building materials, using available raw materials in the environment, and reducing the use of materials polluting the environment during their manufacturing stages, Implementation of a wall product consisting of a group of self-made bricks with interlocking joints that allow dismantling and installation without using other materials for fixing, for example, mortar or nails. Finally The Hypothesis is to Access a construction method that is easy to manufacture and implement, The possibility of using materials with environmentally friendly specifications, Orientation to use modern and innovative means to manufacture temporary and fast facilities easily, Reducing the use of building materials that pollute the environment, such as cement and other building materials.

Key words: Self brick, Flexibility, Interlock, Mould.

1. Introduction:

The innovation in the world turning to ways for developing new methods and systems for construction to reduce pollution and harmful emissions in the environment. Therefore, a set of different ideas are being worked out to design and manufacture a unit with approved technical specifications. It has what it lacks and we have designed a new model of the building unit registered in the industrial model unit for a self-building brick as in Figure (1), which has the following advantages:

- Bonding without using materials for installation in a vertical and horizontal interlock manner in an engineering way that can be disassembled, installed and assembled again.
- It is made of a variety of materials, including recycled materials.
- Can be used to forming walls with different shapes and sizes and form an integrated structure, as shown in Figure (2).

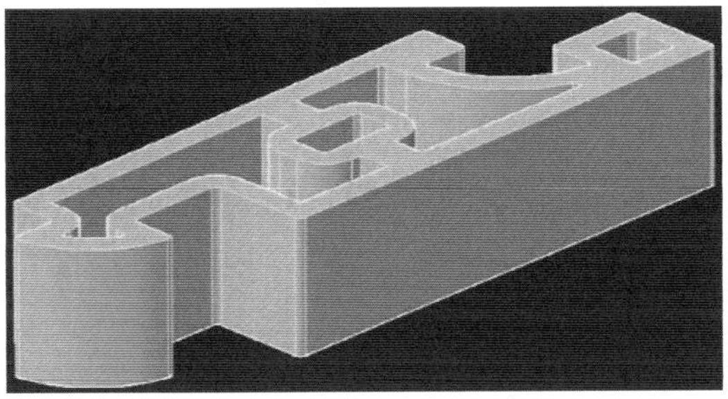

Figure 1. Self-Brick Design

Figure 2. Wall of a group of bricks

- Especially can be used more in temporary structures and quick-established buildings to be implemented in times of crises and disasters, with the aim of achieving sustainable green construction.

2. What is the self-brick

It is a unit with an innovative design in which the architectural thought is combined with the mechanical function, which is represented by the method of interlocking the combined vertical and horizontal interlock in the same unit and represents the architectural basis for dividing the spaces of the building. The materials and nature of their use according to the available in the market, with installation without the need for supporting materials, with the flexibility of dismantling the units and reinstalling them again.

In order to facilitate the transfer of the idea and its clarification, several models were made with different manufacturing methods, some of which are available locally, and those that are available internationally did not reach the local market to meet all the requirements, and each method of manufacturing is determined by changing technical specifications for the final product.

It has the following characteristics:

- Dimension 25 * 12 * 10 cm.
- Produced from local materials.
- Possibility of manufacturing using different materials suitable for recycling.
- Male and female form in one design.
- It can be manufactured using the equipment available from the factories.
- Faster implementation of temporary projects.

3. Architectural analyses of the unit:

Design Idea: An innovative registered design idea that represents an economic and technological breakthrough because it includes keeping pace with the scientific and technical development used in the field of construction globally. The interlock system used can be done automatically later. The design consists of curves aiming to increase the forces of bearing pressure and tension as a result of the distribution of those forces affecting the curve in the design on the entire unit, which increases the resistance to these forces affecting the total, as for the lines, they represent the interconnection surfaces between each brick and another. The brick façade has multiple measurements, as it fits with the existing curve and the height of the brick in global proportions to fit visually when looking at the building, and the vertical assembly unit is located in the middle. The proportions of the sizes have been taken into account with the availability of different production machines to fit with The manufacturing methods available in the local market and the first cost in each manufacturing method that is followed and depends on the two interlock units (male & female) in one model until the shape is meshed vertically and horizontally to form the integrated unit of a structure that can stay in it and deal with it and through it with the external environment as it nourishes the feeling of safety and stability as Figures (3,4,5).

Figure 3. Side interface.

Figure 4. Horizontal projection of the self-brick.

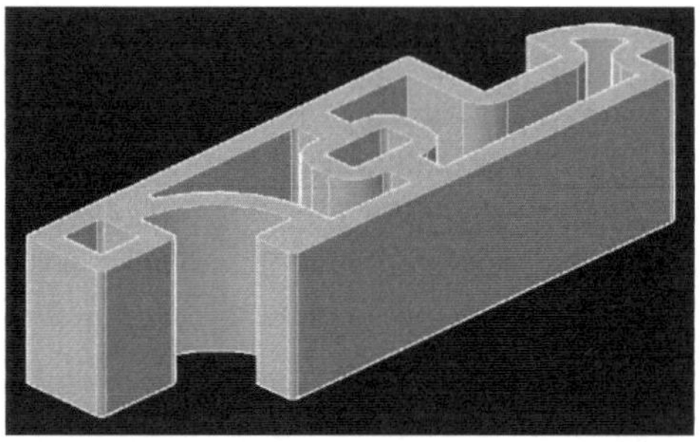

Figure 5. Perspective interface for self-brick.

4. Structural and mechanical analysis of the unit:

Structural system: Vertical elements consisting of sectors of cans made of metal, often as they help in the process of vertical support and linking the elements together, the horizontal installation works to tighten the bonding and make the wall cohesive more strongly it works to increase cohesion.

Connection method: Horizontal meshing with the sectors shown in the design, which depends in essence on curves that add durability and resistance to any motive forces that affect the two units interconnected with each other. Vertical meshing is done by modules made of the same material or using what is shown here, which is the vertical boxes will be prepared with the structural design of the building, it is related to the base of wall on which the building is built, as shown in Figure (6).

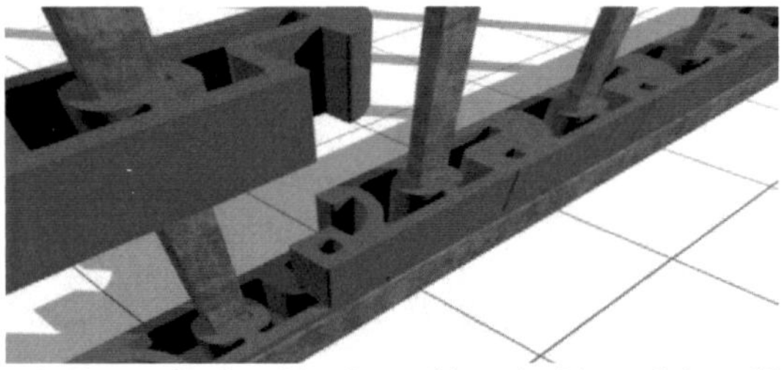

Figure 6. The vertical and horizontal interlocking of the self-brick.

The materials used in the manufacture of the self-brick: The materials vary according to the different sources that are available in the environment, as well as those available in the local market. They are not limited to specific materials, but rather provide opportunities for use of many of the materials subject to studies in specialized research institutes to provide opportunities for practical application of these materials, including but not limited to materials next:

Silt: It is poured into the Mould prepared for it and left to dry, by heating or drying in the sun.

Magnesite cement: It has many specifications and advantages, the most important of which is the economy in the use of fresh water. There are mines of raw materials required for this sector of industries, and the units constructed from these raw materials enjoy being more robust the more they are exposed to the coastal atmosphere.

Recycled plastic: Materials that are a major source of pollution to the environment are reused by chemical means. These types of plastic can be treated in accordance with many requirements, especially ignition conditions, as well as harmful emissions resulting from combustion.

Aluminium: the use of this material by making Moulds for manufacturing through the aluminium extrusion process using special machines for this, thus making sticks that are cut into the target sizes.

5. Possibilities available in the field of self-brick application-n in Egypt:

In the context of the growing interest globally in the adoption and orientation of green construction, especially the shift to the application of the standards of green facilities, which differ from one society to another and from one thought to another, in order to reduce the use of elements that work on environmental pollution Especially cement products that are not recyclable and which need to be dealt with wasting environmental resources that are the lifeblood of water, so the proposed project to design, manufacture and innovate in bricks that depends in its installation on interlocking to reduce the use of supporting materials to make the necessary installation seeks to support this approach to use it in the formation of various facilities and the preparation of technical cadres with support for markets and emerging minds, as they carry the ability to create, innovate and develop, and this provides new job opportunities while seeking to remove obstacles that prevent the benefit of local minds for sustainable local development.

6. Methods of Making Self-Brick:

The brick is manufactured in several different ways depending on the nature of the materials to be manufactured from, and we can summarize some of the use of these methods as follows:

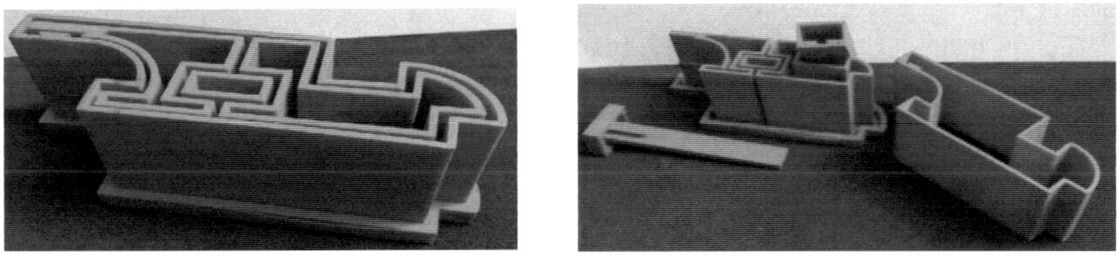

Figure 7. A model of a Mould for pouring in it.

Figure 8. A model that has been poured and taken out of the Mold.

Figure 9. A model of a Mold shape and a sample of the model is cast inside it.

<u>Vibration units:</u> the components are placed in manufacturing Molds and placed on units that operate in the manner of successive vibrations at a frequency, so the components merge and the surface becomes smooth as a result of successive shaking, as is done in the manufacture of floor slab units (interlock).
<u>Manufacturing using moulds (casting forms):</u> the materials are mixed and poured with the molds prepared for that, subjected to pressure processes to crush the components until they withstand high pressures, and among the forms of casting models as shown in Figures (7,8,9).
<u>Extrusion system:</u> In this resort, a new method of manufacturing is used, where polymeric materials or aluminium are used when manufacturing as recycled materials.
3D printing: In this, another design registered in the name of researchers is used using 3D printing technology, as shown in Figures (10, 11).

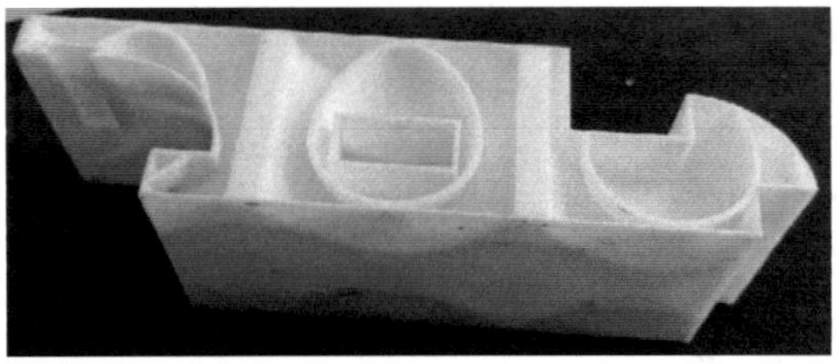

Figure 10. The general figure of the brick model and manufacturing using 3D printing.

Figure 11. The brick model with a segmented unit to illustrate the method of connecting a compass and manufacturing using 3D printing.

7. Other examples:

There are many designs that depend on interlock, but they lack both vertical and horizontal interlock in the market, including types of hollow cement bricks compared to the self-brick that was presented in the research as Figure (12).

a) A model of bricks that are fixed vertically only through voids by placing iron rods in order to increase bonding, and cement is poured into these voids But the horizontal installation is with the second course that will connect the two units to each other, especially those in which the skewers will be placed.

b) A model of bricks that shows only the horizontal fixation of the bricks, and the process of vertical fixation is by placing vertical iron bars with cement in order to increase the bonding.

c) Hollow modular concrete bricks model.

d) The shape of the subject brick, in which the bonding process is carried out vertically and horizontally.

Figure 12(a, b, c and d). Models of types of bricks compared to self-blocking.

It is clear from the following table (1) that the characteristics of different types of bricks compared with self-brick are as following:

Table 1. Comparison between self-blocking and hollow concrete.

Compare	Self-brick	Hollow concrete brick
Advantages And the shape	• Reducing the use of cement by 65%. • Preserving the environment. • You do not need trained labour for installation. • Interlocking without adhesives, nails or welds. • Speed in implementation and does not require skilled labour. • Saving in cost. • The ability to disassemble and install and the ease of movement from one place to another for more than one job and in different formative forms. • No need for finishing.	• Use of cement. • Environmental Pollution. • You need trained labor in construction. • Bonding with mortar and other materials. • Wasted time. • Expensive. • Traditional building methods.

	• Flexibility of design, which allows the possibility of adapting the design to suit the quality of the materials used. • The possibility of using more than one material to make the brick mold according to the capabilities and usage, as well as from recycled materials. • Modern technological construction methods such as 3D printing or by manufacturing a Mold to pour the material inside it.	

8. Economic study of self-brick:

The production unit consists of a self-construction unit that is easy to disassemble and install with a system of vertical and horizontal interlock together that can be traded in the local market compared to the similar in its cost that is the cheapest and fastest in its implementation and production in quantity, re-transportation, use and manufacturing with raw materials from recycled materials that do not pollute the environment.
Market approach to win procurement, develop strategic business and achieve a difficult equation as shown in Figure (13).

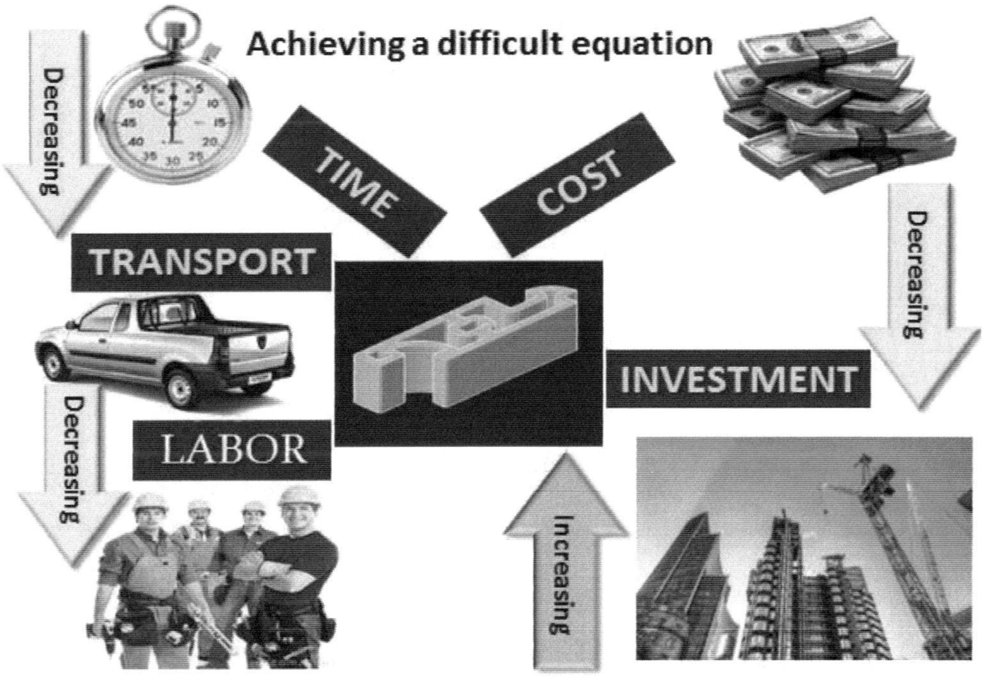

Figure 13. Characteristics of self-brick in the local market.

We can analyze the above compared to the types of commonly used bricks with the following graphs, which show a comparison of the cost per square meter using each type, the comparison between the number of days required for implementation as shown in Figure (15,14), the rates of need for raw materials required to manufacture each type of study as shown in Figure (16) And finally, compare the time required to implement a building with an area of 47 square meters, as shown in Figure (17).

IOP Conf. Series: Earth and Environmental Science 1056 (2022) 012016 doi:10.1088/1755-1315/1056/1/012016

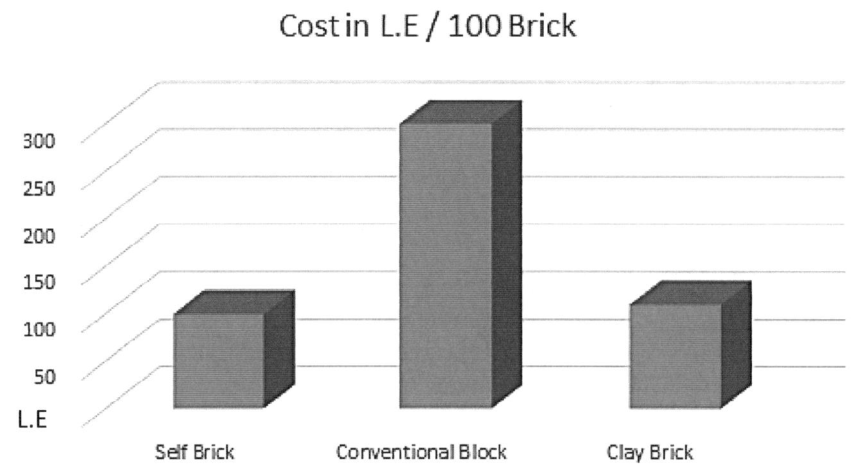

Figure 14. Graph showing the average cost comparison for the different types of bricks in L.E.

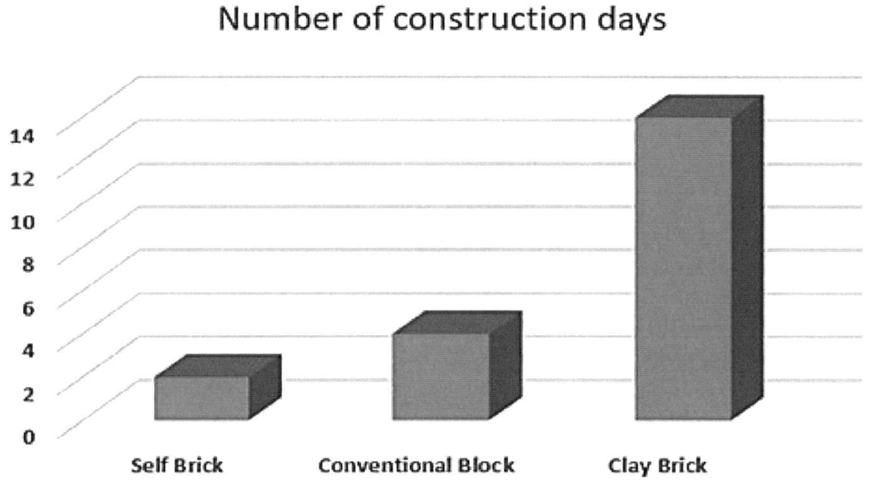

Figure 15. Graph showing the comparison of the average implementation time with different types of bricks in days.

Cost comparison of raw material costs including reinforcement to complete a 47m^2 structure:

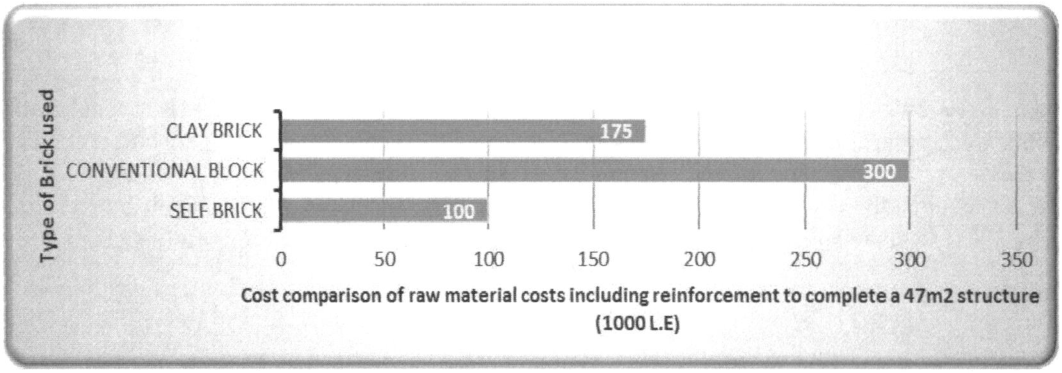

Figure 16. Illustrative statement of the proportions of materials required to implement construction operations using different types of bricks.

Number of working days to complete a 47m² structure:

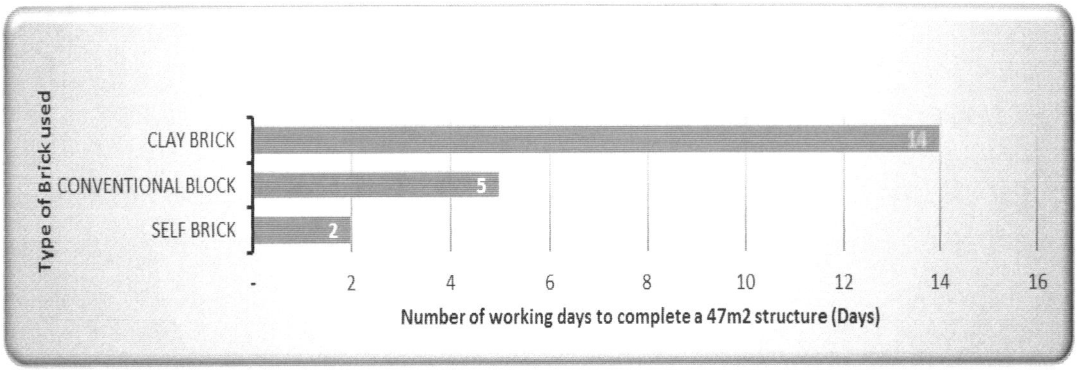

Figure 17. A comparison of the time periods required to implement a one-story building with an area of 47 square meters.

9. Conclusion:

The design is applicable to raw materials processing, casting and vibrating Molds, The self-brick is suitable for the ongoing urban expansion, and for reconstruction works in various regions of the world, Seeking to lose weight and diversify the raw material, The possibility of its acceptance by international bodies and donors in the reconstruction work, Its compatibility with the new global trends under the title of green economy and the preference it includes for environmentally friendly buildings Among the recommendations of the study providing opportunities for practical application of new materials, The assistance of a production engineer to determine the most appropriate method of production, develop a code and guideline for quick establishments, presenting the model to the various research centres with the aim of developing and mechanizing assembly and disassembly.

References

[1] Nicholas J. Garber, Lester A. Hoel. 2018. Traffic and Highway Engineering. 4th Edition. University of Virginia. CENGAGE Learning. ISBN-13: 978-0-495-08250-7.
[2] Barry, R. (Robin). 2009. The construction of building. 7th Edition. Blackwell Science Ltd. Oxford U.K. ISBN 0-632-05261-9 (v. 1).
[3] Ed van Hinte, Marc Neelen, Jacques Vink, Piet Vollaard. 2003. Smart Architecture. 010 Publishers, Rotterdam. ISBN 90-6450-490-3.
[4] Fouad Al-Jumaili and Abdullah Odaibat. 2019. Entrance to the construction of buildings. Available from: https://e3arabi.com/?p=984434https://e3arabi.com/?p=984434 [accessed 29 September 2021].
[5] The American Ceramic Society. 2010. The 16th Annual New England Technical Ceramics Expo. Available from: https://ceramics.org/ceramic-tech-today/construction/the-many-types-of-bricks [accessed 27 October 2021].
[6] Juan Rodriguez. 2021. 5 Types of Materials Used for Bricks. Available from: https://www.thespruce.com/bricks-types-uses-and-advantages-844819 [accessed 15 November 2021].
[7] Campbell, J. W. P. and Pryce, W. 2006. TECHNICAL NOTES on Brick Construction. Available from: https://www.gobrick.com/docs/default-source/read-research-documents/technicalnotes/9-manufacturing-of-brick.pdf?sfvrsn=0 [accessed 23 November 2021].
[8] UKEssays. 2018. Alternative Recycling Materials for Bricks Production Construction Essay. Available from: https://www.ukessays.com/essays/construction/alternative-recycling-materials-for-bricks-production-construction-essay.php?vref=1 [accessed17 December 2021].
[9] Timothy Dale, 2021, 7Types of Brick All DIYers Should Know. Available from: https://www.bobvila.com/articles/types-of-brick/ [accessed 26 December 2021].

IOP Conf. Series: Earth and Environmental Science 1056 (2022) 012017 doi:10.1088/1755-1315/1056/1/012017

Design Approach for Eccentrically Loaded Reinforced Concrete Sections

A E Mohamed[1, *]

[1] Civil Engineering Department, Modern Academy for Engineering and Technology Cairo, Egypt

*Corresponding Author Email: AYMAN.EZZAT@eng.modern-academy.edu.eg

Abstract. The behavior of eccentrically loaded reinforced concrete sections depends on both the eccentricity and load value. The type of failure can be determined according to the eccentricity-thickness ratio and the applied load value. Interaction diagrams can be used for designing eccentrically loaded section, but the section dimensions are required to use these diagrams. According to the section dimensions and the load value, the type of failure and the required reinforcement can be determined from the interaction diagrams. Most design codes controls the design process to ensure ductile failure. But according to the design procedures of interaction diagrams, the failure type can't be controlled simply. This paper aims at finding a design procedure for eccentrically loaded reinforced concrete sections that enable the designer to simply control the design process and the failure mode. The proposed design approach makes the design process more direct and simplified.

Keywords: axial load, eccentric load, failure mode, Interaction diagram, plastic centroid.

1. Introduction

Axially loaded members occur rarely in practice. Most reinforced concrete members are considered as eccentrically loaded members. The behavior of the sections under eccentric load depends on the magnitude of the applied load and the eccentricity [1-3]. Vivo and Rosti examined the behaviour of reinforced concrete sections subjected to eccentric loads [4]. The studies explained that both the section capacity and the failure mode is greatly affected by the eccentricity [1-4]. The same conclusions are obtained for ordinary and high strength concrete. The same results are obtained by F. Peng, F. Alhussainy, M. Chellapandian, I. Geevar, M. Guerin, W. Xue and G. Kabtamu [9 - 15] for different cases.

The design process of eccentrically loaded section depends on the failure mode [1-3]. For small eccentricity, the resulted moment is small compared to the applied load and the column behavior will dominate. If the eccentricity is large enough, the beam behavior is expected [5]. To determine the failure mode; the section dimension, the applied load value and the eccentricity must be known [1-3, 5]. Using the interaction diagrams, the failure mode can be determined and also the required reinforcement ratio. Estimating the section dimensions is required before using interaction diagrams.

Most design codes control the design process to ensure ductile failure [6-8, 22, 23]. To achieve a certain type of failure by using interaction diagrams, many trails can be carried to get the suitable section dimensions. This process is time consuming and is not a suitable one. This paper study the eccentrically loaded rectangular reinforced concrete sections to introduce a design approach that control the section dimensions and the reinforcement ratio according to the required failure mode.

2. Plastic centroid of singly reinforced sections

For singly reinforced section, shown in figure1, the plastic load (P_{ups}) is:

$$\frac{P_{ups}}{f_{cu}\, b\, t} = \frac{\mu\, f_y}{\gamma_s\, f_{cu}} \times \frac{d}{t} + \frac{0.67}{\gamma_c} \tag{1}$$

Taking the moment about the top fiber, then:

$$P_{ups} \times c_{ps} = (C_s \times d) + \left(C_c \times \frac{t}{2}\right)$$

$$\frac{P_{ups}}{f_{cu} b\, t} \times c_{ps} = \frac{\mu\, d\, f_y}{f_{cu}\, t\, \gamma_s} \times d + \frac{0.67}{\gamma_c} \times \frac{t}{2} \tag{2}$$

Simplifying it and from Eq. (1), then:

$$\frac{c_{ps}}{d} = \left[\frac{\mu\, f_y}{\gamma_s\, f_{cu}} \times \frac{d}{t} + \frac{0.335}{\gamma_c}\frac{t}{d}\right] \Big/ \left[\frac{\mu\, f_y}{\gamma_s\, f_{cu}} \times \frac{d}{t} + \frac{0.67}{\gamma_c}\right]$$

where c_{ps} is the distance between the plastic centroid of the singly reinforced section and the most compressed fiber. For $\gamma_c = 1.75$, and $\gamma_s = 1.34$ [6,7,8,22,23] then:

$$\frac{c_{ps}}{d} = \left[0.746\, \frac{\mu\, f_y}{f_{cu}} \times \frac{d}{t} + 0.191\, \frac{t}{d}\right] \Big/ \left[0.746\, \frac{\mu\, f_y}{f_{cu}} \times \frac{d}{t} + 0.382\right] \tag{3}$$

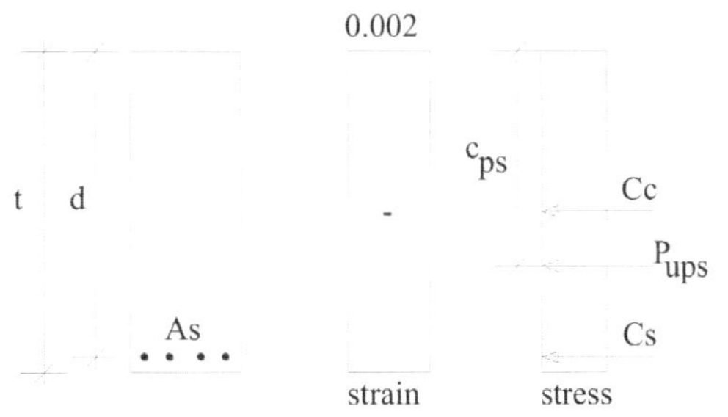

Figure 1. Plastic centroid of singly reinforced sections.

3. Balanced failure of eccentrically loaded singly reinforced sections

For balanced failure of eccentrically loaded singly reinforced section shown in figure2, the neutral axis-depth ratio is given by [6, 18-20]:

$$\frac{c_b}{d} = 690/(690 + f_y) \tag{4}$$

and from equilibrium it can be found that:

$$\frac{P_{ubs}}{f_{cu}\, b\, t} = \left[\frac{369.84}{(690 + f_y)\gamma_c} - \frac{\mu\, f_y}{\gamma_s\, f_{cu}}\right]\frac{d}{t}$$

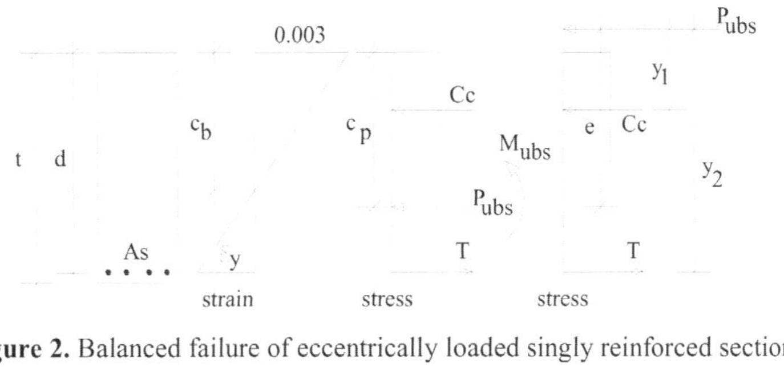

Figure 2. Balanced failure of eccentrically loaded singly reinforced sections.

For balanced failure, $\gamma_c = 1.5$, and $\gamma_s = 1.15$ [6,7,8] then:

$$\frac{P_{ubs}}{f_{cu}\, b\, t} = \left[\frac{246.56}{(690+f_y)} - \frac{0.87\, \mu\, f_y}{f_{cu}}\right]\frac{d}{t} \tag{5}$$

which can be written as:

$$\frac{P_{ubs}}{f_{cu}\, b\, t} = Z_1 \tag{6}$$

where

$$z_1 = \left[\frac{246.56}{(690+f_y)} - \frac{0.87\,\mu\, f_y}{f_{cu}}\right]\frac{d}{t} \tag{7}$$

Taking the moment about the point of application of the eccentric load, then:

$$C_c\, y_1 - T\, y_2 = 0$$

$$\frac{y_1}{y_2} = \frac{\mu\, \gamma_c\, f_y}{0.67\, \gamma_s\, f_{cu}} \times \frac{690+f_y}{552} \tag{8}$$

Since

$$y_2 - y_1 = d - \frac{276}{690 + f_y} d$$

then

$$\frac{y_2}{d} = \left(1 - \frac{276}{690+f_y}\right) \Big/ \left(1 - \left(\frac{1.947\,\mu\, f_y}{f_{cu}} \times \frac{690+f_y}{552}\right)\right) \tag{9}$$

Since

$$y_2 - e = d - c_{ps} \tag{10}$$

$$\frac{\left(1 - \frac{276}{690 + f_y}\right) d}{\left(1 - \left(\frac{1.947\,\mu\, f_y}{f_{cu}} \times \frac{690 + f_y}{552}\right)\right)} - \frac{M_{ubs}}{P_{ubs}} = d - c_{ps}$$

Simplifying and rearranging it, then:

$$\frac{M_{ubs}}{P_{ubs}} = \left[\frac{\left(1-\frac{276}{690+f_y}\right)}{\left(1-\left(\frac{1.947\,\mu\, f_y}{f_{cu}} \times \frac{690+f_y}{552}\right)\right)} - 1 + \frac{c_{ps}}{d}\right] d \tag{11}$$

which can be written as:

$$\frac{M_{ubs}}{P_{ubs}} = z_2\, d \tag{12}$$

$$z_2 = \left[\frac{\left(1-\frac{276}{690+f_y}\right)}{\left(1-\left(\frac{1.947\,\mu\, f_y}{f_{cu}} \times \frac{690+f_y}{552}\right)\right)} + \frac{\left[0.746\frac{\mu f_y}{f_{cu}} \times \frac{d}{t} + 0.191\frac{t}{d}\right]}{\left[0.746\frac{\mu f_y}{f_{cu}} \times \frac{d}{t} + 0.382\right]} - 1\right] \tag{13}$$

The coefficients z_1 and z_2 depend on the reinforcement ratio (μ), the properties of both concrete and steel (f_{cu} *and* f_y) and the depth-thickness ratio (d/t). For different values of these factors, designing tables and charts for the coefficients z_1 and z_2 can be prepared. Table1 and figure3 show an example of these design tables and charts. According to the values of the moment and the applied load, and using the coefficient z_2, the required depth can be estimated from Eq. (12), and then the required width can be estimated using Eq. (6) and the coefficient z_1.

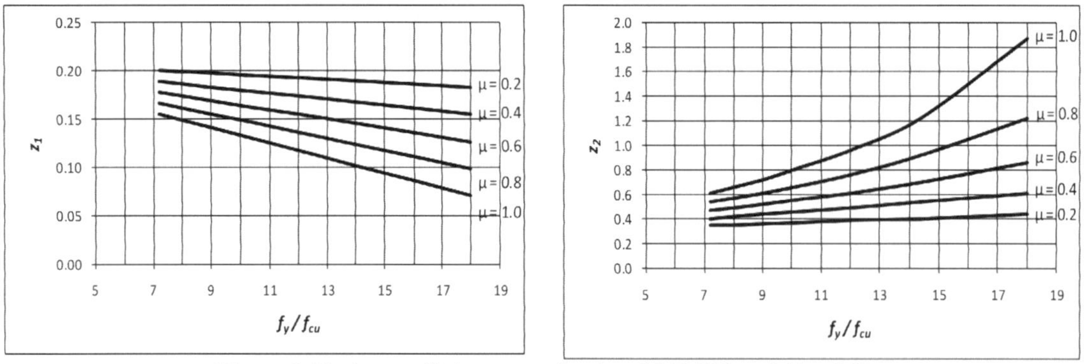

Figure 3. Design coefficients z_1 and z_2 (steel 360/520 and d/t = 0.9).

Table 1. Design coefficients z_1 and z_2 (steel 360/520 and d/t = 0.9)

f_{cu}	z	μ %				
		0.2	0.4	0.6	0.8	1.0
20	z_1	0.183	0.155	0.127	0.099	0.070
	z_2	0.433	0.611	0.855	1.225	1.874
25	z_1	0.189	0.166	0.144	0.121	0.099
	z_2	0.402	0.533	0.698	0.916	1.225
30	z_1	0.193	0.174	0.155	0.136	0.117
	z_2	0.383	0.487	0.611	0.764	0.960
35	z_1	0.195	0.179	0.163	0.147	0.131
	z_2	0.369	0.455	0.555	0.672	0.814
40	z_1	0.197	0.183	0.169	0.155	0.141
	z_2	0.359	0.433	0.515	0.611	0.722
45	z_1	0.199	0.186	0.174	0.161	0.149
	z_2	0.351	0.415	0.487	0.567	0.658
50	z_1	0.200	0.189	0.178	0.166	0.155
	z_2	0.345	0.402	0.464	0.533	0.611

To ensure balanced failure (and to prevent dividing by zero, for Eq. (7)) the following condition must be satisfied:

$$\frac{0.87\,\mu\, f_y}{f_{cu}} < \frac{246.56}{(690 + f_y)}$$

$$\mu_{max} = \left(\frac{f_{cu}}{f_y} \times \frac{283.5}{690+f_y}\right) \tag{14}$$

Table2 shows the maximum reinforcement ratios (for different steel types) for balanced failure of eccentrically loaded singly reinforced sections.

Table 2. Maximum reinforcement ratio for balanced failure of eccentrically loaded singly reinforced sections

Steel type	μ_{max}
240/350	$12.70 \times 10^{-4} \times f_{cu}$
280/450	$10.43 \times 10^{-4} \times f_{cu}$
360/520	$7.50 \times 10^{-4} \times f_{cu}$
400/600	$6.50 \times 10^{-4} \times f_{cu}$

4. Balanced failure of eccentrically loaded doubly reinforced sections

For balanced failure of eccentrically loaded doubly reinforced sections (figure4), since

$$\frac{\varepsilon_s'}{0.003} = \frac{c_b - d'}{c_b} \tag{15}$$

and to ensure yield of compression reinforcement [16,17], then

$$\frac{d'}{d} \leq \frac{(690-f_y)}{(690+f_y)} \tag{16}$$

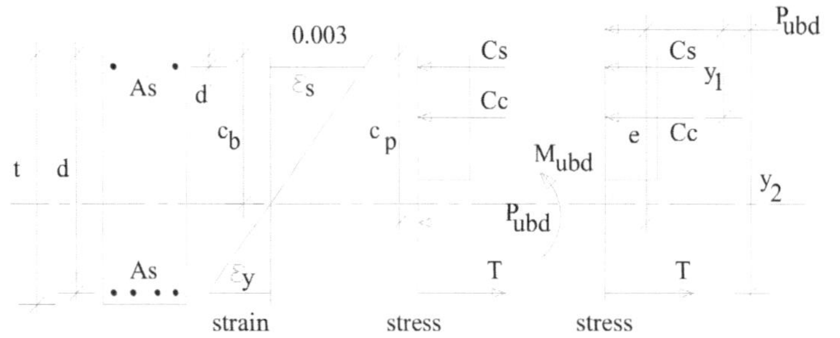

Figure 4. Balanced failure of eccentrically loaded doubly reinforced sections.

According to the equilibrium equations of the section:

$$P_{ubd} = \frac{0.67\, f_{cu}\, b\, a_b}{\gamma_c} + \frac{A_s' f_y}{\gamma_s} - \frac{A_s f_y}{\gamma_s} \tag{17}$$

Assuming that

$$A_s^* = A_s - A_s'$$
$$\mu^* = (1-\alpha)\mu \tag{18}$$

then from Eq. (21)

$$\frac{P_{ubd}}{f_{cu}\, b\, t} = \left[\frac{369.84}{(690+f_y)\gamma_c} - \frac{\mu^* f_y}{\gamma_s f_{cu}}\right]\frac{d}{t} \tag{19}$$

which is the same value of singly reinforced section with reinforcement ratio μ*, then

$$\frac{P_{ubd}}{f_{cu}\, b\, t} = z_1^* \tag{20}$$

$$z_1^* = \left[\frac{246.56}{(690+f_y)} - \frac{0.87\,\mu^* f_y}{f_{cu}}\right]\frac{d}{t} \tag{21}$$

Then the section can be treated as singly reinforced section in addition to the additional moment due to an amount of reinforcement, equals the compression reinforcement, placed in both compression and tension sides (shown in figure5), as following:

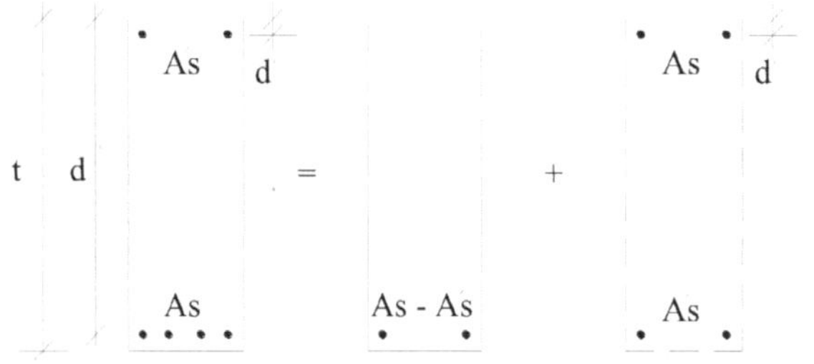

Figure 5. Modification of doubly reinforced section.

a) for singly reinforced section (with reinforcement ratio μ^*):

$$\frac{M_{ubs}}{P_{ubs}} = \left[\frac{\left(1-\frac{276}{690+f_y}\right)}{\left(1-\left(\frac{1.947\,\mu^* f_y}{f_{cu}}\times\frac{690+f_y}{552}\right)\right)} + \frac{\left[0.746\frac{\mu^* f_y}{f_{cu}}\times\frac{d}{t}+0.191\frac{t}{d}\right]}{\left[0.746\frac{\mu^* f_y}{f_{cu}}\times\frac{d}{t}+0.382\right]} - 1\right] d \tag{22}$$

b) the additional moment (due to A_s' added in both sides):

$$\Delta M_u = \frac{\alpha\,\mu\, b\, d\, f_y}{\gamma_s}\times \xi t \tag{23}$$

where *ξ = (d-d')/t*. Since

$$M_{ubs} = z_2^*\, d\, P_{ubd} \tag{24}$$

$$z_2^* = \left[\frac{\left(1-\frac{276}{690+f_y}\right)}{\left(1-\left(\frac{1.947\,\mu^* f_y}{f_{cu}}\times\frac{690+f_y}{552}\right)\right)} + \frac{\left[0.746\,\frac{\mu^* f_y}{f_{cu}}\times\frac{d}{t}+0.191\,\frac{t}{d}\right]}{\left[0.746\,\frac{\mu^* f_y}{f_{cu}}\times\frac{d}{t}+0.382\right]} - 1\right] \tag{25}$$

$$M_{ubd} = z_2^*\, d\, P_{ubd} + \frac{\alpha\,\mu\, b\, d\, f_y}{\gamma_s}\times \xi t \tag{26}$$

from Eq. (20); then

$$\frac{M_{ubd}}{P_{ubd}} = \left[z_2^* + 0.87\,\frac{\alpha\,\mu\, f_y\,\xi}{z_1^*\, f_{cu}}\right] \tag{27}$$

5. Tension failure of eccentrically loaded sections

For tension failure of reinforced concrete sections, the Egyptian Code of practice (ECP-203) [6] limits the neutral axis-depth ratio to 2/3 of the balanced one. Then, the previous equations of balanced sections can be modified using the neutral axis-depth ratio as:

$$\frac{c}{d} = \frac{460}{690+f_y} \tag{28}$$

For singly reinforced sections and using the previous relation of the neutral axis-depth ratio, then:

$$\frac{P_{us}}{f_{cu}\, b\, t} = \frac{246.56}{(690+f_y)\gamma_c}\,\frac{d}{t} - \frac{\mu\, f_y}{\gamma_s\, f_{cu}} \tag{29}$$

$$\frac{P_{us}}{f_{cu}\, b\, t} = z_3 \tag{30}$$

$$z_3 = \left[\frac{164.37}{(690+f_y)} - \frac{0.87\,\mu\, f_y}{f_{cu}}\right]\frac{d}{t} \tag{31}$$

Similarly

$$\frac{M_{us}}{P_{us}} = z_4\, d \tag{32}$$

$$z_4 = \left[\frac{\left(1-\frac{184}{690+f_y}\right)}{\left(1-\left(\frac{1.947\,\mu\, f_y}{f_{cu}}\times\frac{690+f_y}{368}\right)\right)} + \frac{\left(0.746\,\frac{\mu\, f_y}{f_{cu}}\times\frac{d}{t}+0.191\,\frac{t}{d}\right)}{\left(0.746\,\frac{\mu\, f_y}{f_{cu}}\times\frac{d}{t}+0.382\right)} - 1\right] \tag{33}$$

with maximum reinforcement ratio given by the following relations:

$$\frac{0.87\,\mu\, f_y}{f_{cu}} < \frac{164.37}{(690+f_y)} \tag{34}$$

$$\mu_{max} = \frac{189}{(690+f_y)}\times\frac{f_{cu}}{f_y} \tag{35}$$

$$\left(\frac{1.947\,\mu\, f_y}{f_{cu}}\times\frac{690+f_y}{368}\right) < 1 \tag{36}$$

$$\mu_{max} = \left(\frac{f_{cu}}{f_y}\times\frac{189}{690+f_y}\right) \tag{37}$$

Then, for doubly reinforced sections

$$\frac{P_{ud}}{f_{cu}\, b\, t} = z_3^* \tag{38}$$

$$\frac{M_{ud}}{P_{ud}} = \left[z_4^* + 0.87\,\frac{\alpha\,\mu\, f_y\,\xi}{z_3^*\, f_{cu}}\right] d \tag{39}$$

where z_3^* and z_4^* are

$$z_3^* = \left[\frac{164.37}{(690+f_y)} - \frac{0.87\,\mu^*\, f_y}{f_{cu}}\right]\frac{d}{t} \tag{40}$$

$$z_4^* = \left[\frac{\left(1 - \frac{184}{690+fy}\right)}{\left(1 - \left(\frac{1.947\,\mu^*\,fy}{fcu} \times \frac{690+fy}{368}\right)\right)} + \frac{\left(0.746\,\frac{\mu^*\,fy}{cu} \times \frac{d}{t} + 0.191\,\frac{t}{d}\right)}{\left(0.746\,\frac{\mu^*\,fy}{fcu} \times \frac{d}{t} + 0.382\right)} - 1 \right] \tag{41}$$

Figure 6 represents a chart for the design parameters z_3^* and z_4^*. Tables3 and 4 are examples of the design parameters z_3^* and z_4^* and the maximum reinforcement ratio μ_{max}. From these tables and according to the concrete characteristic strength and the reinforcement ratio, the design parameters z_3^* and z_4^* can be obtained. Knowing the applied moment and load and using Eq. (39), the required depth (*d*) can be estimated and then the required thickness (*t*). From Eq. (38), the required breadth can be estimated.

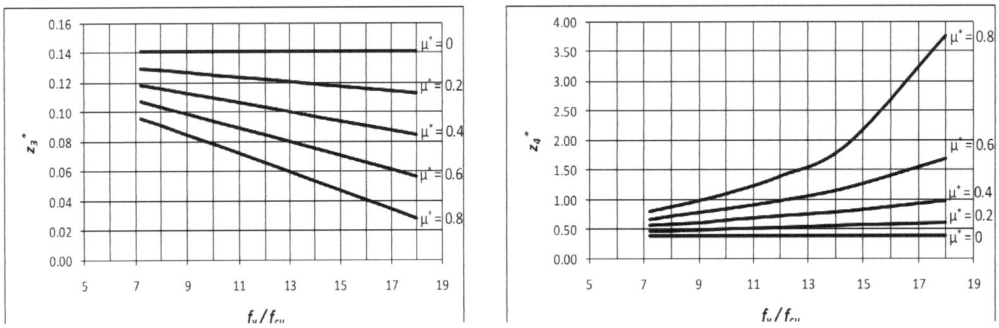

Figure 6. Design coefficients z_3^* and z_4^* (steel 360/520 and d/t = 0.9).

Table 3. Maximum reinforcement ratio for tension failure of eccentrically loaded doubly reinforced section

Steel type	μ^*_{max}
240/350	$8.47 \times 10^{-4} \times f_{cu}$
280/450	$6.96 \times 10^{-4} \times f_{cu}$
360/520	$5.00 \times 10^{-4} \times f_{cu}$
400/600	$4.33 \times 10^{-4} \times f_{cu}$

Table 4. Design parameters z_3^* and z_4^* (f_y = 360 N/mm² and d/t = 0.9)

f_{cu}	z^*	μ^* %				
		0.0	**0.2**	**0.4**	**0.6**	**0.8**
20	z_3^*	0.141	0.113	0.085	0.056	0.028
	z_4^*	0.380	0.613	0.980	1.688	3.768
25	z_3^*	0.141	0.118	0.096	0.073	0.051
	z_4^*	0.380	0.559	0.809	1.200	1.921
30	z_3^*	0.141	0.122	0.103	0.085	0.066
	z_4^*	0.380	0.525	0.715	0.980	1.387
35	z_3^*	0.141	0.125	0.109	0.093	0.076
	z_4^*	0.380	0.502	0.655	0.854	1.131
40	z_3^*	0.141	0.127	0.113	0.099	0.085
	z_4^*	0.380	0.486	0.613	0.772	0.980
45	z_3^*	0.141	0.128	0.116	0.103	0.091
	z_4^*	0.380	0.473	0.582	0.715	0.880
50	z_3^*	0.141	0.130	0.118	0.107	0.096
	z_4^*	0.380	0.463	0.559	0.672	0.809

6. Conclusions

Most design codes controls the design process to ensure ductile failure. But according to the design procedures of interaction diagrams, the failure type can't be controlled simply. The proposed formula for the plastic centroid of singly reinforced sections is a powerful formula in preparing the design approaches of designing eccentrically loaded reinforced concrete sections. The proposed design procedure is more suitable for controlling the design process of eccentrically loaded sections. The proposed design procedure makes it easy to control the failure mode of eccentrically loaded sections. The proposed design approach makes the design process more direct and simplified.

References

[1] R. Park and T. Pauly, "Reinforced concrete structures" John Willy & Sons, New York, 1975.
[2] F. Kong and R. Evans, "Reinforced and prestressed concrete" Springer-Science+Business Media, 1994.
[3] S. Pillai and D. Menon, "Reinforced concrete design" McGraw-Hill publishing company limited, 2003.
[4] L. Vivo and L. Rosti, "Ultimate strength analysis of reinforced concrete sections subjected to axial force and biaxial moment" Computer Methods in Applied Mechanics and Engineering, vol. 166, issue 3-4, pp 261-287, 1998.
[5] W. Chen, "Theory of beam-columns" Vol. 1: In-plane behavior and design, J. Ross, New York, 2008.
[6] Egyptian Code of Practice for the Design and Construction of Concrete Structures ECP 203-2007, Cairo, 2007.
[7] ACI Committee 318, Building code requirements for reinforced concrete ACI 318-05. American Concrete Institute, Detroit, 2005.
[8] Eurocode 2, Design of concrete structures - part 1: general rules and rules for buildings EC-2, Brussels, 2004.
[9] F. Peng and W. Xue, " Reliability analysis of eccentrically loaded concrete rectangular columns reinforced with fiber-reinforced polymer bars" ACI Structural journal, vol. 116, issue 4, 2019.
[10] F. Alhussainy, M. Neaz Sheikh, and N. Muhammad, "P-M interactions of self-consolidating concrete columns reinforced with steel tubes" vol. 116, issue3, 2019.
[11] M. Chellapandian and S. Suriya, "Axial compression-flexure interaction behavior of hybrid fiber-reinforced polymer-strengthened columns" vol. 116, issue2, 2019.
[12] I. Geevar and D. Menon, "Strength of reinforced concrete pier caps—experimental validation of strut-and-tie method" issue1, vol 116, 2019.
[13] M. Guerin, H. Mohamed, B. Benmokrane, A. Nanni, and K. Shield "Eccentric behavior of full-scale reinforced concrete columns with glass fiber-reinforced polymer bars and ties" issue2, vol. 115, March 2018
[14] W. Xue, F. Peng, and Z. Fang "Behavior and design of slender rectangular concrete columns longitudinally reinforced with fiber-reinforced polymer bars" issue2, vol. 115, 2018.
[15] G. Kabtamu and Z. Adil "Interaction Diagram for Slender Concrete Column" LAP Lambert Academic Publishing, 2015.
[16] P. Smith, "Structural design of buildings" Wiley-Blackwell, 1st. edition, 2016.
[17] H. Thomas, "Compression failure mechanisms of composite structures", PN, 1986.
[18] J. Hellesland, N. Challamel, C. Casandjian and C. Lanos " Reinforced concrete beams, columns and frames: Section and slender member analysis" Wiley-ISTE, 1st. edition, 2013.
[19] D. Fanella, "Reinforced concrete structures: Analysis and design" McGraw-Hill Education, 2nd. edition, 2015.
[20] C. McCormac and R. Brown, "Design of reinforced concrete" Wiley, 10th. edition, 2015.
[21] C. Chung, S. Wang, "Computational analysis and design of bridge structures" Routledge; 1st. edition, 2017.
[22] P. Bhatt, T. MacGinley and B. Seng, "Reinforced concrete design to Eurocodes: Design theory and examples" CRC Press, 4th. edition, 2013.
[23] American Concrete Institute, "314R-16 Guide to Simplified Design for Reinforced Concrete Buildings" American Concrete Institute (ACI), 2016.

Behavior of R.C Beams Reinforced with Onsite Manufactured GFRP shear reinforcement

Karim Helmi[1], Alaa M. Morsy[2], Muntaser Blla Abdul Menem[3,*]
[1] Construction & Building Engineering Department, College of Engineering & Technology AASTMT, Egypt
[2] Head of Construction & Building Engineering Department, College of Engineering & Technology, AASTMT, Egypt.
[3] College of Engineering & Technology AASTMT, Egypt

[*]Corresponding Author Email: mntsrazeem70@gmail.com

Abstract. This paper includes an experimental program to study the shear behaviour of concrete beams reinforced in shear with fiber glass reinforced polymer GFRP and carbon fiber reinforced polymer CFRP stirrups manufactured onsite. A total of six specimens were tested to study several parameters including the type of FRP material used to manufacture the stirrups, the angle of inclination of the stirrups and the thickness to bend radius ratio (r/t) of the stirrups. The experimental results showed that the ultimate strength of beams reinforced with carbon fiber stirrups is significantly higher than that of beams reinforced with fiber glass stirrups for the same area, for the same area increasing the diameter increases the failure load and strains in the stirrups while decreasing the thickness decreases the failure load and strains in stirrups, due to the reduction of the stress concentration at the corners of the stirrups and that increasing the inclination angle of the stirrups with the horizontal increases the strain in the stirrups and ultimate strength.

Keywords: Glass fiber-reinforced polymers, Carbon fiber reinforced polymers, Shear behavior, Bend strength (r/t).

1. Introduction

Fiber-reinforced-polymer (FRP) composites have been widely used in strengthening reinforced concrete (RC) beams, slabs and columns in addition to new construction whether as longitudinal or transverse reinforcement. Several studies have been carried out on the shear and flexural behaviour of concrete beams reinforced with FRP, because FRP materials have several advantages such as corrosion resistance, electromagnetic neutrality, high strength to weight ratio, and the ability to be tailored to specific applications [1-3]. Several researchers studied the behaviour of factory manufactured FRP transverse reinforcement in RC beams, for example El-Sayed et al. [4] studied the tail length of the stirrup beyond the bent portion, the stirrup anchorage, the bar diameter, and the embedment length. Issa et al. [5] studied basalt FRP stirrups and concluded that the shear capacity increased when the area of basalt FRP reinforcement increased for the same span to depth ratio (a/d), whereas the shear capacity decreased when the span to depth ratio (a/d) increased.

Factory manufactured FRP stirrups are generally produced using the pultrusion method. While Factory manufactured FRP stirrups have some advantages like higher quality control and the potential for mass production, they usually are more expensive to produce than straight bars and do not have enough flexibility for changes on site. Therefore, several researchers studied carbon Fiber-reinforced-polymer (CFRP) stirrups manufactured on site. Lee et al [6] compared site manufactured stirrups with to factory manufactured FRP and steel stirrups and the researchers concluded that they produce comparable results. In addition, the researchers studied the effect of reinforcement ratios. Other researchers [7, 8] investigated the shear behaviour of concrete beams reinforced with handmade CFRP strip stirrups manufactured on site by wrapping strips cut from carbon fiber fabric and wrapped around a rig or longitudinal reinforcement. The studied parameters included shear reinforcement material type, the cross-sectional geometry of CFRP stirrups, and the inclination angle of stirrups with the beam axis. The researchers concluded that onsite manufactured CFRP stirrups were comparable to factory produces stirrups. Spadea et al [9] studied FRP transverse reinforcement made using the filament winding technique to develop fully automated manufacturing methods for concrete structures reinforced with FRP.

The previous studies concluded that the stress concentrations at the bend have a major effect on the shear strength of beams reinforced with FRP stirrups. Therefore, other studies focused on the effect of thickness to- bend radius ratio (r/t) [10-11] and the studies confirmed the previous findings regarding the reduction in the strength of FRP stirrups due to bending process or shaping FRP into stirrup configuration. The reduction in tensile strength at the bend zone is caused by the localized stress concentration due to curvature, which is exaggerated due to the weakness of the fibers in the transverse direction accompanied by the kinking of the fibers in the innermost layer at the bend. To date there have been no research conducted studying the behaviour of GFRP on site manufactured stirrups.

The aim of this paper is to study shear behaviour of R.C beams reinforced with onsite manufactured Glass Fiber-Reinforced Polymer (GFRP) stirrups, which to the knowledge of the authors was not studied to date. The main objective of this research is to study the thickness-to-bend radius ratio (r/t), the effect of the installation of angle stirrups and the type of FRP material.

2. EXPERIMENTAL PROGRAM

2.1 Details of Test Specimens

In total six concrete beams were tested under four-point loading, the beam's dimensions were 2.4 m long, 0.15 m width, and 0.3 m height. The parameters studied in this paper were the effect of stirrup inclination, the effect of FRP material type and the effect of thickness to bend radius ratio (r/t) of the stirrups. The specimens included a control specimen "C" that did not have any transverse reinforcement for comparison.

The specimens were coded (W/X/Y/Z) as show figure 1. Code W indicates stirrups manufactured using glass (G) or carbon (C). Code X indicates width the stirrups in cm, Y indicates number of layers used to manufacture the stirrups and Z indicates angle of installation of the stirrups with the longitudinal bars. In all specimens, the stirrups were constructed by wrapping FRP strips cut to the desired width from commercially available unidirectional fabric directly around the main reinforcement while using four steel stirrups to support the main reinforcement in place, except for specimen (G/1.5/5/90/25) which was wrapped around 25 mm reinforcement bars and then installed on the main reinforcement of the specimen. The stirrups had a spacing of 150 mm for all specimens. Details of the specimens is presented in table1.

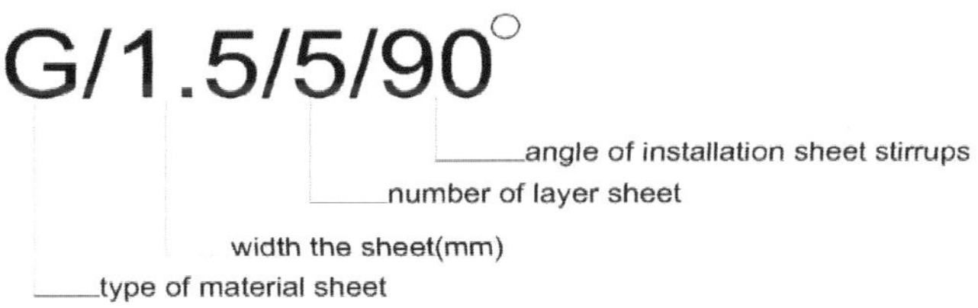

Figure 1. Coding of specimens.

Table 1. Details of all specimens.

	Stirrup material	Width of sturrups (cm)	Number of layers	Stirrup angle	stirrup thickness(mm)	r/t	First crack load (kN)	Failure load (kN)
C	-	-	-	-	-	-	80	140
G/1.5/5/90	Glass	1.5	5	90	2.25	7.11	80	220
G/2.5/3/90	Glass	2.5	3	90	1.85	8.65	80	180
G/1.5/5/60	Glass	1.5	5	60	2.25	7.11	80	240
G/1.5/5/90/25	Glass	1.5	5	90	2.25	11.11	100	240
C/1.5/5/90	Carbon	1.5	5	90	1.95	8.2	80	280

2.2 Test setup and instrumentation

All specimens were tested under 4 point bending, the span was 2.2 m and the distance between loadings points was 0.4 m. The shear span-to-depth ratios was a/d≈3 and was kept constant for the all specimens. A dial gauge was installed in middle of the span to measure deflection. Two strain gauges were installed on the third stirrup from the right side of the beam and the second stirrup from the left side of the beam as show figure 2.

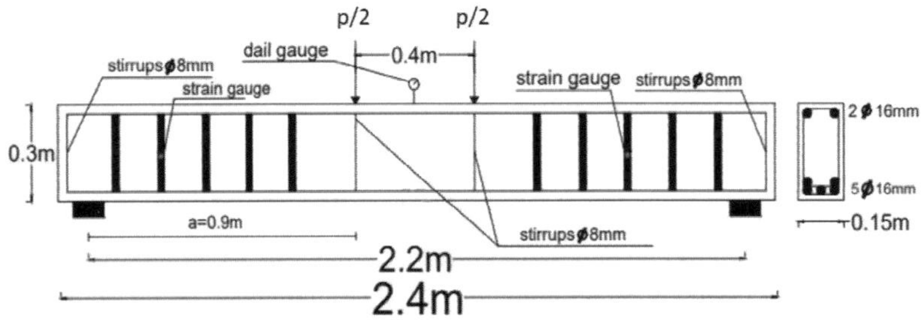

Figure 2. Detail specimens and measurement tools.

2.3 Details of material

5 lower and 2 upper, 16 mm steel bars grade B350DWR [12] were used as the longitudinal reinforcement. The concrete mix was designed to achieve a compressive strength of 35 MPa the mix proportions are provided in Table 2. The sheets used in manufacturing the stirrups were glass fibers "Sika Wrap Hex 430G" and carbon fiber sheets "Sika Wrap Hex 230c" in the form of dry unidirectional flexible fabric. The properties of the CFRP & GFRP sheets as provided by the manufacturer are listed in Table 3. A two components epoxy resin KEMAPOXY 150 (Epoxy)" was used to as the matrix for the FPR stirrups. The flexural strength of the epoxy resin as provided by the manufacturer is (20 – 40 N / mm^2), the compressive strength is (180 – 210 N / mm^2), and the tensile strength is (15 – 25 N / mm^2).

Table 2. Detail of the concrete mix design.

Cement (Kg/m^3)	Water (Liter/m^3)	Coarse aggregate (Kg/m^3)	Fine aggregate (Kg/m^3)
410	193	1160	645

Table 3. Show the properties of CFRP & GFRP sheets.

	Fiber type	Elongation at break %	Density g/cm^3	Fabric thickness mm	Tensile strength of fibers N/mm^2	Modulus of Elasticity Gpa
Sika Wrap Hex 230c	High strength carbon fibers	1.7	1.82	0.129	4000	225
Sika Wrap Hex 430G	glass fibers	2.7	2.56	0.168	2500	70

3. Results and discussion

3.1 Failure mode

All the beams failed in shear tension as a result of the fracture of the stirrups at the corners due to the concentration of stress as shown in figure 5, except for beam (C/1.5/5/90) which failed in flexure due to the yielding of bottom reinforcement and crushing of concrete at the top of the beam in the middle as shown in figure 6.

The mode of failure of specimens (C, G/1.5/5/90 and G/2.5/3/90) was a single large diagonal crack at an angle of about 50^0 . The crack initiated from the support and extended to the load exertion point as show figure 3. The mode of failure of specimens (G/1.5/5/90/25 and G/1.5/5/60) was a double large diagonal crack in addition to small cracks appearing in the middle as shown figure 4.

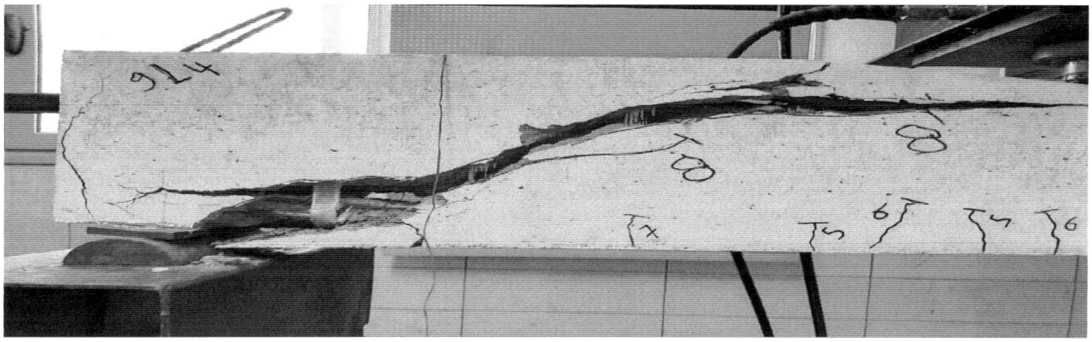

Figure 3. (G/2.5/3/90) mode failure.

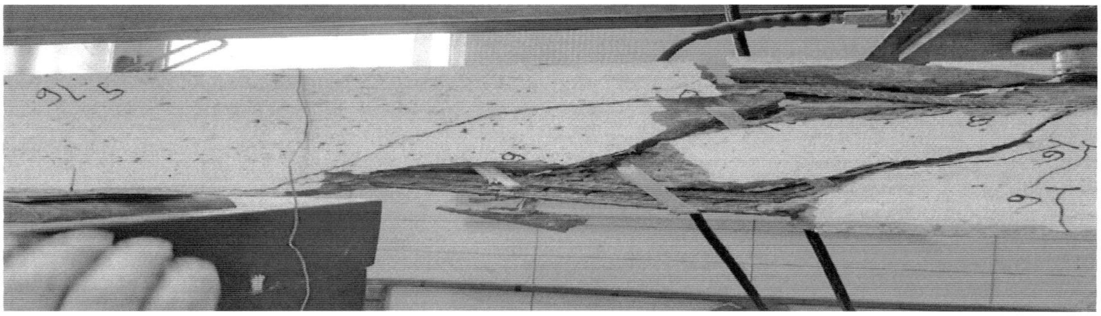

Fig 4. (G/1.5/5/60) mode of failure.

Figure 5. Fracture the sheet from the corners beam (G/1.5/5/90/25).

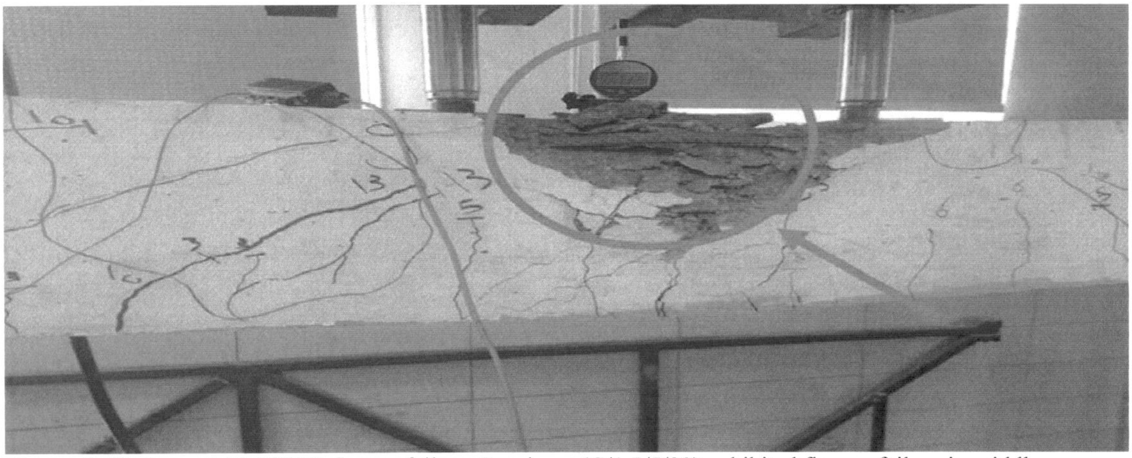

Figure 6. The beam (C/1.5/5/90) flexure failure. Specimen (C/1.5/5/90) exhibited flexure failure in middle.

3.1.1 Effect of inclination stirrups

Comparing specimens (G/1.5/5/90) and (G/1.5/5/60) the specimen with vertical stirrups failed at a load 36% higher than the control beam while the specimen with inclined stirrups failed at a load 41% higher than the control beam. The first crack for the control beam C occurred at a deflection 9.16 mm and the maximum deflection was 12.5mm. The first crack for specimens (G/1.5/5/90) and (G/1.5/5/60) occurred at a deflection 6.97 mm, and 6.5 mm respectively and the maximum deflection was 20.3mm, and 20.2mm respectively. The ultimate deflection in beams (G/1.5/5/90) and (G/1.5/5/60) increased by (38%) compared to the maximum deflection of the control as show in figure 7. The angle of inclination of the stirrups did not significantly affect the stiffness of the beams. The maximum strain for beam (G/1.5/5/90) was 0.0122 mm/mm at a load of 220 kN and for beam (G/1.5/5/60) was 0.014 mm/mm at a load of 240 kN, as shown in figure 9, indicating a slightly higher strain at failure for the inclined stirrups.

3.1.2 Effect of type FRP material

Comparing specimens (G/1.5/5/90) and (C/1.5/5/90) the specimen with fiber glass stirrups failed at a load 36% higher than the control beam while the specimen fiber carbon stirrups failed at a load 50% higher than the control beam, a significantly larger load. However, a direct comparison cannot be made because the mode of failure was different. The first crack in the specimens (G/1.5/5/90) and (C/1.5/5/90) occurred at a deflection of 6.97 mm, and 8.74 mm respectively and the maximum deflection was 20.3mm, and 24.1mm respectively. The ultimate deflection in beams (G/1.5/5/90) increased by (38%) compared to the maximum deflection of the control and (C/1.5/5/90) increased by (48%) compared to the maximum deflection of the control. It could be also observed from figure 7 that specimen C/1.5/5/90 had a slightly higher stiffness compared to specimen G/1.5/5/90. The maximum strain for beam (G/1.5/5/90) was 0.0122 mm/mm at a load of 220 kN and for beam (C/1.5/5/90) was 0.0046 mm/mm at a load of 280 kN, as shown in figure 9. The lower strain in the carbon fiber stirrups is due to the higher stiffness of the carbon fibers compared to the glass fibers. A direct comparison cannot be made due to the difference in failure mode.

3.1.3 Effect of thickness-to- bend radius ratio (r/t)
Comparing specimens (G/1.5/5/90), (G/1.5/5/90/25) and (G/2.5/3/90) specimen (G/1.5/5/90) failed at a load 36% higher than the control beam, specimens (G/1.5/5/90/25) failed at a load 41% higher than the control beam while the specimens (G/2.5/3/90) failed at a load 22% higher than the control beam, indicating for the same area increasing the diameter increases the failure load while decreasing the thickness decreases the failure load. The first crack in the specimens (G/1.5/5/90), (G/1.5/5/90/25) and (G/2.5/3/90) occurred at a deflection of 6.97 mm, 8.5 mm and 3.17 respectively and the maximum deflection was 20.3mm, 19.8mm and 9.15mm respectively. The ultimate deflection in beam (G/1.5/5/90) increased by (38%) compared to the maximum deflection of the control and in beam (G/1.5/5/90/25) increased by (36%) compared to the maximum deflection of the control and in beam (G/2.5/3/90) decreased by (25%) compared to the maximum deflection of the control as show in figure 8. The maximum strain for beam (G/1.5/5/90) was 0.0122 mm/mm at a load of 220 kN and for beam (G/1.5/5/90/25) was 0.01 mm/mm at a load of 240 kN and for beam (G/2.5/3/90) was 0.006 mm/mm at a load of 180 kN. It is clear from figure 10 that increasing the r/t ratio decreases the stress concentration at the bends allowing for higher strain of the stirrups at failure, thus increasing the effectiveness of the FRP material.

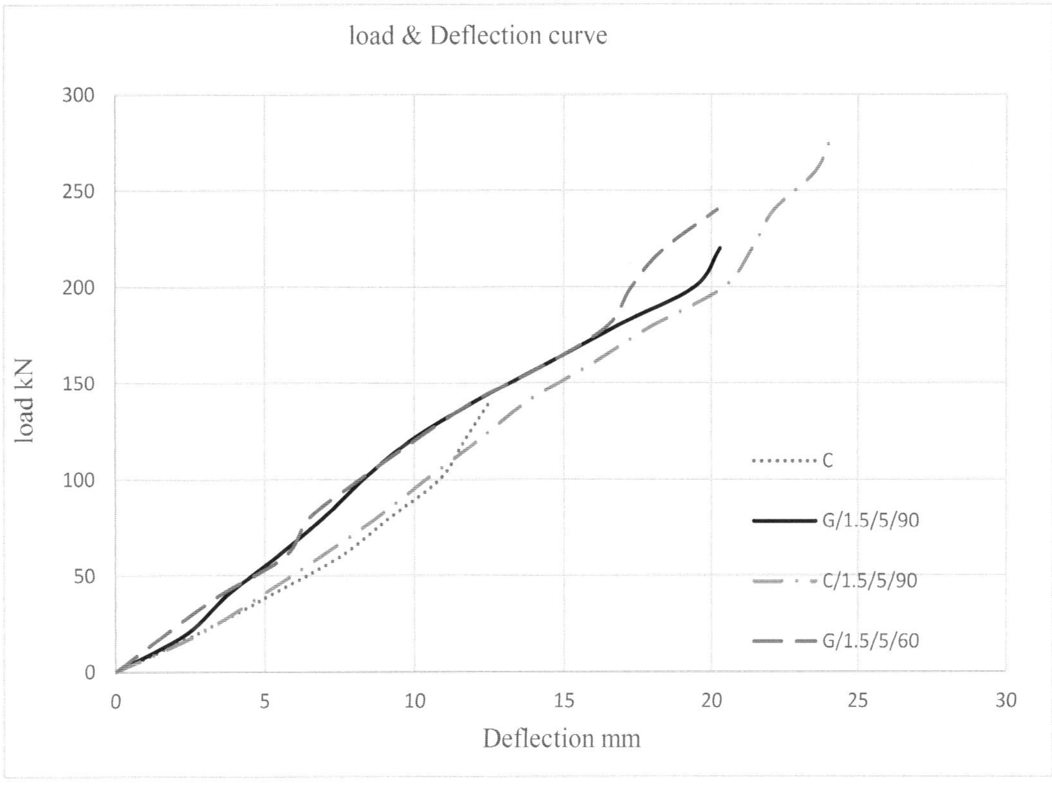

Figure 7. Load & Deflection curve (G/1.5/5/90), (G/1.5/5/60), (C/1.5/5/90) and control beam.

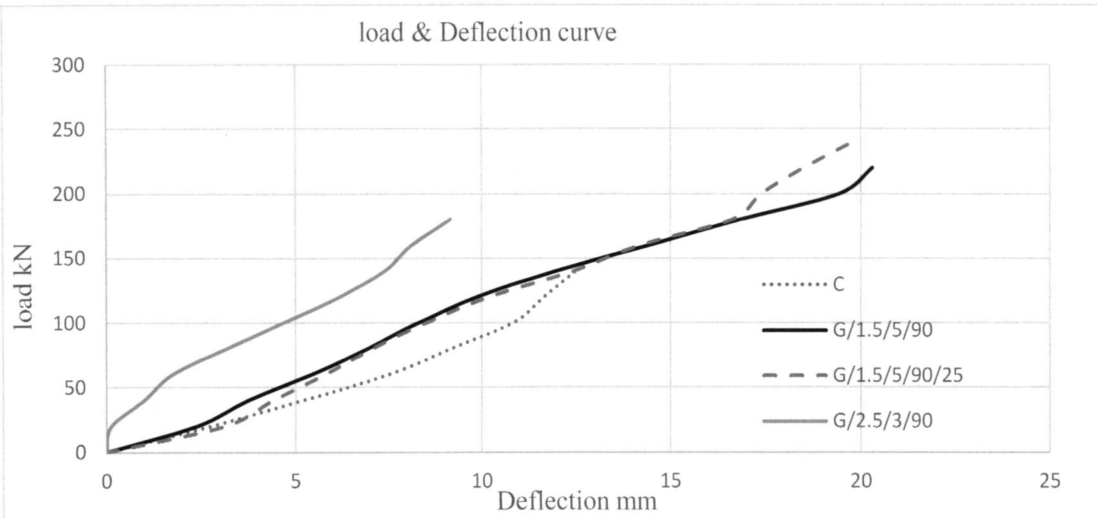

Figure 8. Load & Deflection curve (G/1.5/5/90), (G/1.5/5/90/25), (G/2.5/3/90) and control beam

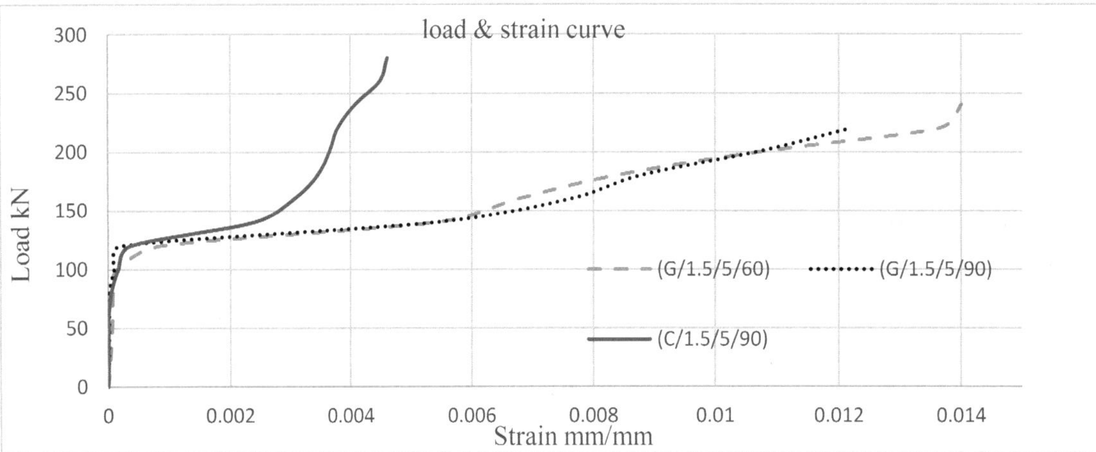

Figure 9. Load & strain curve (G/1.5/5/90), (G/1.5/5/60) and (C/1.5/5/90).

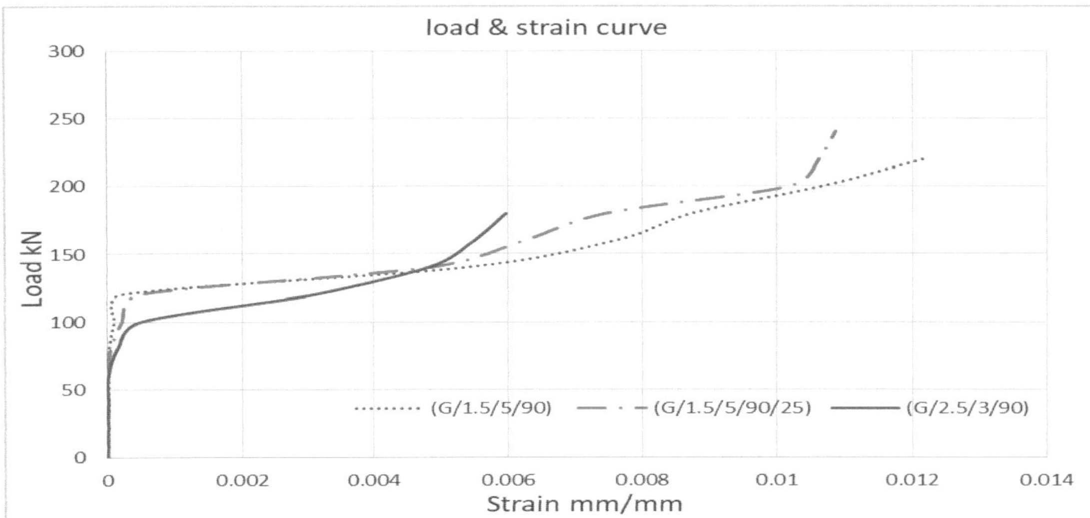

Figure 10. Load strain curve between (G/1.5/5/90), (G/1.5/5/90/25), and (G/2.5/3/90).

Conclusions

The present study showed that replacing steel stirrups with hand-made GFRP stirrups is feasible. According to the behavioural observations and obtained results of experimental of six reinforced concrete beam, it can be concluded that the ultimate strength of beams reinforced with carbon fiber stirrups is significantly higher than that of beams reinforced with fiber glass stirrups for the same area. For the same area, increasing the diameter increases the failure load and strains in the stirrups while decreasing the thickness decreases the failure load and strains in stirrups, due to the reduction of the stress concentration at the corners of the stirrups. Finally, increasing the inclination angle of the stirrups with the horizontal increases the strain in the stirrups and ultimate strength

References

[1] Fakhreddine Djeddi, Youcef Ghernouti, Yazid Abdelaziz and Li Alex. Strengthening in flexure–shear of RC beams with hybrid FRP systems: Experiments and numerical modelling. Journal of Reinforced Plastics and Composites 2016, Vol. 35(22) 1642–1660.
[2] C. Lee; S. Lee; and S. Shin. Shear Capacity of RC Beams with Carbon Fiber-Reinforced Polymer Stirrups with Rectangular Section. American Society of Civil Engineers 2015.
[3] A. Sivandi-Pour, A. Dalvand. Experimental Assessment of Using Composite Ties in Shear Behavior of Reinforced Concrete Beams. Iranian Journal of Structural Engineering 2017 4(1) 29-37.
[4] El-Sayed AK, El-Salakawy E, Benmokrane B. Mechanical and structural characterization of new carbon FRP stirrups for concrete members. J Compos Constr 2007; 11(4):352–62.
[5] Mohsen A. Issa, F.ASCE; Thilan Ovitigala, Ph.D., P.E., M.ASCE; and Mustapha Ibrahim. Shear Behavior of Basalt Fiber Reinforced Concrete Beams with and without Basalt FRP Stirrups. Journal of composite for construction volume 20 Issue 4, 2016.
[6] C. Lee; S. Lee; and S. Shin. Shear Capacity of RC Beams with Carbon Fiber-Reinforced Polymer Stirrups with Rectangular Section. American Society of Civil Engineers 2015.
[7] Ye Yuan, Zhenyu Wang. Shear behavior of large-scale concrete beams reinforced with CFRP bars and handmade strip stirrups. Composite Structures 227 (2019) 111253.
[8] Muhammad Tahira, Zhenyu Wanga, Kanwar Majid Alia, Haytham F. Isleem. Shear behavior of concrete beams reinforced with CFRP sheet strip stirrups using wet-layup technique. 2352-0124/ © 2019 Institution of Structural Engineers. Published by Elsevier.
[9] Saverio Spadea, Ph.D; John Orr, Ph.D., C.Eng; Antonio Nanni, Ph.D., P.E., F.ASCE; and Yuanzhang Yang. Wound FRP Shear Reinforcement for Concrete Structures. American Society of Civil Engineers 2017.
[10] Lee C, Ko M, Lee Y. Bend strength of complete closed-type carbon fiber-reinforced polymer stirrups with rectangular section. J Compos Constr 2013; 18:4013022.
[11] Ehab A. Ahmed1 ; Ahmed K. El-Sayed2 ; Ehab El-Salakawy3 ; and Brahim Benmokrane. Bend Strength of FRP Stirrups: Comparison and Evaluation of Testing Methods. J. Compos. Constr., 2010, 14(1): 3-10.
[12] Egyptian code of practice for design and construction of concrete structures. No, ECP 203-2020.

Parametric Studies on Dynamic Analysis of Blast-Loaded Retrofitted RC Columns

Mahmoud K. Ameen[1], Fouad B. A. Beshara[2], and Youssef M. H. Hammad[2]

[1] Engineering Authority for Egyptian Armed Forces
[2] Civil Engineering Department, Faculty of Engineering (Shobra), Banha University

[*] Corresponding Author Email: mahkhayri@gmail.com

Abstract. The majority of public and private facilities, existing today, were not originally designed to withstand blast loads rising from terrorist attacks. Hence, such facilities require effective countermeasures to minimize losses inflicted on property and people in the event of terrorist bombing. Retrofitting of main structural elements is very effective to increase blast resistance. In this paper, comprehensive parametric studies for the dynamic response of blast-loaded RC columns are performed using the proposed nonlinear coupled model and AUTODYN software. Many material and reinforcement parameters are considered in the paper such as concrete compressive strength, steel yield stress, longitudinal steel ratio and transverse stirrups ratio. Blast-loading parameters such as using different explosive weights, changing the stand-off distance of explosion, are also studied. Finally, the retrofitting parameters for blast-loaded columns, such as using different retrofit material types like the Carbon Fiber Reinforcing Polymers (CFRP), Glass Fiber Reinforcing Polymers (GFRP) and Steel Reinforcing Polymers (SRP) with different thicknesses for each material type, are considered. Important conclusions are drawn for the sensitivity of dynamic response and failure modes of RC columns under blast loading.

Keywords: Blast loading, Reinforced concrete columns, Finite element modelling, Dynamic nonlinear analysis, AUTODYN, Parametric Studies.

1. Introduction

Hazard assessments are continuously updated and increased blast resistance may be desired for specific structures. Implementing a standoff distance between a building and a potential hazard is not always possible, especially in urban environments. A different approach proposes to retrofit existing buildings to increase their blast resistance. Several experimental and theoretical studies [1-7] are reported for the response of as-built and retrofitted RC columns under blast loading. A complete literature survey is given in [1] which indicates that retrofitting with advanced composite materials, enhances significantly the lateral resistance of columns subjected to blast loading.

In [1], the details of modelling and validation studies are given for nonlinear dynamic analysis of blast-loaded retrofitted RC columns. A three-dimensional solid element and a connecting linear element are used for spatial discretization of concrete and steel reinforcement, respectively. A strain rate- and history-dependent behavioral model is proposed for the progressive failure of concrete under explosion. The proposed model is an improved version of RHT model and takes into account the different material non-linearities and strain rate sensitivity in compression and tension. Steel is modeled as a rate-dependent elastoplastic material with strain hardening. For the retrofitting layers, the FRP and SRP composite material in the layered solid element is modeled as a linear elastic orthotropic material. The proposed model has been implemented in the AUTODYN program [8]. Several validation and sensitivity studies have been successfully performed [1] for retrofitted blast-loaded RC columns.

This paper is concerned with parametric studies on the performance of RC columns under blast loading. The parametric studies were held on a concrete column specimen, shown in Figure 1, and previously tested by Crawford et al. [5]. It was numerically modelled in program AUTODYN, to validate his results [1]. Accordingly, due to the good agreement between both the experimental results and our numerical results as shown in Figure 2, the Crawford's specimen is kept as the reference column

C1 for the parametric studies. The load-time function used in the computer analysis is the triangular function with duration of 0.2 msec. Many material and reinforcement parameters are considered in the paper such as. concrete compressive strength, steel yield stress, longitudinal steel ratio and transverse stirrups ratio. Blast-loading parameters such as using different explosive weights, changing the stand-off distance of explosion, are also studied. Finally, the retrofitting parameters for blast-loaded columns, such as using different retrofit material types like CFRP, GFRP and SRP with different thicknesses for each material type, are considered.

2. Material Parameters

2.1 Effect of Concrete Compressive Strength

To study the effect of concrete compressive strength on the dynamic response of blast-columns, three RC columns with different compressive strengths were considered and labelled as C1, C2 and C3. The compressive strength for the three columns was respectively 31.2 MPa, 25 MPa, and 37.4 MPa. The displacement-time history is shown in Figure 3 for the three different cases. The solid-line curve without markers represents the original validated case C1, the curve with circular markers represents the column C2 with 25% strength less, while the triangle marked curve represents the column C3 with 20% strength higher. The original case C1 reached a maximum displacement of 362 mm and the residual deflection is 250 mm. For column C2, the maximum deflection and residual displacement are 384 mm and 265 mm. For column C3, the maximum displacement and residual deflection are respectively 344 mm and 238 mm. Compared with C1, the maximum displacement ratio is 1.06 for C2, and 0.95 for C3. The little effect of concrete compressive strength on displacement is due to the fact that the flexural resistance of RC column under lateral blast loading is slightly affected by concrete strength.

The damage shapes of different columns are shown in Figure 4 at the end of dynamic analysis. For the original column C1, it is obvious that the upper half of the column consists of many concrete deteriorated parts, the concrete is almost crushed near both the upper and lower supports. For column C2, the concrete is also deteriorated near the ends of the column and some parts are entirely crushed and the steel cage is completely seen clearly. The column side facing the blast become less damaged due to increasing the compressive strength. Also, some damaged parts are noticed in the middle of the upper half of the column. For column C3, the deteriorated parts are noticed to be decreased specially near the supports, some crushed concrete is seen but far less than the last two cases.

2.2 Effect of Yield Stress of Steel

The effect of steel yield stress on the dynamic response of blast-columns was illustrated by three RC columns with different yield stress levels. These columns were labelled as C1, C4 and C5. The yield stress for the three columns was respectively 420 MPa, 360 MPa, and 500 MPa. The different steel yield stresses give different displacement-time history curves which are shown in Figure 5. The original validated case 420 MPa is represented by the solid line, the 360 MPa case is represented by the curve with circular markers, while the 500 MPa case is represented by the triangle marked curve. The original case C1 reached a maximum displacement of 362 mm and the residual deflection is 250 mm. A maximum displacement and residual deflection of 387 mm and 268 mm respectively, was experienced by column C4, while a maximum displacement of 326 mm, and residual deflection of 225mm is reached by column C5. This reflects that the C4 case reached a 7% higher displacement, while C5 reached a lower displacement by 10%. The moderate effect of steel yield stress on displacement is due to the fact that the flexural resistance of RC column under lateral blast loading is moderately affected by the longitudinal steel reinforcement.

At the end of dynamic analysis, the damage shapes of different columns are shown in Figure 6. For the original column C1, the upper half of the column consists of many concrete deteriorated parts, the concrete is almost crushed near both the upper and lower supports. For column C4, the concrete is also deteriorated near the ends of the column and some parts are entirely crushed and the steel cage is completely seen clearly. As for the steel cage it was bent and had a slight deflection in both the longitudinal and transverse steel bars but with no breakage. The column side facing the blast become more damaged due to decreasing the steel yield stress. Also, some damaged parts are noticed in the middle of the upper half of the column. For column C5, the deteriorated parts are noticed to be decreased specially near the supports, some crushed concrete is seen but far less than the last two cases.

3. Reinforcement Parameters

3.1 Effect of Steel Longitudinal Ratio

For dynamic analysis under blast, three columns were considered and labelled as C1, C6 and C7. The longitudinal steel ratio for the three columns was respectively 1.1%, 1.5%, and 2%. Figure 7 shows the displacement-time history for the three different cases of longitudinal steel ratio. The solid-line curve without markers represents the original validated case C1, the curve with circular markers represents the column C6 with 36% strength higher, while the triangle marked curve represents the column C7 with 81.8% strength higher. The original case C1 reached a maximum displacement of 362 mm and the residual deflection is 250 mm. For column C6, the maximum deflection and residual displacement are 326 mm and 225 mm. For column C7, the maximum displacement and residual deflection are respectively 286 mm and 198 mm. Compared with C1, the maximum displacement ratio is 0.9 for C6, and 0.79 for C7. It was noticed that in maximum displacement column C6 reached a decrease of 10.2% while C7 was lower by 26.6%. The significant effect of longitudinal steel ratio on displacement is due to the fact that the flexural resistance of RC column under lateral blast loading is highly affected by the longitudinal steel reinforcement.

The damage shapes of different columns are shown in Figure 8. At the end of dynamic analysis, it was obvious that by increasing the longitudinal steel ratio, the experienced damage by the column specimen was decreased. For the original column C1, it is obvious that the upper half of the column consists of many concrete deteriorated parts, the concrete is almost crushed near both the upper and lower supports and the steel cage was exposed. For column C6, the concrete is also deteriorated near the ends of the column and some parts are entirely crushed and the steel cage is completely seen clearly. The column side facing the blast become less damaged due to increasing the steel reinforcement. Also, some damaged parts are noticed in the middle of the upper half of the column. For column C7, the deteriorated parts are noticed to be decreased specially near the supports, some crushed concrete is seen but far less than the last two cases.

3.2 Effect of Transverse Stirrups Ratio

Three RC columns with different transverse steel ratios were considered and labelled as C1, C8 and C9. The steel stirrup ratio (ρ) for the three columns was respectively 0.28%, 0.15%, and 0.3%. The displacement-time history is shown in Figure 9 for the three different cases of transverse steel ratio. The solid-line curve without markers represents the original validated case C1, the curve with circular markers represents the column C8 with 26% stirrups ratio less, while the triangle marked curve represents the column C9 with 59% stirrups ratio higher. The original case C1 reached a maximum displacement of 362 mm and the residual deflection is 250 mm. For column C8, the maximum deflection and residual displacement are 380 mm and 263 mm. For column C9, the maximum displacement and residual deflection are respectively 300 mm and 205 mm. Compared with C1, the maximum displacement ratio is 1.05 for C8 and 0.82 for C9. An increase of 4.97% for C8 and a decrease of 17% for C9, was observed at the maximum displacement. The positive effect of steel stirrups ratio on displacement is due to the fact that the shear resistance of RC column under lateral blast loading is quite affected by the transverse steel reinforcement.

Figure 10 shows the damage shapes of different columns at the end of dynamic analysis. For the original column C1, it is obvious that the upper and lower half of the column consists of many concrete deteriorated parts, the concrete is almost crushed near the supports. For column C8, the concrete is also deteriorated near the ends of the column and some parts are entirely crushed and the steel cage is completely seen clearly specially in the lower half, casing most of the deflection. The column side facing the blast is more damaged due to decreasing the stirrups reinforcement. Also, some damaged parts are noticed in the middle of the column. For column C9, the deteriorated parts are noticed to be decreased specially near the supports, some crushed concrete is seen but far less than the last two cases.

4. Blast Loading Parameters

4.1 Effect of Explosive Weight

To study the effect of Explosive weight on the dynamic response, three RC columns were considered and labelled as C1, C10 and C11. The explosive weight for the three columns was respectively 750Kg, 1000Kg, and 350Kg. For all cases, the Stand-off distance was constant, and equals 4.5 m. Figure 11 shows the displacement-time history curves under three different explosive weights. The solid-line curve without markers represents the original validated case C1, the curve with circular markers represents the column C10 with 33% weight higher, while the triangle marked curve represents the column C11 with 47% weight less. The original case C1 reached a maximum displacement of 362 mm and residual deflection is 250 mm. For column C10, the maximum deflection and residual displacement are 488 mm and 338 mm. For column C11, the maximum displacement and residual deflection are respectively 139 mm and 95 mm, making an increase in maximum displacement of 35% for column C10, and a decrease of 61.6% for C11. Compared with C1, maximum displacement ratio is 1.35 for C10 and 0.38 for C11.

The predicted damaged state for the three columns at the end of analysis is shown in Figure 12. For the original column C1, it is obvious that the upper half of the column consists of many concrete deteriorated parts, the concrete is almost crushed near both the upper and lower supports. For column C10, the concrete is massively damaged at the upper half and the concrete was blown off, the steel cage is badly deformed, as both the longitudinal and transverse steel. The column side facing the blast become more damaged due to increasing the explosive weight. For column C11, the deteriorated parts are noticed to be negligible specially near the supports, minor cracks spreading in concrete is seen, the longitudinal steel bars is found to be slightly bent, with no considerable damage to the transverse steel with no breakages in the steel cage, the overall damage is far less than the last two cases.

4.2 Effect of Stand-off Distance

To study the effect of the Stand-off distance on the dynamic response, three RC columns were considered and labelled as C1, C12 and C13. The Stand-off distance of the simulated exploding charges for the three columns was respectively 4.5 m, 5 m, and 4 m. For the three cases, the equivalent TNT weight was constant and equals 750 kg. The displacement-time history is shown in Figure 13 for three different cases of Stand-off distance. The solid-line curve without markers represents the original validated case C1, the curve with circular markers represents the column C12 with 22% longer Stand-off distance, while the triangle marked curve represents the column C13 with 22% shorter Stand-off distance. The original case C1 reached a maximum displacement of 362 mm and the residual deflection is 250 mm. For column C12, the maximum deflection and residual displacement are 284 mm and 195 mm. For column C13, the maximum displacement and residual deflection are respectively 445 mm and 308 mm, a decrease of 21.5% for C12 and an increase of 22.9% for C13, was observed at the maximum displacement. Compared with C1, the maximum displacement ratio is 0.78 for C12 and 1.23 for C13. The significant effect of the Stand-off distance on displacement is due to the fact that the resulted pressure and impulse of blast is highly affected by the distance of the explosive from the column.

Figure 14 illustrates the different damage shapes of tested columns at the end of dynamic analysis. For the original column C1, it is obvious that there is a considerable lateral deflection in the column at mid-height, the upper half of the column consists of many concrete deteriorated parts, and the concrete is almost crushed near both the upper and lower supports. As for the reinforced steel, the steel cage was found to be bent considerable with no breakage found. For column C13, the concrete is also deteriorated near the ends of the column and some parts near the lower support are crushed and the steel cage is completely seen clearly. The column side facing the blast becomes less damaged due to decreasing the blast pressure. But the overall damage is far less that of C1 column. For column C12, the deteriorated parts are noticed to be decreased specially near the supports, some crushed concrete is seen but far less than the last two cases.

5. Retrofit Material Parameters

5.1 Effect of CFRP Thickness

Three RC columns with different retrofitting material thickness were considered and labelled as C1, C14 and C15. The thickness of CFRP on the three columns was respectively 0 mm, 3 mm, and 6 mm. The displacement-time history is shown in Figure 15 for three different CFRP thicknesses. The solid-line curve without markers represents the original validated case C1, the curve with circular markers represents the column C14 with 3mm thick CFRP, while the triangle marked curve represents the column C15 with 100% thicker CFRP. The original case C1 reached a maximum displacement of 362 mm and the residual deflection is 250 mm. For column C14, the maximum deflection and residual displacement are respectively, 224 mm and 155 mm. For column C15, the maximum displacement and residual deflection are respectively 145 mm and 100 mm. Compared with C1, C14 column experienced a decreased percentage of 38% while C15 experienced a decreased percentage of 60%, compared to the reference case of C1 column. In case of comparing maximum displacement, the ratio is 0.62 for C14 and 0.4 for C15. The significant effect of CFRP thickness on displacement is due to the fact that flexural resistance of RC column under lateral blast loading is significantly affected by thickness of CFRP.

At the end of dynamic analysis, the damage shapes of different columns are shown in Figure 16. For the original column C1, the upper half of the column consists of many concrete decomposed parts, the concrete is crushed near both the upper and lower supports. For column C14, damage to concrete is almost negligible, due to the CFRP protection to the concrete, there is a partial tear of the retrofitting material on the side facing the blast. Also, some damaged was noticed in the reinforcement steel, as some bars and ties where bent, but no breakage was found. For column C15, the damage to CFRP material was greatly reduced, only a very slight distortion was found. The concrete is seen to sustain no damage.

5.2 Effect of GFRP Thickness

The dynamic response of blast-loaded GFRP-retrofitted RC columns is significantly affected by the thickness of wrapping layers. Three RC columns with different GFRP thicknesses were considered. These columns were labelled as C1, C16 and C17. The thickness of GFRP on the three columns was respectively 0 mm, 3 mm, and 6 mm. The displacement-time history is shown in Figure 17 for the three different cases of GFRP thicknesses. The solid-line curve without markers represents the original validated case C1, the curve with circular markers represents the column C16 with 3 mm thick GFRP, while the triangle marked curve represents the column C17 with 100% thicker GFRP. As mentioned before, the original case C1 reached a maximum displacement of 362 mm and the residual deflection is 250 mm. For column C16, the maximum deflection and residual displacement are 253 mm and 175 mm. For column C17, the maximum displacement and residual deflection are respectively 174 mm and 121 mm. It was calculated that C16 column experienced a decreased percentage of 30% while C17 experienced a decreased percentage of 52%, compared to the refinance case of C1 column. Compared with C1, the maximum displacement ratio is 0.7 for C16 and 0.48 for C17. The significant effect of

GFRP thickness on displacement is due to the fact that the flexural resistance of RC column under lateral blast loading is significantly affected by the thickness of GFRP.

Figure 18 indicates the shape of the damage sustained by the columns at the end of dynamic analysis. For the original column C1, the upper half of the column consists of many concrete decomposed parts, the concrete is crushed near both the upper and lower supports. For column C16, damage to concrete is significantly decreased, due to the GFRP protection to the concrete, there is a tear of the retrofitting material on the other side of the blast, and the concrete surface was exposed and had some cracks as it sustained the damage. Also, some damage was noticed in the reinforcement steel in the form of steel bars and ties where bent, but no breakage was found. For column C17, the damage to the GFRP material was greatly reduced, only a very slight tear was found. The concrete is seen to sustain the damage but the concrete surface had minor cracks.

5.3 Effect of SRP Thickness

In order to study the effect of retrofitting the column by SRP wrapping layers, three RC columns with different SRP thicknesses were considered and labelled as C1, C18 and C19. The thickness of SRP was respectively 0 mm, 3 mm, and 6 mm. The thickness was doubled for specimen C19 compared to C18. For the three different cases of SRP retrofitting cases, the displacement-time history is shown in Figure 19. The solid-line curve without markers represents the original validated case C1, the curve with circular markers represents the column C18 with 3 mm thick SRP, while the triangle marked curve represents the column C19 with 100% thicker SRP. As mentioned before, the original case C1 reached a maximum displacement of 362 mm and residual deflection is 250 mm. For column C18, the maximum deflection and residual displacement are 235 mm and 153 mm. For column C19, the maximum displacement and residual deflection are respectively 163 mm and 113 mm, resulting the following enhancement to the columns resistance, as the maximum displacement the column experienced decreased by 35% for C18 while C19 experienced a decreased percentage of 55%, all is compared to the refinance case of C1 column. Compared with C1, the maximum displacement ratio is 0.65 for C18 and 0.45 for C19. The significant effect of SRP thickness on displacement is due to the fact that flexural resistance of RC column under lateral blast loading is significantly affected by the thickness of SRP.

Figure 20 shows the damage shapes of different columns at the end of dynamic analysis. For the original column C1, it is obvious that the upper half of the column consists of many concrete deteriorated parts, the concrete is almost crushed near both the upper and lower supports. For column C18, damage to concrete is almost negligible, due to the SRP protection to the concrete, there is a partial tear of the retrofitting material on the side facing the blast. Also, some damaged was noticed in the reinforcement steel, as some bars and ties where bent, but no breakage was found. For column C19, the damage to the SRP material was significantly reduced, only a very slight distortion was found. The concrete is seen to sustain some damage. An overall conclusion is that, as the thickness of the SRP cover increases, the maximum mid-height displacement decreases. In addition, the damaged concrete parts are nearly negligible, and the concrete of the column side facing the blast also experience almost no damage, while the SRP absorbed all the damage. Also, it was noticed that the damage dealt to the SRP wrapping is greatly reduced as the thickness increased.

6. Conclusions

From the comprehensive parametric studies for the dynamic response of blast-loaded RC columns, the following major points are concluded:

- For RC columns subjected to blast loading, the displacement at mid-height, increases significantly with the increase of explosive weight (W_{exp}), and with the decrease of stand-off distance (SOD). Increasing W_{exp} leads to the concrete being blown of, and a significant deformation to steel cage. For smaller SOD, the steel cage was found to be bent considerably.

- For blast-loaded RC columns, the mid-height displacement decreases moderately with the increase of concrete compression strength and steel yield stress. In addition, the damage and concrete deterioration all over the specimen decrease due to the enhancement of flexural strength and stiffness.
- For blast-loaded columns, the critical displacement decreases moderately with the increase of longitudinal steel ratio, and stirrups ratio. Also, the concrete deterioration was decreased slightly especially at the middle of the column for the longitudinal steel ratio increasing case, and near the supports for the transverse steel ratio increasing case.
- The dynamic response of blast-loaded retrofitted RC columns decreases with the increase of thickness of retrofit layers of advanced composite materials. Also, the concrete damage of retrofitted column decreases significantly with nearly no deterioration. As well, increasing the thickness of FRP materials, decreases slightly the partial tearing damage.

References

[1] Ameen M. K., "Dynamic Analysis of Retrofitted RC Columns under Blast Loadings", Banha University, Shoubra Faculty of Engineering, Dep. of Civil Engineering, PhD. Thesis, Cairo, Egypt 2019.

[2] Carriere M. D., "Steel Reinforced Polymer Strengthening of Reinforced Concrete to Resist Blast Loads", M.Sc. Thesis, Faculty of the Royal Military College of Canada. 2006.

[3] Berger J. O., Heffernan P. J., and Wight R. G., "Blast Testing of CFRP and SRP Strengthened RC Columns", WIT Trans. Built Environ, Vol. 98, pp. 95–104, 2008.

[4] Vappera, M., Lasn, K., "Blast protection of concrete columns with thin strips of GFRP overlay", El-Seiver, Structures, V.25, PP. 491-499, 2020.

[5] Kadhom, B., " Blast Performance of Reinforced Concrete Columns Protected by FRP Laminates", PhD Thesis, University of Ottawa, Canada, 2016.

[6] Jacques E., Lloyd A., Imbeau P., Palermo D., and Quek J., "GFRP-Retrofitted Reinforced Concrete Columns Subjected to simulated Blast Loading", J. Struct. Eng., Vol. 141, no. 11, pp. 1–13, USA, 2014.

[7] Crawford J. E., Malvar L. J., Morrill K. B., and FerrittoJ. M., "Composite Retrofits to Increase the Blast Resistance of Reinforced Concrete Buildings", the Tenth International Symposium on Interaction of the Effects of Munitions with Structures, Vol. 10, pp. 1–25, 2001.

[8] Century Dynamics, AUTODYN User and Theory Manual, 2016.

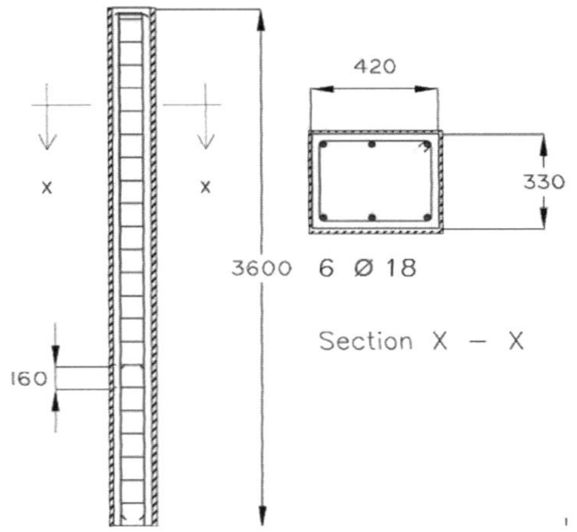

Figure 1. Details of Rectangular Column Specimen [7].

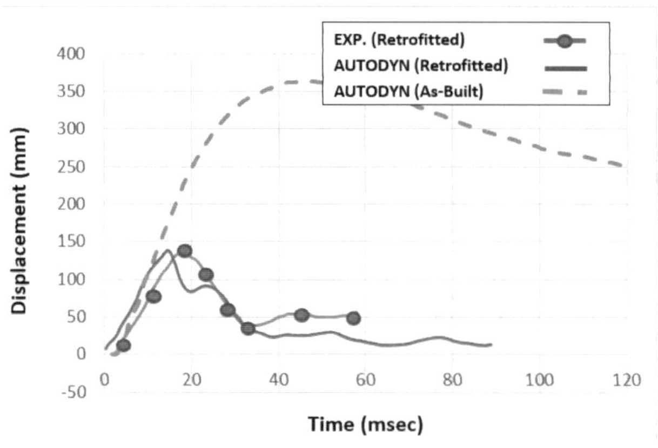

Figure 2. Mid-Height Displacement-Time History for Test for Rectangular Column.

IOP Conf. Series: Earth and Environmental Science 1056 (2022) 012019 doi:10.1088/1755-1315/1056/1/012019

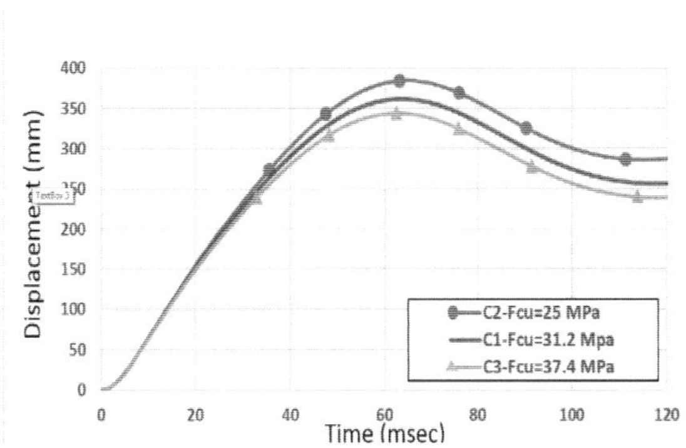

Figure 3. Displacement – time history for different concrete compressive strengths.

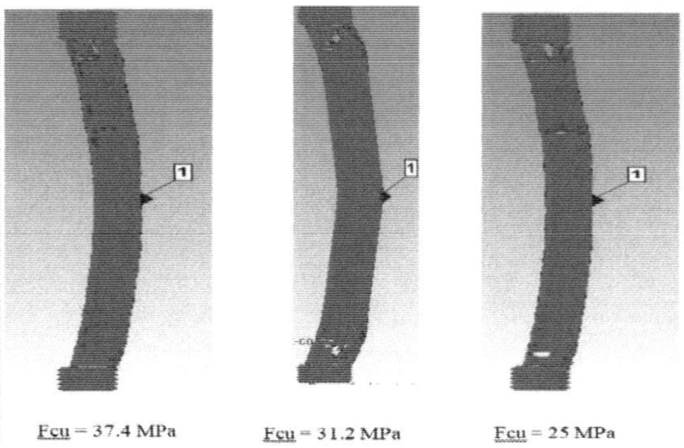

Figure 4. Damaged shape columns for different concrete compressive strengths.

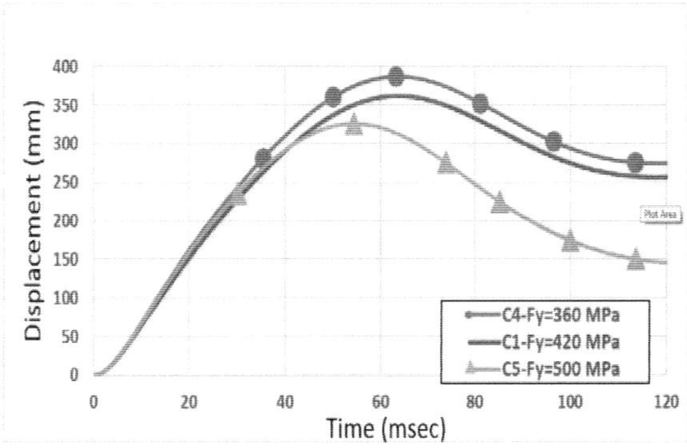

Figure 5. Displacement – time history for different steel yield strengths.

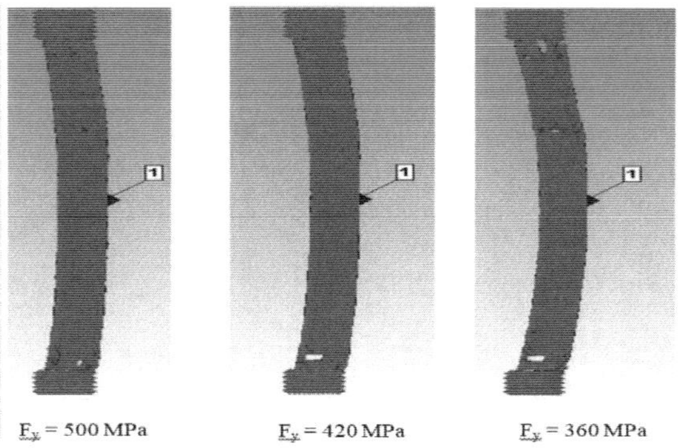

Figure 6. Damaged shape columns for different steel yield stresses.

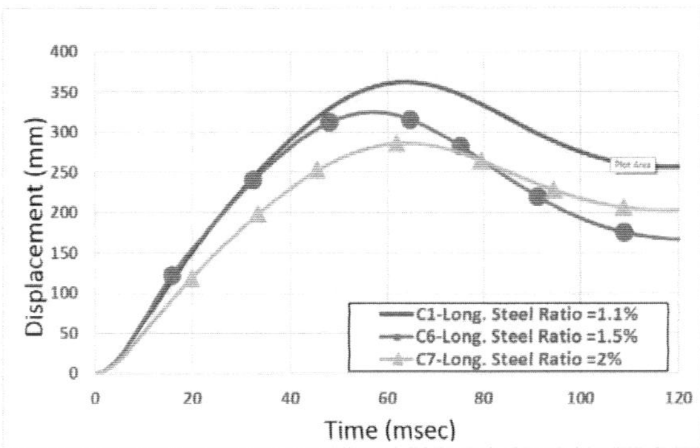

Figure 7. Displacement – time history for different longitudinal steel ratios.

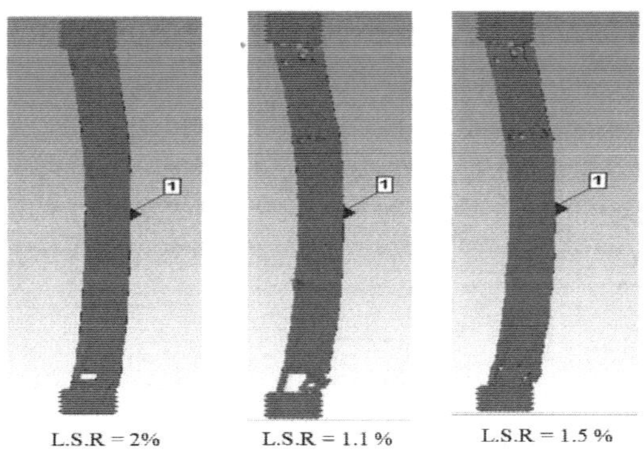

Figure 8. Damaged shape columns for different longitudinal steel ratios.

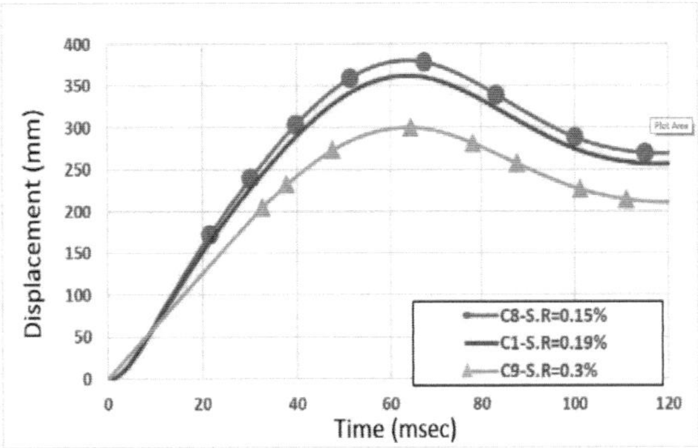

Figure 9. Displacement -time history for different transverse steel ratios.

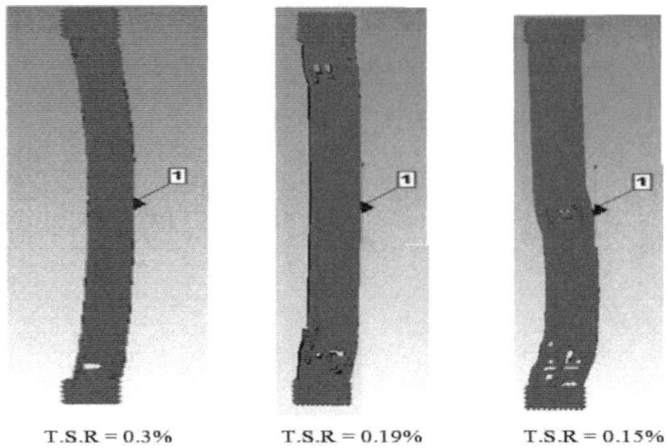

Figure 10. Damaged shape columns for different transverse steel ratios.

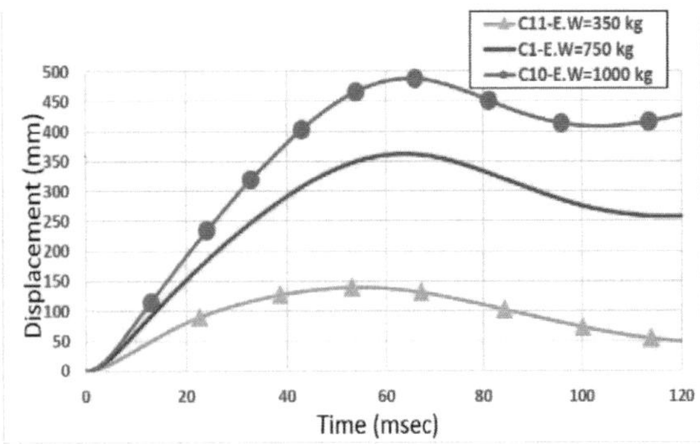

Figure 11. Displacement – time history for different explosive weights.

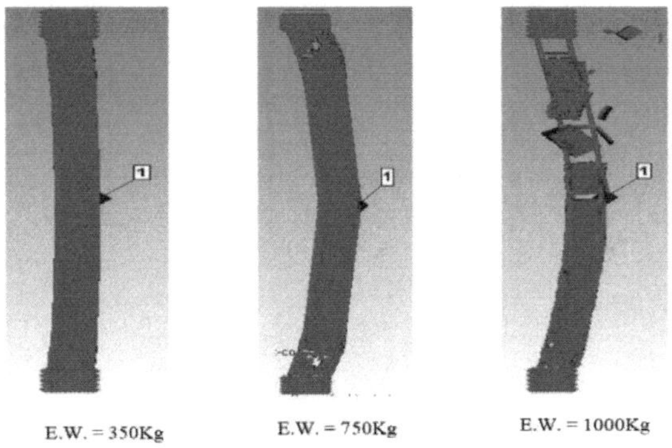

Figure 12. Damaged shape columns for different explosive weights.

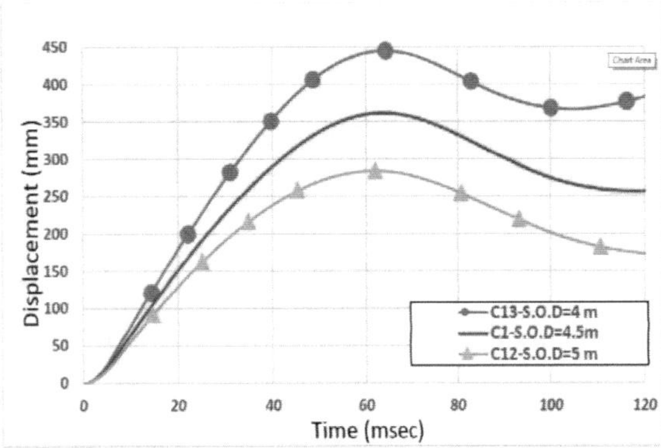

Figure 13. Displacement – time history for different stand-off .distances

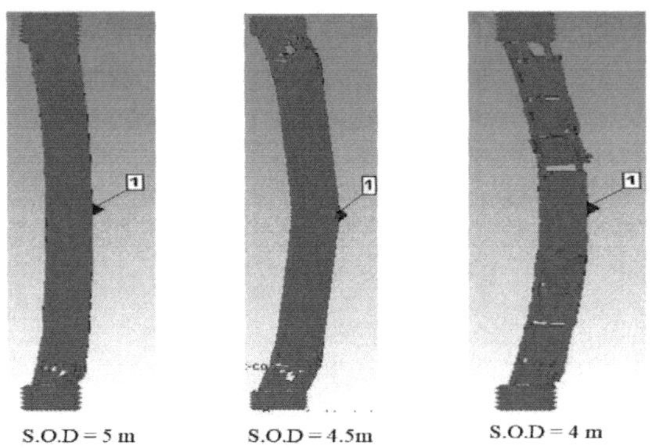

Figure 14. Damaged shape columns for different stand-off distances.

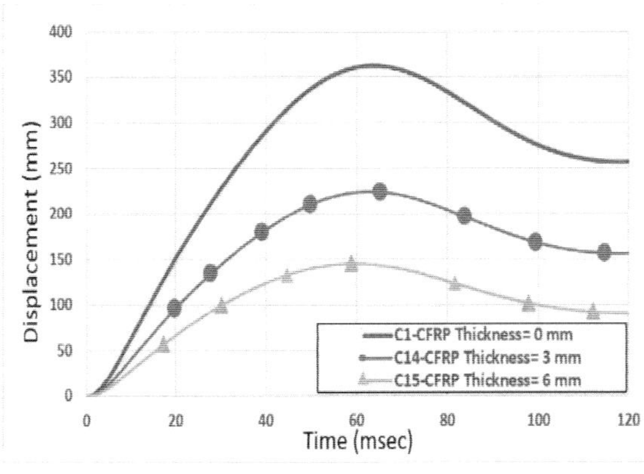

Figure 15. Displacement – time history for different CFRP. thicknesses

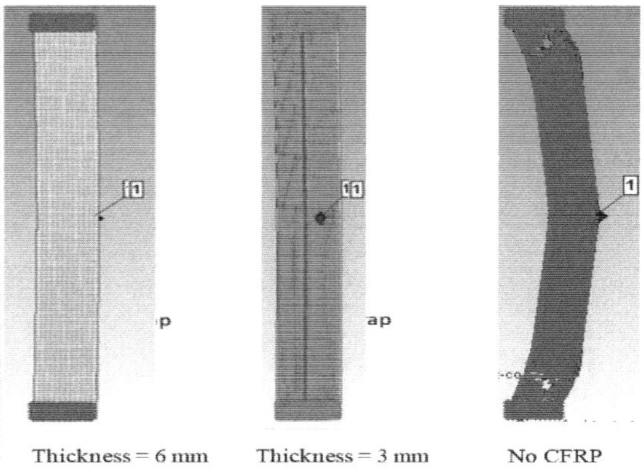

Figure 16. Damaged shape columns for different CFRP thicknesses.

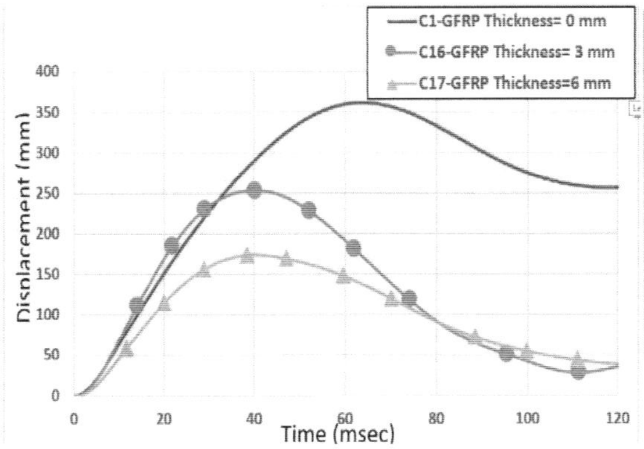

Figure 17. Displacement – time history for different GFRP thicknesses.

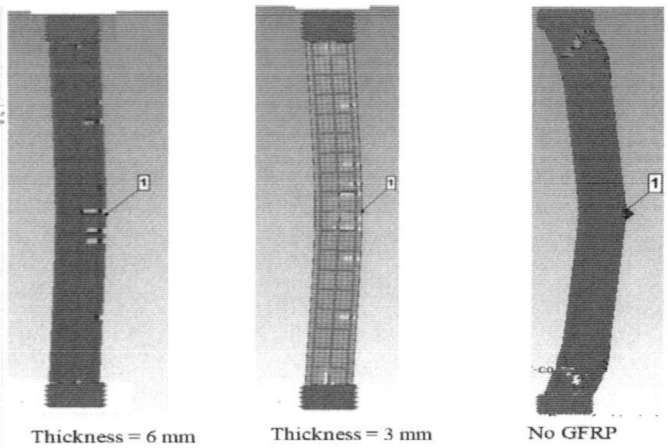

Figure 18. Damaged shape columns for different GFRP thicknesses.

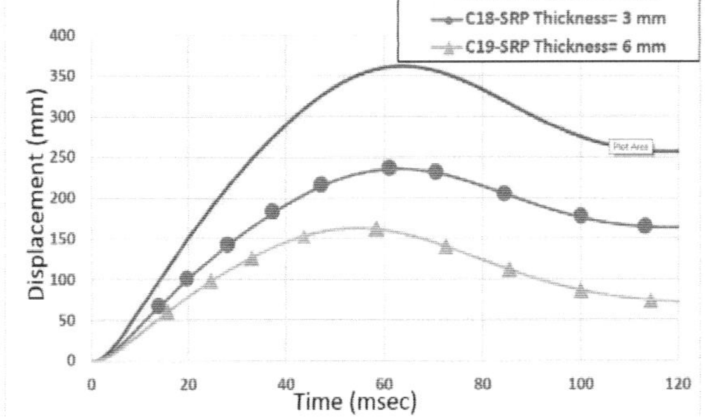

Figure 19. Displacement – time history for different SRP thicknesses.

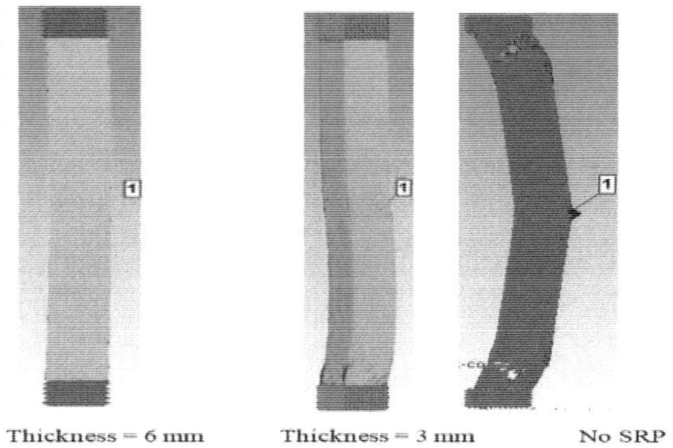

Figure 20. Damaged shape columns for different SRP thicknesses.

Properties, Durability and Fields of Application of Polymer Concrete

Mounir Kamal[1], Zeinab A. Etman[1], Noha M. Soliman[1,*], Mustafa R. Elbheary[3]

[1] Civil engineering department, faculty of engineering, Menoufia University, Egypt.
[2] Higher Institute of Engineering and Technology, Menoufia, Egypt.
[3] Civil engineering at the Arab contractors, Osman Ahmed Osman Company,

[*] Corresponding Author Email: dr-noha-soliman@sh-eng.menofia.edu.eg

ABSTRACT. Polymer concrete is a type of special concretes which casting by using the polymeric resin replaces the Portland cement/water binder of conventional concrete. The aim of this paper is investigated the hardened properties and the micro structure of polymer concrete. The main variables of this study were the type of epoxies [Polymer modified repair mortar (sika top122) and epoxy (Euxit 50)], epoxy to aggregate percentage and sand to dolomite ratio and. Twenty-four concrete mixes were cast to carrying out the experimental program. Compression, tensile, flexure and bond strength were conducted to evaluate the properties and impact loads. At sika top122 use water had a constant ratio according to sika top 122 equal to 35% with all mix from 1:1 to 1:16 in sixteen concrete mixes. At Euxit 50 divided into three groups in eight concrete mixes, first group had four mixes with fine aggregate (natural sand) and Euxit 50 and without added any water and coarse aggregate. The second and third groups components were coarse aggregate (natural dolomite), fine aggregate (sand) and Euxit 50 (welded polymer material) with variable mix ratios. Each group mixes had the same Euxit 50 to aggregate (coarse and fine) 1:5 and 1:6 for second and third group respectively. Get the best mix ES4 with high properties Compression, tensile, flexure and bond strength were 1,040.00, 44.60, 262.5 and 14.95 respectively.

Keywords: Polymer concrete, solidified properties, micro structure.

1. Introduction

Polymeric Concrete is a special type of concrete that can be obtained by adding polymeric materials that as a sealant for the spaces between the aggregate grains, which represent from 6 to 25% of the weight of concrete. It is sometimes called plastic concrete or resin concrete. Added polyester and epoxy gave the most important features as high-pressure resistance, high tensile stress, highly resistant to shrinkage, corrosion resistance, water permeation, sulfate resistance and density range from 1970 to 2350 kg/m^3 [1]. The mechanical properties of polymer-modified light weight aggregate concrete were investigated. And the compressive strength increased at 7-days and density by 31% and 10 %, respectively compared to ordinary concrete [2]. The polymer and cement in the mortars was in the range of 34 to 86% of the weight of concrete. Decreasing in the elastic modulus of polymer mortar compared with cement mortar causing reducing cracking resulting from shrinkage [3]. Pascal et al [4], analyzed the mechanical behavior of polymer modified mortars (PMMs). Different types of polymers (latex, styrene, and butadiene) were used. Polymer to cement (P/C) ratio was (0, 5, 7.5, 10, 12.5, 15) % by weight. The results show that 15% was the optimum ratio can be used, increasing in the flexural strength as P/C ratio increases by 10 % was recorded. four formulations of polymeric concrete (PC) casted with hardened epoxy resins. When the polymer ranged from 6 to 16 % of weight the polymeric concrete was exposed to temperatures up to 250°C. The results show that when the epoxy polymeric concrete was exposed to 225°C, a significant loss in compressive strength by 70 and 98% for PC and OC. This was due to the thermal oxidative decomposition of the epoxy polymer and the disintegration between the aggregate and the binder [5].

Adding the nano-clay to an epoxy in polymeric concrete doesn't guarantee that all mechanical qualities will improve. As a result, different amounts of nano clay can be used to improve different mechanical properties. This implies that the amount of nano clay used affects the behavior of the produced material under various loading circumstances. As a result, the nano clay amounts should be determined based on the desired qualities. Mechanical testing revealed 12.5 percent, 13.3 percent, 7.4 percent, and 25.5 percent increases in tensile and flexural moduli, compressive strength, and fracture toughness. However, the strain to failure was reduced by 21% and the tensile strength by 25% [6].

Using epoxy polyurethane acryl concrete (EPUAC) was investigated to produce polymer concrete. A quantity of 70% aggregates and 30% polymer were selected for preparing epoxy polyurethane acryl. The compressive and flexural strength were 52.5 and 11.52 N/mm^2.The high compression resistance 52.5 N/mm^2.It was found that the bond force between concrete and pull-out stress for a plain bar is 2.89 N/mm^2.A strong relationship between adhesion EPUAC with cement concrete is 4.7 N/mm^2 and these values are smaller than in the case of PEC, which the minimum values were 3.52 and 5.6 N/mm^2, respectively [7]. Polymer modified slurries (PMM) was investigated to reduce polymer addition without losing the properties. P/C ratios of 0, 1 and 3% were used. A fixed sand to cement ratio of 3, and a water to cement ratio of 0.5. The bending test results showed that the graded polymer distribution resulted in higher flexural strength [8].

The optimum mixture design for polymer polysulfide concrete was presented with a new polymer additive (polysulfide), which the freezing and thawing reduced the strength and had little effect on the relative modulus of elasticity. Over 300 cycles of freezing and thawing caused a problem in terms of compressive strength and get 41.4MPa in compressive strength [9]. Polyester polymer concrete (PPC) under low strain and compare polymer concrete (PC) were studied. Stress rates from 0.00033 to 0.15, the rate at 0.00033 was 52.03 MPa. A maximum 40% increase in tensile strength of the neat polyester, the compressive strength of PPC increases by 25% while the elastic modulus did not change significantly. The flexural strength was 973 kg/cm^2 [10].

The influence of latex, cement, aggregates, air content, polymer-cement ratio, anti-foam and processing methods were found. The tensile strength is a slight increase ranging from 2.2 to 8.54% for 2.5 and 15%, respectively [11]. The effect of using basalt fibers and clay nano particles as epoxy polymer were studied on the properties of polymer reinforced concrete. Also, the effect of temperature (50°C, 100°C, 150°C, 200°C and 250°C) was observed. Use 2% basalt and 10% of nano-fibers, the flexural and tensile strength of increased by 4.8 and 35%. Increasing in the compressive by flexural strength by 7 and 27%, respectively when adding 3% by weight of basalt core between PC and basalt fiber. Weight losses and reduction in compressive strength were recorded at 250°C [12].

The mechanical properties of polymer were studied as rapid repair of concrete structures. Different polymer ratio of mix was 10, 12 and 14% of the weight of concrete and two different types of aggregates ranging between 4.75 to 9.5 mm and 9.5 to 19 mm were used. The results showed that when the temperature was decreased from 25°C to 15°C, the mechanical properties of PC improved, and that when the temperature was increased from 25 to 65°C, it negatively affected the mechanical properties. The Taguchi method determined the design of the mixture with a polymer ratio of 14%, which contains aggregates from 9.5 to 19 mm, and tested at a temperature of 15 °C, the best mixture in terms of compressive, tensile strength and bending [13]. The mechanical properties and durability of sustainable unsealed and confined geopolymer for GPC exposed to sulfuric acid under constant and periodic loads were investigated. Increase in strength was 8.1 to 9.1% and 8.4 to 8.9% for GPC. GPC is more resistant to sulfuric acid attack than the OPCC analogue for both confined and unconfined GPC samples based on this result SEM images. Acid exposure was reduced by 41% compared to the control samples (unstrained and not exposed to acid). The corresponding of unconfined OPCC was 53% against the reduced strength of GPC [14].

The polymeric concrete according the resin to filler ratios were studied. Severe separation occurred at less than 40% of the filler between them resulting in reduced tensile and bending and reduced ductility. The optimal resin ratio was 30:70 and 60: 40 for the irregular and uniform distribution of the totals respectively at aggregated ratio 1:1.35 were obtained good balance between performance and cost. Density gradually increased from 1732 to 1869 kg/m^3, high compressive and tensile strength were 500 and 160 kg/cm^2 respectively [15].

2. Research Significance
This paper presents the properties and durability of polymer concrete mix table. According to resin used as cement, the high mechanical properties and more durable against sulfate and acid exposure. Also, this loads to use the polymer concrete in special application such as prestress concrete, nuclear power plants, industrial and electrical structure, prefabricated structures such as acid concrete tanks, inspection rooms drainage pipes, water insulation and drainage works. It's Use as concrete repair mortar for repairing damages in concrete structures like beams, piles, slabs, posts, pipes, precast elements, etc.

3. Experimental Program

3.1. MATERIALS.

3.1.1. Fine aggregate (sand).
Natural sand was used as fine aggregates. Fineness modulus of sand is 2.5. Its characteristics satisfy the Egyptian Standard Specifications E.S.S. 1109/2007 [16].

3.1.2. Coarse Aggregates (Dolomite).
Natural dolomite was used as a natural coarse aggregate. The dolomite has a maximum nominal size of 10 mm and specific gravity 2.64 t/ m^3. Water absorption of the used dolomite was 0.37 per unit by weight. Its characteristics satisfy E.C.P 203/2018 [17] and its grading satisfy ASTMC33- 03 [18].

3.1.3. Water.
Tap water without special taste, smell, color, or turbidity was used for mixing and curing procedures according to the Egyptian code of practice E.C.P 203/2018 [17].

3.1.4. Steel Bars.
High tensile steel rebars (St. 52) of Ø8 mm diameter was used to evaluate the bond strength. The main properties of steel type used are satisfying E.S.S 262/2011 [19], the yield and ultimate strength of steel used were 365& 521 MPa respectively.

3.1.5. Sika Top 122 (Polymer modified repair mortar).
Sika top 122 is a two-component high performance, cement based, polymer modified, fiber reinforced, structural concrete repair mortar. Suitable for horizontal, vertical and overhead applications. which the polyamide fibers are added {A (liquid substance) polymer and B (powder substance) cement} when they are mixed together and as a result of mixing them together, they produce mortar suitable for restoring and leveling the concrete surface. The mechanical, physical and chemical properties of the sika top 122 used to show in table (1).

3.1.6. Euxit 50 (epoxy consisting of two solvent free compounds).
The product of the base epoxy is low viscosity and it is multi-advantage over the rest of the epoxy types, as its uses are many, which is called pure epoxy. Euxit 50 is consists of from two compounds A(liquid substance) and B(liquid substance).The mechanical, physical and chemical properties of the Euxit 50 used shown in table (2).

3.2 MIX PROPORATIONS AND MIXING METHOD.
Concrete mixtures were cast to evaluate the properties of polymer concrete the experimental program divided to main two group according to the polymer types. Sika top 122 and Euxit 50 used as polymer types. 24 mixes were cast to evaluate the aim of the paper. Frist prepare the materials used individually. Second, mix the two compounds of epoxy material. Third, adding the aggregate (fine then coarse) while

striping the mixture at least ten minutes with constant speed. The mix process was drawn using electric hand mixture. Finally, purring the concrete in the forms of samples. The table 3 shows the mix proportions of concrete mixes using sika top 122 and Euxit 50.

3.2.1 Tested Specimens
Mechanical Properties tests were conducted on the standard samples according the Egyptian Code of Practice (E.C.P203/2018) [17].Cubes (70×70×70mm), cylinders (50×100mm) and prisms (40×40×160mm) were used to obtain compressive, tensile and flexural strengths, respectively, to evolute conduct bond strength, standard cylinders (50×100mm) with embedded high tensile steel rebars (St.52 with 8mm×120mm), and micro structure test was tested to determine structure composition and the main chemical elements compotes of this polymer concrete. While Plates (200×200×20mm) of this polymer concrete were tested to know its behavior in impact load.

4. Test Results Mechanical Properties for Polymer Concrete

4.1. Test results mechanical properties for polymer concrete using sika top 122.

4.1.1 Compressive strength.
The results show that increasing the sika top 122: sand ratios up to 1:6 increase the compressive strength then increasing sika top 122 decrease the compressive strength. Figure (1) shows the optimum compressive strength occurred at the mix SS6 which cast using sika top 122 to sand ratio 1:6 at 28 days, this compressive strength increases by 70 % compared to the control mix.

4.2. Test results mechanical properties for polymer concrete using Euxit 50.

4.2.1. Temperature Characteristics of Fresh Concretes.
At the first minutes of preparing the mixes the temperature was measured for the mixtures. Increasing the aggregate to Euxit 50 ratios lead to decrease the temperatures of mixes, and increasing the dolomite to sand ratios lead to decrease the temperatures of mixes as shown in figure (2). Which the decreasing in the temperatures occurred due to decrease the quantity and the ratios of polymer ratio which increase the temperature of the reaction.

4.2.2. Density Properties for Euxit 50 Polymer Concrete Samples.

Figure (3) shows the density properties of this polymer concrete. The increasing the aggregate ratios to Euxit 50 increases the density of this polymer concrete. While the density is increasing by 26.7% by using the ratio of Euxit 50 to aggregate 1to 6 to 1 compared to the control mix which the ratio of Euxit 50 to aggregate 1to 1 to 1. As a result of increasing the weight of the aggregate in the mixture and reducing the weight of the Euxit 50. Because dolomite and sand have similar density, the increasing the density is 1.7% of the mix ED63 compared to the mix ES6.

4.2.3. Compressive strength.
Figure (4) shows the compressive strength at 3, 7 and 28 days for all mixes using Euxit 50 in polymer concrete. The best compressive strength is occurred at mix ES4. The compressive strength increases by (46%, 25% and 44.4%) compared to the control mix ES1 at 3, 7 and 28 days. But increasing the ratio of aggregate to Euxit 50 by 5:1 and 6:1 decrease the compressive strength. Using dolomite in this polymer concrete decrease the compressive strength at the same ratio of Euxit 50 to aggregate, in the mixes ED52 and ED53, the compressive strength decreases by 9% and 27.3% respectively compared to ES6 at 28 days ages. Figure (5) shows that, when samples are exposed to different temperatures 50, 75 and 100°C, they negatively affect the compressive strength and reduce it. When the temperature rises to 100°C, the effect is more and the compressive strength decreases by 44.6%, 63.6%, 44%, 55%, 55%, 60% and 29.4% for mixes ES4, ES5, ES6, ED52, ED53, ED62 and ED63 respectively compared themself at 25°C.

4.2.4. Tensile Strength.

The tensile strength is increased by increasing the sand to Euxit 50 ratios as shown in figure (6). The tensile strength increases by 54.9%, 43.1%, 174.5%, 25.5%, 25.5% and 15.7% for mixes ES5, ES6, ED52, ED53, ED62 and ED63 respectively compared to mix ES1. But it decreases by 11.8% for mix ES4 compared to mix ES1.

4.2.5. Flexure Strength.

The flexural strength is increased by increasing the sand to Euxit 50 ratios as shown in figure (7). The flexural strength increases by 9.6%, 51.7%, 15.8%, 1.7% and 17.1% for mixes ES4, ES5, ES6, ED53, and ED63 respectively compared to mix ES1. But it decreases by 3% and 18.8% for mix ED52 and ED62 compared to mix ES1.

4.2.6. Bond Strength.

The bond strength is increased by increasing the sand to Euxit 50 ratios as shown in figure (8). The bond strength increases by 194.1%, 117.6%, 94.1%, 206% and 53% for mixes ES6, ED52, ED53, ED62 and ED63 respectively compared to mix ES1. But it decreases by 3% and 18.8% for mix ES4 and ES5 compared to mix ES1.

Note: The irregularity of the different mechanical properties with the compressive strength is visible in this form of cast and tested concrete because the samples that give the maximum compressive strength differ from those that give both tensile, flexural and bond strength.

4.2.7. MICRO STRUCTURE (scanning electron Microscope (SEM) (EDX).

The cubes or small slices (5×5mm) or less to fit the root the surface of the samples is wrapped in gold until their surface is conducted to be easily tested under the electronic microscope, which depends on the electronic beam's interaction with the sample surface atoms inside the microscope. As shown in fig (23), A scanning electron microscope test was done to examine the microstructure of concrete mixes. The microstructure of concrete mixes was examined by conducting a scanning electron microscope test on broken specimens after indirect tensile strength tests at 28 days. In this test, samples are cut into small cubes are slices (5×5 mm) or less to fit the device room. In this test, the samples were enlarged to × 12000 as showed in figures (9 to 15).

In the micro structure, the highest mass of carbon and oxygen are at ratio 1:6 of epoxy (Euxit 50) to aggregate and 1:2 of sand to dolomite. And, the best mass of silicon is at ratio 1:4 epoxy (Euxit 50) to aggregate. And the highest mass of calcium is at ratio 1:5 epoxy (Euxit 50) to aggregate and using 1:1 of sand to dolomite 1:1 as showed in figure (16).

4.2.8. Impact

The impact test was carried out by installing the sample in the shock device and falling a mass weighting half a kilogram free fall under the influence of weighing only on the cast tiles (20×20×2 cm) and calculating the number of blows until the appearance of cracks and calculating the blows at the break of the sample.
Figure (17) shows that, the impact energy increases by 103.8% and 253.8% when increasing the sand ratio at mixes ES5 and ES6 compared to ES4. But impact energy decreases by 6% and 65.4% when increasing the dolomite to sand ratios at mixes ED52 and ED53 compared to ES5. And impact energy decreases by 83.5% and 87.2% when increasing the dolomite to sand ratios at mixes ED62 and ED63 compared to ES6.

5. Conclusions

The following conclusions can be drawn as follow:

- The characteristics of Euxit 50 are better than that of sika top 122, Which can be used to form a good mortar. But Euxit 50 is used to form excellent concrete.
- At 3 and 7days using ratios 1:4, 1:5, 1:6 epoxy (Euxit 50) to aggregate by weight the compressive strength increased by 93%, 50% and 20%, respectively compared with the plain concrete.

- At 28 days using ratios 1:4, 1:5, 1:6 epoxy (Euxit 50) to aggregate by weight the compressive strength increased by 160%, 110% and 108% respectively (the best mix is at epoxy to aggregate 1:4).
- At 3 day using ratios 1:5, 1:6 epoxy (Euxit 50) epoxy to aggregate by weight at sand to dolomite (1:1 and 1:2 respectively) the compressive strength decreased by 15%, 52%, 31%, and 42% respectively compared with the epoxy to aggregate 1:4.
- At 7 day using ratios 1:5, 1:6 epoxy (Euxit 50) to aggregate by weight at sand to dolomite (1:1 and 1:2 respectively) the compressive strength decreased by 5%, 50%, 30%, and 35% respectively compared with the epoxy to aggregate 1:4.
- At 28 day using ratios 1:5, 1:6 epoxy (Euxit 50) to aggregate by weight (1:5 & 1:6) at sand to dolomite (1:1 and 1:2 respectively) the compressive strength decreased by 23%, 38%, 35%, and 35% respectively compared with the epoxy to aggregate 1:4.
- The best ratio for using ratio epoxy (Euxit 50) to aggregate is 1:4 to create best compressive strength
- The tensile strength was increased by (200%), while using ratio 1:5 of epoxy (Euxit 50) to aggregate. But the flexure strength and bond strength were increased by (300%), while using ratio 1:6 of epoxy (Euxit 50) to aggregate.
- The best compressive strength is at using ratio 1:4 of epoxy (Euxit 50) to aggregate, which increased by (62%, 62% and 15 %) at different temperature (50 ℃, 75 ℃ and 100 ℃) respectively. However, the compressive strength decreases at these temperatures while using other ratios of epoxy (Euxit 50) to aggregate.
- In the micro structure, the highest mass of carbon and oxygen are at ratio 1:6 of epoxy (Euxit 50) to aggregate and 1:2 of sand to dolomite. And, the best mass of silicon is at ratio 1:4 epoxy (Euxit 50) to aggregate. And the highest mass of calcium is at ratio 1:5 epoxy (Euxit 50) to aggregate and using 1:1 of sand to dolomite 1:1.
- Using ratio 1:5 of epoxy (Euxit 50) to aggregate at sand to dolomite 1:1 is the best mix on Energy Impact at First Crack (241). And the ratio 1:6 of epoxy to aggregate has the highest value of Energy Impact at Failure (466).

References

[1] Mounir Kamal, "polymer concrete" M.Sc. Thesis, Menoufia University, Egypt, 1977.
[2] J. A. Rossignolo and M. V. C. Agnesini, "Mechanical properties of polymer-modified lightweight aggregate concrete," Cem.Concr.Res., vol. 32, no. 3, pp. 329–334, 2002.
[3] M.M.Al-Zahrani, M. Maslehuddin, S. U. Al- Dulaijan, and M. Ibrahim, "Mechanical properties and durability characteristics of polymer- and cement-based repair materials, Cem. Concr. Compos, vol. 25, no. 4–5, pp. 527–537, 2003.
[4] S. Pascal, A. Alliche, and P. Pilvin, "Mechanical behavior of polymer modified mortars," Mater. Sci. Eng. A, vol. 380, no. 1, pp. 1–8, 2004.
[5] E. Oussama, G. Elham, M. Valérie, and B. Ouezdou, "Mechanical and physical properties of epoxy polymer concrete after exposure to temperatures up to 250 ° C," Constr. Build. Mater., vol. 27, no. 1, pp. 415–424, 2012.
[6] M. M. Shokrieh, A. R. Kefayati, and M. Chitsazzadeh, "Fabrication and mechanical properties of clay / epoxy Nano composite and its polymer concrete," J. Mater., vol. 40, pp. 443–452, 2012.
[7] L. Agavriloaie, S. Oprea, M. Barbuda, and F. Luca, "Characterization of polymer concrete with epoxy polyurethane acryl matrix," Constr. Build. Mater., vol. 37, pp. 190–196, 2012.
[8] X. Kong, C. Wu, Y. Zhang, and J. Li, "Polymer-modified mortar with a gradient polymer distribution. Preparation, permeability, and mechanical behavior," Constr. Build.Mater., vol. 38, pp. 195–203, 2013.
[9] S. Hong, H. Kim, and S. Park, "Optimal mix and freeze-thaw durability of polysulfide polymer concrete," Constr. Build. Mater. vol. 127, pp. 539–545, 2016.
[10] M.M. Shokrieh, S. Rezvani, and R. Mosalmani, "Mechanical behavior of polyester polymer concrete under low strain rate loading conditions, Polymer Testing, Vol. (63), September 2017, pp.596-604 DOI:10.1016/j.polymertesting.2017.09.015.
[11] H. Berkak, Z. Makhloufi, M. Bederina, A. Demdoum, and Y. Hadj Aissa, "Physico-mechanical characterization of composite cementitious mortars modified by resin containing antifoam," MATEC Web Conf., vol. 149, pp. 1–6, 2018.
[12] M.H. Niaki, A. Fereidoon, and M. G. Ahangari, "Experimental study on the mechanical and thermal properties of basalt fiber and Nano clay reinforced polymer concrete, Compos.Struct.vol.191, no. January, pp. 231 – 238,

2018.
[13] K. Jafari, M. Tabatabaeian, A. Joshaghani, and T. Ozbakkaloglu, "Optimizing the mixture design of polymer concrete: An experimental investigation," Constr. Build. Mater., vol. 167, pp. 185–196, 2018.
[14] R. Alzeebaree, A. Çevik, B. Nematollahi, and J. Sanjayan, "Mechanical properties and durability of unconfined and confined geopolymer concrete with fiber reinforced polymers exposed to sulfuric acid," vol. 215, pp. 1015–1032, 2019.
[15] W. Ferdous et al., "Optimal design for epoxy polymer concrete based on mechanical properties and durability aspects," Constr. Build. Mater, vol. 232, p. 117229, 2020.
[16] E.S.S.1109/2008, "Aggregates for Concrete," Ministry of Industry, Cairo, Egypt, 2008.
[17] E.C.P.203/2018, "Egyptian Code of Practice: Design and Construction for Reinforced Concrete Structures," Research Centre for Houses Building and Physical Planning, Cairo, Egypt, 2018
[18] ASTM.C-33, "Aggregates," American Society for Testing and Materials ASTM International, Philadelphia, USA, 2003.
[19] E.S.S.262/2011.Egyptian Standard Specification for Steel Bars. Cairo, Egypt, 2011.
[20] Sika Egypt-constraction chemicals-swiss quality, (2022), "Product Data sheet for sika top 122", https://egy.sika.com/en/construction/waterproofing/waterproofing-bridgedecks/cementitious-rendering/sikatop-122.html. April 2022, Version 01.02.
[21] SwissChem- Costruction Chemicals, (2019), "Product data sheet for Euxit 50", https://www.swisschem.com/en/products/resin-mortars-flooring-products/, 2019 swisschem Group.

Table 1. Mechanical, physical and chemical properties of the sika top 122, [20].

Property	Value	Property	Value
Density at 20 c Specific density	A liquid (1.02 kg/L) B powder (1.2 kg/L) (A+B) (2.2 Kg/L)	mechanical resistance (after 28 day) - pressure resistance - bending resistance	150 – 200 gm 40 -50 n/mm2 12 – 15 n/mm2 1.5 – 3 n/mm2
Hardness	7 days after last face	- resistance to consistency between concrete	
static elastic modulus	20000 n/mm2	- resistance to consistency between concrete	
Mixing ratio by volume	1:3	consumption rate	2.15 kg/l
validity	12 months from the date of production	Coefficient of resistance to the spread of carbon dioxide	> 4000
the color	A liquid (white) B powder (grey) Mix (grey)	Important considerations	Application temperature not less than 8 Minimum thickness of layer = 3 mm Maximum thickness of layer = 20 mm

Table 2. Mechanical, physical and chemical properties of the Euxit 50, [21].

Property	Value	Property	Value
Density at 20°C - Specific density	✓ A liquid (1.02 kg/L)- ✓ B powder (1.2 kg/L) ✓ (A+B) (2.2 Kg/L)	Density at 20°C Mixing ratio by weight	300 cm poise 3:1
validity	12 months the date of production	Mixing ratio by volume	5:2
- Compressive Strength (after 28 day) - bending resistance (after 28 day) - resistance to consistency between concrete - resistance to consistency between concrete	520 kg/cm². 400 -500 kg/cm². 120 – 15 kg/cm². 15 – 30 kg/cm².	Mixing ratio by volume Mixing ratio by weight Single-face consumption rate the color	1:3 1:6 150 – 200 gm transparent
Hardness	7 days after last face	Viscosity at 20 C cps	300
Layer Thickness Substrate Temperature	Minimum 5 mm / maximum 20 mm +8 °C minimum; +30 °C maximum	Specific gravity at 20 C (g/cm³) Pencil hardness	1.05 2 H
Operating at a temperature at 10 °C	1.5 Hour	Compression (N/mm²)	85
Operating at a temperature at 20°C	1 Hour	Bending property (N/mm²) E-module (N/mm²)	45 2800
Operating at a temperature at 30 °C	¼ Hour	Breaking tension (%)	4
Hardness	7days after last face	Compression at 15 N/ mm²) Volume shrinkage (%)	0.5 3.5

Table 3. The mix preparations of the total mixes.

Name of polymer	Code mix	Materials Ratio		Code mix	Materials Ratio	
		Poly./ Agg.	Poly./W		Poly./ Agg.	Poly./W
Sika top 122	SS1	1:1	0.35	SS9	1:9	0.35
	SS2	1:2		SS10	1:10	
	SS3	1:3		SS11	1:11	
	SS4	1:4		SS12	1:12	
	SS5	1:5		SS13	1:13	
	SS6	1:6		SS14	1:14	
	SS7	1:7		SS15	1:15	
	SS8	1:8		SS16	1:16	

Name of polymer	Code mix		Materials Ratio	
			Poly./Agg.	S/ D
Euxit 50	1st group	ES1	1:1	1:0
		ES4	1:4	1:0
		ES5	1:5	1:0
		ES6	1:6	1:0
	2nd group	ED52	1:5	1:1
		ED53	1:5	1:2
	3th group	ED62	1:6	1:1
		ED63	1:6	1:2

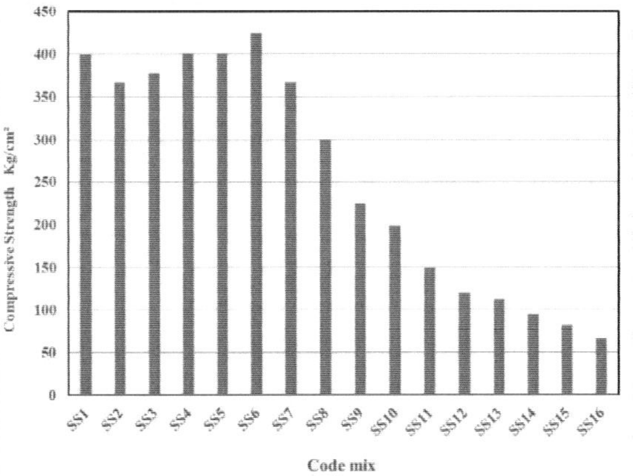

Figure 1. The Compressive Strength of All Mixes Using Sika top 122.

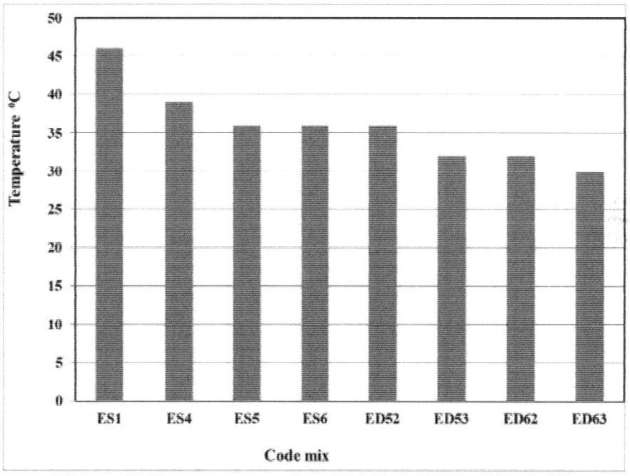

Figure 2. Temperatures the All Mixes Using Euxit 50.

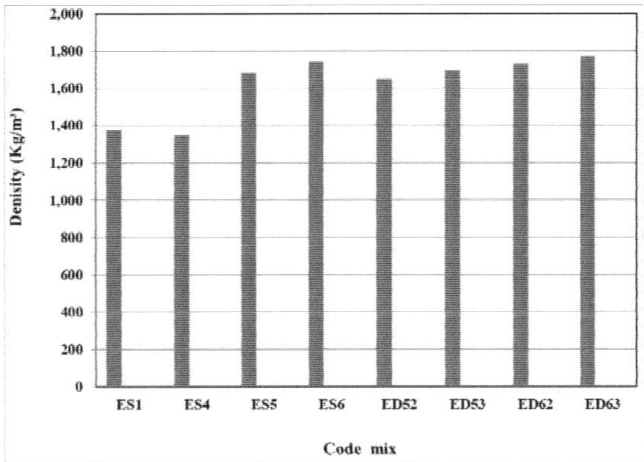

Figure 3. The Density of the All Mixes Using Euxit 50.

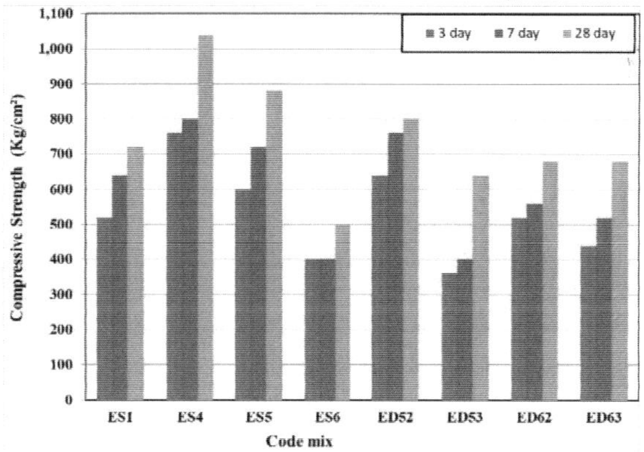

Figure 4. The Compressive Strength of the All Mixes Using Euxit 50 at 3, 7 and 28 Days.

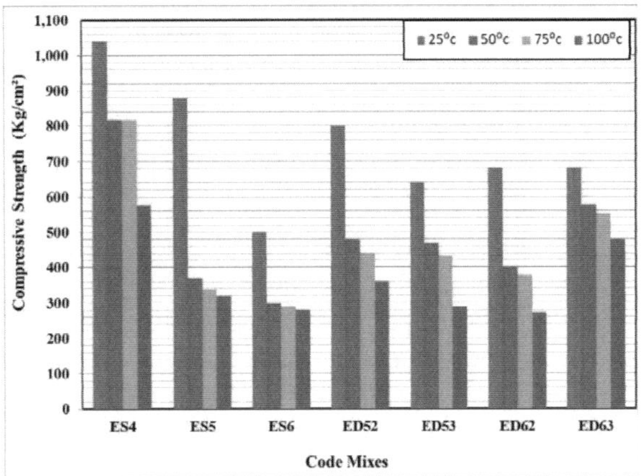

Figure 5. The Compressive Strength of the Mixes Using Euxit 50 Curing at Different Temperatures.

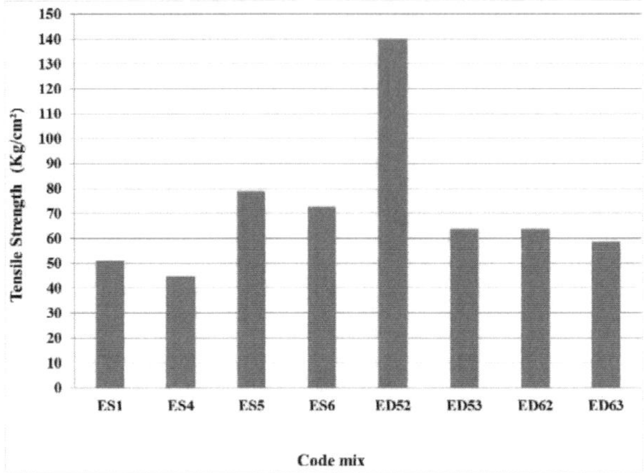

Figure 6. The Tensile Strength of the All Mixes Using Euxit 50 at 3, 7 and 28 Days.

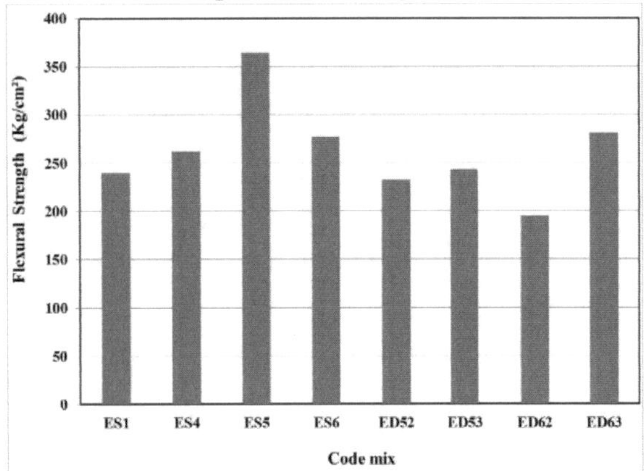

Figure 7. The Flexural Strength of the All Mixes Using Euxit 50 at 28 Days.

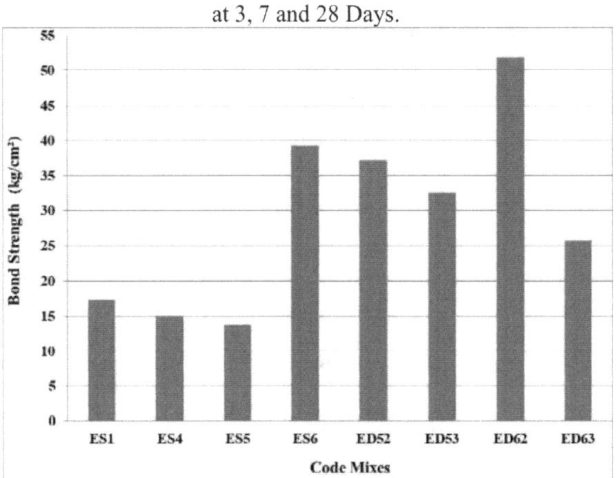

Figure 8. The Bond Strength of the All Mixes Using Euxit 50 at 28 Days.

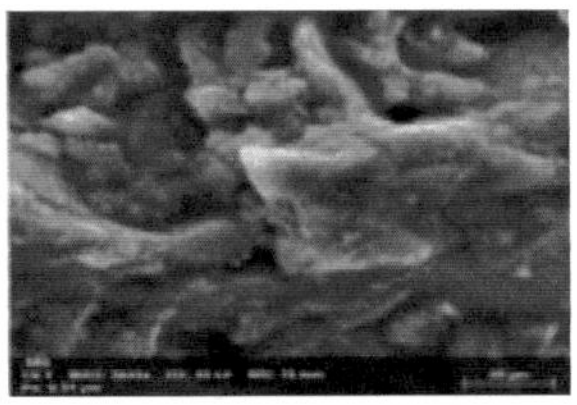

Figure 9. Micro structure at mix ES4.

Figure 10. Micro structure at mix ES5.

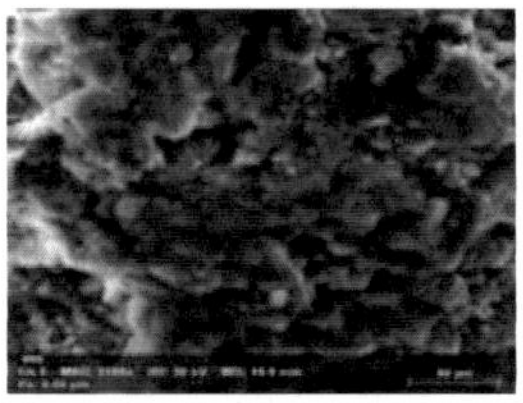

Figure 11. micro structure at mix ES6.

Figure 12. Micro structure at mix ED52.

Figure 13. Micro structure at mix.

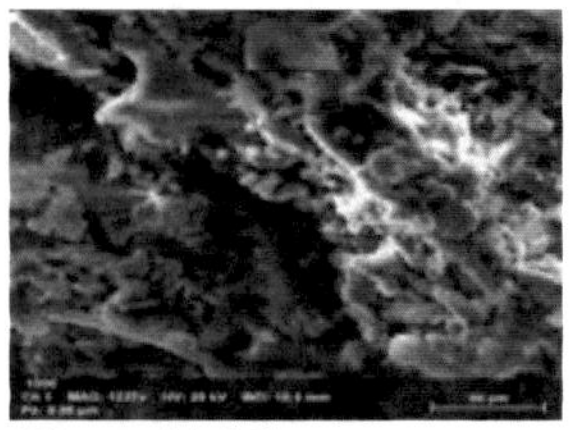

Figure 14. Micro structure at mix.

ED53. ED62.

Figure 15. Micro structure at mix ED63.

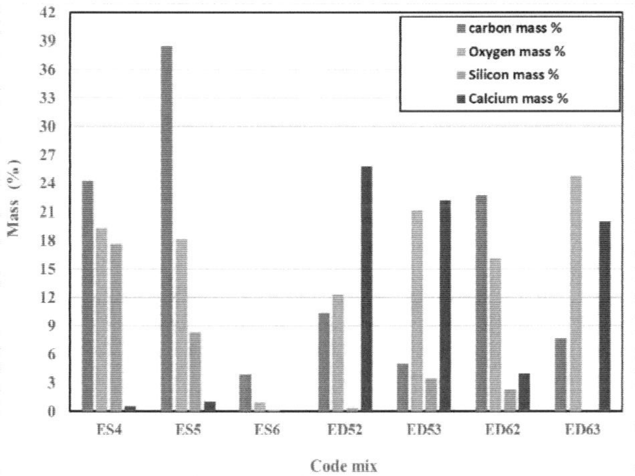

Figure 16. Comparison of carbon, oxygen, silicon and calcium ratios of mass for the mixes.

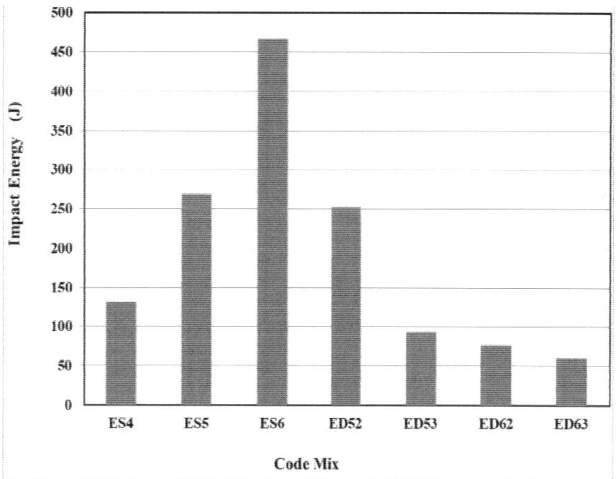

Figure 17. The Impact Energy of the All-Plates Casting Using Euxit 50 at 28 Days.

Building envelope optimization technique to increase energy efficiency of office building in Egypt

Ayah Mohamed Ramadan[1, *]

[1] Faculty of Engineering, Modern Academy for Engineering and Technology, Cairo, Egypt

[*]Correposnding Author Email: ayahezzat54@gmail.com

Abstract. Sustainable design is a design strategy that aims to improve environmental quality as well as the quality of a building's internal atmosphere by limiting negative consequences on the structure and the surrounding environment. Several new building envelope technologies and concepts have been presented as solutions for enhancing indoor comfort and lowering environmental impact during building life cycles. In Egypt, the high sensitivity to temperature fluctuations resulted in an increase in power usage to meet the necessary cooling loads. Energy simulations are a common way to improve a building's energy efficiency. Design Builder program with the Energy Plus simulation engine was used for energy analysis of the building. This was confirmed by comparing the energy efficiency performance of selected case studies of buildings with and without a sustainable development approach. In two parametric series, an existing prototype of a residential unit module will be utilized to evaluate energy performance, co2 emission and thermal comfort. It has been discovered that a building envelope with a variety of characteristics may provide the inhabitants with the highest level of residential comfort. In comparison to similar-sized structures, the building façade decreases energy usage by 80% related to CO_2 emissions, according to this study.

Keywords: Sustainable design, Energy Plus, Energy efficiency, Thermal comfort.

1. Introduction

The exterior environment and the inside working area are separated by the building envelope. Building skins with smart features help to improve the implementation of buildings. Extreme energy usage is pushing the need for conservation efforts around the world, with buildings accounting for more than 30% of global energy consumption [1]. This article discusses new and current multi-layer structures that can benefit from smart skin technologies that can be integrated with closed systems. The purpose of implementing a smart approach is to lower building energy usage while also improving the quality of life. The value is in providing intelligent buildings, that is, buildings designed or remodelled utilising intelligent (intelligent design) and technology solutions, both in intelligent architecture (intelligent shell) and in equipment (intelligent systems). There is interaction between the user and the environment (smart people) [2].

As an alternative to attaining building sustainability through sustainable envelope design, the paper suggests that the building envelope be made sustainable. However, as far as building sustainability is concerned, it is necessary to investigate the effects of the environment on the building envelope. The ideas and requirements for building envelope sustainable performance and building sustainability were outlined in Figure 1.

These are interrelated domains were identified using the implications for sustainable development, [3]:

- Future principle emphasises intergenerational equity and the need to maintain natural resources for future generations by recycling resources and using transparent techniques.
- A basic element that characterises sustainable development is the importance of public participation in making sustainable design decisions. It comprises including building and construction specialists in the decision-making process for long-term sustainability.
- The environment prioritised ecosystem protection, energy efficiency, and resource conservation in future construction and envelope development.

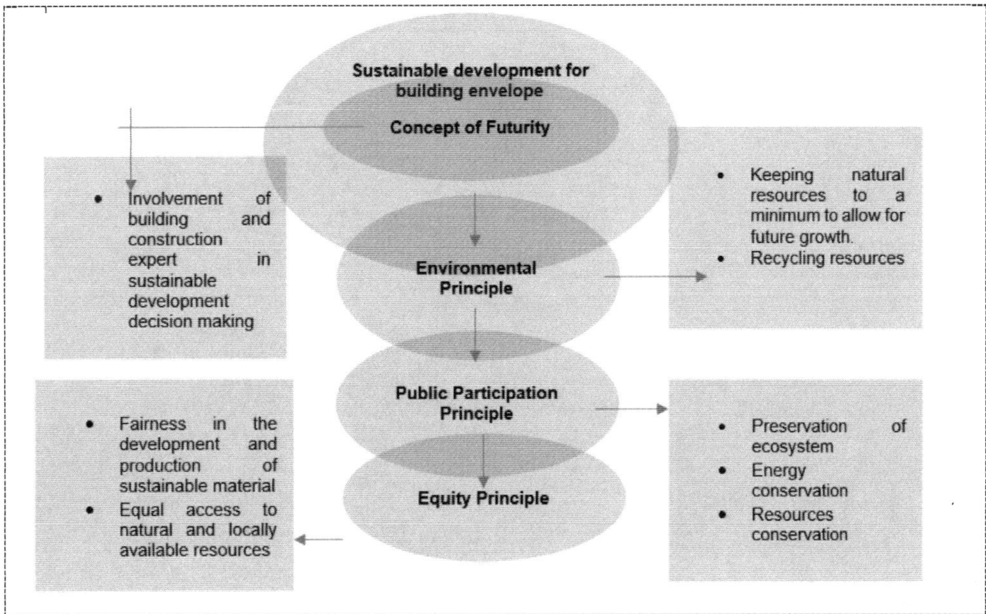

Figure 1. Modified model of sustainable development for building envelops [4].

2. Research problem

Based on the risks of global warming, one of the main challenges for architects today is to minimize the energy consumption of a building and maintain the proper indoor atmosphere at the same level. Most modern office buildings built in Egypt are not designed to leave buildings with large glass surfaces resulting in huge energy consumption, electricity bills and carbon emissions compared to the local climate shown in Figure 2. [5].

Figure 2. Office buildings in Cairo, Egypt with curtain wall façades facing west and south.

The regulating and protective functions of the building envelope are depicted in Fig. 2. These building envelope regulation activities aid in the achievement of building sustainability. Wind, rain, temperature differences, vapour pressure differences, industrial pollutants, solar radiation, and soil temperature are all factors that can cause damage to a building's envelope [6].

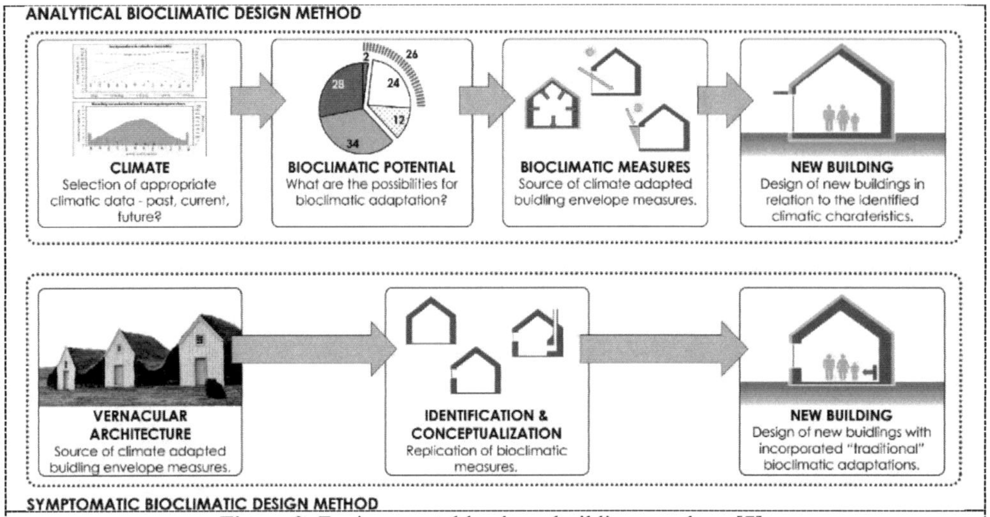

Figure 3. Environmental loads on building envelope [7].

Furthermore, as clarified in Fig. 2, the building envelope protects the indoor environment of the building by providing regulatory functions such as temperature management, moisture control, and indoor air quality control against environmental impacts on the building system [8].

3. Green walls

Today's first green wall system, called "plant brick", was invented in 1938 by Professor Stanley Hart white of the University of Illinois. The green wall can be placed on the inner or outer wall of the building. Lighting and orientation are key to the life of the green wall and the overall health, growth and appearance of the plant material. These walls can create a private, aesthetically pleasing living space while reducing noise from the surroundings. They became popular only when Patrick Blanc, often referred to as the "Father of the Green Wall," designed his first vertical garden for the Museum of Science and Industry in 1988. Bran's design was so innovative that it quickly became world-famous for its green walls and has been popular ever since [9].

- Vertical Displays
- Trellised Walls
- Modified Trellised Walls
- Planting Pockets
- Engineered Modular Systems

Figure 4. Green wall types [10].

4. Building Envelope Classification

To understand the behaviours of building components, it is necessary to classify different building types to determine the right combination for location and climate. The integument of a building is usually divided into two parts: [11]

- Single skin façade building envelope
- Double or multiple skin façade building envelope

Types of double skin façade

Box DSF: The first form of facade used in construction is the box DSF. Vertical divisions are used to divide the façade horizontally across the building. The front of the building is separated into windowsills. High degrees of heat and sound insulation are possible with this material. It is the only type of architecture that can incorporate this feature inside a regular rectangular aperture.

Corridor DSF: A cavity or open region between the exterior and interior façades that is separated horizontally via each ground. It is available and comprehensive enough to be used as a provider platform. Herbal and mechanical ventilation are both options. Within the exterior façade layer, close to the ground and ceiling, there are air consumption and extraction apertures. Benefits of hall facades include the fact that they no longer limit the height of buildings. [12].

Figure 5. Double skin façade types [13].

5. Energy Plus Simulation tools

Energy Plus simulation tools are a total of energy simulation building programs established by the US. Department of Energy. And because it is incorporated in the Design-Builder, it is therefore installed during the Design-Builder installation. Design-Builder is the elderly, easiest to employ, with powerful graphical UI to Energy Plus [14].

6. Methodology

A Methodology to Improve Office Buildings Energy Performance and Thermal Comfort Using Double Green Skin Facade systems is considered the largest zone in Egypt which includes the desert region.

6.1. Building Description

According to the simulation procedure, the results of Double Green Skin Facade systems simulation will be showed and explained on three main sequential stages (as shown in Fig 6,7):

- Simulation of the Case-study Models "Base-Case"& (DGSF), as shown in Fig 6.
- Double Green skin technique results with a 0.5 m cavity corridor in between both skins.
- Double Green skin technique results with a cavity in between of width. Over than 1.0 m – (Large Type cavity corridor in front of the offices), as shown in Fig 6,7.

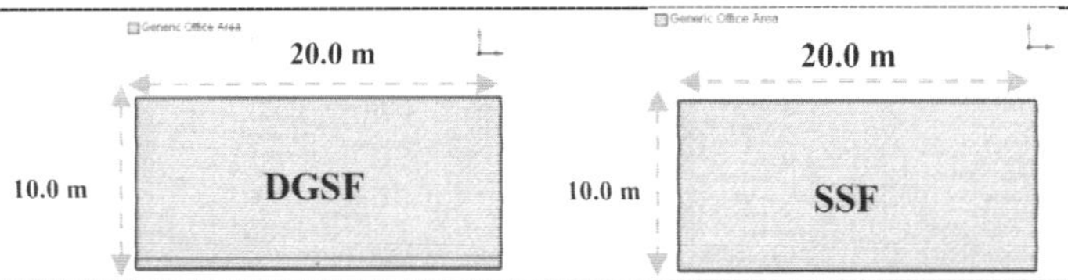

Figure 6. Architecture plans of Single Skin Façade (SSF) & Double Green skin façade (DGSF).

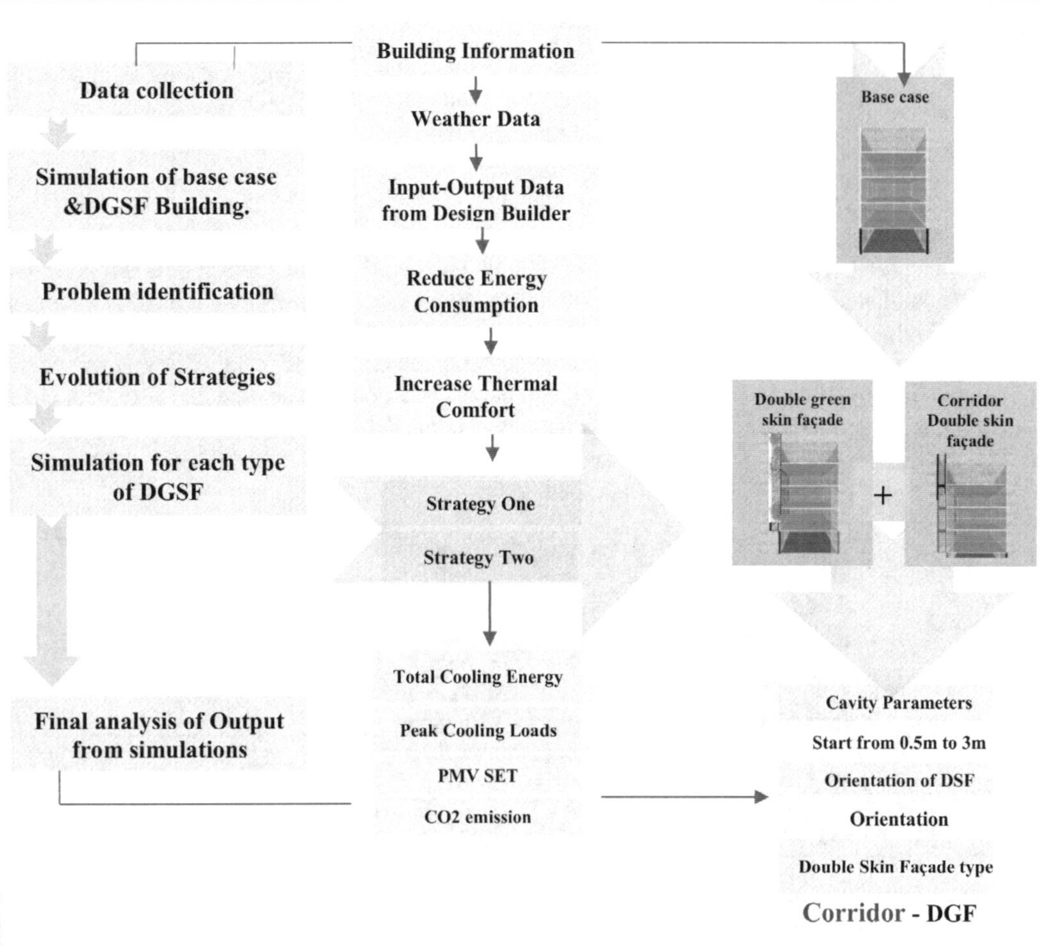

Figure 7. Simulation Methodology of the Practical Research.

6.2. Effect of Air Cavity of DGSFs Façade Configuration

Simulating the different 'Air-Cavity' widths for the (Corridor) DGSF and base scenario (small = 0.50,1.00 m, Medium = 1.50,2.00 m, Large = 2.50,3.00 m.), with the same fixed values for the other parameters (SSF).

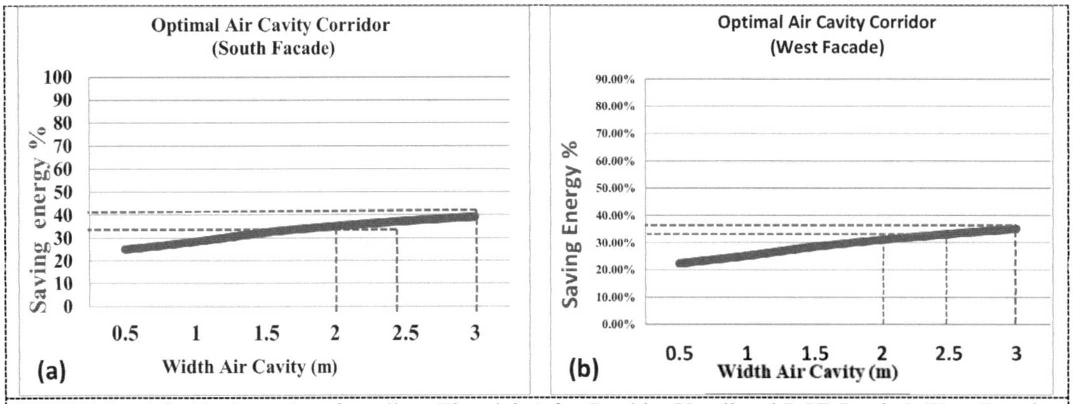

Figure 8. Optimal Air Cavity of Cooling (Electricity) for Corridor Ventilated DGF South & West Facade.

Figure 8(a) shows that air cavity widths of (0.5,1.0,1.5,2.0,2.5, and 3.0) m on the south façade achieve and save energy of (24,28,32,35,37, and 40 percent) accordingly. On the other hand, it was a different story on the western facade (22 percent ,25 percent ,28 percent ,31 percent ,33 percent & 35 percent) fig.8(b). Both figures clearly show that air gaps ranging from (50 to 100 cm) saved the least amount of energy and air gaps ranging from (50 to 100 cm) saved the most (150 to 200 cm). While air gaps greater than 2.00 m saved energy, the increase was not particularly noticeable.

6.3. Effect of Corridor Double Green Skin Facade on-air cavity width and orientation

It's clear that all the different cavity widths led to energy saving and directly a decrease in cooling loads and started to stabilize at 2.0 m cavity widths of the hybrid system. Effective Air Cavity Width, the small size is (0.50,1.00 m), Medium (1.50,2.00 m) and large widths (over 2.00 m) however the Medium 'Air-Cavity' widths are the most effective and economic for saving energy than other considering to the climatic zone and other parameters of the study.

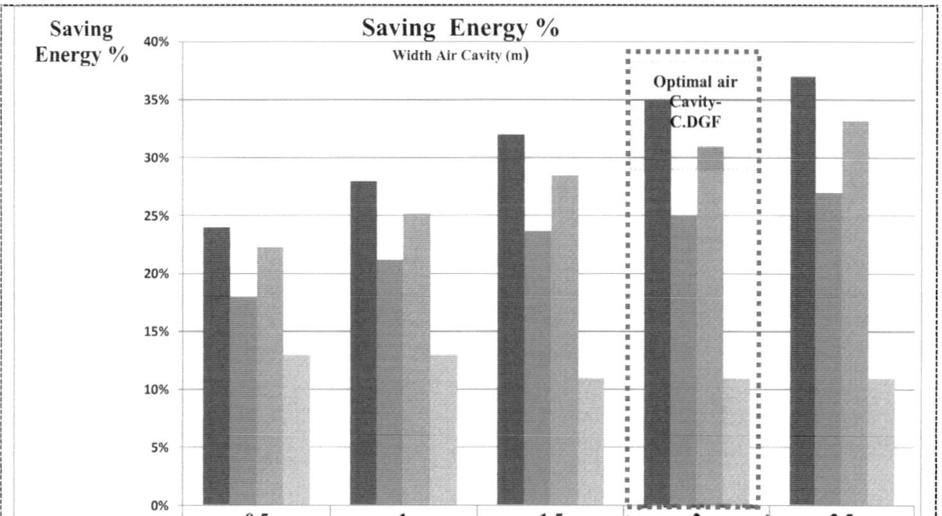

Cavity Depth(m)	0.5	1	1.5	2	2.5
South saving	24%	28%	32%	35%	37%
East saving	18%	21.20%	23.70%	25%	27%
West saving	22.30%	25.14%	28.47%	31%	33.20%
North saving	13%	13%	11%	11%	11%

Figure 9. Optimal Air Cavity of Saving energy for Corridor Green DSF Façade.

Figure 9 summarizing the total monthly cooling loads for the four orientations (East, West, North and South) facades for corridor Green DSFs with low, medium and large air cavity width with annual demand for cooling loads indicated that their energy saving at 2.0 m width was (25 %,31%,11% and 35%) respectively.

6.4. Effect of Corridor Double Green Skin Facade orientation with optimum air cavity width on Annual Cooling Demand

From the graphical output, it can be clearly seen that the total cooling energy used by the building in applying the Corridor DGSF, it was clear that the Annual Sensible Cooling for the south and west façade performance was the highest with a 2.0 m cavity width compared to the base case (SSF), while the east and north achieve lesser variation in annual sensible cooling compared to the SSF at the south façade, as shown in figure 10.

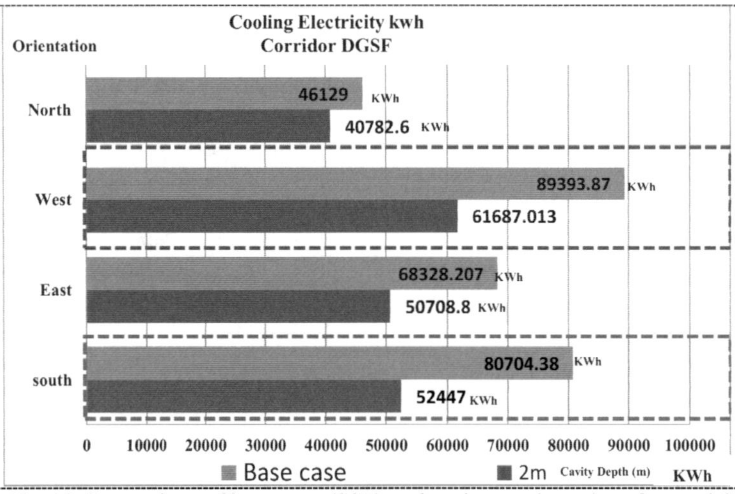

Figure 10. (a) Comparison of base case (SSF) and optimum air cavity of annual Sensible Cooling for Corridor DSF for the four orientations.

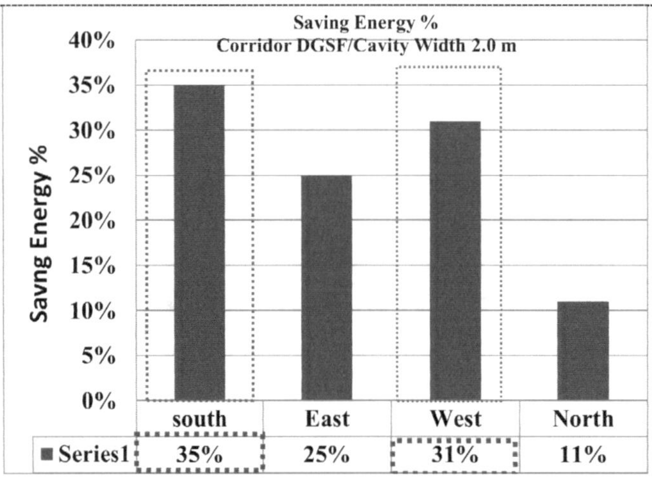

Figure 11. Saving Energy % for Corridor DSF orientation.

Figure.11 Shows the Effect of applying Corridor DGSF on Saving Energy with a 2.0m air cavity width, it reached up to 35%, 31%, 21%, and 11% for the south, western, eastern and northern orientation respectively.

6.5. Shading device for Corridor Double Green Skin Facade.
The annual energy consumption of the Corridor Double Green Skin Facade examined within the office area of the building was accomplished, in order to determine which Double-Skin Façade configuration and cavity airflow system is achieving the highest degree of efficiency in relation to energy consumption with the use of horizontal shading device, Fig .12.
The graph shows that the zone sensible cooling with the optimum air cavity width of Corridor DGSF with a horizontal shading device oriented to the south achieved best enhancement of sensible cooling of 9000(kWh) while Corridor DGSF without shading device the zone sensible cooling increased to 18000 (kWh) compared to the base case model which have the highest value of zone sensible cooling of 25000 (kWh) on July as shown in Fig 12,where it was clear that the horizontal shading devices was more efficient in the winter seasons, as the suns angle is quite lower and its impact is quite high.

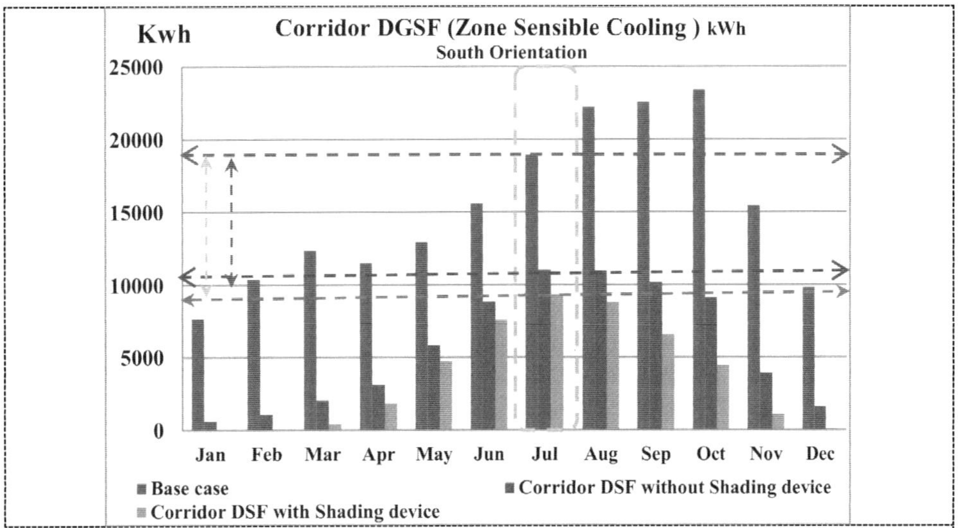

Figure 12. Monthly results of Zone Sensible Cooling kWh of an office building with a Shading of Corridor DGSF.

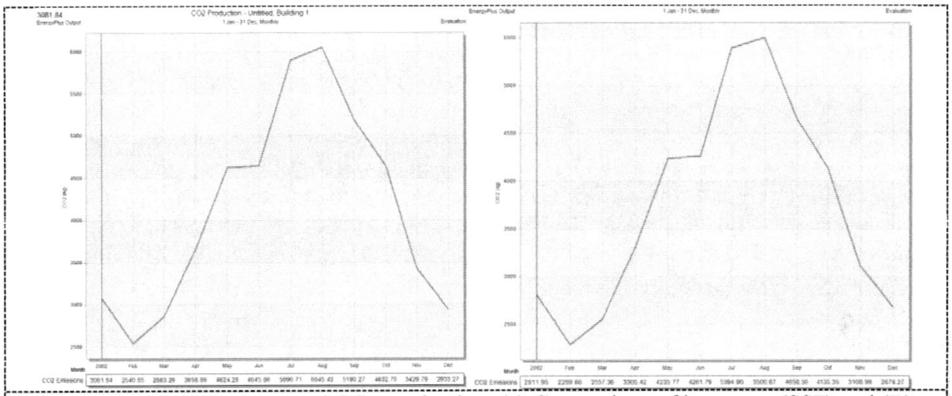

Figure 13. Building's annual CO_2 production, (a) Comparison of base case (SSF) and (B) Corridor DGSF.

Figure 13 show the development in the reduction of CO_2 emission can be adapted by applying Green Double Skin Façade as a source of energy.

7. Conclusion

The investigation, analysis, and optimization of Double green façade system essential factors in office buildings using Configuration types was the major purpose of the research. In 'hot-arid' conditions, practical study has proven that Double Green Facade (DGF) is the most effective of the Double Skin Facade (DSF) varieties, with the following significant impacts: Application of (DGF) decreases the total energy consumptions by more than (50% till about 75%) with respect to correspondent "Base-Case" (SSF) Façade type.

Environmental impact: The configuration of a glazed office building facades in 'Hot-arid' climates like Egypt (Greater-Cairo) has the main impact on reducing/increasing the building's energy demand, depending on the building location, orientation, and occupancy the façade's double green skin façade configuration is dependable for up to (60%) of the cooling loads needed.

References

[1] Egyptian Electricity Holding Company (2013). Annual Report 2011/2012, Arab Republic of Egypt, Ministry of Electricity and Energy.

[2] Aksamij A 2009 Perkins+Will Res. J. Context based design of double skin facades: climatic consideration during the design process vol 1 (1) p 54–69.

[3] Duran E, Jorge A, Paucar C, and Alberto A (2010). Sustainability model for the vallecula lake in Puebla, Mexico: combining system dynamics and sustainable urban development.

[4] Irene L, and Robert T (2007). Examining the role of building envelopes towards achieving sustainable buildings. In: Horner M, Hardcastle C, Price A, and Bebbington J (Eds), International Conference on Whole Life Urban Sustainability and its assessment, Glasgow.

[5] Sayed A, and Fikry M (2019). Impact of glass facades on internal environment of buildings in hot arid zone, Alexandria Engineering Journal vol 58 p1063–1075.

[6] EEH., Egyptian Electricity Holding Company Annual Report, (2020).

[7] Pajek L, and Košir M (2018). Implications of present and upcoming changes in bioclimatic potential for energy performance of residential buildings Build Environ vol 127 p 157–172.

[8] J. Iwaro and A. Mwasha, (2013). The impact of sustainable building envelope design on building sustainability using Integrated Performance Model. International Journal of Sustainable Built Environment, (2) 153–171.

[9] Ottelé M 2011 Comparative life cycle analysis for green façades and living wall systems Energy Build vol 43 p 3419–29.

[10] Wong NH 2010 Thermal evaluation of vertical greenery systems for building walls. Build Environ vol 45 p 663–72.

[11] Hazem A, Ameghchouche M, and Bougriou C (2015). A numerical analysis of the air ventilation management and assessment of the behaviour of double skin facades Energy Build vol 102 p 225–236

[12] Radhi H, Sharples S, and Fikiry F (2013). Will multi-facade systems reduce cooling energy in fully glazed buildings? A scoping study of UAE buildings, Energy Build vol 56 p 179–188.

[13] Rahmani B, Kandar M, and Rahmani P (2012). How double skin façade's air-gap sizes effect on lowering solar heat gain in tropical climate? World Appl. Sci. J. vol 18 (6) p 774–778.

[14] DesignBuilder (2019). DesignBuilder Software Ltd). Available [Online] at: https://designbuilder.co.uk/.

Soundscape Assessment for Indoor Spaces: A case of the Higher Education Environment

Ayat K. Kamal[1], Mostafa R. Ismail[2], Mohammed S. Mayhoub[1]

[1]Architecture Engineering Department, Al-Azhar University, Cairo, Egypt
[2]Architecture Engineering Department, Ain Shams University, Cairo, Egypt

*Corresponding Author Email: ayatkalaf@azhar.edu.eg

Abstract. As a concept, an indoor soundscape is an innovative approach to building environments with a shape that "sounds good," meeting the preferences and needs of the occupants and established according to their unique indoor soundscape perception. This paper presents the assessment of the indoor soundscape regarding objective acoustic parameters such as the equivalent noise levels (Leq) and the reverberation time (RT), and subjective method including a questionnaire survey (n = 403) in three buildings of Engineering in Al-Azhar university. The data analyses were conducted using SPSS v.21 statistical software. Overall, results show that indoor assessment is associated with five factors that are: acoustic environment factors, demographics and social factors, sound sources, duration and time of measurement, and non-acoustic parameters. The conceptual framework showed the relations between the acoustic environment, built environment, and auditory perception of university students and staff. Various buildings present a series of acoustic problems based on different activities that change the perception of users, so the outcomes feature the importance in the future considering the type of activity as a key factor in assessing the effects of noise on humans.

Keywords: Soundscapes of buildings, human Perception, learning space environment, acoustic environment, indoor soundscape factors.

1. Introduction

Because we spend most of our modern lives indoors, the soundscape is one of the most influential factors influencing human comfort and health. Sound landscape studies are a new integrative area of research that emerged in the late 1960s when Murray Schafer studied acoustic ecology in a project investigating the interaction of individuals and sound [1]. Soundscape research brings new knowledge in the field of noise control. It is utilized to assess environmental sounds, not just by means of high the sound pressure level is yet additionally by understanding the connections among sound sources, context, and space users [2]. The International Organization for Standardization (ISO) defines a soundscape as "*an acoustic environment that is perceived or experienced and/or understood in context by one or more persons*" [2]. This definition centers on context; hence, the concept of a soundscape for the most part applies to indoor environments. Standard ISO/TS 12913-2:2018 has as of late reported different data collection methods for comprehensive evaluations of acoustic experiences in contexts, such as guided interviews, sound walk, questionnaires, binaural measurements, and non-participatory observations [3]. Soundscape research is quite possibly the most well-known point contemplated in the field of acoustics. To study soundscape, five principle indoor assessment factors are present in the literature. This topic mainly deals with soundscape perception and acoustic quality that is categorized into the quantitative and qualitative aspects of soundscape assessment.

Assessing sound environments needs to be approached in several ways. As shown in Figure 1, in addition to determining the assessment factors of the building and its soundscape, the evaluation of the sound environment using various objective and subjective methods is decisive for carrying out acoustic investigations in closed buildings. As for indoor environments, their soundscape assessment is influenced by five main built entity factors, acoustic environment factors, demographics and social factors, sound sources, duration and time of measurement, and non-acoustic parameters. Before examining or assessing the acoustic properties of the context, these factors ought to be alluded to and presented. The main purpose of this paper is to feature ' quantitative and qualitative aspects and factors of soundscape assessment' that ought not to mix up with noise control methods. While objective metrics and parameters are essential in deciding guidelines and tangible results, there is a need to move from a physical perspective to a more psychological one. Users with their sensory experience are the key to better information about auditory environments. Other topics covered in this article are the various discussion topics that focus on the study of soundscape and acoustic quality.

2. Methodology

2.1. Case Study

2.1.1. Case study setting. The three buildings in the selected case study are located in the capital of Egypt, Cairo, in Al- Azhar University. The study was carried out in the faculties of engineering for girls "Building A" that include architecture, planning, computer, and electricity departments and for boys "Building B" that include architecture, planning, and petrol departments and "Building C" that include electricity and computer departments (Fig. 1). The buildings comprise longitudinal constructions of low height with a maximum of three floors mostly made of a painted brick wall and tile or wood floor and windows of aluminium frames and thin glass. The Three buildings have multiple classrooms, laboratories, studios, and lecture halls.

Figure 1. Study area, *at Al-Azhar University.*

In terms of the two-dimensional formal organization, Building C has a U-shaped plan, whereas both Building A and B are rectangular in plan. In Building A, the studios were connected to private and close-ended corridors causing additional circulation access from another point as well as the main circulation of the building. The lobby area is linked with the main stair and is near the lecture halls in Building A and C and is near the studios in Building B. Building C has an atrium void, the atrium is located above the foyer space and near to the classrooms, with a height of two floors. The location of the atrium and the lobby is important for sound analysis. In addition to the varying architectural characteristics, materials properties such as absorption coefficients and the room furniture should consider in the sound environment analysis for indoor soundscape studies.

2.1.2. Participants, Data was gathered through the questionnaire. Students and teaching staff voluntarily participated in the questionnaire for the lecture halls (n = 182), studios (n = 132), classrooms (n = 49), and laboratories (n = 40). The demographic distribution of students and teaching staff was recorded as 208 males (51.6%) and 195 females (48.4%). Of the total, 382 students were between 20 and 23 years of age, and 21 lecturers were between 24 and 45 years of age, since the selected rooms are used for elective courses.

2.2. Methods

2.2.1. Acoustic environment. To understand the relationship between the physical acoustic environment and people's perception in the selected educational facilities, the equivalent noise levels (Leq) was measured. The TENMARS IEC 61672 sound level meter was used for the measurements. The measurement point was located one meter from the main reflective surfaces and 1.5 meters from the grounds (in consideration of the sitting position of the receiver, as referred to in the American National Standard. The measurements all occurred in December 2021. The process consisted of the following: from 09:00 AM to 04:00 PM during active hours, a group of sound level measurements was taken every half hour and the mean value was calculated from 10 measurements per group. The interval between each measurement was five seconds [4]. There were great variances between the different functions of the rooms and their soundscapes when they were compared to each other. In order to reveal such differences, sound measurements were carried out on 65 typical rooms. The reverberation time (RT) was calculated by the Sabine formula in an unoccupied condition [5], assuming all rooms without tables and benches. The results of the equivalent noise levels (Leq) measurements show that there was a significant difference in terms of sound levels between the four types of rooms. The calculating reverberation time (RT) shows that none of the selected spaces meet the reverberation time requirements, duo to the recommendation of the standers [6,7] that recommended reverberation time values for rooms designated for education are between 0.6 and 0.8 (see Appendix).

2.2.2. Data collection, the questionnaire survey was used to collect data, analysed with the SPSS v 21. statistical analysis software. The questionnaire (prepared in Arabic) evaluates the soundscape over several topics: perception, annoyance, sound environment quality, and interpretation of the sound environment (as indicated in ISO 12913-1 [2]). Semantic descriptors are used in the survey to provide the most comprehensive information regarding soundscapes with a five-point scale ranking system. Data collected through questionnaires were evaluated using statistical software IBM SPSS Statistics v.21. First, we checked the reliability of the survey using Cronbach's alpha value. Then, as the number of participants exceeded 30, the data were analysed using the Kolmogorov-Smirnov test of normality to control for a normal distribution. A frequency analysis was performed to understand the distribution. The results of the Kolmogorov-Smirnov normality test indicated that the data were not normally distributed ($p < 0.05$). Mann-Whitney U test is a nonparametric statistical test that has been chosen to assess the difference between every two conditions using different spaces. Finally, to analyse dependencies between variables “Spearman's Rho correlation” is used.

3. Results and Discussion

3.1. Results

3.1.1. Questionnaire survey. The normality of the data was checked using the Kolmogorov-Smirnov test, and in all cases, data followed a normal distribution. Internal reliability scores were calculated for all ranked question groups. For the questions on factorial space usage, Cronbach's α value is 0.823. For the questions on factorial sound source perception, Cronbach's α value is 0.792. For the questions on factorial sound source annoyance, Cronbach's α value is 0.815. Finally for the questions on sound environment quality, Cronbach's α value is 0.746.

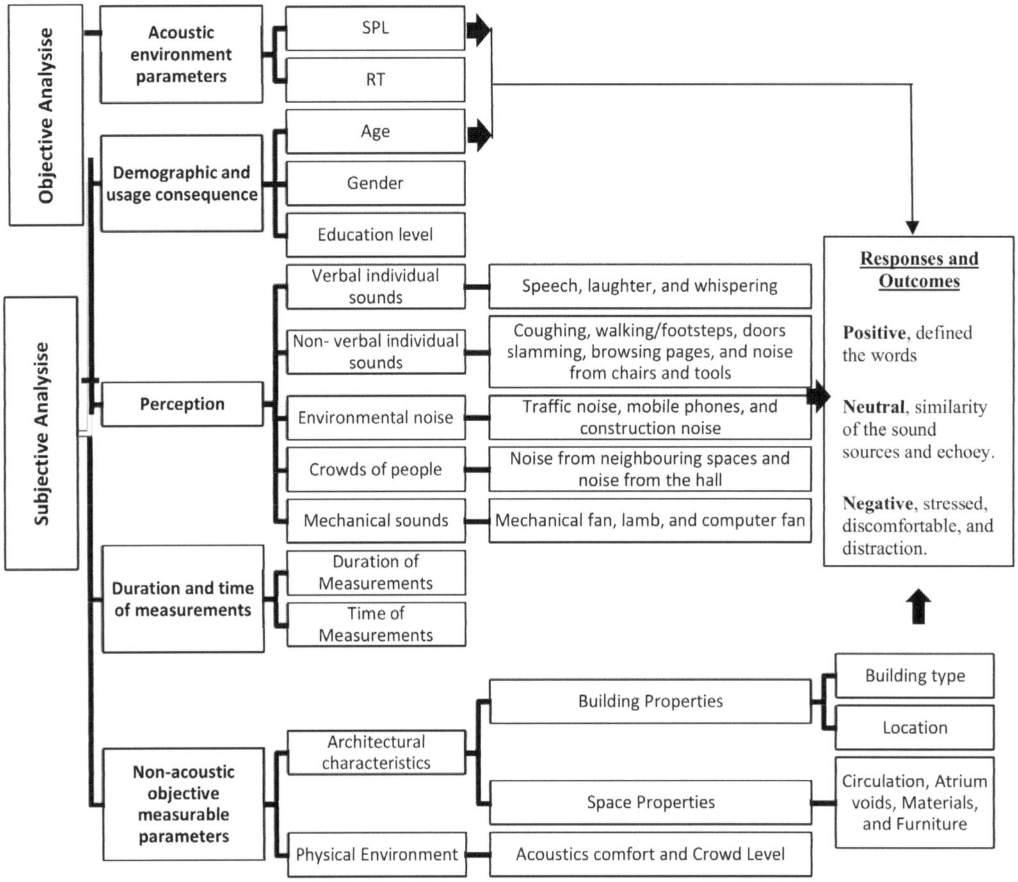

Figure 2. The conceptual framework for learning spaces.

3.1.2. Acoustic environment parameters. To examine the relationships between floor level and sound pressure level (SPL) in all three buildings, the nonparametric statistical test "Spearman's Rho correlation" is used. The statistical analysis shows that there is a significant negative correlation between floor levels and SPL ($r_s = -.566$, $p < .001$), highlighting the tendency for sound pressure level to increase with a decrease in the floor level and near to the lobby. In addition, one-way analysis of variance (ANOVA) is used to indicate the differences in the means of the variables. A Bonferroni Post hoc tests were also used for further analysis to determine the level of significance (p). ANOVA test results with Bonferroni post hoc test show differences in the means of the variables (sound pressure level - SPL for different floors - SPL for different room types). Floor level had a statistically significant effect on the change in sound pressure level ($F = 10.273$, $p < .001$). The effect of room type on the change in sound pressure level was also statistically significant ($F= 115.737$, $p= .005$). All floors and both types of rooms (halls and classrooms) were determined to be significant when Bonferroni adjusted for the number of comparisons. Statistical tests and their results highlight that floor and crowd level (here considered the number of students) as one of the physical factors that influence sound pressure levels. Although humans in the learning space are expected to act as absorbers, human-related activities were found to increase the analysed parameter SPL.

3.1.3. Demographic and social factors. Demographic factors are an important factor that influences any auditory perception study and therefore one should keep in mind that factors such as age, gender, and education level are key elements that influence auditory perceptions in indoor spaces, although the influence may vary with different types of indoor spaces and sounds. Regarding age, the result of the Mann-Whitney U Test, the difference between common sound sources perception of students age between 18-20 and students age between 21-23 were found significant in slamming doors, browsing pages, tools noise, mobile phone, traffic noise, and noise from neighboring spaces and in coughing, slamming doors, mechanical fan, lamb, construction noise, and noise from neighboring spaces for sound source annoyance. The difference between common sound sources perception of students age between 18-20 and teaching staff age between 24-45 were found significant in traffic noise ($p < 0.05$, $U = 1021.5$, $Z = -2.326$) and construction noise ($p < 0.05$, $U = 740$, $Z = -3.380$). While the difference between common sound sources annoyance of them were found significant in computer / projector ($p < 0.05$, $U = 773$, $Z = -3.237$) and mechanical lamb ($p < 0.05$, $U = 964$, $Z = -2.115$). At last, the difference between common sound sources perception of students age between 21-23 and teaching staff age between 24-45 were found significant in traffic noise ($p < 0.05$, $U = 1617$, $Z = -3.411$), construction noise ($p < 0.05$, $U = 1601$, $Z = -3.129$), and noise from neighbouring spaces ($p < 0.05$, $U = 1432$, $Z = -3.639$). While the difference between common sound sources annoyance of them were found significant in speech inside the room ($p < 0.05$, $U = 1899$, $Z = -2.352$), walking/ footsteps ($p < 0.05$, $U = 1872.5$, $Z = -2.385$), and computer / projector sound ($p < 0.05$, $U = 1666.5$, $Z = -3.029$).

Regarding the effects of gender, the difference between common sound sources perception of both males and females were found significant in 13 sound sources. While the difference between common sound sources annoyance of both males and females were found significant in ten sound sources as shown in Figure 3. In terms of education level, significant relationships were detected between academic level and both perception and annoyance from sound sources. The difference between common sound sources perception of students and teaching staff were found significant in traffic noise ($p < 0.05$, $U = 2809.5$, $Z = -2.700$), construction noise ($p < 0.05$, $U = 1690.5$, $Z = -4.585$), and noise from neighboring spaces ($p < 0.05$, $U = 2232$, $Z = -3.558$). While the difference between common sound sources annoyance of them were found significant in noise from tools, mobile phone, computer / projector sound, mechanical fan, lamb, and traffic noise. The respondents' educational levels correlated with their soundscape evaluations. In other words, lecturers had different perceptions of acoustical environments and tend to be less annoyed by noise, it could be attributed to the fact that when in the learning spaces, a lecturer's attention is focused on their work. One interesting result is in the approach to health and mood. They have a significant effect on the perception of the soundscape. To be illustrate, a student suffering from a health condition such as a migraine or during the exam even waking up all night working on a project, was likely to have a negative mood and would not tolerate irrelevant noises.

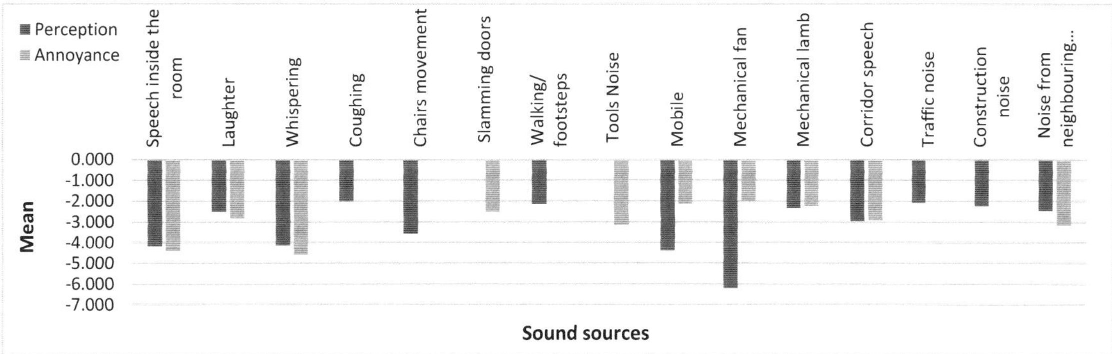

Figure 3. The Sound sources perception and annoyance according to gender.

3.1.4. Sound sources. To identify the recognized sound, participants were asked to determine the frequency of hearing of 17 sound sources in their lessons from not at all to in all my lesson indicating sound source perception, then the participants evaluate the sound sources from unpleasant to pleasant indicating sound source annoyance. The sound sources investigated include individual verbal sounds (e.g., speech, laughter, etc.), individual non-verbal sounds (i.e., coughing, walking/footsteps, etc.), environmental noise (e.g., sounds of birds, traffic noise, etc.), crowds of people (e.g., noise from neighbouring spaces, passers-by, etc.), and mechanical sounds (e.g., elevators, air conditioners, etc.). Among the perceiving sound sources, chairs movement sound was the most frequently mentioned in the Studios environment, construction noise in the lecture classroom, while speech inside the room was mostly referred to in the lecture hall and laboratory environments Figure 4. In addition, sound sources were also crucial in terms of the auditory perception of participants. The same sound sources could be perceived differently in different activities. Generally, students and teaching staff annoyance to hear sounds and negatively affected if the task in question requires concentration because any sounds were perceived as distraction noise, even if the students did not annoy to hear them in different room environments when they were doing different activities.

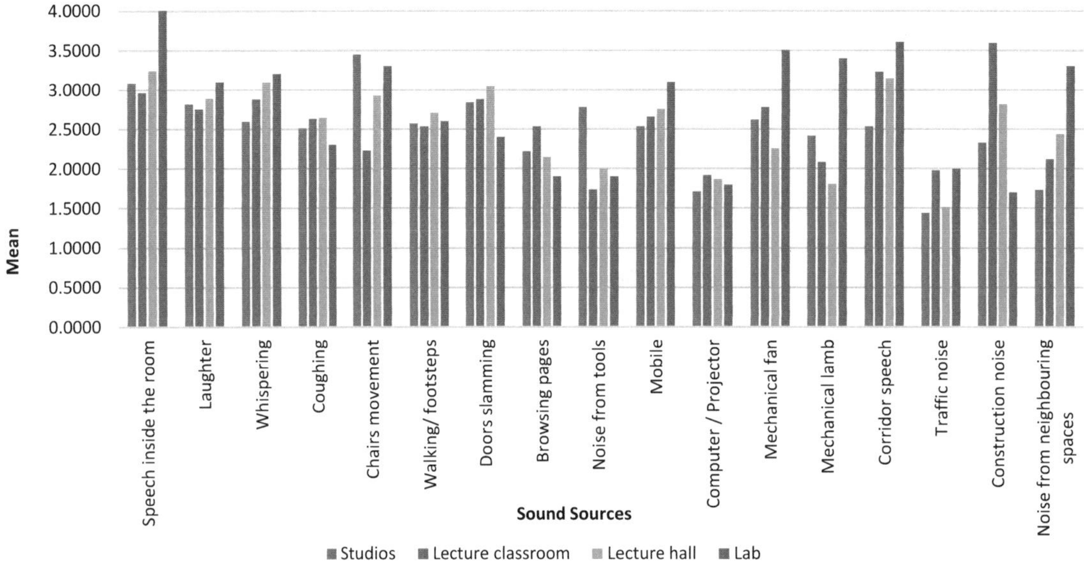

Figure 4. The frequency of the Sound sources.

As for sound sources annoyance, chairs movement, speech inside the room, corridor speech, and door slamming were ranked as significantly more annoying by participants in studios. Corridor speech, construction noise, chairs movement, and door slamming were ranked as significantly more annoying by participants in lecture classrooms. Speech inside the room, construction noise, corridor speech, and whispering were ranked as significantly more annoying by participants in lecture halls. Finally, corridor speech, mechanical fan, noise from neighbouring spaces, and chairs movement were ranked as significantly more annoying by participants in the laboratory Figure 5. According to the results of the Mann-Whitney U Test, the difference between common sound sources of studios and lecture classrooms environments was found significant in corridor speech ($p < 0.05$, $U = 2235.5$, $Z = -3.407$), traffic noise ($p < 0.05$, $U = 2513$, $Z = -2.394$), construction noise ($p < 0.05$, $U = 2222$, $Z = -3.452$), and noise from neighbouring spaces ($p < 0.05$, $U = 2741.5$, $Z = -1.637$). The difference between common sound sources of studios and lecture hall environments was found significant in chairs movement, walking/ footsteps, doors slamming, noise from tools, computer / projector, mechanical fan, and mechanical lamb. Apart from coughing and noise from tools sounds, all sound sources significant difference in perception of sound sources between common of lecture hall and laboratory environments were found.

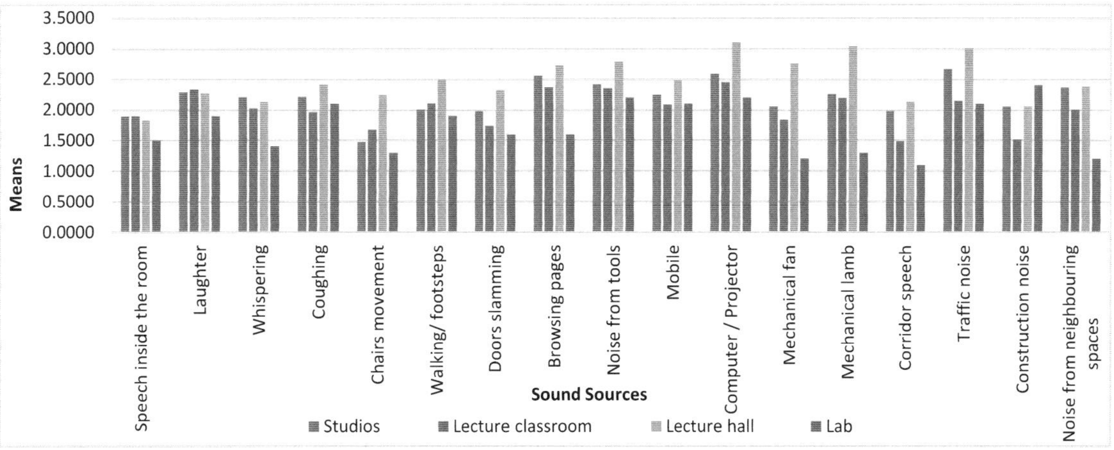

Figure 5. Sound sources annoyance.

3.1.5. Duration and time of measurements. Time is an essential factor in indoor soundscape assessment, mainly according to two dimensions: the duration of the data collection process and the time when the measurements and analysis are made. Regarding the duration of measures, questionnaires were designed to take from ten to 15 minutes. Before undertaking any soundscape assessment, participants spend from 30 to 60 minutes inside the rooms before beginning data collections to adjust to the learning spaces sound environment. Both measurements and analysis made to understand the acoustic conditions of a space. The measurements and analyses conducted on multiple days (multiple times each day). The survey time distributed across the day as usual occupancy period i.e., lecture time to sample coherent representation of the space including the varied sound environment and sources depending on the usage and function during the opening hours. Soundscapes differ over time. As individuals change their activities, their perception of the soundscape as restorative can also vary, particularly in terms of fascination and extent ratings. It was found that at noon, there was a significant increase in SPL values in the three buildings. SPL values at 9:12 AM were between 59.1 dB and 76 dB, while SPL values at 1:4 PM were between 60 dB and 83.2 dB. According to the results of the Mann-Whitney U Test, the difference between common SPL at 9:12 AM and 1:4 PM was found significant ($p < 0.05$, $U = 683.5$, $Z = -2.516$).

3.2. Discussion

3.2.1. Non-acoustic objective measurable parameters. Architectural characteristics relationships change the perception of sound and the space that produces the soundscape. These factors were circulation, vertical voids, space location, layout, furniture, and materials. They impacted the quality of the acoustic environment and, consequently, sound perception during the lecture. To illustrate, in the lecture halls and studios in Buildings A and B, students reported that, due to the location of their rooms, they always hear voices in the lobby in front of the main staircase or corridor speeches during lecture hours. The atrium in Building C causes unwanted changes and acoustic formation, resulting in high sound pressure on all floors. Therefore, the built environment can lead to changes in the sound perception of the acoustic environment. Another important part of the conceptual model is the coping method. Students improved some solutions to achieve their ideal acoustic conditions as coping methods helped them deal with existing situations. Coping methods such as closing the door, wearing headphones in the studio, and isolation. Studios and lecture halls were spacious but have a high Level of the crowd, while only lecture halls have the ability to find the way around via sound.

3.2.2. Responses and Outcomes. The method of "semantic differential" is used also to test participant's responses towards learning spaces' acoustic environments quality. According to the results of the Mann-Whitney U Test, the difference between common negatively connotated adjectives of Building A and Building B was found significant in noisy, chaotic, diverse, blurred, unpleasant, stressed, discomfort, monotonous, and distraction. The difference between common negatively connotated adjectives of Building B and Building C was found significant in noisy, chaotic, and stressed. The difference between common negatively connotated adjectives of Building A and Building C was found significant in echoey, unpleasant, stressed, discomfort, monotonous, and distraction. In other words, the participants have positive, neutral, and negative responses towards learning spaces' acoustic environments. Positive responses were exemplified as defining the words of the speakers in Building B and C. The neutral responses were exemplified as defining the words of the speakers in Building A, the similarity of the sound sources in Building B, and the room's echoey in Building C. The most negative responses were stressed, discomfort, and distracted in all three buildings.

4. Conclusion

This paper demonstrates the association between objective and subjective measures and the growing interest in indoor soundscapes to improve the performance of learning spaces. Therefore, it is important to incorporate today's auditory perception-sensitive approaches to architecture and sound engineering into the larger goal of the "well-being" index. The study was conducted in four types of rooms, studios, lecture halls, lecture classrooms, and laboratories across the three engineering colleges. Such research will help identify ways to indirectly reduce noise and improve the soundscape of a venue. Five main assessment factors for indoor soundscapes were identified as: acoustic environment, demographic and social factors, sound source, duration and time of measurements, and non-acoustic parameters, ignoring one of these factors would result in limited analysis and therefore not a complete understanding of the soundscape. Interpretation of the soundscape can be significantly affected if the context of the sound is consistent with the environment or suitable for its function. In general, different room types present a range of acoustic problems according to different activities, so when studying the interaction between humans and their acoustic environments (e.g., activities) in different buildings, the research must investigate the integration of additional variables. Non-acoustic objectively measurable parameters can also significantly affect the sound evaluation of indoor spaces. These include architectural characteristics and the physical environment. In indoor soundscape analyses, such characteristics should be regarded as potential confounding variables. The conceptual framework shows that positive perceptions of the soundscape contributed to defining the speaker's words in both male buildings, but neutral responses were also exemplified as echo and similarity, while negative responses to the soundscape Perception caused stress, discomfort, and distraction in all three buildings. This is related to the fact that increasing the positive perception of the soundscape will improve learning acoustics quality and classroom performance, which are the main goals of educational spaces.

Acknowledgements

We appreciate the support of the students of faculty of engineering and thank the heads of the Departments and teaching staff for their effort and cooperation during conducting the questionnaire.

References

[1] R. M. Schafer, *The tuning of the world.* New York, NY: Knopf, 1977.
[2] ISO. ISO 12913-1:2014(E) Acoustics - Soundscape - Part 1: Definition and conceptual framework. Geneve: Switzerland; 2014.
[3] ISO. ISO—Acoustics—Soundscape Part 2: Data Collection, *Geneva: ISO,* 2018.
[4] ANSI. American National Standard Acoustical Performance Criteria, Design Requirements, and Guidelines for Schools, Part 1: Permanent Schools. Acoust 2010.
[5] Sabine, W. C. (1964). Collected papers on acoustics. Reprint of the Harvard University Press in 1922 (Dover Publications, New York, NY).
[6] The Egyptian Environmental protection law No.4/1994 and its amendment law No. 9/2009.
[7] ISO 3382. (1997). Acoustics — Measurement of the reverberation time of rooms with reference to other acoustic parameters.

Passive measures for energy-efficient economic class housing units in national housing projects

Walaa S.E. Ismaeel[1, *]

The British University in Egypt, Department of Architecture, Faculty of engineering, El-Sherouk City, Egypt.

*Corresponding Author Email: Walaa.Salah@bue.edu.eg

Abstract. This study investigates passive building design measures and how they may differ according to the surrounding context. This particularly stands as an important consideration for economic class housing units which are seen as repetitive cluster units of the same architectural design plan. These units have a typical design that does not respond to their contexts, hence, the study proves with evidence that the building performance varies from one city to another, taking three cities Cairo, Alexandria and Aswan of different climatic zones. This is carried through a series of computer simulations using a validated software program to investigate the effect of adding passive design measures on building performance e.g. thermal walls, double glazing system, and shading devices. The results verified the significant difference in building energy performance in the three studied cities. It also showed that shading devices are a common passive design measure that contributed towards reducing energy consumption in the prototype building in the three cities. This showed the necessity of integrating proper passive design features in economic class housing units because they represent a repeated prototype building.

Keywords: Building energy performance, Economic class housing units, Energy-efficiency, National housing projects, Passive measures.

1. Introduction

The current global goal is to reduce the climate change and global warming problems caused by Green House Gas (GHG) emissions, with a focus on CO_2. International agreements, such as the Kyoto Protocol and the Intergovernmental Panel on Climate Change (IPCC), set targets and unify goals for emission reduction by imposing emission control strategies. The IPCC presented methods to achieve an 80% reduction in carbon emissions by 2050 from the 1990 base emissions [1].

Buildings accounted for 39% of end-use energy consumption and 42% of carbon emissions in 2009, according to the US Energy Information Administration. The building sector cause growing environmental impact, particularly in terms of energy consumption, raw material exploitation, and water use, resulting in solid waste and carbon dioxide emissions [2–4].

Developments of the green building industry account for the efficiency of materials and resources throughout different project phases and requires adopting an integrated design approach [5–7]. This has a long term benefit in managing the significant increase in energy costs, decreasing the environmental impact by reducing GHG emissions, and enhancing the overall value of green buildings. The Egyptian government has set a goal of reducing CO_2 emissions by 20% by 2030 through implementing policies, incentives, and methods that promote energy efficiency and reduce emissions [8]. According to Attia et al. (2017), in 2010/2011, residential buildings in Egypt utilised roughly 51 TWh and 7 Mtoe energy, accounting for a percentage of 41.10 of the overall power consumption. Considering that Egypt has over

20 million residential flats and over 6 million mixed-use apartments, it's clear that the residential sector needs rapid attention. It accounts for 50% of carbon emissions, and these emissions are constantly increasing. This requires efforts from all concerned parties [9]. Hence, concerns about building performance are increasingly becoming an integral part of the building sector. This aims at reaching a sustainable building performance that result in minimal environmental impact [10]. This is particularly important in governmental national housing projects (GNHP) which include prototypical designs repeated in many cities in Egypt. Hence, a thorough investigation of the design implications can lead to significant change for the large number of units executed each year. Thus, the authors argue that applying passive measures to improve the building performance in this type of project is essential but draws attention to the suitability of the measures employed to the prevailing climatic conditions. Hence, this paper studied prototype building of implemented GNHP, tested and compared building performance to the original state. Eventually, this presented a set of design recommendations for similar large scale projects. Therefore, the results can be quite beneficial for future projects that will take into consideration the importance of climate responsive building units.

2. Methodology

The research method was explained in the following sequence of steps shown in Fig. (1). This started with a thorough review of the literature to trace the development of GNHP and how this reflected the design of individual residential units in three different cities of different climatic zones: Cairo, Alexandria and Aswan. Then, recent case study projects were modelled and their energy consumption was simulated using the Design builder software program. Following that, a series of passive design measures were applied to investigate their effect on reducing the building's energy consumption for each city.

The data-gathering phase comprised international and national contexts. Energy-efficient design guidelines were investigated from international green building rating systems e.g. Leadership in Energy and Environmental Design (LEED) and Building Research Establishment's Environmental Assessment Method (BREEAM). This was integrated with a latter step searching local building codes and the national Green Pyramid rating system (GPRS). This pointed to recommendations to apply passive design measures, particularly interventions of applying thermal walls, double glazing systems, and shading devices. Then, the later step investigated national guidelines for designing GNHP to be able to select representative case study buildings. A simulation method was followed to test the combinations of recommended passive design measures and pinpoint optimum design scenarios. Eventually, the research hypothesis was tested and a set of recommendations were proposed for planning for GNHP projects.

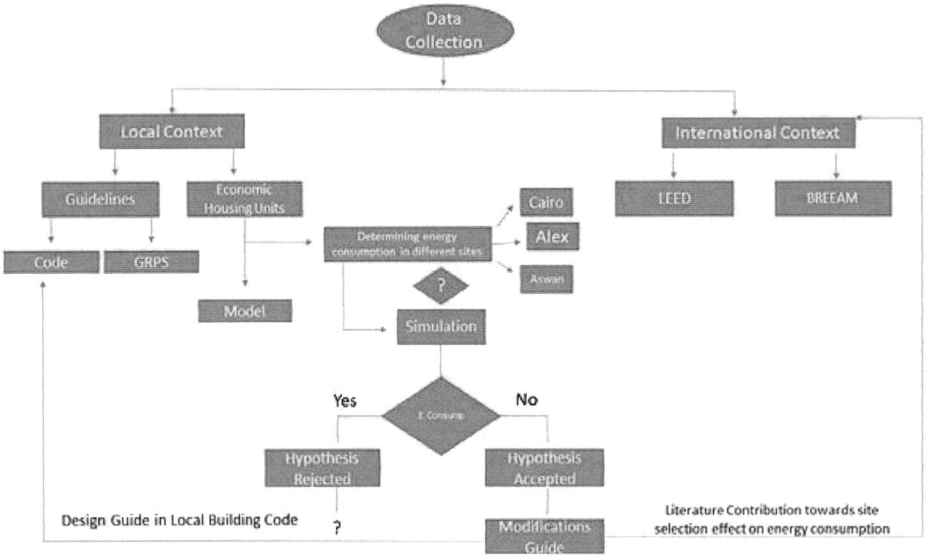

Figure 1. Research methodology.

3. Literature review

A primary review of existing literature conducted over the last ten years utilising the Science Direct database has resulted in the revision of books, journals, working papers, and technical reports. The following keywords were used in the web search; Building performance, Economic Class Housing Units, Energy efficiency, National housing projects and Passive design to determine previous experiences and draw scientific contribution directions This demonstrated that this is rising attention for applying passive design measures to reduce buildings' energy consumption.

Passive design is a technique used in the design of buildings to improve architectural design used while reducing consumption of energy to enhance overall thermal comfort for users. Passive design can be applied in most buildings to improve design procedures while ensuring the results will accomplish less consumption of energy while ensuring that it will not influence climate change. The major aim of passive design is to remove any active mechanical operations while ensuring the comfort of users [11,12]. During the past few years, most designers and architects in Egypt did not take into consideration the use of certain techniques to reduce impacts on the environment. Passive design strategies are not applied in most architectural designs including the use of proper lighting, natural lighting, decreasing the use of air conditioning by providing proper ventilation system, use of shading systems to ensure the comfort of users and finally providing better orientation for the building [13].

Passive cooling is defined as one of the energy efficiency techniques that can reduce the economic impact for buildings. Several studies in hot weather conditions demonstrated that high energy efficiency could be accomplished through the use of passive cooling techniques. Therefore, passive cooling techniques and systems accomplished greater attention by most designers, architects and engineers. In addition, multiple stages are needed to reach such passive cooling where the initial step is to concentrate on designing the building to reduce and decrease the energy needed such as taking into account the orientation of the building and building layout. Any building must be planned with the regions and climate characteristics in mind. The passive cooling design can be improved by reducing the internal gain of heat within the building [14,15]. Because the site can be affected by a variety of circumstances, including economic factors and zoning rules, site design and planning are considered to be important contributors in reducing internal gain. The first step in reducing heat gain is to properly manage solar radiation. The second stage in the basic principle of passive cooling is to change the building's model, such as the types of windows, shading systems, and wall or roof materials. To ensure the design of various parts in the building, the thermal design must be effectively handled. The building envelope is one of the techniques available for estimating heat modulation in the building during the day, estimating inside temperature, and reducing heat gain inside the structure throughout the day. In most cases, the building envelop is the most important aspect in reducing heat loss, especially in the winter. On the other hand, the third stage of passive cooling focuses on eliminating any additional heat in the building through the design of a building layout that can eliminate any heat inside the building which is not needed. The elimination of additional heat in the building mainly relies on two major conditions, the first one is the availability of an adequate environment to sink the heat, while the second one is to provide proper thermal couple between building and sink to ensure transfer of heat and that the temperature is well preserved. Various processes can be applied to offer certain savings in the consumption of energy [16].

Climate-response building units: improper decisions might influence the required sustainability features. The adequate selection of sites can play a major role in indicating the required performance in terms of energy [17].

Building Orientation: Daylighting and solar gain are the main factors that are directly related to the proper orientation of buildings. The orientation can define the potential exposure to solar radiation at which can offer greater use of natural daylight, thus offering extensive save in the used energy for artificial lighting [18].

Energy-efficient electric appliances: The proper use of energy appliances can provide several benefits in terms of reducing utilities. Even though that the initial installation cost of such appliances can be higher than the basic ones, but it can offer extensive savings and provide needed comfort for users [16].

Ventilation and Heating systems: The reduction in the heat loads within any building can allow the installation of fewer and smaller cooling and heating systems. The significance of ventilation systems that has high quality can be introduced during the design stage, but the major concern is green buildings [19].

4. Case study

Egypt belongs to the hot-arid region, with an average annual temperature of 18 degrees and the maximum possible temperature might reach 45 degrees. The Egyptian Typical Meteorological Year Authority stated that the annual average temperature in the city of Cairo was recorded to be 22.4 degrees while the highest average temperature was recorded to be 35.4 degrees in the summer, and 18 degrees in the winter. Therefore, passive design and climate response building units are major parameters that should be taken into account in building design with further considerations for site conditions [20–22]. The case study discusses governmental economic housing units that are currently being assembled in multiple new settlements in different cities in Egypt to ensure affordable living places for youth. The basic design of each building is quite similar with 3 different zones of land that covered an area of a 300-meter square. This is shown in Fig. (2). In addition, the chosen models had 2 various architectural designs to provide the capability of changing it towards better efficiency in terms of design and to reach higher and better outcomes. Thus, the same floor area was applied in 2 various architectural plans, the initial one as a single unit and the second plan was designed as twin units on the shorter side of the land. Typically, each building included 4 floors starting from the ground floor, and the other 3 upper floors. Three different locations were chosen such as Cairo, Alexandria, and Aswan. Each location has different characteristics at which can yield different energy performance where one of them has moderate weather conditions, hot weather conditions, and humid weather conditions. Other parameters are briefly described in Table (1).

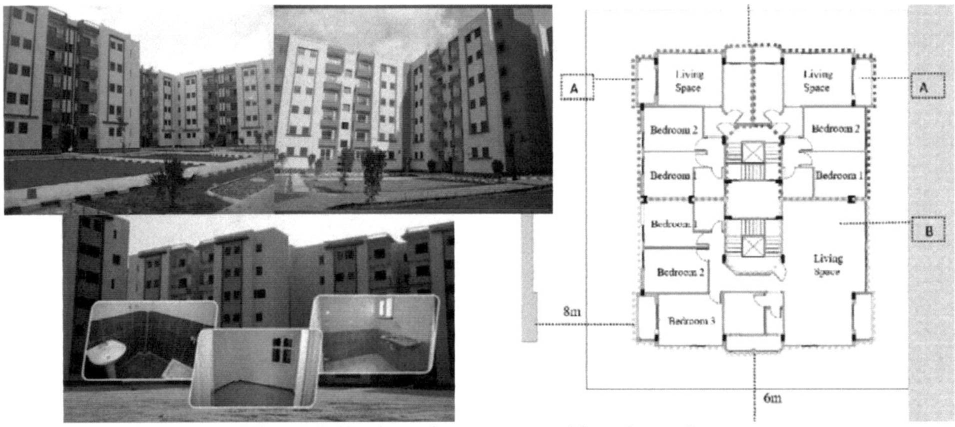

Figure 2. Design of governmental housing units.

Table 1: Properties of the case study building.

Location	Three alternatives: **Cairo** 30.0074° N, 31.4913° E, 250 m above sea level **Alexandria** 30.8761° N, 29.7426° E, -1 below sea level **Aswan** 24.0889° N, 32.8998° E, 194 m above sea level
Orientation	Front Elevation facing West
Floor to Floor Height	3.00 m
Occupancy Density	5
Floor Area	240 (Gross):240m^2 (Ground Floor), 240m^2 (First Floor), 240m^2 (Second Floor),
Window to Wall Ratio	30 %
Exterior Walls	0.2 m Sandstone (dense) +0.02 m Cement mortar (medium) +0.25 m Bricks (medium) +0.02 m Cement mortar (medium) + 0.01 m Plaster (lightweight)

Roof	Depending on the roof type applied either extensive green roof, sloped, or cool reflective roof.
Infiltration	1.25 air changes per hour (ACH) (ground floor), 0.75 ACH (typical floor).
HVAC system type	Residential system direct expansion (DX) Air conditioner (AC fan coil).
Light Power Density	21 W/m² (ground floor), 13 21 W/m² (typical floor).
Window	0.004 m tinted glass + 0.012 m air gap +0.04 m tinted glass.

5. Research Method

The Design builder software program was used to carry the simulation to estimate the energy performance of the building. It offers the capability to investigate different passive techniques that could be applied to this building e.g. shading effect and takes into account the influence of adjacent buildings. The building was oriented towards the east direction. In addition, the simulation analysis was applied to estimate cooling load and indoor temperature- focusing on the upper floor plan for similar as it is subject to the greatest solar radiation. Then, the building was divided into several thermal zones at which can aid in obtaining full information regarding the use of energy and provide more reliable and efficient data for thermal simulation. It was divided into 4 main sections: northeast, northwest, south-east and south-west. The simulation is shown in Fig. (3).

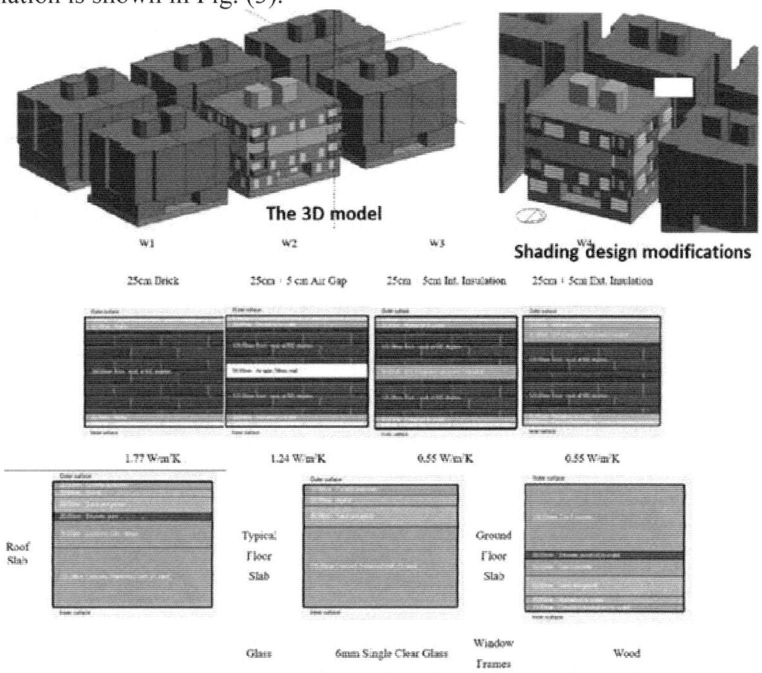

Figure 3. Modelled urban unit and wall design alternatives.

The passive measures tested were:

Roof Insulation: the roof of any building is normally the highest one to be exposed to direct solar radiation. Enhancing the U-value for roofs played a major step towards upgrading the thermal performance of the building and ensuring that it will have proper thermal insulation. Thus, the design of this roof was changed to include an insulation material i.e. Expanded Polystyrene Standard with a thickness of 5 centimetres.

Types of Walls: material selection for the type of wall have a huge influence on the consumption of energy through transferring the heat into the building [23], consequently 4 different types of walls were tested as exterior building materials. Each type was compared with the base wall which was the use of a red brick wall. Each wall was designed to include 3 various layers, the first one was constructed using a brick surface and had a thickness of 10 inches. the second layer is an air layer with 5 inches of thickness. The third layer was a bricklayer and had a dimension of 10 inches.

Types of Glazing: Glazing is the main element responsible for around 30 percent of energy load within the building, and has a huge influence on the thermal comfort of users because it is mainly related to indoor air temperature [12,13,24]. In the case study, the height of the glazing system was 1.5 meters and represent 30 percent of the façade area and they were made from aluminium window frames. Thus, 3 alternatives of glazing were tested: single tinted (blue) 6 mm of 5.77 W/m^2K, Double clear 13 mm with air gap and of 2.66 W/m^2K and Double tinted (blue) 6 mm with 13 mm air gap and of 2.66 W/m^2K.
Window Shading: Several studies showed the positive support and influence of shading components on energy gain especially in hot and dry weather conditions [12]. This should achieve a balance between thermal performance and daylight in the building [13]. The model was simulated to include vertical shading fins on the east and west window faced facades and horizontal overhangs on the southern ones. Passive measures are described in Table (2).

Table 2. Describing the proposed passive measures.

Design Aspects	Thermal Wall	Glazing	Shading devices
Description	Changing the layers of the wall to reduce thermal transfer	Testing single and double glazing with different specifications	Protect windows from direct solar radiations
Design procedure	2 bricklayers, and 1 air layer that divides them	Include an insulation layer of air or argon gas between the two layers of glazing	Shading devices can be made of wood, or aluminium which takes the width and depth of the window depending on its location

6. Results & discussion
Comparing the three proposed passive measures (Shading devices, Thermal walls and Double glazing) for the building model in the three cities is shown in Fig. (4). Comparing their Total site energy in Fig. (4.1) and Net source energy in Fig. (4.2) shows that the same building unit has various energy consumption in different locations i.e. highest in Aswan, followed by Cairo then Alexandria. In addition, it indicates that the use of shading devices is an efficient passive measure in all three locations, which represent different climatic zones.

7. Conclusion
This study discussed passive measures for energy-efficient economic class housing units in national housing projects. It showed that for the same model, the energy consumption of the typical unit differs according to the prevailing climatic condition. This was clear when comparing the energy consumption for the unit in three cities in Egypt which belonged to different climate zones. It also showed that some design guidelines should be incorporated in these building models to reduce their energy consumption e.g. the use of thermal mass, energy-efficient alternatives of glazing systems and proper design of shading devices. Hence, the design of these major scale housing projects should not be repeated across different cities and site locations but should follow a climate-responsive design approach. Furthermore, implementing passive design features is a must to ensure an energy-efficient design. Proper design of shading devices was found to result in the least energy consumption when tested in all three climatic zones, hence, it is one of the recommended design guidelines, followed by energy-efficient glazing and using thermal mass for exterior walls. Hence, a mix of these passive measures shall reduce energy consumption compared to the base case scenario. The outcome of this study is useful for designers of national housing governmental projects. The results of the simulation can further be developed to measure and validate findings indoors to make sure that individual building units perform as expected.

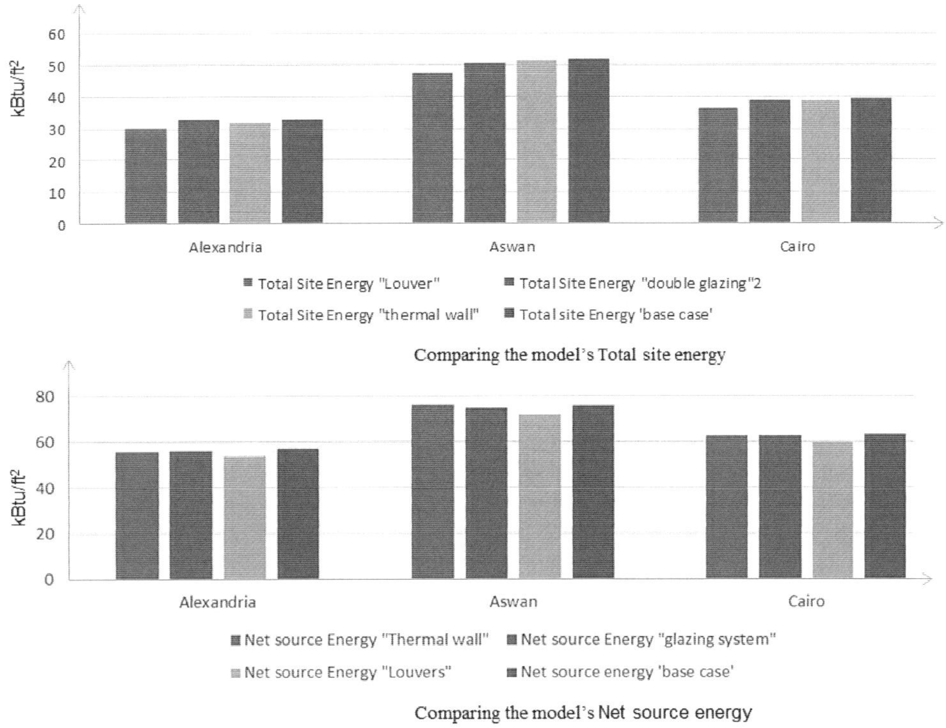

Figure 4. Comparing the building energy performance of the three proposed alternative measures.

References

[1] IPCC. Climate Change 2014: Mitigation of Climate Change: Contribution of Working Group III to the Fifth Assessment Report of the Intergovernmental Panel on. Cambridge Univ Press … 2014:1132 pp.

[2] Elsayed MA, Ismaeel WSE. Environmental assessment for major development projects: A case study "Qattara Depression." J Clean Prod 2019;215:522–33. https://doi.org/10.1016/j.jclepro.2018.12.301.

[3] Morsi DMA, Ismaeel WSE, El Hamed AEA, Othman AAE. Applying LCA-BIM integration for a sustainable management process. ARCOM 2020 - Assoc Res Constr Manag 36th Annu Conf 2020 - Proc 2020:416–24.

[4] Morsi DMA, Ismaeel WSE, El-Hamed AEA. Life cycle assessment of a residential building in Egypt: A case study. IOP Conf Ser Mater Sci Eng 2020;974:7–9. https://doi.org/10.1088/1757-899X/974/1/012028.

[5] Ismaeel WSE. Drawing the operating mechanisms of green building rating systems. J Clean Prod 2019;213:599–609. https://doi.org/10.1016/j.jclepro.2018.12.115.

[6] Ismaeel WSE. An integrated model for energy-efficient building opening design. ARCOM 2020 - Assoc. Res. Constr. Manag. 36th Annu. Conf. 2020 - Proc., 2020, p. 576–84.

[7] Ismaeel WSE. An integrated model for energy-efficient building practices. 9th Int. Conf. Softw. Inf. Eng. (ICSIE 2020)-accepted Pap., 2020, p. 204–8. https://doi.org/10.1145/3436829.3436859.

[8] GOPP. General Organization for Physical Planning 2017.

[9] Attia S, Hamdy M, Ezzeldin S. Twenty-year tracking of lighting savings and power density in the residential sector. Energy Build 2017. https://doi.org/10.1016/j.enbuild.2017.08.041.

[10] Dabaeih M, Kenawy I, Ismaeel WSE, El-Sayed MA. Carbon mapping for residential low carbon retrofitting. Adv. Technol. Sustain. Syst., vol. 4, 2017, p. 79–91.

[11] Mahmoud S, Ismaeel WSE. Developing sustainable design guidelines for roof design in a hot arid climate. Archit Sci Rev 2019;0:1–13. https://doi.org/10.1080/00038628.2019.1665984.

[12] M. ElBatran R, Ismaeel WSE. Applying a parametric design approach for optimizing daylighting and visual comfort in office buildings. Ain Shams Eng J 2021;12:3275–84. https://doi.org/10.1016/j.asej.2021.02.014.

[13] Elakkad N, Ismaeel WSE. Coupling performance-prescriptive based daylighting principles for office buildings: Case study from Egypt. Ain Shams Eng J 2021;12:3263–73. https://doi.org/10.1016/j.asej.2020.09.030.

[14] Moussa RR, Ismaeel WSE, Solban MM. Energy generation in public buildings using piezoelectric flooring tiles ; A case study of a metro station. Sustain Cities Soc 2022;77:103555. https://doi.org/10.1016/j.scs.2021.103555.
[15] Ismaeel WSE, Elsayed MA. Sustainable sites in two generations of city development using GIS-MCDM and LEED LT and SS categories. J Clean Prod 2022;330:129782. https://doi.org/10.1016/j.jclepro.2021.129782.
[16] Peuportier B, Thiers S, Guiavarch A. Eco-design of buildings using thermal simulation and life cycle assessment. J Clean Prod 2013;39:73–8. https://doi.org/10.1016/j.jclepro.2012.08.041.
[17] Lutzkendorf T, Lorenz DP. Using an integrated performance approach in building assessment tools. Build Res Inf 2006;34:334–56. https://doi.org/10.1080/09613210600672914.
[18] Yun G, Yoon KC, Kim KS. The influence of shading control strategies on the visual comfort and energy demand of office buildings. Energy Build 2014. https://doi.org/10.1016/j.enbuild.2014.07.040.
[19] Chen X, Yang H, Lu L. A comprehensive review on passive design approaches in green building rating tools. Renew Sustain Energy Rev 2015;50:1425–36. https://doi.org/10.1016/j.rser.2015.06.003.
[20] Ismaeel WSE. Using SITES for Sustainable Rehabilitation of Cultural Heritage Landscape Areas; Case of Giza Zoo. Adv. Technol. Sustain. Syst., vol. 4, 2017, p. 1–9. https://doi.org/10.1007/978-3-319-48725-0.
[21] Ismaeel WSE. Systems thinking and modelling for sustainable site selection. In: In: Gorse, C and Neilson CJ (Eds), editor. 35th Annu. ARCOM Conf., Leeds, United Kingdom: Associations of researchers in construction management; 2019, p. 437–44.
[22] Ismaeel WSE. Sustainable site selection using system dynamics ; case study LEED-certified project. Archit Eng Des Manag 2021;0:1–19. https://doi.org/10.1080/17452007.2021.1889955.
[23] Walaa S.E. Ismaeel. Material selection for sustainable buildings. ARCOM, 2021, p. 784–93.
[24] Maskarenj MS, Avasare M, Ghosh PC. Analysis of Plastic Optical Fiber Based Daylight System Suitable for Building Applications. Appl Mech Mater 2014. https://doi.org/10.4028/www.scientific.net/amm.492.101.

Thermal Performance Assessment for Stacking Shipping Containers as habitable spaces: Cairo Case Study

A S Hassan[1], C R Botros[1], M N Kamal[1], N M Morsi[1] and Y S Elfeky[1,*]

[1] Department of Architecture, Faculty of Engineering, Ain Shams University, Cairo, Egypt

*Corresponding Author Email: yasmeen.sayed19@eng.asu.edu.eg

Abstract. The world is moving towards reuse, recycling, and low economic buildings in the AEC industry. Among these concepts, the reuse of shipping containers (SC) in architecture has seen a remarkable increase in recent times. Few studies have focused on the thermal performance of different stacking configurations using multiple SCs in hot climates to achieve the most suitable thermal performance and testing different passive design techniques. This study aims to know the most effective stacking configuration as a passive technique not tested before using SC as residential spaces in Cairo by analyzing its effect on their internal thermal performance. This paper presents a comparative analysis of four simulation models as separate residential units, including a control unit, an uninsulated SC, and three types of different configurations, three units horizontally connected, three units vertically connected, and the last model consists of three units connected both horizontally and vertically. The thermal performance of these units is tested as uninsulated SCs in the hottest and coldest days (23 August and 20 January, respectively) in the year using Design-Builder software. The paper concludes that the best configuration of SCs models that contribute to improving SC's thermal performance in the hot climate is the vertical stacking configuration.

Keywords: shipping containers, stacking, thermal performance, hot climate, residential unit.

1. Introduction

The world is moving towards the concepts of reuse and recycling in the AEC industry. Among these concepts, the reuse of shipping containers (SC) in architecture has witnessed a remarkable increase in recent times. SC is characterized by a solid and durable structure, besides, their uniformity and modularity, which makes these prefabricated units suitable for being treated and reused economically in less time than the traditional concrete structures [1]. On the other hand, in terms of energy conservation, the energy used to melt and recycle (Sc) is 20 times the energy used to reuse them [2]. Therefore, SC provides a great opportunity to provide architectural spaces suitable for multiple human activities.

Egypt has many seaports on the coasts of the Mediterranean and the Red Sea, where there is a surplus of SCs, due to the increase in imports over exports. This surplus represents a promising potential that could help provide rapid solutions for the housing problem, especially in Cairo, to cope with the increasing demand for residential buildings resulting from the growing population there [3]. Also, researchers raised concerns over the habitability of SC regarding its environmental performance, additional costs, and construction obstacles [4].

While assessing the thermal performance of container-based buildings in Port Said, a hot and humid region, through conducting a comparative analysis of six simulation models; including a conventional building as a base model, an uninsulated SC, and four externally insulated SCs with four different thermal insulation materials, Ghada had found that thermal insulation is indispensable in SCs reused for habitation [4]. Also, Fulhan Antony conducted a similar field experiment to measure the thermal performance of an uninsulated container and two insulated containers of the same size through the indoor and outdoor temperature and wind speed values. The results recorded that insulated containers have an increase in the internal temperature following the change in external temperature by a noticeable time [5].

Moreover, other studies have shown that the different designs for the SC could affect achieving energy efficiency and thermal performance of using shipping containers in residential buildings in the Egyptian coastal area. Ashraf Al-Mokadam had suggested six design models, based on using SCs, arranged together horizontally in different ways according to the design criteria of residential buildings while considering the general requirements of economic housing design in Egypt. He had predicted the environmental performance of each design type by using simulation programs and concluded the most suitable design achieved the highest environmental performance [3]. According to the literature, there is a potential in exploring the effect of arranging multiple uninsulated SC of not only horizontal but also vertical arrangements to achieve the most effective configuration thermal performance. The aim is to know the most effective stacking configuration using SC as residential spaces in Cairo by analyzing its effect on their internal thermal performance.

2. Container Architecture

Container architecture began in the USA in 2006 in a two-story house, the first home of its kind that met all building code requirements. Also, in 2005, 1,000 student housing units were constructed in Amsterdam in twelve five-storey container buildings, [2] and there are many housing examples of reused containers as shown in Figures 1 and 2. Various countries also presented innovative models for reusing containers in many projects, including fast-food restaurants. It is also used as shops, exhibition halls, and waiting areas [7].

In Egypt, the idea of using old shipping containers is not new, but its use is very limited. Most of the uses are for the construction of labour housing on construction sites, small shops, and kitchens for restaurants - Figures 3 and 4 - in open spaces.

Figure 1. Redondo beach house - 2008 [8].

Figure 2. Chemin Brochu – 2007 [9].

Figure 3. The driveway by the waterway restaurant complex 2021 [10].

Figure 4. The headquarters of Cubix [10].

3. Methodology

This paper has proposed three different stacking configurations each containing three residential units using Shipping Containers. The study focuses on the effect of stacking of SCs in their uninsulated basic state without any HVAC systems, mechanical ventilation, and zone equipment. Then, a comparative simulation assessment has been presented by using the Design-Builder program according to their environmental performance on the hottest day (23 August) and the coldest day (20 January) for each of the four models.

The research is divided into three basic phases to study the effect of the types of stacking, whether horizontally or vertically, on the thermal performance inside the Shipping containers (SCs) when they are used as residential spaces:

- Phase one: Designing a residential unit model for one or two people, consisting of one shipping container.
- Phase two: Suggesting three models for stacking configurations to express the basic walls, floors, and ceiling connections of the containers and compare them to the single unit (the base case).
- Phase three: Building models and simulating their thermal performance using the Design-Builder program and extracting environmental data for each unit.

3.1. Phase one: Design parameters of (SCs) as a residential model

3.1.1. Structure details.

- The residential unit was designed from a standard 40ft High Cube Containers to represent a studio space because its height corresponds to the appropriate height of a housing unit. And its dimensions are shown in Table 1 which is suitable for the accommodation of one or two people.
- The unit's area is 30 m^2.
- Each unit contains a seating area, a sleeping area, an open-plan kitchen, and one bathroom.
- Openings: as shown in the horizontal plan in Figure 1; a door (90 cm wide), and a window on the opposite side (120 cm wide).

Table 1. Dimensions of base case container used in the simulation.

	Length (m)		Width (m)		Height (m)	
	Internal	external	Internal	external	Internal	external
40 ft High Cube Containers	12	12.2	2.34	2.40	2.71	2.89

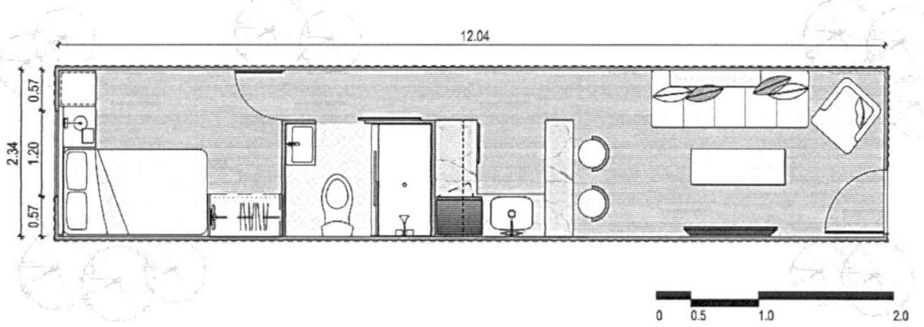

Figure 5. The suggested model floor plan, (By authors).

3.1.2. *Materials details.* Walls are made of Corten steel sheets 2 mm thick, one layer, no insulation and U-value 5.88 (W/m^2-K), and ceilings are made of steel deck 2 mm thick, one layer, no insulation and U-value 7.14 (W/m^2-K). The floor is made of 25 mm thick Plywood with U-value 2.29 (W/m^2-K). Windows are made of single glazed blue glass with a thickness of 6 mm with U-value 5.79 (W/m^2-K), and a wooden frame U-value 3.63 (W/m^2-K).

3.2. Phase two: Models configurations

- Each design model contains three units distributed horizontally, vertically, or both as these are the basic configurations, as shown in Figure 2.

IOP Conf. Series: Earth and Environmental Science 1056 (2022) 012024 doi:10.1088/1755-1315/1056/1/012024

- The units are stacked in the manner intended for stacking containers on cargo ships, so there is no need for any additional structure.
- The configurations are designed to test different ways of connecting the units' faces as follows: Model (A): Control unit, model (B): Walls connections, model (C): Ceiling-floors connections, model (D): Walls and ceiling-floors connections.

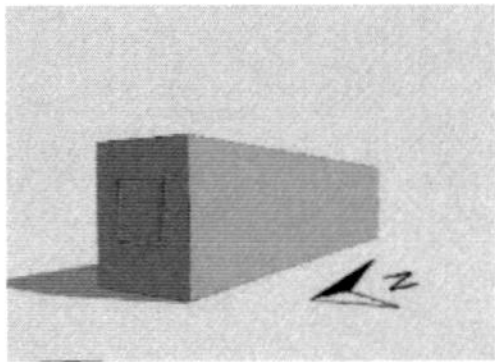

(a) Model (A) Control Unit.

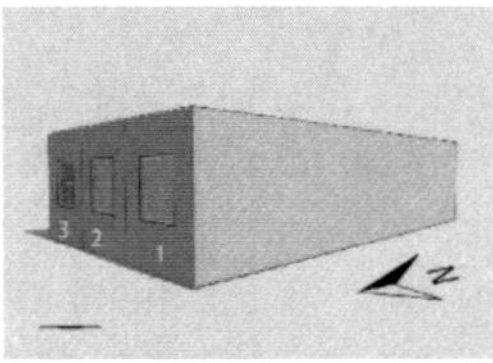

(b) Model (B).

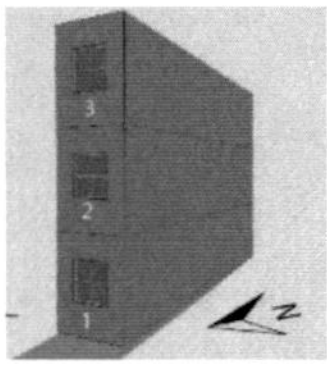

(c) Model (C).

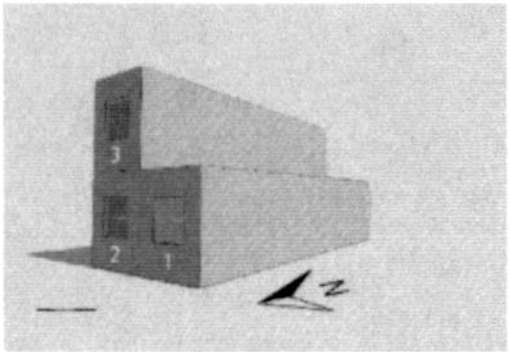

(d) Model (D).

Figure 6. The suggested configurations, (By authors).

3.3. Phase three: Simulation of suggested models

Design-Builder analyzes the four models by:

- Building the control unit, entering the materials data, and setting model design setting as shown in table 2.
- Creating the three configurations composed of 3 units.
- Orientation of the four models so that the opening of the second window (width 120 cm) is on the north side.
- Choosing the weather data for Cairo Governorate available in the program.
- Extracting the average hourly readings of Operative temperature (°C) - Pierce PMV – Relative Humidity (%) for the hottest day (23 August) and the coldest day (20 January) for each of the four models.
- Comparing outputs according to simple ASHRAE standards. [6]

Table 2. Model design setting.

Window orientation	North
Window percentage	17%
Window dimensions	1.2m*1.2m
Activity level of simulation (template)	Domestic lounge
Occupancy density (people/m²)	0.0188
Occupancy schedule	average hourly
Fresh air supply (I/S-person)	10.00
Heating setpoint (°C) Cooling setpoint (°C)	12 28
Metabolic rate (met)	1.00
Clothing level (clo)	Summer clothing: 0.50 Winter clothing: 1.00
Air velocity (m/s)	0.14
HVAC template	No active cooling systems
Mechanical ventilation	Non
Natural ventilation	Through opening

4. Results

Tables 3 and 4 refer to the recorded results of operative temperature, pierce predicted mean vote (PMV), and relative humidity; by simulating the average of total hours of both the hottest and coldest day of the year, respectively.

Table 3. Results of the four different models throughout the hottest day (23 August).

Model (A)	Unit A1			Mean
Operative temperature	36.57	-	-	36.57
Pierce PMV SET	2.64	-	-	2.64
Relative Humidity	40.45%	-	-	40.45%
Model (B)	**Unit B1**	**Unit B2**	**Unit B3**	**Mean**
Operative temperature	37.37	37.56	37.09	37.20
Pierce PMV SET	2.86	2.93	2.82	2.87
Relative Humidity	37.61%	36.25%	37.82%	37.23%
Model (C)	**Unit C1**	**Unit C2**	**Unit C3**	**Mean**
Operative temperature	35.30	35.5	36.13	35.64
Pierce PMV SET	2.52	2.55	2.57	2.55
Relative Humidity	40.03%	40.14%	41.15%	40.44%
Model (D)	**Unit D1**	**Unit D2**	**Unit D3**	**Mean**
Operative temperature	36.39	35.41	36.32	36.04
Pierce PMV SET	2.69	2.58	2.60	2.62
Relative Humidity	39.21%	38.94%	40.86%	39.67%

Table 4. Results of the four different models throughout the coldest day (20 January).

Model (A)	Unit A1			Mean
Operative temperature	15.69	-	-	15.69
Pierce PMV SET	-0.77	-	-	-0.77
Relative Humidity	58.77%	-	-	58.77%
Model (B)	**Unit B1**	**Unit B2**	**Unit B3**	**Mean**
Operative temperature	16.14	16.33	16.03	16.16
Pierce PMV SET	-0.74	-0.72	-0.75	-0.74
Relative Humidity	56.55%	55.21%	56.78%	56.18%
Model (C)	**Unit C1**	**Unit C2**	**Unit C3**	**Mean**
Operative temperature	15.97	15.66	15.29	15.64
Pierce PMV SET	-0.77	-0.80	-0.83	-0.80
Relative Humidity	56.30%	57.55%	60.06%	57.97%
Model (D)	**Unit D1**	**Unit D2**	**Unit D3**	**Mean**
Operative temperature	16.09	16.23	15.48	15.93
Pierce PMV SET	-0.75	-0.75	-0.81	-0.77
Relative Humidity	56.16%	54.47%	59.32%	56.65%

4.1. Operative Temperature

4.1.1. *For the hottest day (23 August).* According to the results of measuring operative temperatures for the four models (figure 7), unit “C1” recorded the least temperature (35.30°) between all units, including the control unit A (36.57°), while “B2” recorded the highest temperature of 37.56 °.

IOP Conf. Series: Earth and Environmental Science 1056 (2022) 012024 doi:10.1088/1755-1315/1056/1/012024

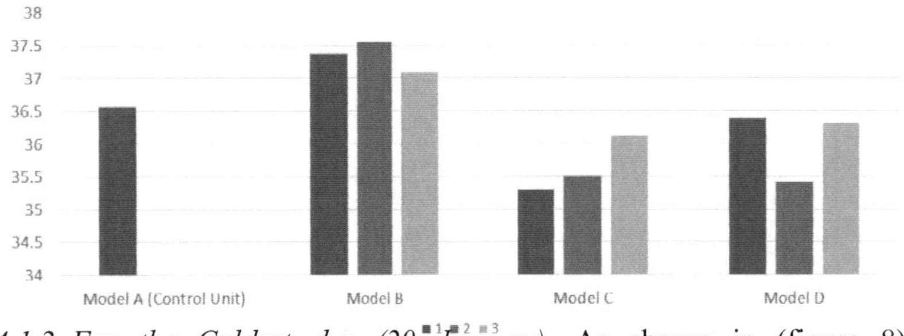

Figure 7. Comparison of operative temperature for each unit of the four simulation models (23 August).

4.1.2. For the Coldest day (20 January). As shown in (figure 8), unit "C3" recorded the lowest temperature of 15.29°; however, unit "B2" had the highest operative temperature of 16.33° while the control unit "A" recorded 15.69°.

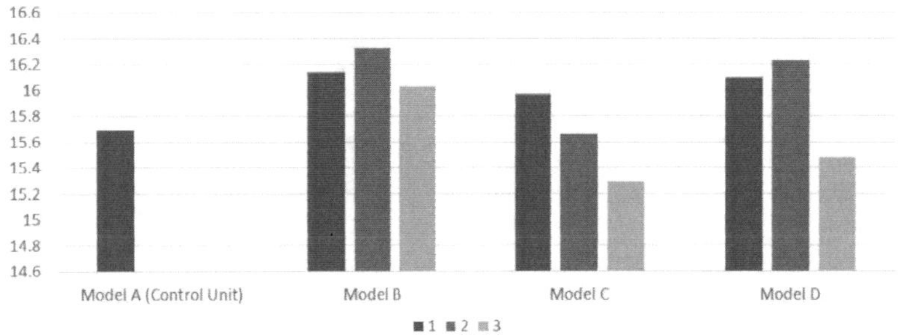

Figure 8. Comparison of operative temperature for each unit of the four simulation models (20 January).

4.2. Pierce PMV SET

4.2.1. For the hottest day (23 August). In figure 9, the PMV of the four models, with unit "C1" has the PMV closest to the comfort zone, 2.52, and the unit "B2" has a PMV of 2.93 furthest from the comfort zone. These PMV results show all the units to be in the warm to the hot zone. However, compared to the control unit "A" we see that the units "C1, C2, C3, D2, and D3" all have better PMV scores.

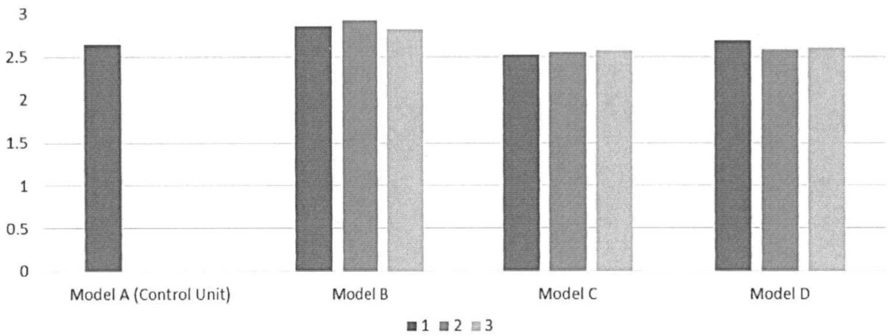

Figure 9. Comparison of pierce PMV for each unit of the four simulation models (23 August).

4.2.2. For the Coldest day (20 January). In figure 10, all the units have a PMV between 0 and -1 which is slightly cool, but within the comfort zone, with unit "B2" being the closest to the comfort zone with a PMV of -0.72 and unit "C3" being the furthest away with a PMV of -0.83, they are both relatively close to the control unit "A" with PMV -0.77.

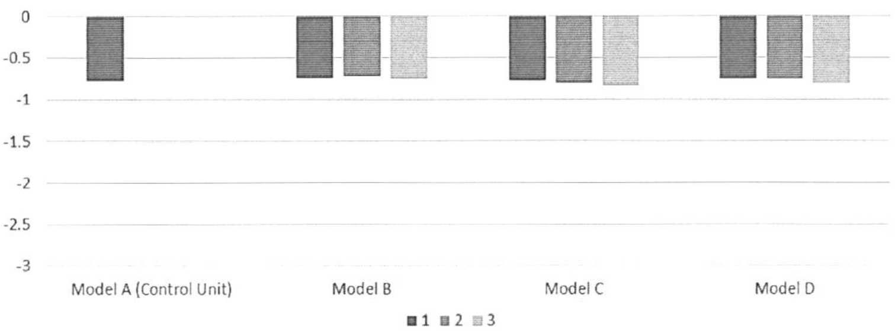

Figure 10. Comparison of pierce PMV for each unit of the four simulation models (20 January).

5. Discussion

On the hottest day (23 August), the comparative analysis in table 5 represents units` operative temperatures are weighted from the lowest (1) to the highest (9). The units highlighted with yellow have higher operative temperatures than the control unit. For the relative humidity, the units with an acceptable range have weighted 1, and 2 is given to the units out of the range. For the PMV values, the weight of the unit is from 1 to 9, with "1" being the closest to the neutral state ASHRAE sensation scale. The units highlighted with yellow have worse values than the control unit. Thus, all units of model C decreased in their operative temperature and an acceptable range of relative humidity (40%). The PMV in all models was recorded hot on the ASHRAE thermal sensation scale even when stacking is applied, but at the same time, vertical stacking helped in decreasing the PMV value compared to the control unit. As a result, vertical stacking is the most effective configuration during hot days.

Table 5. Comparative analysis of the 3 models on Hottest Day (hourly average).

	Operative Temperature			Relative Humidity			PMV		
	Model B	Model C	Model D	Model B	Model C	Model D	Model B	Model C	Model D
Unit 1	8	1	5	2	1	2	8	1	6
Unit 2	9	3	2	2	1	2	9	2	4
Unit 3	7	4	6	2	1	1	7	3	5
Total	24	8	13	6	3	5	24	6	15

All the nine units are weighted from 1 to 9 and arranged from the most affected unit by stacking during the hottest day (Weighted 1) to the least affected unit (weighted 9).

On the coldest day (20 January), the comparative analysis in table 6 represents units` operative temperatures are weighted from the highest temperature (1) to the lowest (9). The units highlighted with yellow have lower operative temperatures than the control unit. The relative humidity and PMV values are as mentioned before in table 4. All units of Model B have the highest operative temperature, with a high range of relative humidity (55 - 57%). For the PMV in all models, they recorded an acceptable range recorded from 0 (neutral) to - 1 (slightly cool). Thus, according to the comparative analysis in table 5, where units are weighted according to their effect by stacking them horizontally, vertically, and both horizontal and vertical stacking (Model B, C, and D respectively) and the control unit readings, horizontal stacking is most effective during cold days.

Table 6. Comparative analysis of the 3 models on Coldest Day (hourly average).

	Operative Temperature			Relative Humidity			PMV		
	Model B	Model C	Model D	Model B	Model C	Model D	Model B	Model C	Model D
Unit 1	3	6	4	1	1	1	2	4	3
Unit 2	1	7	2	1	1	1	1	5	3
Unit 3	5	9	8	1	2	1	3	7	6
Total	9	22	14	3	4	3	6	16	12

All the nine units are weighted from 1 to 9 and arranged from the most affected unit by stacking during the coldest day (Weighted 1) to the least affected unit (weighted 9).

6. Conclusion

This paper presents a thermal performance analysis of Shipping Containers (SC) in Cairo, Egypt, (a hot climate). The selection of stacking configurations in this study is based on the three basic walls, floors, and ceiling connections. All units used are in their basic form without any insulation material. This paper has conducted a simulation study of four models, including a base model with a standalone SC unit, three SC units connected horizontally, three SC units connected vertically, and three SC units connected both horizontally and vertically. The results conclude that stacking influences thermal performance, especially in hot weather conditions. Vertical stacking works best in Summer (Model C), while horizontal stacking works best in winter (Model B). As Cairo is in a hot climate zone, and the three models have a relatively good PMV value in winter, then we advise that the vertical stacking has the best thermal performance.

Although in vertical stacking, the hourly average results showed a drop in both operative temperature and PMV values during the hottest day simulation. The results of the top unit in vertical stacking were lower than the control unit in its operative temperature. However, it still needs additional measures to reach thermal comfort.

This paper suggests having more research on taking advantage of shipping containers structure benefits like stacking up to 9 units without additional supports and how this can help in reducing costs of their treatment to be low-cost habitable spaces. And also studying different stacking configurations coupled with different insulation materials.

Acknowledgments

The authors wish to acknowledge guidance from Professor Ahmed ElKhateeb, and Eng. Aya Shaheen. The assistance, and encouragement from our colleagues: Mohamed Gadallah, Mohamed Prince, and Hady Hany.

References

[1] Tarek El-Kashef, (2010). Air Temperature and Relative Humidity Effects on Thermal Comfort in Egypt, (In Arabic). A Workshop for Environmental Issues in the Arab World, Cairo, Egypt.

[2] Islam H, Zhang G, Setunge S and Bhuiyan M A 2016, Life cycle assessment of shipping container home: A sustainable construction. *Energy Build* **128** 673–85.

[3] Elmokadem A, Eleinen O M A, Aly R H and Ezz-eldien D A 2019 Shipping Containers as Flexible Components for Sustainable Buildings in Coastal Zone - Egypt. *Int. J. Innov. Res. Sci. Eng. Technol.* **8**(6) 6814–22.

[4] Elrayies G M 2017 Thermal performance assessment of shipping container architecture in hot and humid climates. *Int. J. Adv. Sci. Eng. Inf. Technol.* **7**(4) 1114–26.

[5] Anthony F 2018 An Assessment of Thermal Performance of office units built with shipping containers around Lagos. *Int. J. Sci. Eng. Res.* **9**(2) 44–72.

[6] Turner SC, Paliaga G, Lynch BM, Arens EA, Aynsley RM, Brager GS, et al. 2010 Thermal Environmental Conditions for Human Occupancy American society of heating, refrigerating and air-conditioning engineers. Int. J. Refrig. pp 2–6

[7] Gehan Fouad, (2017). Ideas for Small Enterprises among the International Practice of using Containers in Sustainable Design, (In Arabic), International Design Journal, 7(1), 279-290.

[8] Pierre Morency Architecte [Internet]. Pierremorencyarchitecte.com. 2021 [cited 30 November 2021]. Available from: http://www.pierremorencyarchitecte.com/

[9] Redondo Beach Container House - House Tour - Bob Vila [Internet]. Bobvila.com. 2021 [cited 30 November 2021]. Available from: https://www.bobvila.com/slideshow/house-tour-redondo-beach-container-house-2445#redondo-beach-house

[10] Qubix | Home [Internet]. Qubixstudios.com. 2021 [cited 30 November 2021]. Available from: https://www.qubixstudios.com/

The Impact of Street Trees on a Typical Urban Canyon in Eastern Cairo Region

W M Elbardisy[1,*], M A Salheen[1], M Fahmy[2,3]

[1] Faculty of Engineering, Ain Shams University, Abbaseya, Cairo 11517, Egypt
[2] Department of Architecture Engineering, Military Technical Collage, Kobry El-Kobba, Cairo 11662, Egypt
[3] Head of Research Excellence Center for Urban Environment and Sustainability Governance, Military Technical Collage, Kobry El-Kobba, Cairo 11662, Egypt

[*]Corresponding Author Email: welbardisy@eng.asu.edu.eg

Abstract The trend of upgrading urban areas in Egypt is expanding across the region for better urban mobility patterns. Meanwhile, the concerns towards climate change and creating resilient communities are evolving to avoid the dramatic consequences on urban environments. Street trees offer great benefit in climate moderation, particularly in hot arid regions as in Cairo. In this work, different spatial arrangements and locations for evergreen, deciduous trees and palms at NW street canyon are tested using ENVI-met numerical simulation tool. It is aimed to maximize the cooling effect of tree shade on street facades and attain better pedestrian thermal comfort in peak summer times. This ultimately gives better guidance for urban planners and policymakers towards an improved planting design pattern. Comparing simulation results, it is concluded that evergreen trees located at the building edge can provide up to 6°C cooler walls and up to 3.8°C when placed at the building centre. Whilst other tree species have lower cooling effects. Also, these trees have a great role in attaining better thermal comfort levels for pedestrians at the street level. It is recommended to wisely choose suitable tree species and arrangements to maximize the cooling benefits in the urban environment.

Keywords: Trees, Urban Microclimate, Cairo, Thermal comfort, ENVI-met, Air temperature, Solar radiation, Street canyon.

1. Introduction

Cairo, known for its hot arid climate according to Köppen-Geiger climate classification [1], is falling towards the risk of Urban Heat Island (UHI) [2]. This, in turn, warms the urban environments, threatens human health with longer heat waves resulting in thermal stress, and impacts the water quality [3], [4].Thus, mitigating and adapting the urban climate is paramount to reduce the risks of UHI patterns. Trees, in specific, among other vegetative elements have promising strategies towards UHI mitigations and adaptation in cities [5]. Attaining a better pedestrian thermal comfort in urban microclimates through solar irradiance reduction and shade provision [6]from one hand, and cooling and moisturizing air in the other hand through its canopy via evapotranspiration effect [7]. Its reduction to air temperature could vary from 0.1°C as in research conducted in Indiana, United States, 0.9°C at Australia, and up to 5.6 °C at the dense context in Bangalore, India [8]. The reduction percentage highly depends on the street orientation and the study site location [8] and tree percent [9] and [10].

Seizing the opportunity of the major urban development occurring all over Cairo for better urban mobility pattern and wellbeing, this research aims to contribute towards developing insights about mitigation strategies to face this phenomenon since there is no specific retreat. The research attempts to specify the suitable tree species and the arrangement pattern which on one hand maximize human thermal comfort at the street level, and the other hand moderate temperature and radiation ranges at the building facades and its front.

2. Methods and tools

In the present study, a numerical simulation ENVI-met V5, [11] was conducted to simulate an urban street canyon in Egypt for evaluating the different vegetation arrangements. The outcome of ENVI-met simulations has been verified in different research as in [12] and [13]. The output of ENVI-met simulation were verified previously through research conducted in the same urban street [9]. Meternorm [14] tool was used to generate context weather file since onsite weather station is not available. Through analyzing the weather data on Ecotect tool [15], it was found that 20th of July represents the peak and average summertime.
For assessment and test, an urban street canyon is selected to give insight on developing a suitable urban street tree with minimum cost and maximized benefits. This, in turn, is a guide for landscapers and policy makers for the upcoming urban upgrading developments. The trees used in the model are numerically modelled and its LAI and LAD data were adopted and generated from Fahmy et al work on investigating the trees [16–18] and the palm's data was adopted from ENVI-met library [11].

2.1. Urban canyon

This study examined El Shaheed Sayed Zakareya urban street at El Nozha district (30◦06′N and 31◦22′E). The street is oriented to the Northwest, 46° direction. The study section is 1 km and 100m wide with a sky View Factor range of 0.8 to 0.9, see figure 1.Most of the buildings located on the street are residential buildings with commercial use at the ground. Buildings are skeleton structures painted in light colored mortar plaster. Facades are oriented towards NE and SW orientation. Pedestrian and walkways are covered with yellow and red colored interlock with albedo=0.5, while the car traveling way are of an asphalt surface with albedo=0.5.

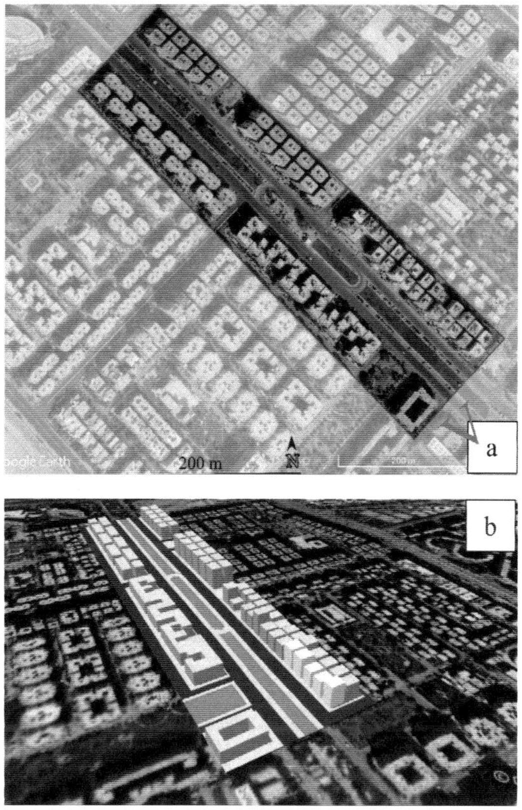

Figure 1. a) Satellite image for urban street marked in red rectangle, captured from Google earth on 22nd Nov. 2021. (b) Perspective view for the studied urban street with the current surface materials, author.

2.2. Street tree species and scenarios

Following the previous researches in the Egyptian context; [9], [16], [19], [20], three tree types are introduced in the study. Deciduous, Cassia Nodosa H=15, W=13), Evergreen Ficus Nitida H=15, W=9, and Palm Phoenix H=15, W=9) are numerically modelled using Albero tool, a plugin in ENVI-met software[21]. The trees are placed on the pedestrian walkways in front of the buildings in the NE and SW directions. Smaller scaled trees are used in the NE direction due to the limited space available H=5, W=3. For these trees, two arrangement patterns are tested; firstly, trees are placed at the building edge, p1 and p3, and secondly, trees are placed in the building centre p2. Trees in the middle median are linearly arranged on both sides based on the finding of Elbardisy et al. [9], Figure 2 a ,b, c ,and d. The vegetative cover percent is 4, 4, 8 for evergreen, palm, and deciduous tree, respectively for the edge and centre tree arrangement. In total six introduced simulation patterns are conducted. Additionally, the current planting pattern and the no vegetation simulation scenario are modelled for evaluation. Table 1 sums the input data used for simulation in ENVI-met and table 2 lists the abbreviations used for each scenario in ENVI-met.

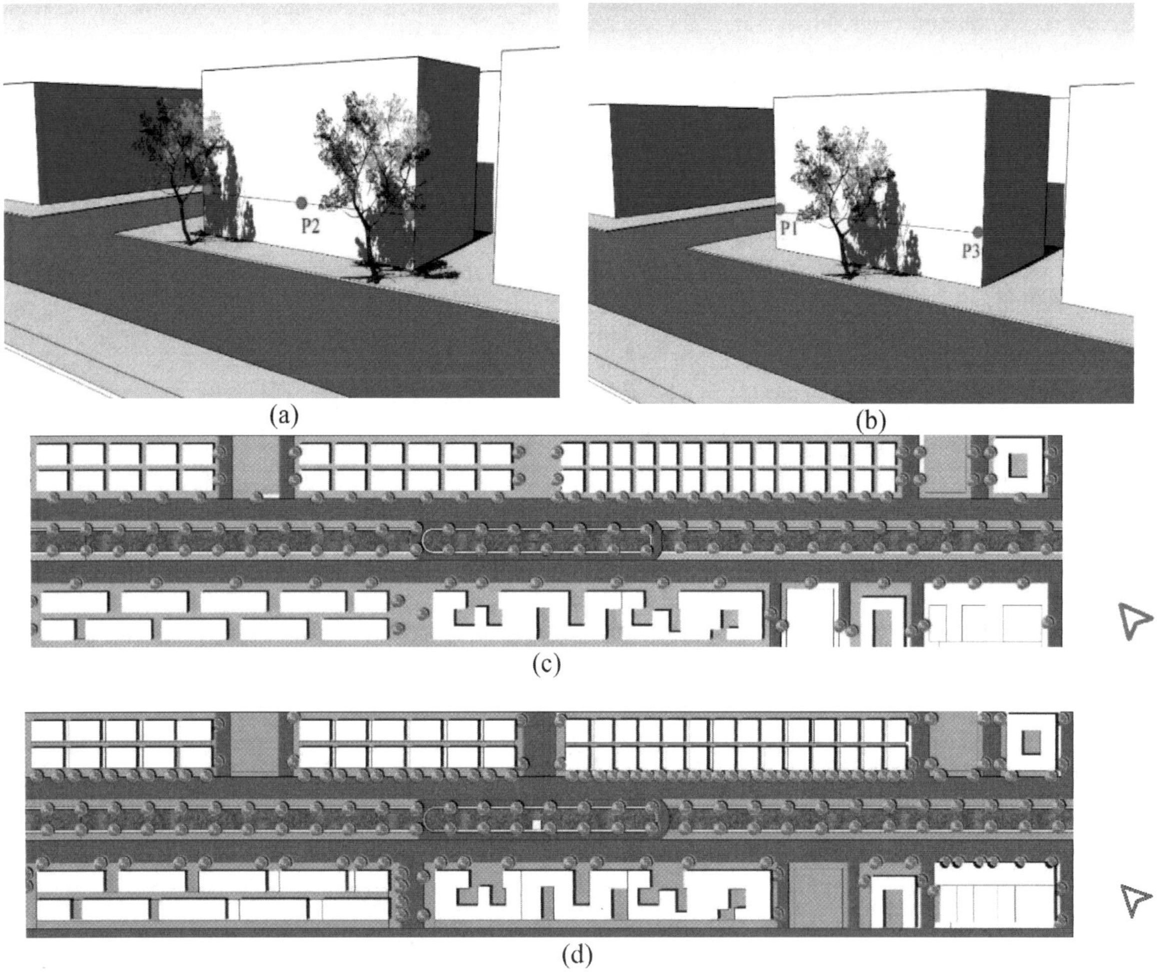

Figure 2: the proposed spatial arrangement of street trees along the buildings; (a) trees are placed on the middle of the buildings, three study points are selected p1 and p3 are in the building edges, while point 3 is in the middle of the building (b) trees are placed on the buildings' edge, three study points are selected p1 and p3 are in the building edges, while point 3 is in the middle of the building. (c) the layout of mid building tree pattern, and (d) the layout of edge building tree pattern.

Table 1. Data used for simulation in ENVI-met.

Input Parameter	Value(s) used
City location	Cairo, Egypt (Lat: 30.06°; Long: 31.25°)
Simulation day	20th July 2020
Simulation duration	24 h
Model resolution	42*191*12 m
Climate type	Hot arid
Windspeed and direction	3 m/s at 240°
Temperature	Min:26.4°C; Max.:39.70°C (forced sampling)
Relative humidity	Min:27%; Max: 59%
Cloud cover	Default setting in ENVI-met
Material albedo	Grass=0.2; yellow tiles=0.5; asphalt= 0.2
Building wall and roof materials	Default settings in ENVI-met

Table 2. The abbreviation/coding used in ENVI-met simulation and analysis

Scenarios	Value(s) used
No VG	No vegetation is present
Current	The existing tree pattern
EDGE_EG	Evergreen are placed in the center of the building
EDGE_DC	Deciduous are placed in the center of the building
EDGE_PLM	Palms are placed in the center of the building
MID_EG	Evergreen trees are placed at the center of the building
MID_DC	Deciduous trees are placed at the center of the building
MID_PLM	Palms are placed at the center of the building

3. Results

This research examines the impact of three commonly used tree species, which vary in their canopy and size, on pedestrian thermal comfort, solar irradiance reduction at an urban canyon in El-Nozha district. ENVI-met urban simulations were conducted on summer design day 20th of July analysed through Ecotect tool [15] after generating a context weather file through Meteronorm [14] and validation to the ENVI-met outputs through the author's publication [22]. Periodic assessments were conducted every 2 hrs. starting from 12:00 till 16:00 for the horizontal layers at 1.5m above the ground since it is the mean human height of neck and nose [23]. Also, assessments were on the vertical urban layer at the building façade (indicated previously) at NE and SW street sides. The following analyses and findings are divided into two sections. Firstly, the shade pattern and radiation generated on the ground surfaces and facades. Secondly, pedestrian thermal comfort generated across the different trees pattern and microclimate parameters change.

3.1. Shade pattern and radiation

Throughout ENVI-met simulations, the shade flag upon NE and SW buildings is computed for the different spatial arrangements and tree species. According to the sun's azimuth, diagram present in figure 3, the SW building is gradually shade starting from 12:00 giving the opportunity to shade its façade and the front pedestrian path as well, in contrast to the NE building behaviour. Regarding the tree arrangements and species, evergreen, deciduous, and the hybrid arrangements performs better in shade provision when compared to palms and no vegetation scenario, figure 3.

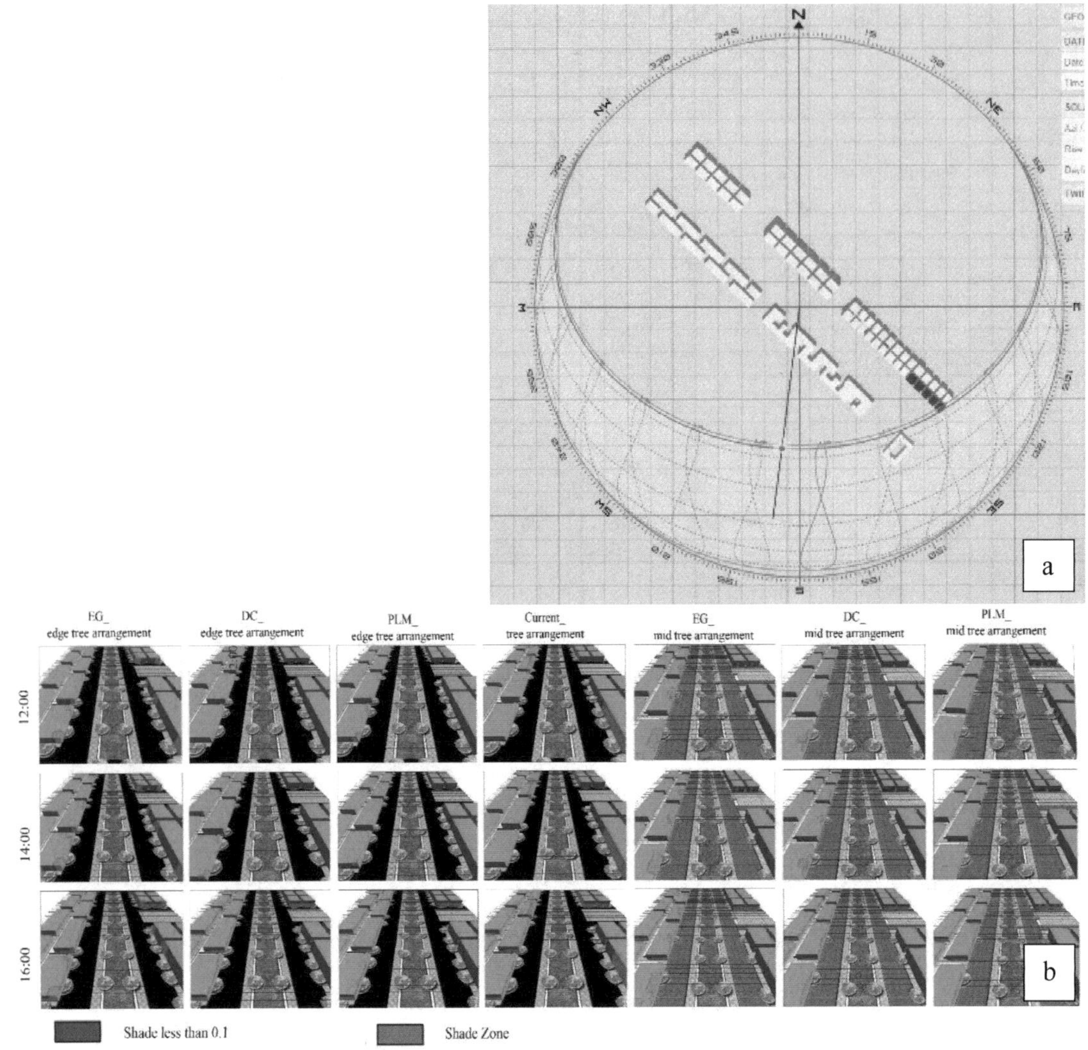

Figure 3:a)The sun path analysis for the study site for the design day, b) shade altered on the buildings at 12:00, 14:00, and 16:00 time series for the different tree patterns generated through Leonardo plugin in ENVI-met. The blue color resembles to no shade or low shade.

3.1.1. *Shortwave radiation on Facades*

Aligning with the shade pattern generated in the SW façade, Evergreen trees have a distinguished effect on intercepting shortwave radiation for both SW and NE facades regardless its spatial arrangement [24], see Fig.4&5. This ,in turn, led to a variation of wall temperature of the façade [25], [26]. For example, the edge evergreen arrangement on the SW façade lowered SW radiation by 85 w/m^2 and wall temperature by 6°C when compared to the no vegetation scenario midday. While at 14:00 and 16:00, wall temperature reduced by 5° and 3°C when trees are placed at the building edge and by 3.7,3,2.4°C, respectively, arranged in the building centre. The solar radiation blockage depends on the trees' canopy character; the percentage of the leaves in each branch, density of leaf, and foliage percent[3]. Thus, no significant change in palm scenarios radiation on façade is merely witnessed since its max. reduction on SW façade at 12:00 is 8 w/m^{2,} 5 w/m^2 at 14:00 and 16:00 for centred tree arrangement and minor change in the wall temperature by 0.2 over the day. The same findings are present for edge palm arrangements in both facades. Meanwhile, the middle palm arrangement lowered wall temperature to 0.2 and 1.5 at 14:00 and 16:00.

On contrary, deciduous trees, after evergreen trees, play a significant role in radiation reduction on façade at the NE. especially at 14:00 and 16:00 LST reaching 551 and 595 w/m^2 and minor decrease at SW direction. Unexpectedly, the mid deciduous trees arrangement elevated wall temperature to around 46 at 12:00, 47.6 at 14:00 and 45.5 at 16:00 by 0.1 and 0.2° C and these values are much higher than no vegetation scenario. this illustrates that the 8% canopy coverage have a great influence on trapping the air between the tree and the façade [19] when placed in the facade front. However, sufficient space is present for the air to flow in edged arrangement, thus, temperature and wall temperature is decreased by 0.9 and 0.7 at 14:00 and16:00.

Although evergreen and deciduous for both arrangements perform well, the edged arrangements outweigh the mid tree arrangement at the building since larger façade portion is shaded. For NE building facade, although smaller sized trees were introduced, a reduction effects are clearly seen, especially in the evergreen. Additionally, the hybrid arrangement of trees has an acceptable role in ameliorating wall temperature; the same reduction rate, around 0.7 °C is reduced for both facades along the day.

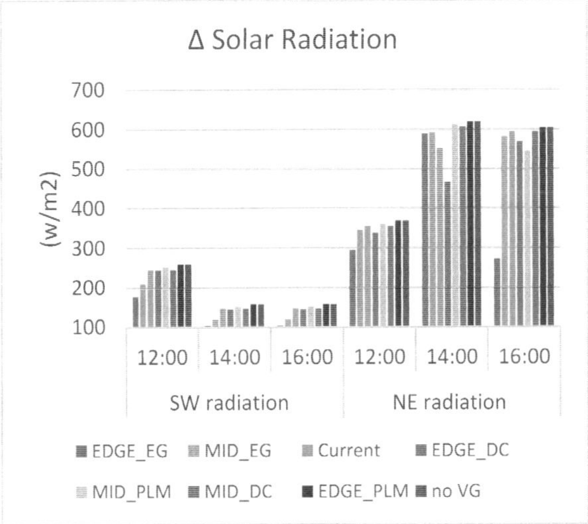

Figure 4. The Shortwave radiation reaching the facades over the day for the different tree patterns.

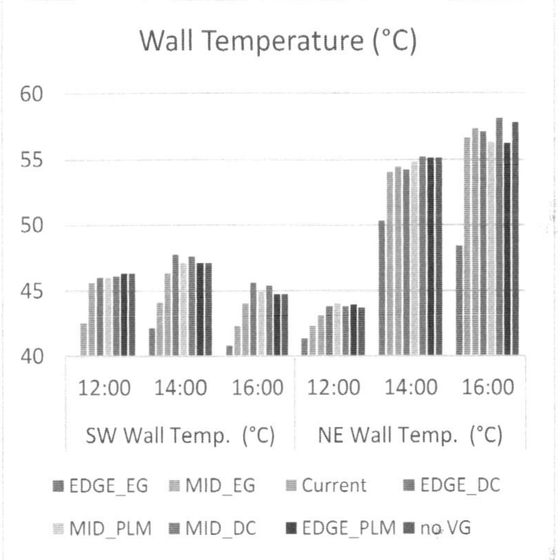

Figure 5. Wall temperature of facade along the day for the different tree arrangement.

3.1.2. Mean Radiant temperature on ground surfaces

From the graph in Fig.6, it is clear that all the introduced design scenarios dramatically reduce the radiant temperature regardless the speices and its arrangemet [9]. The canopy of the tree with its foliage are the key factor in attenuating solar radiation and reducing the solar heat gain on the urban surfaces [28].
Edge evergreen arrangement at the building edge amilorates the radation falling into the ground to the lowest values reaching to 48.77, 50.71, and 47.6°C, respectively at 12:00, 14:00 and 16:00. A lower reduction is also present when trees are placed in the middle of the building reaching to 52.3,56, and 51.3 at the same timeseries.The performance of the present existing random pattern, and palm regardless the spatial pattern along the day reached to 45,47 at 10:00, 50,52 at 12:00, and 50 at 16:00, repectively. This finding resembles that the geomtry and evergreen trees have the capacity of amilorting the extreme radition released in the context [10].

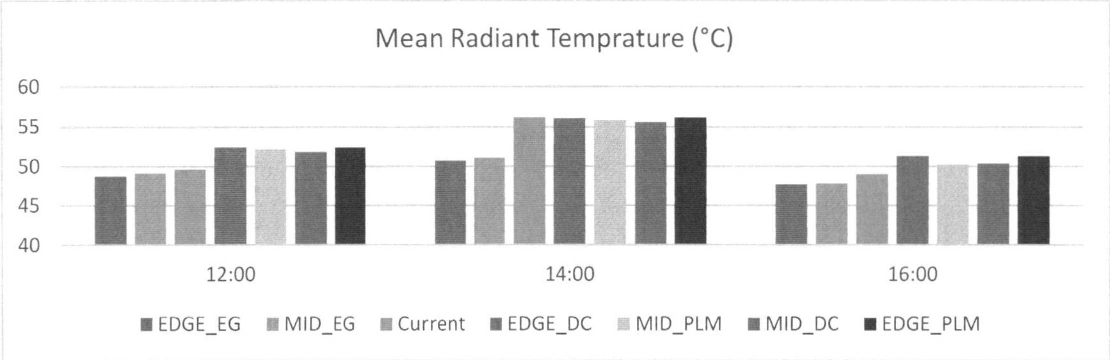

Figure 6. Mean Radiant Temperature across 12:00, 14:00 16:00 time series.

3.2. Thermal comfort

Differently than other tree types, evergreen spatial arrangement attains the best pedestrian thermal comfort throughout the whole day, shown in fig.7. No significant impact of deciduous trees on enhancing the human comfort level, and this finding is similar to the work of Elbardisy et al. [9]. The performance of the palm trees is like the deciduous tree for both the edged tree arrangement and the centred arrangement. Human thermal comfort as discussed by [29] reached its maximum when the space is shaded with high wind flow and the worst case when no wind pattern is present with no shade. Additionally, the evapotranspiration effect the leaves does, moisturize the air affecting the thermal comfort index as well [30]. Since the context wind flow is relatively low and the vegetative density in the deciduous tree is high (almost 8% for both arrangements), the airflow is blocked and there is no window for the air to flow. This could happen with the evergreen arrangements; however, the vegetative cover density is lower 4% since the canopy size is smaller than the deciduous tree. Although palms have no impact in increasing the humidity values (nearly constant), its canopy performance in shielding the radiation is low, thus, minor impact on thermal comfort.

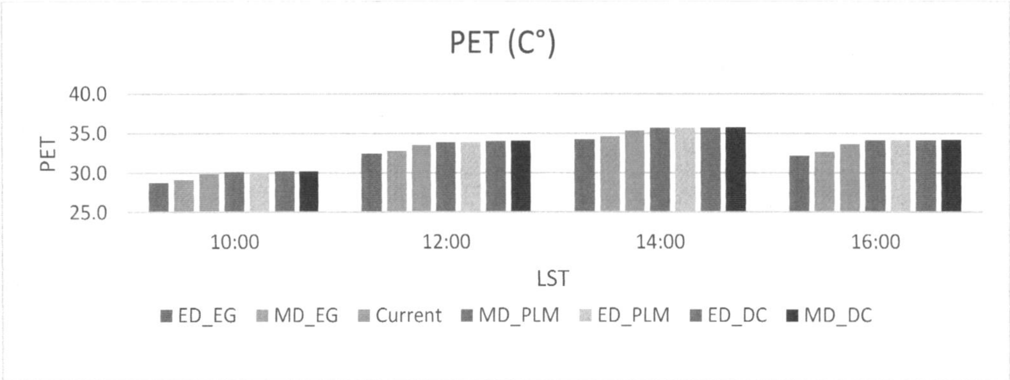

Figure 7. PET ranges for the different tree arrangements 12:00, 16:00, and 16:00.

4. Conclusions

This research has found that planting evergreen species in its full size, in specific Ficus Nitida, at Northwest Street canyon with 0.8 Sky view factor is typically most effective for cooling and shading building's wall, especially if trees are placed at the edge of the building. A slight lower cooling and shading patterns are witnessed when these trees are placed in the middle of the building. This also attains better pedestrian thermal comfort, when compared to other tree species and arrangements. Other tree species, also, impacts the urban environment but in a lower range, since the solar radiation is blocked or reduced through the canopy character, leaf, and foliage percentage. Therefore, a minor effect is witnessed with palm scenarios regardless of the spatial arrangement. However, its presence is vital since it has minor impact on the wind pattern and enhances the space aesthetics'.

Hybrid tree arrangements: intermix between evergreen, deciduous, and palm are also promoted since it offers benefits to the urban residential environments. It is better to place evergreen trees in front of the buildings individually with no overlap to give a window for the wind to flow and cool the surrounding. For better results matured sized trees are placed at NE and SW facades edges and do not exceed 4%. Additionally, a lower percent of deciduous trees and palms can be added for spaces aesthesis noting that matured deciduous trees are not recommended to be placed at the building centre.
This research is considered one of the pioneering attempts to incorporate with the ongoing planning and upgrading patterns present in Cairo. This in turn will help and provide better guidance to urban planners and policy makers in long term planning strategies for enhanced urban environments and wellbeing.

References

[1] M. Kottek, J. Grieser, C. Beck, B. Rudolf, and B. Rubel, "WORLD MAPS OF KÖPPEN-GEIGER CLIMATE CLASSIFICATION," *Meteorol. Z*, 2006. .
[2] S. Huq, S. Kovats, H. Reid, and D. Satterthwaite, "Reducing risks to cities from disasters and climate change," *Environ. Urban.*, vol. 19, no. 1, pp. 3–15, 2007.
[3] M. Fahmy, S. Mahmoud, I. Elwy, and H. Mahmoud, "A review and insights for eleven years of urban microclimate research towards a new Egyptian era of low carbon, comfortable and energy-efficient housing typologies," *Atmosphere (Basel).*, no. 3, 2020.
[4] Z. Wu and L. Chen, "Optimizing the spatial arrangement of trees in residential neighborhoods for better cooling e ff ects : Integrating modeling with in-situ measurements," *Landsc. Urban Plan.*, vol. 167, no. July, pp. 463–472, 2017.
[5] S. Teshnehdel, H. Akbari, E. Di Giuseppe, and R. D. Brown, "Effect of tree cover and tree species on microclimate and pedestrian comfort in a residential district in Iran," *Build. Environ.*, vol. 178, no. May, p. 106899, 2020.
[6] A. Peeters *et al.*, "A decision support tool for calculating effective shading in urban streets," *Urban Clim.*, vol. 34, no. July, p. 100672, 2020.
[7] E. Ng, *Designing High-Density Cities for Social and Environmental Sustainability*. UK: Earthscan, 2010.
[8] R. Sanusi, D. Johnstone, P. May, and S. J. Livesley, "Microclimate benefits that different street tree species provide to sidewalk pedestrians relate to differences in Plant Area Index," *Landsc. Urban Plan.*, vol. 157, pp. 502–511, 2017.
[9] W. M. Elbardisy, M. A. Salheen, and M. Fahmy, "Solar irradiance reduction using optimized green infrastructure in arid hot regions: A case study in el-nozha district, cairo, egypt," *Sustain.*, vol. 13, no. 17, 2021.
[10] W. M. El-Bardisy, M. Fahmy, and G. F. El-Gohary, "Climatic Sensitive Landscape Design: Towards a Better Microclimate through Plantation in Public Schools, Cairo, Egypt," *Procedia - Soc. Behav. Sci.*, vol. 216, no. October 2015, pp. 206–216, 2016.
[11] M. Bruse, "ENVI-met," 2021. [Online]. Available: http://www.model.envi-met.com/hg2e/doku.php?id=apps:biomet_pmv.
[12] X. Yang, L. Zhao, M. Bruse, and Q. Meng, "Evaluation of a microclimate model for predicting the thermal behavior of different ground surfaces," *Build. Environ.*, vol. 60, pp. 93–104, 2013.
[13] M. Fahmy, H. El-hady, M. Mahdy, and M. F. Abdelalim, "On the green adaptation of urban developments in Egypt ; predicting community future energy efficiency using coupled outdoor-indoor simulations," *Energy Build.*, vol. 153, pp. 241–261, 2017.
[14] "Meteonorm Version 8 - Meteonorm (en)." [Online]. Available: https://meteonorm.com/en/meteonorm-version-8. [Accessed: 30-Jun-2021].
[15] "Autodesk - Autodesk Ecotect Analysis." [Online]. Available: http://wayback.vefsafn.is/wayback/20091118210953/http://usa.autodesk.com/adsk/servlet/pc/index?id=12602821&siteID=123112. [Accessed: 30-Jun-2021].
[16] M. Fahmy, Y. Ibrahim, E. Hanafi, and M. Barakat, "Would LEED-UHI greenery and high albedo strategies mitigate climate change at neighborhood scale in Cairo, Egypt?," *Build. Simul.*, vol. 11, no. 6, pp. 1273–1288, 2018.

[17] M. Fahmy, S. Sharples, and M. Yahiya, "LAI based trees selection for mid latitude urban developments: A microclimatic study in Cairo, Egypt," *ElSevier*, vol. 45, no. 2, pp. 345–357, Feb. 2010.

[18] M. Fahmy, M. Mahdy, H. El-Hady, Y. Ibrahim, and M. F. Abdelalim, "Simulating the carbon foot print reduction of an arid urban form applying canopy layer green coverage under climate change scenarios," *Int. J. Appl. Eng. Res.*, vol. 12, no. 22, pp. 12906–12913, 2017.

[19] R. O. Shata, A. H. Mahmoud, and M. Fahmy, "Correlating the sky view factor with the pedestrian thermal environment in a hot arid university campus plaza," *Sustain.*, vol. 13, no. 2, pp. 1–25, 2021.

[20] I. Elwy, Y. Ibrahim, M. Fahmy, and M. mahdy, "Outdoor microclimatic validation for hybrid simulation workflow in hot arid climates against ENVI-met and field measurements," *Energy Procedia*, vol. 153, no. November, pp. 29–34, 2018.

[21] ENVI-met, "ENVI-met - Decode urban nature with ENVI-met software," 2021. [Online]. Available: https://www.envi-met.com/. [Accessed: 30-Jun-2021].

[22] E. Cuce, P. M. Cuce, F. Sher, T. Bali, and I. Altin, "The role of plants in temperature regulation of external walls: An experimental and numerical research," pp. 1–10, 2016.

[23] T. A. L. Martins *et al.*, "Impact of Urban Cool Island measures on outdoor climate and pedestrian comfort: Simulations for a new district of Toulouse, France," *Sustain. Cities Soc.*, vol. 26, pp. 9–26, 2016.

[24] Q. Zhao, E. A. Wentz, and A. T. Murray, "Tree shade coverage optimization in an urban residential environment," *Build. Environ.*, vol. 115, pp. 269–280, 2017.

[25] H. Takebayashi, "Thermal Environment Design of Outdoor Spaces by Examining Redevelopment Buildings Opposite Central Osaka Station," 2019.

[26] Q. Zhao, J. Yang, Z. Wang, and E. A. Wentz, "Assessing the Cooling Benefits of Tree Shade by an Outdoor Urban Physical Scale Model at Tempe , AZ," pp. 1–20, 2018.

[27] F. Salata, I. Golasi, R. De Lieto, and A. De Lieto, "Urban microclimate and outdoor thermal comfort . A proper procedure to fit ENVI-met simulation outputs to experimental data," *Sustain. Cities Soc.*, vol. 26, pp. 318–343, 2016.

[28] C. Bartesaghi Koc, P. Osmond, and A. Peters, "Towards a comprehensive green infrastructure typology: a systematic review of approaches, methods and typologies," *Urban Ecosyst.*, vol. 20, no. 1, pp. 15–35, 2017.

[29] C. Hsieh, F. Jan, and L. Zhang, "A simplified assessment of how tree allocation , wind environment , and shading affect human comfort," *Urban For. Urban Green.*, vol. 18, pp. 126–137, 2016.

[30] M. A. Ruiz, M. B. Sosa, E. N. C. Cantaloube, and M. A. Cantón, "Urban Climate Suitable configurations for forested urban canyons to mitigate the UHI in the city of Mendoza , Argentina," 2015.

Energy efficiency, carbon emissions, and thermal comfort comparisons between conventional and proposed clustered open courtyard housing using CEB blocks

M. Fahmy[1, 2, *], S. Mahmoud[1], H. Abdelkhalik[1], M. Abdelalim[3], M. Elshelfa[1]

[1]Architecture Engineering Department, Military Technical College, Cairo, Egypt
[2] Head of Research Excellence Center for Urban Environment and Sustainability Governance, Military Technical College, Cairo, Egypt
[3] Architecture Engineering Department, Prince Sultan University, Riyadh, Saudi Arabia

[*]Corresponding Author Email: md.fahmy@mtc.edu.eg

Abstract. This research aims to assess a compact economic housing model design proposal using walls constructed of compressed earth blocks as a nucleus for sustainable urban housing in Egypt. This research also aims to evaluate the effect of the direction of the proposed compact building on the building's internal thermal comfort, energy consumption, and carbon dioxide emissions. According to previous studies, residential buildings represent about 50% of land uses in Egyptian cities, consume about 19% of the energy produced, and are responsible for 10% of carbon dioxide emissions. All previous environmental studies commended the thermal performance of courtyard buildings in the hot-arid climate of which Egypt is a part. As a result, the authors directed many previous studies towards new housing patterns in Egyptian cities using court-yarded clustered housing to improve thermal performance, decrease energy consumption, and reduce carbon dioxide emissions. This research compares two residential buildings using design-builder simulation. The two buildings have the same housing unit area of 90 m^2 to represent the social housing unit area in Egyptian cities. The first comparative building is a conventional residential building surrounded by solar radiation from all sides and uses walls constructed of solid cement bricks (SCB). The second comparative building is a cluster open court-yarded building using walls built of Compressed Earth Blocks (CEB). Results show that both the clustered open courtyard housing model whose main axis is in the (N-S) direction and the use of CEB has improved all of the examined measures.

Keywords: Housing, CEB, Clustered Court-Yarded Planning, Sustainability.

1. Introduction.

New Egyptian cities have developed over time since the emergence of the first generation of Egyptian cities in 1975 (10th of Ramadan City), through the second generation in 1985 (Badr City), then the third generation in 1997 (Shorouk City) [1] and finally the fourth-generation cities in 2013 (the new administrative capital). Hence, sustainability began to be viewed as an essential component of construction determinants in Egypt [2]. Thus, the efficiency of energy consumption and the reduction of carbon dioxide emissions resulting from it were considered direct elements of achieving environmental sustainability. Therefore, reducing and rationalizing energy consumption has become a national priority and goal in Egypt. According to the report of the Egyptian Ministry of Electricity, the total electric energy sold for the year 2019/2020 was about 151.9 billion kWh, and the total consumption of residential buildings was only 62.8 billion kWh, a rate of 41.4% and an increase of 1.8% for the year

2018/2019 [3]. Therefore, the construction sector, especially residential communities, possesses a great opportunity to reduce energy consumption and thus reduce carbon dioxide emissions (the main cause of global warming). Therefore, researchers have strived towards designing sustainable buildings in which the energy used to adapt the internal spaces of the building is rationalized, and the external urban climatic environment of the building is optimized to adapt to the hot climatic conditions resulting from the expected climatic changes [4, 5]. According to global environmental reports, it is estimated that the air temperature will increase by 4 to 6.4 by the end of this century [6]. As a result of the global population increase, which is expected to reach 70% by 2050, the quest to reduce energy consumption and improve the environmental performance of buildings and the built environment to reduce carbon emissions and increase the green area [7] is an imperative that cannot be overlooked [8]. However, the planners ignored the use of architectural blocks previously used in the local architecture. The unique design of these buildings reflects the character of Egyptian architecture, which is proportional to the climatic conditions to a large extent and proved its ability to adapt to the hot, dry climate in most Egyptian regions [9, 10]. They used a separate modernist model of the blocks and were repeated in the urban environment to achieve short-term economic gains. And they ignored all the research that called for rationalizing energy consumption or improving the local climate.

Furthermore, some essential rules are overlooked, such as: ignoring the orientation of the streets, building ratios, choosing white landscape raw materials, arranging vegetation cover or using different shading applications [11, 12], or setting organizing laws that oblige the implementers of buildings to make a cover for buildings so that it is not less than 25 cm or even study the proportions of the window openings or their direction according to the function of the void. This prompted the residents of the buildings to use air conditioning [13] while they were inside to achieve thermal comfort, which led to a substantial increase in the rates of electrical energy consumption and thus carbon dioxide emissions. Therefore, the authors opt to 1) use of new housing patterns from Egyptian cities using courtyard clustered housing to reduce the area exposed to sunlight [5]; 2) use of green building materials that have high thermal-insulating properties and possesses low embodied energy; and 3) and the use of surface paints that reduce the absorption of surfaces to sunlight and the use of heat galleries to create shaded pedestrian paths [14] and adding different and innovative window shading applications to create a more comfortable urban environment. Also, the use of vegetation cover solutions[15, 16], green roofs [17, 18], and the composition of the urban fabric [19, 20] to create an architectural and urban environment that provides comfort to humans through a passive urban design [21, 22], which is what the authors called (the housing of Egyptian fifth-generation cities), [5].

2. Methodology

This research aims to spotlight 1) the effect of applying suggested passive compact design; 2) using compressed earth blocks (CEB) as low energy consumption wall materials; 3) the effect of orientation of the main axis of the building; and 4) using different window shading applications, on reducing energy consumption and carbon dioxide emissions and improving indoor thermal comfort, hence conducting a comparative analysis of different housing patterns and different building materials as It is shown in figure 1.

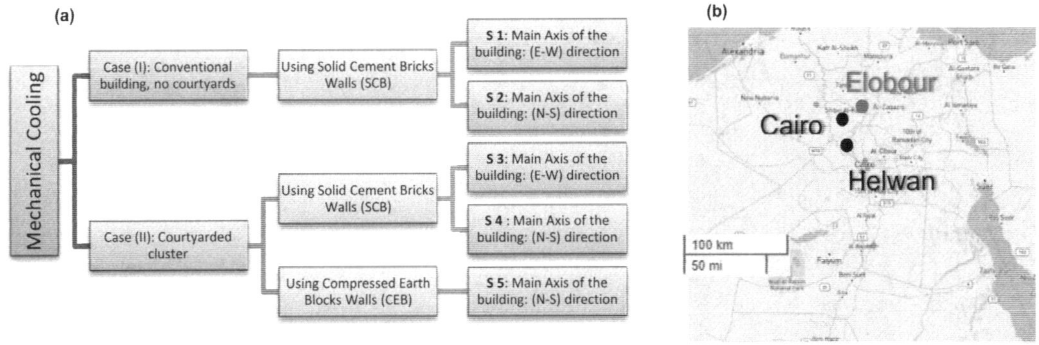

Fig 1. (a) The five simulation alternatives; (b) the location of El Obour El Jadida, Egypt, and the nearest weather stations derived from OpenStreetMap, 2021.

(a) The case study was carried out on an urban settlement in El Obour El Jadida, Cairo using a design-builder simulation on two different types of buildings. One is a conventional already used, and the other is a proposed passive design. (b) Weather data inputs for the simulation of the case study location for the new transit site (30.19° N, 31.45° E) were obtained from the nearest new transit weather station.

As shown in Figure 2 (a), case (I) is a conventional design building example currently used for new cities in Egypt. Case (II) represents a passive design for an open courtyard clustered attached housing typology. Both buildings consist of five floors (ground floor + four typical floors) with the same area of the housing unit (90 m2), representing the housing unit's area at the economic level in Egyptian cities. The simulation was carried out in five directions and divided into five different simulation alternatives, according to building materials used for walls and the orientation of the buildings, as previously shown in Figure 1 (a) and figure 2 (c)

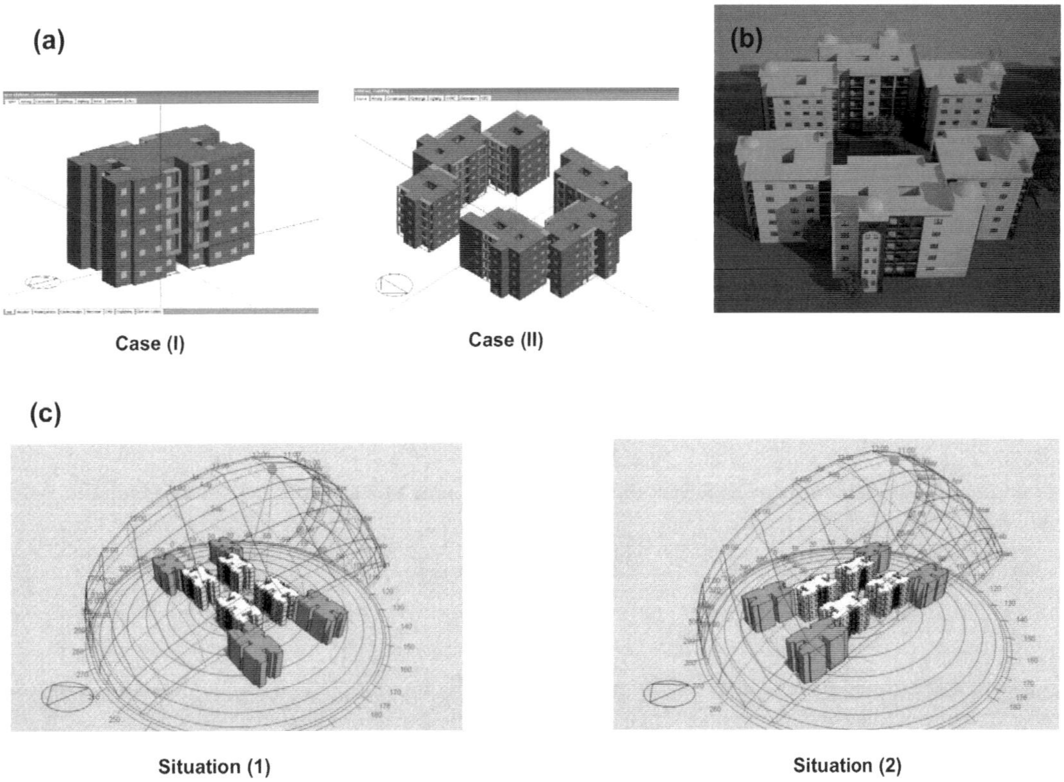

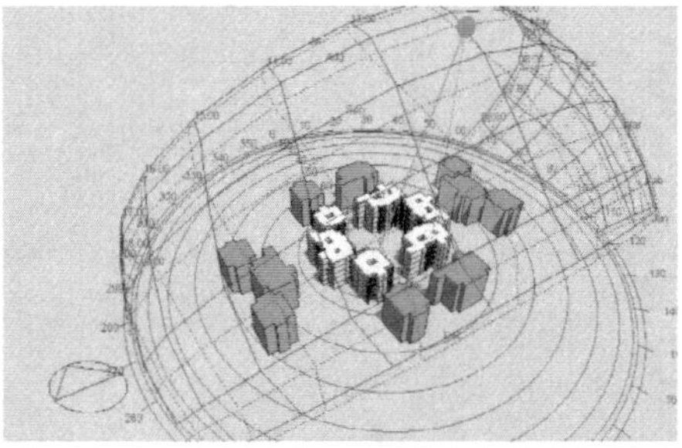

Situation (3-5)

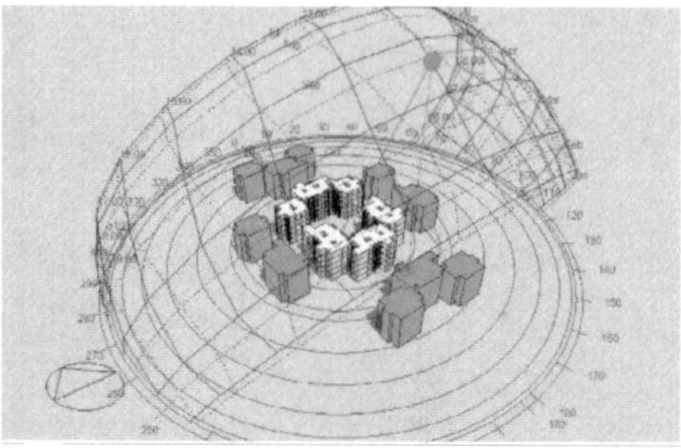

Situation (4-5)

(d)

Fig. 2. (a) The graphical user interface of Design Builder software showing the models of Case (I) and Case (II); (b) The 3D model for the open clustered housing of the 5th Generation of Egyptian Cities suggested by authors to generate clustered court-yarded attached housing of Case (II); (c) 3d model for five situations of alternative simulations with sun path diagram in 15-august and orientations; (d) The 3D model of a preliminary design alternative combining more than one cluster.

Table 1. Material constructions' layers used in Design Builder simulations.

Construction	Layers
Ground floor	0.2 cast concrete, 0.02m bitumen felt sheets, 1m earth common, 0.1m cast concrete, 0.06 sand, 0.02m cement plaster mortar, and 0.01m ceramic glazed tiles.
Typical floors	0.025m plaster cement mortar, 0.12m cast concrete, 0.05m sand, 0.02m cement plaster and 0.01m ceramic glazed tiles.
Roof floor	0.02m cement plaster, 0.12m reinforced concrete, 0.1m sand, 0.02m cement plaster, 0.02m bitumen felt/sheet, 0.05m foam (heat insulation), 0.14m cast concrete, 0.06m sand, 0.02m cement plaster and 0.02m cement tiles.
SCB wall	0.025m cement plaster, 0.12m / 0.25m / 0.38m solid cement bricks and 0.025m cement plaster.
CEB wall	0.025m cement plaster and 0.25m / 0.38m compressed earth blocks.

All materials used in the construction of a roof, the typical floors, the ground level, and the walls of the buildings were the same in all the simulation alternatives. Walls of SCB with an external thickness of 25 cm and internal thickness of 12 cm. but In the fifth comparative case (S5), walls of CEB units with an exterior thickness of 25 cm and inner thickness 12 cm, the same parameters were fixed for all the alternatives within the simulation. Table 1 shows the detailed structural layers for all construction and building works. Table 2 shows the thermal properties of the building units of the building walls, including the external building envelope and insulation materials, which were added to the simulation engine.

Table 2. Thermal characteristics of walls and fenestration materials.

	Walls	
Thermal Data	Solid cement bricks (SCB)	Compressed earth blocks (CEB)
Thermal conductivity (W/m. K°)	1.6	0.548
Density (Kg /m^3)	1800	1629
Specific heat (J/Kg. K°)	880	826
	Fenestration	
Thermal Data	6 mm clear glass	
Total solar transmission (SHGC)	0.71	
Light transmission	0.65	
U-value (W/m^2. K°)	5.76	

3. Results and discussion

The simulation outputs of overall energy consumption and carbon dioxide emissions were analyzed to compare the aforementioned alternatives and show the effect of using an open courtyard cluster on reducing energy consumption rates and carbon dioxide emissions. Total monthly and annual readings were normalized per unit area. In comparison, overall energy consumption combines lighting, heating, cooling loads. Case (S4), the open courtyard cluster and its main axis in (N-S) direction, achieve better performance in energy consumption and carbon dioxide emissions by saving 6% in energy consumption and 6% in carbon dioxide emissions, compared to the first case that uses the conventional building its main axis in (E-W) direction as shown in Figure 3 and Figure 4.

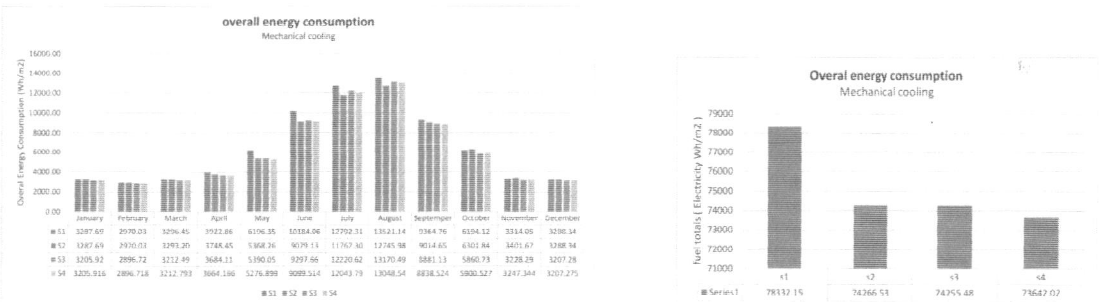

Fig. 3. Monthly and annual energy consumption of the first four simulations alternatives in (kWh/m^2).

Considering the annual energy consumption and carbon dioxide emissions in the previous four cases, which all use walls of (SCB). Figures 3 and 4 show that the first case (S1), which uses a conventional design with the main axis in (E-W) direction consumes more electrical energy with a total of 78.33 kWh/m^2 and thus produces more amount of CO_2 emissions of 47.46 tons/m^2 compared to the annual consumption of the fourth case (S4) which uses Open court yarded cluster with the main axis in (N-S) direction consumes less electrical energy totaling 73.64 kWh/m^2 and thus produces less amount of carbon dioxide emissions 44.62 ton/m^2. The results of the annual energy consumption and carbon dioxide emissions throughout the year for Building (S5) that uses an open cluster courtyard with the main axis in (N-S) direction and using walls built of compressed earth blocks (CEB) are shown in figure (5). It was found that the total energy consumption has been reduced to 65.9 kWh\m^2, which is 10.5% lower compared to Building (S4) that uses the same building and in the same orientation that consumes annual energy of 73.64 kWh\m^2. Moreover, the carbon emission was 39.95 tons, a decrease of 10.5% compared to Building (S4), in which its carbon dioxide emissions is 44.6 tons.

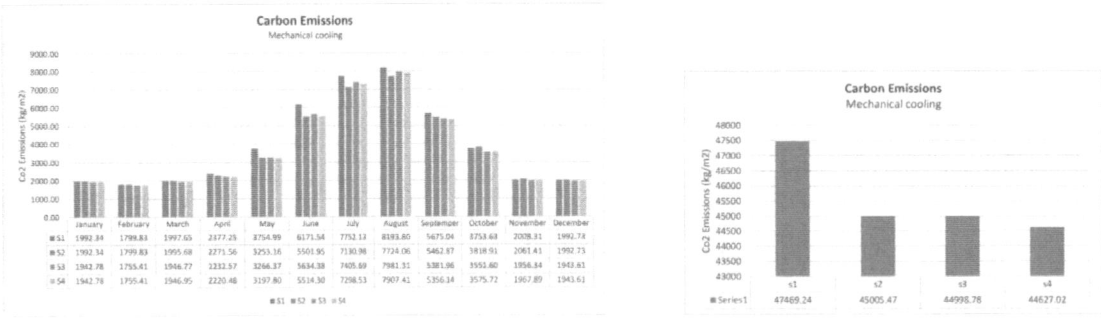

Fig. 4. Monthly and annually CO_2 production of first four simulation alternatives in (kg).

Fig. 5. Annual simulation outputs of (a) total energy consumption (left); and (b) CO_2 production among (S4) and (S5), (right).

4. Conclusion

This paper gives support for the use of the proposed model for designing sustainable urban housing, which links urban planning and architectural design, drawing the ideas from local architecture that has proven its effectiveness in creating an urban and architectural environment commensurate with the hot climatic conditions of the region to regulate sustainability measures the Fifth generation city housing in Egypt. Also, this research demonstrates the benefits of using a group of passive design applications in both directions (individual building and urban form) to reduce energy consumption and carbon dioxide emissions. The simulation was conducted using Design builder software and climatic data from the nearest weather station to Elobour Elgededa to distinguish between Conventional social detached buildings and open courtyard cluster attached pattern typologies, construction system, and envelope material alternatives. The results show the importance of passive design strategies in improving sustainability measures, especially when combining the design with low energy and cost materials. The results led to the choice of Cluster open courtyard design with CEB walls materials to reduce the average energy consumption and cooling loads and carbon dioxide emissions by 16.5% compared to the conventional design that uses SCB and achieves better results in the values of Indoor thermal comfort. Looking at the 3-D models of Figure No. 2, the open courtyard cluster design and sustainable construction bricks (CEB) have improved energy consumption and internal thermal comfort and sheds light on a new and distinct housing style with a different character, especially when it comes to looking at the cluster as a neighborhood housing design unit that generates a compact urban pattern instead of Parcel that generates highrise dot urban patterns. And last but not least, another aspect of the design appears, "the character."

Acknowledgments

This publication is part of the work of Environmental Design Research Group, Architecture Engineering Department of the Military Technical College of Egypt and has been funded by Science, Technology and Innovation Funding Authority Project ID STDF45851; Center of Research Excellence for Urban Environment and Sustainability Governance whose head; professor Mohammad Fahmy is the PI of the above-mentioned project ID STDF45851.

References.

[1] National Organization of Urban Harmony, "New Egyptian cities," 2009.

[2] Ministry of Housing, "Urban design, planning and division of 25 urban and service projects in 11 new cities," 2020.

[3] E. E. H. Company, "Egyptian Electricity Holding Company, Annual Report 2019/2020," 2020.

[4] M. Fahmy, Y. Ibrahim, E. Hanafi, and M. Barakat, "Would LEED-UHI greenery and high albedo strategies mitigate climate change at neighborhood scale in Cairo, Egypt?," *Building Simulation,* vol. 11, pp. 1273-1288, 2018.

[5] M. Fahmy, S. Mahmoud, I. Elwy, and H. Mahmoud, "A Review and Insights for Eleven Years of Urban Microclimate Research Towards a New Egyptian ERA of Low Carbon, Comfortable and Energy-Efficient Housing Typologies," *Atmosphere,* vol. 11, p. 236, 2020.

[6] IPCC, "Climate Change 2014: Synthesis Report," 2014.

[7] U. Berardi, "The outdoor microclimate benefits and energy saving resulting from green roofs retrofits," *Energy and Buildings,* vol. 121, pp. 217-229, 2016.

[8] M. J. Holmes and J. N. Hacker, "Climate change, thermal comfort, and energy: Meeting the design challenges of the 21st century," *Energy and Buildings,* vol. 39, pp. 802-814, 2007.

[9] A. Aboelata and S. Sodoudi, "Evaluating the effect of trees on UHI mitigation and reduction of energy usage in different built-up areas in Cairo," *Building and Environment,* vol. 168, p. 106490, 2020.

[10] O. M. Galal, H. Mahmoud, and D. Sailor, "Impact of evolving building morphology on microclimate in a hot arid climate," *Sustainable Cities and Society,* vol. 45, p. 102011, 2020.

[11] I. Elwy, Y. Ibrahim, M. Fahmy, and M. Mahdy, "Outdoor microclimatic validation for hybrid simulation workflow in hot arid climates against ENVI-met and field measurements," *Energy Procedia,* vol. 153, pp. 29-34, 2018.

[12] D. M. A. Hamdan and F. L. d. Oli, "The impact of urban design elements on microclimate in hot arid climatic conditions: Al Ain City, UAE," *Energy & Buildings,* vol. 200, pp. 86-103, 2019.

[13] M. Fahmy, M. Mahdy, and M. Nikolopoulou, "Prediction of future energy consumption reduction using GRCenvelope optimization for residential buildings in Egypt," *Energy and Buildings,* vol. 70, pp. 186-193, 2014.

[14] E. Alexandri and P. Jones, "Temperature decreases in an urban canyon due to green walls and green roofs in diverse climates," *Building and Environment,* vol. 43, pp. 480-493, 2008.

[15] EPA, "Reducing Urban Heat Islands: Compendium of Strategies; Trees and Vegetation," 2009.

[16] T. E. Morakinyo, K. K. L. Lau, C. Ren, and E. Ng, "Performance of Hong Kong's common trees species for outdoor temperature regulation, thermal comfort, and energy-saving," *Building and Environment,* vol. 137, pp. 157-170, 2018.

[17] EPA, "Reducing Urban Heat Islands: Compendium of Strategies; Cool Roofs," 2009.

[18] M. Santamouris, "Cooling the cities–a review of reflective and green roof mitigation technologies to fight heat island and improve comfort in urban environments," *Solar Energy,* vol. 103, pp. 682-703, 2014.

[19] O. Al-Hafith, S. B. K, S. Bradbury, and P. d. Wilde, "The Impact of Courtyard parameters on its shading level An experimental study in Baghdad, Iraq," *Energy Procedia,* vol. 134, pp. 99-109, 2017.

[20] M. Fahmy, H. Kamel, H. Mokhtar, I. Elwy, A. Gimiee, and Y. Ibrahim, "On the Development and Optimization of an Urban Design Comfort Model (UDCM) on a Passive Solar Basis at Mid-Latitude Sites," *climate,* vol. 7, p. 1, 2019.

[21] A. Allouhi, Y. E. Fouih, T. Kousksou, A. Jamil, Y. Zeraouli, and Y. Mourad, "Energy consumption and efficiency in buildings: current status and future trends," *Journal of Cleaner Production,* vol. 109, pp. 118-130, 2015.

[22] M. Razzaghmanesh, S. Beecham, and C. J. Brien, "Developing resilient green roofs in a dry climate," *Science of The Total Environment,* vol. 490, pp. 579-589, 2014.

Comparative Analysis of Occupant Thermal Comfort, Energy Consumption, and CO_2 Emissions for an Educational Building in Cold Climate: A Case Study in Canada

S. Mahmoud[1,*], Mahmoud M. Saad[1], M. Elshelfa[1], H. Abdelkhalik[1], M. Fahmy[1,2]

[1] Architecture Engineering Department, Military Technical College, Cairo, Egypt
[2] Head of Research Excellence Center for Urban Environment and Sustainability Governance, Military Technical College, Cairo, Egypt

*Corresponding Author Email: sherifahmed679@gmail.com

Abstract. Global climate change and global warming capture the attention of planners seeking more sustainable buildings to reduce dependence on fossil fuels and minimize energy consumption. Consequently, architects have developed buildings over the years and applied more sustainable strategies to cope with the environment, either in the shape of the building, the design of openings, or the use of more sustainable materials. These strategies provide more thermal comfort for long-term users. As a result, it is necessary to know the exact effect of using these strategies on thermal performance and energy consumption inside buildings. Therefore, researchers have developed many energy simulation tools and programs to provide a clear view of the nature of the building's thermal performance and energy consumption based on data inputs about local weather situations of the building. Thus, we must know the ability of the software to model different design strategies and be sure that the results are valid. This paper aims to make an environmental simulation of an educational building in Montreal, Canada, using Design-Builder simulation and compare it with the as-built equivalent thermal zones applying different glazing materials, which is the main element of the building facade to enhance the thermal comfort with the lowest energy consumption and CO_2 emissions. This paper shows that using double glass low E 6/6mm Air gap is better for the case study building in cold climate weather that increases thermal comfort for the users, reduces energy consumption by 18.63%, and also reduces CO_2 emissions by 18.57% per year.

Keywords: Architecture, educational buildings, Design-Builder, simulation, sustainability, environment, energy efficiency, thermal comfort, energy consumption, CO_2 emissions.

1. Introduction

The built environment variation is a complex process determined by many factors. Some of these factors are the conditions of outdoor weather, the heating for various indoor heat sources, and indoor and outdoor ventilation. So, it is required to use computer simulation methods to predict the built environment conditions, such as temperature and humidity changes for indoor over time, real-time energy consumption for HVAC systems, and the energy that is required by the environment control of the building over the year [1]. Computer simulation methods can also be used to calculate human thermal comfort which affects by four main environmental factors represented in indoor air temperature, radiant temperature, relative humidity, and air velocity [2]. As the requirements for the quality of the built environment increase and the focus on building energy efficiency is growing, building simulation has become one of the necessary tools for design, valuation, and analysis. Building comfort control consumes most of the building energy represented in heating, ventilation, and air conditioning (HVAC) [3]. Energy efficiency means less energy is used for heating, cooling, and lighting. It also means integrating different methods to ensure energy conservation. The concept of "energy-efficient" buildings has immediate implications on regulations, the economy, energy demand, and the environment. To evaluate absolute energy-efficiency standards for an energy-efficient building: Efficient equipment and materials appropriate for the location and conditions must be used in the building; amenities and services appropriate to the building's intended use must be provided; and having low energy use compared to other, similar buildings [4]. Using simulation programs is the best way to help engineers make the design process faster, avoid major design defects, and increase efficiency. A comparison of a wider range of design variants is available, leading to more optimal designs. Engineers will also understand the consequences of design decisions better after using simulation programs, which increases the effectiveness of the engineering design process [5]. The Interest in computer-based building energy simulations in the design process is growing in North America, the main factor for this change is the growing awareness of sustainable design practices among building owners and policymakers, as a result of that, a greatly increased number of so-called 'green' construction. In fact, in North America, the US Green Building Council's LEED (Leadership in Energy and Environmental Design) rating system has established itself as an industry standard to demonstrate the 'greenness' of a building. Most projects rated under LEED require an energy model to establish the number of energy efficiency credits for which the project is LEED eligible [6]. Design-Builder simulation software is an integral graphical user interface that is developed for a dynamic simulation program (Energy Plus) of energy consumption for buildings. At any stage of the process of any design, it can be applied to optimize the design and evaluation by supplying the performance data, and before the entire design is determined in the initial design stage it can be carried out [1]. Comparative capability is the other function of Design-Builder simulation models. The different active and passive strategies effectively reduce energy consumption in terms of comparison of energy consumption and Thermal comfort, and the more effective solutions could be selected. This paper opts to investigate two main topics: 1) Demonstrate the accuracy of the Design-Builder to build a model of an existing educational building and compare the simulation output energy consumption with the actual one; and 2) How to make the building more sustainable, consume less energy, reduce carbon dioxide emissions and increase thermal comfort for the users.

2. Methodology

The methodology is comprised two subsections: 1) as-built drawing of the case study educational building and; 2) Data-entry for Design-Builder and performing simulation. The first subsection demonstrates the as built-case study model using the same activity and material. The second subsection shows the steps of calibration and the data entry for the Design-Builder tool in the simulations' versions.

2.1. As-built case study simulation

The case study is an educational building exists Montreal, Canada which is composed of ground floor, fifteen above-grade floors, and fourteen typical floors as shown in (Figure 1), with a footprint area of 5272 m^2. The building is directed to the north. Six different simulations were performed as follows:1) one simulation for the base case; and 2) five simulations for various modified scenarios applied to the same building to enhance passively the thermal comfort for the building internal spaces, minimize the summer months' air temperature, and keep the building warmer in winter. The main strategy that will be modified in each simulation to maximize passive thermal optimization will be the exterior walls because 95% of them contain glass areas (curtain walls and windows).

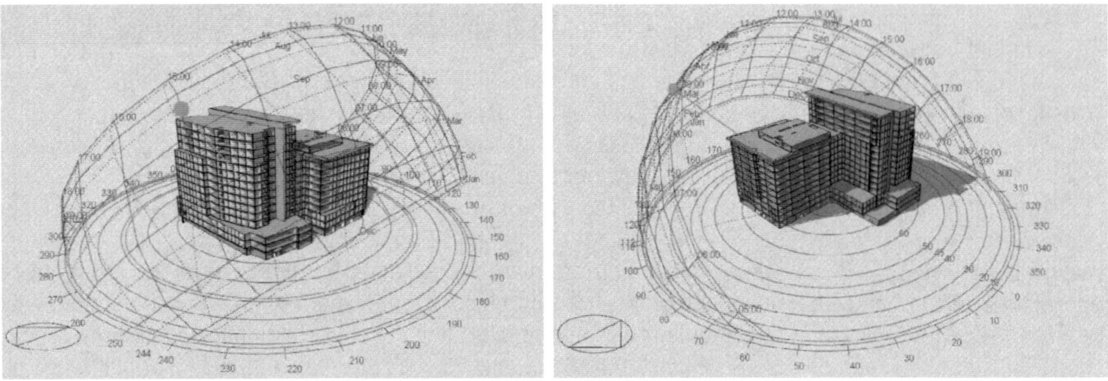

Figure 1: The case study model in Design-Builder.

2.2. Design-Builder data entry calibration and simulation

The required simulations were performed using Design-Builder software for the base case and the five scenarios. Three primary tabs in the software interface, in which setting through all the simulations were kept constant: the activity tab, the HVAC tab, and the construction tab. In the activity tab, the occupancy schedule of was set to be educational classrooms with to occupancy intervals; from 8 am to 3 pm, and from 5 pm to 10 pm. In the environmental control, the cooling setpoint temperature is set to be 23 degrees Celsius, and the setback cooling temperature is set to be 28 degrees Celsius, the heating setpoint temperature is set to be 20 °c, and the setback heating temperature is set to be 12 degrees Celsius. In the HVAC tab, the HVAC template is set to be VAV, Water-cooled Chiller, Air-side HR, and Outdoor air reset, the template (no natural ventilation with heating and cooling) was selected, both are set as electricity from the grid. In the construction tab, the external walls are set to be 250 mm aerated concrete block, with 20 mm cement plaster from inside, and 5mm Aluminium cladding from outside, the internal walls are set to be Lightweight 2 x 25mm gypsum plasterboard with 100mm cavity. The opening tab is the main tab that was changed in every scenario as shown in (Table 1).

Table 1. Simulation scenarios.

Simulation Scenarios	**Base case**	**Modified Case No.1 (S1)**	**Modified Case No.2 (S2)**	**Modified Case No.3 (S3)**	**Modified Case No.4 (S4)**	**Modified Case No.5 (S5)**
openings	Single glass Clear 6mm	Single glass Reflective 6mm	Single glass low E 6mm	Double glass Clear 6/6mm Air gap	Double glass Reflective 6/6mm Air gap	Double glass low E 6/6mm Air gap

3. Results and Discussion

The three main factors that we can use to measure the performance of the building are thermal comfort, energy consumption, and CO_2 emissions. Thermal comfort is measured by the Fanger PMV unit, which lies between ±3 whereas zero is the best thermal comfort. ISO 7730 suggests that PMV should be kept between ranges of ± 0.5 for a good standard of comfort [7]. The result of simulation for thermal comfort and comparison between different scenarios is shown in (Table 2) and (Figure 2), which shows that thermal comfort is the best in Modified Case No.2 (S2) using Single glass low E 6mm at nine months from January to May and from September to December. Energy consumption simulation results are measured by kWh unite and represented in electric consumption as shown in (Table 3) and (Figure 3), which shows that the lowest electric consumption is in Modified Case No.5 (S5) using Double glass low

E 6/6mm Air gap but it is the second-best amount in thermal comfort. CO_2 emissions simulation results are measured by kg as shown in (Table 3) and (Figure 3), which shows that the lowest CO_2 emissions is also in Modified Case No.5 (S5) using Double glass low E 6/6mm Air gap.

Table 2. Thermal comfort (Fanger PMV).

Date/Time	Single Clr. (Base Case)	Single Ref. (S1)	Single low E. (S2)	Double Clr. Air gap (S3)	Double Ref. Air gap (S4)	double low E. Air gap (S5)
January	-1.801012	-1.793306	-1.556308	-1.677675	-1.740677	-1.647184
February	-1.762247	-1.800261	-1.501108	-1.63382	-1.752702	-1.614242
March	-1.452235	-1.602264	-1.144656	-1.297549	-1.546003	-1.293508
April	-2.196045	-2.451227	-1.785284	-1.984209	-2.361709	-1.980758
May	-1.244426	-1.619749	-0.883342	-1.072276	-1.54852	-1.095246
June	-0.2804434	-0.6994104	-4.68E-02	-0.1993167	-0.6741727	-0.2552693
July	-6.13E-02	-0.4160591	0.1107053	-0.01267735	-0.4082153	-0.06795435
August	-0.1903887	-0.5151409	-4.69E-03	-0.1233214	-0.4934928	-0.1680381
September	-0.9430318	-1.289944	-0.6380191	-0.8051034	-1.232481	-0.828703
October	-0.796973	-0.9395702	-0.5173995	-0.6483321	-0.8713551	-0.6346694
November	-1.375543	-1.410569	-1.119432	-1.229025	-1.338127	-1.195175
December	-1.799126	-1.785086	-1.567609	-1.678284	-1.731524	-1.645652

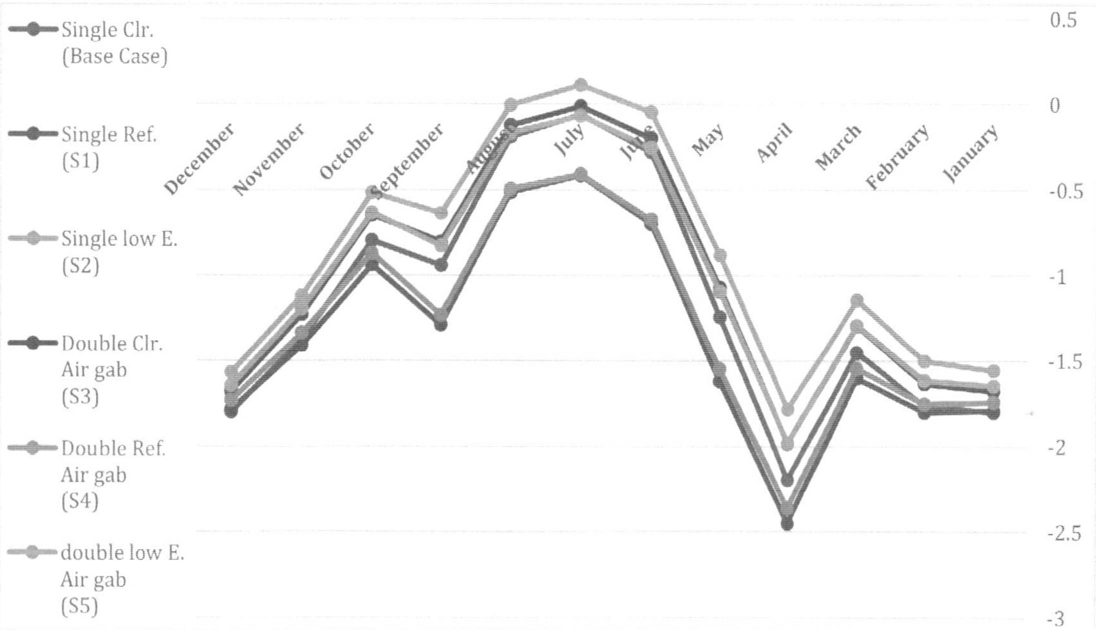

Figure 2. Thermal comfort (Fanger PMV).

Table 3. Electric consumption.

Date/Time	Electric consumption (kWh)					
	Single Clr. (Base Case)	Single Ref. (S1)	Single low E. (S2)	Double Clr. Air gap (S3)	Double Ref. Air gap (S4)	double low E. Air gap (S5)
January	2411208	2486614	2013834	1917015	2095401	1816218
February	1936169	2097104	1608760	1546938	1788424	1485455
March	1292304	1509059	1040767	1041000	1308680	1016587
April	825366.4	941502	674537	690152.4	837943.3	673543.4
May	526220.2	553286.3	493355.4	492159.2	519377.1	483934.9
June	507320.8	460538.3	521325.6	505137.9	451471	492400.9
July	665444.8	588575.3	691216.4	665844.3	577787.2	645781.6
August	657222	583417.4	677048.1	654760.1	570352.1	634392.3
September	469515.3	469276.3	461739.5	455457.5	451420.3	446890.1
October	682492.3	750472.6	581601.7	593368.3	680568.6	578562.7
November	1162253	1201487	953281.4	959443.1	1049231	913521.6
December	2134776	2177023	1765624	1705882	1843277	1611248
Total	**13270291.8**	**13818355.2**	**11483090.1**	**11227158**	**12173932.6**	**10798535.5**

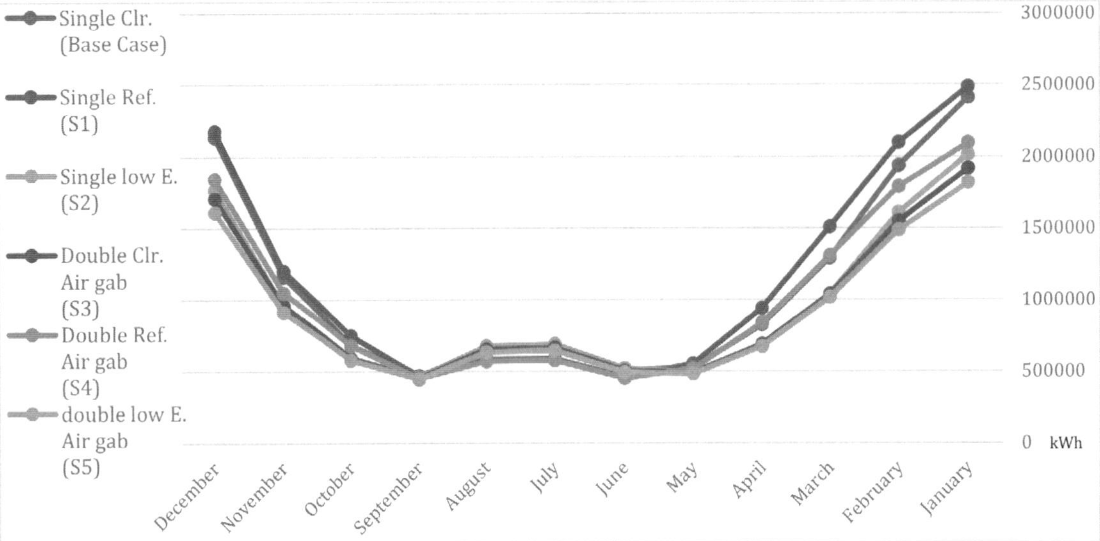

Figure 3. Electric consumption.

Table 4. CO_2 Emissions.

Date/Time	CO_2 Emissions (kg)					
	Single Clr. (Base Case)	Single Ref. (S1)	Single low E. (S2)	Double Clr. Air gap (S3)	Double Ref. Air gap (S4)	double low E. Air gap (S5)
January	1463273	1508969	1222465	1163792	1271894	1102710
February	1175129	1272655	976719.1	939255	1085595	901996.6
March	785038.3	916392.2	632607.5	632748.8	794962.7	617954.4
April	502163.8	572541.9	410761.3	420224.2	509785.4	410159.1
May	320971.7	337373.7	301055.6	300330.7	316824.8	295346.8
June	309248.9	280898.8	317735.8	307926.1	275403.9	300207.4
July	405341.7	358758.8	420959.3	405583.8	352221.3	393425.8
August	400269.6	355544.1	412284.2	398777.7	347626.5	386434.8
September	286427.9	286283.1	281715.8	277908.9	275462.3	272717.1
October	415672.4	456868.5	354532.8	361663.3	414506.7	352691.2
November	706227.5	730003.4	579591.1	583325.1	637736.3	555496.7
December	1295665	1321267	1071960	1035756	1119017	978407.4
Total	**8065428.8**	**8397555.5**	**6982387.5**	**6827291.6**	**7401035.9**	**6567547.3**

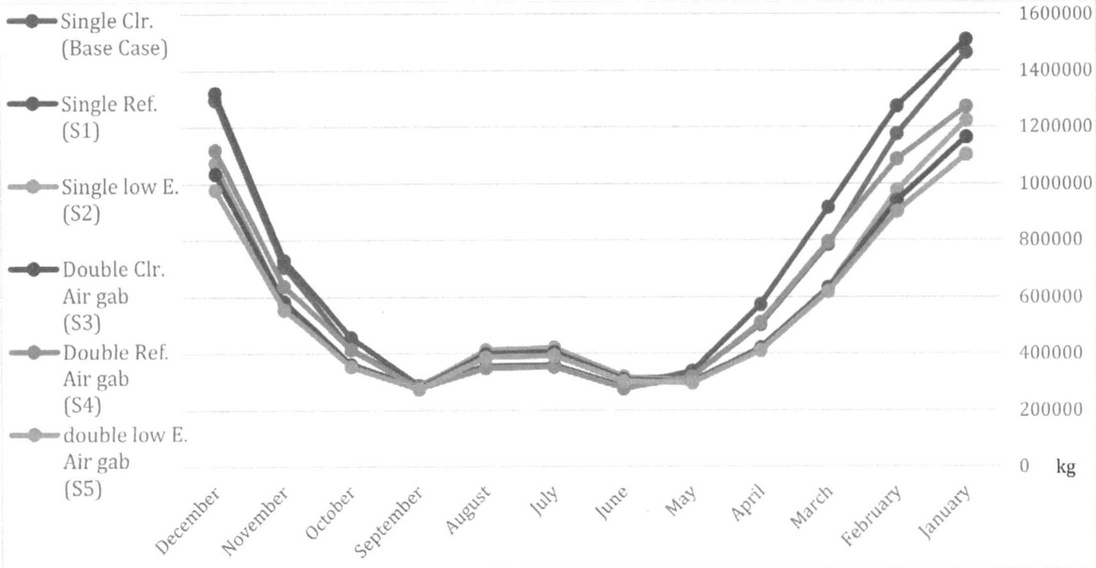

Figure 4. CO_2 Emissions.

4. Conclusions

Design-Builder is an application that applicable to architects, HVAC engineers, energy consultant companies, universities, and some research institutions to make the building more sustainable, consume less energy, reduce carbon dioxide emissions, increase thermal comfort for the users, and even can be carried out in the initial design stage before the entire design is determined. In our case study, we used Design-Builder to build a model of an existing educational building in Montreal, Canada (cold climate weather) and compare the simulation outputs for six scenarios with different glazing material in each one to know which glazing material give us more reduction in energy consumption and carbon dioxide emissions and increase thermal comfort for the users. In Modified Case No.5 (S5) using Double glass low E 6/6mm Air gap, there is a reduction in energy consumption represented in electric consumption by 18.63% per year compared to the base case in which we used single clear glass as a glazing material. In the same case, there is also a reduction in carbon dioxide emissions by 18.57% per year. The best thermal comfort results in two cases (S2) and (S5) when we used single and double low E glass, which mean that the most suitable type that can be used as glazing material in our case study according to simulation outputs is Double glass low E 6/6mm Air gap.

References

[1] Lin Zhang, (2014), Simulation Analysis of Built Environment Based on Design Builder Software, Sichuan College of Architectural Technology, Deyang, Sichuan, China.
[2] Zahiri S, Altan H, (2012), Evaluation of Thermal Conditions in a Secondary School Building in Middle East: A Comparison Study of Computed and Measured Conditions, In Asim 2012, the 1st Asia conference of international building performance simulation association
[3] Frauke Oldewurtela, Alessandra Parisio, Colin N. Jonesc, (2012), Use of model predictive control and weather forecasts for energy-efficient building climate control, Energy and Buildings 45 (2012) 15–27.
[4] Bhagyesh S Pawar, Prof. G.N. Kanade, (2018), Energy Optimization of Building Using Design Builder Software, Warananagar Kolhapur, India.
[5] Augenbroe, G. (2002), Trends in Building Simulation, Georgia Institute of Technology College of Architecture, Atlanta, Georgia USA
[6] Holly A. Wasilowski, Christoph F. Reinhart, (2009), Energy Simulation of an Existing Building Using Customized Internal Load Schedules and Weather Data as Opposed to Default Assumptions, Harvard University, Graduate School of Design Cambridge, MA, USA.
[7] Can Ekici., (2013), A REVIEW OF THERMAL COMFORT AND METHOD OF USING FANGER'S PMV EQUATION, Turkish Standard Institution (Gebze Calibration Laboratory), Turkey.

Challenges for Implementing Environmental Impact Assessment method for construction projects in Egypt

Rodina Abd El-Raouf Lotfy[1, *], Walaa S.E. Ismaeel[1]

The British University in Egypt, Department of Architecture, Faculty of engineering, El-Sherouk City, Egypt

[*]Corresponding Author Email: Rodina172723@bue.edu.eg

Abstract. Assessing the environmental impact of buildings is a prime concern for many countries to abide by international agreements. Various studies have shown that the construction industry has significant implications for the consumption of energy, raw material utilization, and carbon dioxide emissions. Recent Law 9/2009 in Egypt necessitates performing an Environmental Impact Assessment (EIA) for major construction projects. There is a rising number of projects doing the EIA every year. Nevertheless, this cannot be considered a common practice till the present time and many projects are still struggling to perform the EIA studies. Hence, this study investigates the common challenges for implementing EIA for major construction projects in Egypt through a comprehensive literature review and survey among local practitioners. The results showed that the dilemma lies in the segregation of the EIA process along with the project's phases. It also indicates problems associated with the limited availability of data and insufficient practitioners' competencies and skills to perform the required assessment for certain impact categories.

Keywords: Construction projects, Environmental Assessments, Environmental Impact Assessment, Mitigation, Project Management.

1. Introduction

Sustainable indicators, measurements, and certifications are gaining importance in the building industry to comply with the increasing market demands for environmental stewardship [1]. The environmental assessment process aims at providing a comprehensive account of the effect of the building/project/plan or policy on the surrounding environment e.g., area, water, energy, raw materials and resources, and potential causes of pollution, including emissions and waste production. This aims at minimizing the expected environmental impacts as shown in figure 1. Hence, comprehensive strategies and several tools and methods have been developed to provide robust means for carrying out environmental assessments- noting the Environmental Impact Assessment (EIA). It is noted that the decision-making process for major construction projects requires an environmental assessment process for all parameters which may disrupt the natural ecosystem [2–4]. Nevertheless, practitioners are still struggling to perform an EIA for their major construction projects. Hence, the study performs a thorough literature review to pinpoint aspects associated with the successful implementation of the EIA world widely. These were used to design and construct an online questionnaire and structured interviews among local practitioners to investigate critical aspects for implementing EIA in Egypt.

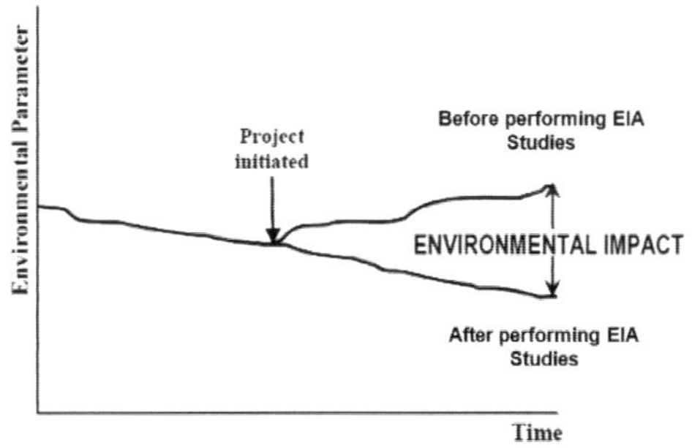

Figure 1. EIA studies can reduce projects' environmental impact [5].

2. Literature review

A review of extant literature was conducted through Science Direct databases from 2016 to 2021. This has culminated in the revision of several books, journals, publications, and technical documents. Construction projects, project management, energy efficiency, Environmental Assessment, and Environmental Impact Assessment were among the keywords utilized in the web search. As illustrated in Figure 2, there is a gradual rise in the global interest in the subject.

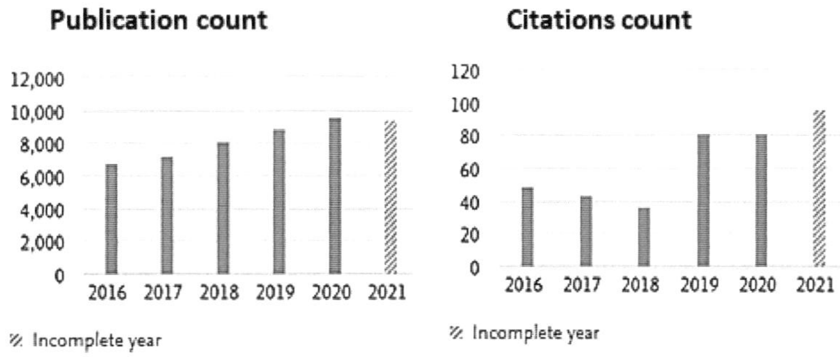

Figure 2. Recorded publications for environmental impact assessment, Scopus directory of research publications, accessed 11 November 2021.

2.1. *Environmental impacts*

Human activities/ development leads to changes in the surrounding environment, alongside the natural changes inherent in the functioning of the environmental system. Man-induced changes in environmental conditions may have positive or negative consequences, which are called "Effects" and imply changes in the natural environment accompanying "Impacts" which imply the outcomes of these changes [6,7]. An impact can be described in terms of its magnitude and significance. This can be later classified into mid-point and endpoint impact categories [8]. Areas of protection may be grouped according to the receptor type (human health, ecosystems, and resources), the urgency of depletion (High, medium, and low), geographic scales (International, regional, and local), and impact medium (air, water, and soil). Additionally, the former includes natural resource diminution (water usage, extraction of minerals, land uses, and limited and finite energy sources). Air impacts include air acidification, photochemical oxidation, and ozone layer depletion. The terrestrial/ aquatic impacts include aquatic eco-toxicity and water eutrophication. Climate effects include climate change as a result of the global warming phenomenon. Human health includes human toxicity, respiratory inorganics, and ionizing radiations [9].

2.2. *Environmental Impact Assessment*

The Environmental Impact Assessment is a standardized approach that aims for the identification, prediction, evaluation, and mitigation of the prospective impacts of the anticipated projects. This investigation should take place for planning strategies, schemes, or legislations on the physical, chemical, biological, economic, and social elements related to the environment right before making major critical decisions and pledges. Firstly, it was established as a non-binding framework in "the National Environmental Policy Act (NEPA)" in 1969, subsequently, it became law in 1971 [10]. It can be considered a part of an environmental planning development tool [11]. Moreover, it provides an early warning process against possible environmental disruption. It also serves as a preventative technique, allowing for the early assessment of harmful and beneficial environmental consequences and the adjustment of those impacts to tolerable ranges. This necessitates a steady transition away from unrestrained consumption and toward efficient natural resources conservation and management. The EIA procedure is focused on three fundamental principles: First, 'Integrity', the EIA process should fulfil internationally acknowledged standards and norms of the practice. Second, 'Utility', it should supply abundant and accurate data for the process of decision-making. Third, 'Sustainability', it should culminate in the application of environmental precautions and safety measures that are suitable to mitigate and attenuate major negative impacts and preclude the undesirable waste of resources and conservation efforts.

The most important aims of the EIA process are resources preservation, wastes reduction, by-product recoveries, efficient utilization of tools, and long-term sustainable growth. It also sets guidelines for urban development projects established within an existing urban area or the new cities constructed outside the existing urban boundaries, and any touristic zone with special nature [12,13].

The main purpose of the EIA reports is to present a comprehensive account of all environmental impacts of a construction project that are likely to have huge impacts on the natural environment as well as addressing all predicted impacts across the planning, design, and construction phases of the developing projects. EIA offers data on negative environmental impacts as well as anticipating dangers associated with activities during the projects' development phases and assisting in finding the areas where threats could be mitigated. In addition to that, it aims for proposing alternatives to planned projects and applying both management strategies and mitigation techniques. Besides, it results in a coherent EIS (Environmental Impact Statement) that communicates the findings to non-experts in the construction sectors as decision-makers on the significance of the probable impacts and their attributes. Hence, it concentrates on measures that could have a substantial environmental impact. Moreover, it enables governmental authorities, corporate executives, and other concerned individuals to comprehend the predicted environmental impacts of the proposed decisions and strive to make smart decisions that sustain the environmental quality for the upcoming generations. Lastly, the EIA process aims to implement a sustainable development strategy, avoid negative environmental impacts following the execution of the construction projects while promoting integrated economic, social, and environmental development [14].

The types of developments that require an EIA vary from one country to another, yet it should cover the following categories: industry and mining-irrigation, transport and communication, projects in sensitive areas, projects on which there are public complaints, massive roads and transportation networks, airports and ports improvements, power plant construction, dam and reservoir construction, and large residential projects [2,15,16].

An EIA entails several steps as shown in figure 3. The first step is 'Screening', which is the process of determining whether the project requisite an EIA (the very first crucial decision in the EIA process) and if so, the determination of the level of environmental evaluation expected, as well as detecting environmental concerns. Because of the vast variety of projects and operations that could be subjected under EIA, some sort of screening procedure is required to establish the proper scope, procedure, and category of environmental assessment (EA). This allocates the project under consideration into classes based on the project's type, area, sensitivities, and size, as well as the nature and extent of its prospective environmental implications. However, it should be highlighted that overuse of the EIA screening procedure overruns expenses and causes time delays for practitioners [17].

The second step is 'Scoping' which is a crucial phase in the EIA preparation process. It analyses more and deeper issues that are most anticipated as a massive concern during the EIA and removes those that aren't. A thorough scoping study reduces the risk of adding or omitting elements that should be considered in an environmental impact assessment [18].

Insofar the preliminary assessment is essential as it examines the needs for environmental coverages and offers background information to decision-makers. It aids in the identification of criteria that must be fulfilled in order to attain sustainable development. The following data is required for a competent preliminary assessment: a broad description and explanation of the project, environment legislation, an estimation of important impacts on the environment, a brief description of the region of effect as well as contingency plans to deal with potential consequences [19,20].

Mitigation is the process of planning and implementing actions to reduce, mitigate, or eliminate undesirable consequences of a project, task, or action on people or the environment. Mitigation can return one or even more environmental conditions to their pre-impact condition. if not feasible, it can restore the fundamental qualities of natural environments. The avoidance, eradication, diminution, or management of a project's negative impacts is also known as mitigation. To mitigate dire implications, there could be possible flexible modifications to make changes to the planning, designing, and construction methods, editing the setting, and management of the project. As a result, certain impact types may be restored to the before construction phase, or at the very least, the environment's basic features will be restored [21–23].

The usage of visual tools to identify, explore, and graphically display the EIA results in a readable manner to communicate findings to decision-makers may include the use of Checklists, Simple cause-effect matrices, Graphs and flow charts, Environmental cartography or map overlay, Networks, and the Geographic information systems.

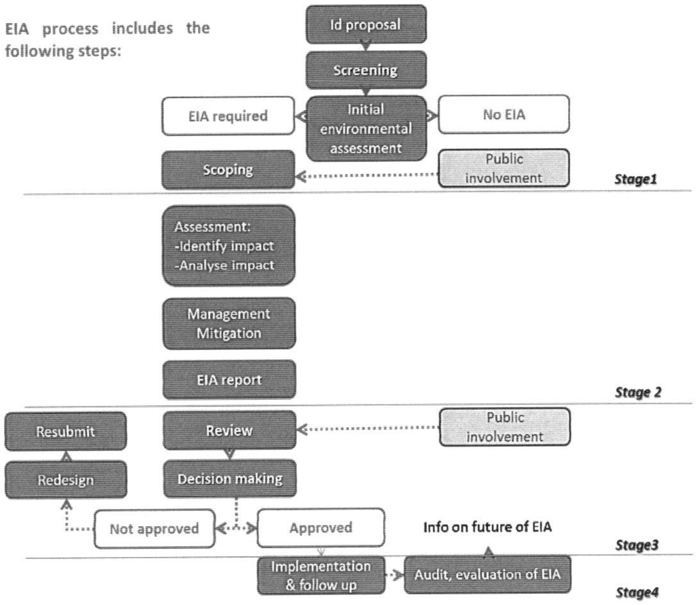

Figure 3. Steps of EIA process [24].

2.3. *EIA in Egypt*

Egypt has been a signatory to nationally and internationally environmental conventions, agreements, and collaborations since 1936. There are several national laws for safeguarding and preserving the environment such as Law 4/1994 and Law 9/2009 for the environment, Law 93/1962 for water waste discharges, Law 38/1967 for General Sanitation, and Law 12/2003 for Labour Works. An EIA is required for new constructions projects, as well as expanding or renovating existing establishments. It stipulates that "the Competent Administrative Authority (CAA) or the licensing authority" for projects that require licenses, must submit an EIA for assessment to "the Egyptian Environmental Affairs Agency EEAA". Following 60 days, the EEAA issues its findings and opinions. A Permission to proceed and continue can be provided if the project has been approved by the EEAA [25].
The purposes of EIA in Egypt are to ensure that the environmental criteria of pertinent Egyptian agencies and applicable international treaties are met or exceeded. Moreover, it should analyse and evaluate critical elements in the present circumstances as well as determining the kind, nature, and severity of any adverse environmental impact during the first phases of execution. from the beginning of the project, it is essential to investigate and propose feasible and applicable mitigation methods to reduce the project's environmental impacts. In addition to that, it targets to guarantee that all the practitioners who may be involved in the project are adequately addressed and that effective communication networks are developed along with the evaluation process and maintained throughout the project's life cycle.

The categories of projects which should implement an EIA are defined by the activities of the project, its location, the amount of natural resource consumption, and energy usage. According to the EEAA [25], the evaluation of urban growth is one among the other eight sectors for which EIA guidelines are presented. The others are “1) Industrial premises, 2) Touristic facilities, 3) Companies operating in mines and quarries and the production of building materials, 4) Enterprises engaged in exploratory drilling, extractions, refinement, storing, and transportation, 5) Electricity generation and distribution facilities, 6) All construction projects, such as plants for the treatment and recycling of wastewater or agricultural drainage water, irrigation projects, highways, bridges, barrages, tunnels, airports, seaports, railway stations, and others, 7) Any other project that is likely to have a significant environmental impact and for which the EEAA will issue a decree with the consent of the relevant administrative body”. For immovable structures that use thermal fuels and have emission levels that surpass the allowed threshold, the form of energy and fuels consumed should be well-studied, and 8) Projects that rely on nuclear power to run their operations.

The EEAA has devised thorough EIA principles, which encompasses the following pillars: 1) defining project aims for the EIA system, 2) Outlining the EIA procedures and regulations, 3) Categorizing the projects as per their impact on the environment and evaluation level as listed in Table (1).

Table 1. Project’s Categories.

1) White Projects	Low impacts projects that can usually be approved based on a simple screening stage
2) Grey Projects	significant environmental impact projects, where a more thorough screening and testing process must be completed by the practitioner. A detailed EIA investigation of recognized impacts or consequences may be demanded in certain circumstances
3) Black Projects	Projects that require a complete, comprehensive EIA analysis because of their potential serious adverse environmental implications for example: Fuel, concrete, and chemicals industries

According to (Law 4/1994) [25], the EIA principles involve requirements for: 1) optimum air temperature and pressure and allowable levels of air pollution, 2) apply certain planning conditions for developments along or adjacent to the coastline, 3) drainage of Liquid Wastes, 4) Solid and Hazardous Waste Management, 5) Noise Emissions and Ambient Noise Levels.

3. Method

This paper adopted a mixed-method approach that incorporated both qualitative and quantitative methodologies. To begin, a review of the literature is made by providing substantial background knowledge on the EIA system, especially in Egypt, project management tools, and mitigation strategies. Moreover, a survey questionnaire was carried out and involved 46 contributors in the field of the construction industry with a variety of their backgrounds (Architect, Environmental Engineers, Project managers, ...) as shown in figure 4. A total of 46 responses were collected. The survey encompassed 13 closed questions in the form of checkboxes and linear scales to evaluate practitioners' familiarity with the EIA process, phases, and tools as well as the communication of EEAA for the aid within obtaining available data. In addition to that, the survey queries the participants about the challenges they face during the integration of EIA through the project. Also, it tackles other pillars including management, mitigation strategies, and the inevitability of their contribution of implementing the EIA to projects' phases.
The Survey was available online using the following link: shorturl.at/nzIP3 and it was open from 21st of November 2021 to 30th of November 2021.

4. Results & discussion

The survey received responses from professionals with a variety of backgrounds as illustrated in figure 4. Firstly, 19 of them declared that not all the practitioners are familiar with the EIA process as shown in figure 5. As for incorporating the EIA process through the construction project phases, 43.5% recommended that it should be integrated with all phases, while 41.3% of respondents suggested integrating it only in the project planning and definition phase. Referring to the tools, they used checklists the most and simulation models the least. Additionally, it was an average level to contact the EEAA. Moreover, respondents stated that they relied on the internet, governmental database, and the Housing and Building Research Institute database for data-sourcing as shown in figure 6. Moreover, figure 7 presented the most significant challenges that face practitioners, which were data availability and limited skills and tools. 56.5% stated that it was important to implement management and mitigation strategies during the project phases. In fact, 19 individuals believed that the EIA process could support other green building rating systems which makes it an important process for sustainable development for all sectors, and that all disciplines should improve their knowledge of EIA.

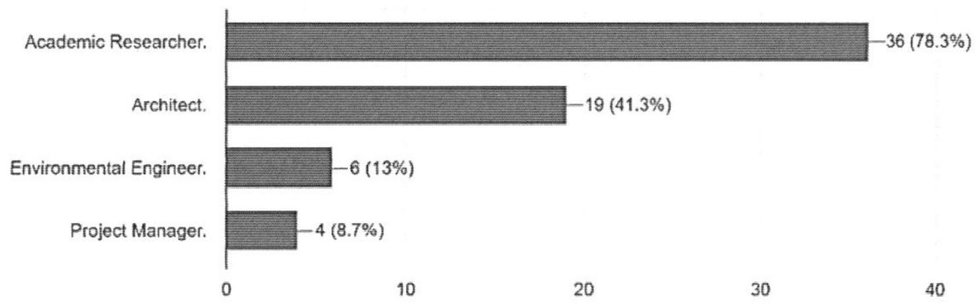

Figure 4. Participant's background.

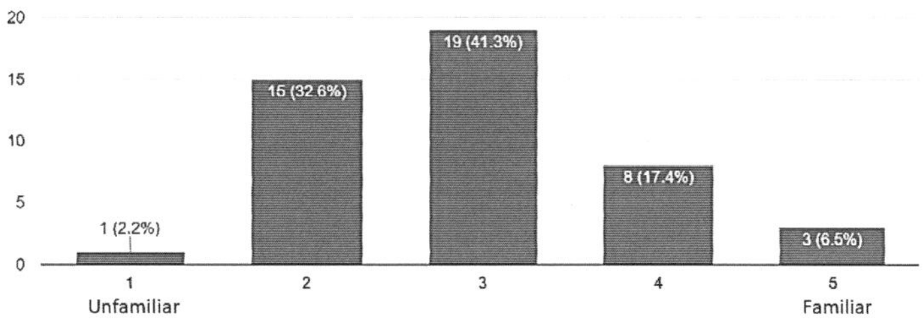

Figure 5. Familiarity of practitioners with the EIA process.

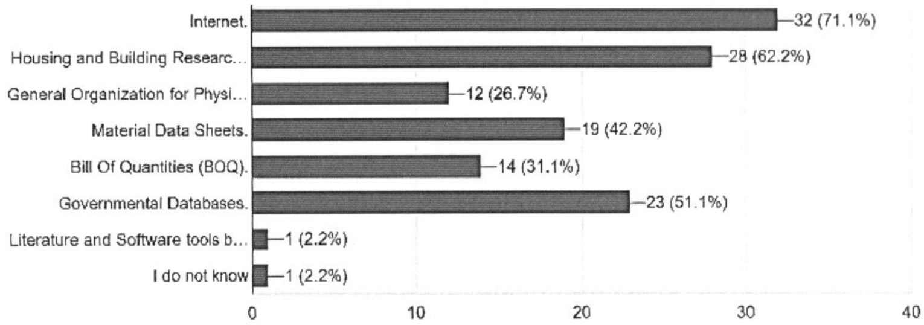

Figure 6. Sources of Data used in EIA Reports.

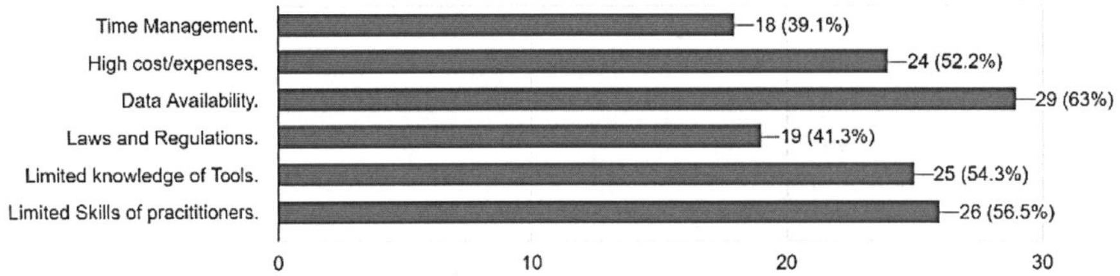

Figure 7. Challenges of implementing the EIA.

5. Conclusion

Preserving the environment and maintaining a sustainable project is a colossal challenge taken under consideration by regular evaluation of the environmental impact of construction projects. This study aimed at investigating the main challenges for implementing EIA in construction projects in Egypt. The results pointed out these challenges and potentials as follows: lack of data availability and reliability, limited practitioners' skills and knowledge of available tools, high expected associated expenses with poor time available to perform the assessment. This suggests that a formalized EIA system framework should be implemented to help practitioners handle some of the obstacles they face and encourage more environmental sustainability. Also, it required better integration along with project development phases from early stages to be able to avoid delays and change orders. Similarly, the knowledge about the EIA tools and methods should be integrated as core science for undergraduate and postgraduate studies to minimize the gap between study and practice and prepare cadres able to adopt responsive methods of design and construction. Additionally, the government's role to ensure data availability and reliability is a must to carry the EIA process within planned time and cost constraints.

References

[1] Ismaeel WSE. Drawing the operating mechanisms of green building rating systems. Journal of Cleaner Production 2019;213:599–609. https://doi.org/10.1016/j.jclepro.2018.12.115.

[2] Momtaz S, Kabir SMZ. Evaluating the effectiveness of environmental impact assessment system in developing countries: The need for an integrated holistic approach. Evaluating environmental and social impact assessment in developing countries, 2013, p. 5–28. https://doi.org/10.1016/B978-0-12-408129-1.00002-4.

[3] Morgan RK. Environmental impact assessment: the state of the art. Impact Assessment and Project Appraisal 2012;30:5–14. https://doi.org/10.1080/14615517.2012.661557.

[4] Ortiz O, Castells F, Sonnemann G. Sustainability in the construction industry : A review of recent developments based on LCA. Construction and Building Materials 2009;23:28–39. https://doi.org/10.1016/j.conbuildmat.2007.11.012.

[5] El-sayed HK. Environmental Impact Assessment Comparative Study Case study: United States, United Kingdom, New Zealand, Egypt. Cairo University 2001.

[6] Munn RE. Environmental impact assessment. Scope 5. Principles and procedures. Environmental Pollution (1970) 1976;10:236. https://doi.org/10.1016/0013-9327(76)90053-7.

[7] Lee N. Environmental impact assessment: a review. Applied Geography 1983;3:5–27. https://doi.org/https://doi.org/10.1016/0143-6228(83)90003-6.

[8] Ismaeel WSE. Midpoint and endpoint impact categories in Green building rating systems. Journal of Cleaner Production 2018;182. https://doi.org/https://doi.org/10.1016/j.jclepro.2018.01.217.
[9] Park WJ, Kim R, Roh S, Ban H. Analysis of major environmental impact categories of road construction materials. Sustainability (Switzerland) 2020;12:1–18. https://doi.org/10.3390/su12176951.
[10] Kang HJ. Development of a systematic model for an assessment tool for sustainable buildings based on a structural framework. Energy and Buildings 2015;104:287–301. https://doi.org/10.1016/j.enbuild.2015.07.015.
[11] Elsayed MA, Ismaeel WSE. Environmental assessment for major development projects: A case study “Qattara Depression.” Journal of Cleaner Production 2019;215:522–33. https://doi.org/10.1016/j.jclepro.2018.12.301.
[12] Arts J, Caldwell P, Morrison-Saunders A. Environmental impact assessment follow-up: good practice and future directions — findings from a workshop at the IAIA 2000 conference. Impact Assessment and Project Appraisal 2001;19:175–85. https://doi.org/10.3152/147154601781767014.
[13] van Buuren A, Nooteboom S. Evaluating strategic environmental assessment in The Netherlands: content, process and procedure as indissoluble criteria for effectiveness. Impact Assessment and Project Appraisal 2009;27:145–54. https://doi.org/10.3152/146155109X454311.
[14] Nita A, Fineran S, Rozylowicz L. Researchers’ perspective on the main strengths and weaknesses of Environmental Impact Assessment (EIA) procedures. Environmental Impact Assessment Review 2021;92. https://doi.org/10.1016/j.eiar.2021.106690.
[15] Wood C. Environmental impact assessment in developing countries. International Development Planning Review 2003;25:301–21. https://doi.org/10.3828/idpr.25.3.5.
[16] Marara M, Okello N, Kuhanwa Z, Douven W, Beevers L, Leentvaar J. The importance of context in delivering effective EIA: Case studies from East Africa. Environmental Impact Assessment Review 2011;31:286–96. https://doi.org/10.1016/j.eiar.2010.10.002.
[17] Cilliers DP, Retief FP, Bond AJ, Roos C, Alberts RC. The validity of spatial data-based EIA screening decisions. Environmental Impact Assessment Review 2022;93:106729. https://doi.org/10.1016/j.eiar.2021.106729.
[18] Adelowo OO, Akinlabi IA, Fagade OE. Environmental impact assessment of Attenda abattoir, Ogbomoso southwestern Nigeria on surface and groundwater quality using geo-electrical imaging and microbiological analysis. Environmental Monitoring and Assessment 2012;184:4565–74. https://doi.org/10.1007/s10661-011-2285-2.
[19] Jasch C. Environmental performance evaluation and indicators. Journal of Cleaner Production 2000;8:79–88. https://doi.org/10.1016/S0959-6526(99)00235-8.
[20] Huertas-Olivares C, Norris J. Environmental Impact Assessment. Environmental Monitoring and Assessment 2013;1:423–97.
[21] Morgan RK. Environmental impact assessment: the state of the art. Impact Assessment and Project Appraisal 2012;30:5–14. https://doi.org/10.1080/14615517.2012.661557.
[22] Senécal P, Goldsmith B, Conover S, Workshop_Participants II, Sadler B, Brown K. Principles of environmental impact assessment, best practice. INTERNATIONAL ASSOCIATION FOR IMPACT ASSESSMEN 1999:1–4.
[23] Suwanteep K, Murayama T, Nishikizawa S. Environmental impact assessment system in Thailand and its comparison with those in China and Japan. Environmental Impact Assessment Review 2016;58:12–24. https://doi.org/10.1016/j.eiar.2016.02.001.
[24] Ismaeel WSE, Elsayed MA. The Interplay of Environmental Assessment Methods; Characterising the Institutional Background in Egypt. Journal of Environmental Assessment Policy and Management 2018;20:1850003-1-1850003–27. https://doi.org/10.1142/S1464333218500035.
[25] EEAA. Egyptian Environmental Affairs Agency n.d. https://www.eeaa.gov.eg/en-us/home.aspx (accessed December 25, 2021).

Effect of Bentonite Content on Hydraulic Conductivity of Sand-Bentonite Mixtures Used in Landfill Liners as an Alternative to Clay Liner in Egypt

Amr Barrima[1,*], Ibrahim M Mashhour[1], Nabil H Amer[1]

[1] Construction and Building Department, Arab Academy for Science and Technology and Maritime Transport, Smart Village, Giza, Egypt

[*]Corresponding Author Email: Amrbarrima@gmail.com

Abstract: Compacted clays are commonly utilized as landfill liners due to their impermeable properties, however, the availability of natural clay soil may be difficult or prohibitively expensive in some regions thanks to the expense of transportation, the local soil availability or regulation. As an alternative hydraulic barrier, a compacted mixture of sand with a low percentage of bentonite (6% to 14%) was used. The purpose of this research is to investigate the applicability of bentonite available in the Egyptian market for controlling the hydraulic conductivity of sand-bentonite mixtures to be used in landfill liners. The study was divided into two phases: first, laboratory tests were conducted to identify the optimum bentonite percentage and bentonite types, using local commercial bentonite products. Then, in two distinct landfill liners layers, full-scale measurements were obtained. The experimental program demonstrated that Bentonite produced and found in the Egyptian market can be utilized in landfill liners with an acceptable quality control procedure to prevent any outcome deviation.

Keywords: Waste containment liner systems, Landfill, Sand Bentonite mixtures, Atterberg limits, Hydraulic conductivity.

1. Introduction

Waste should be disposed in way that does not harm the environment to any extent. Therefore, it is very important to be aware of the things we recognize as waste, in order to identify sources of danger to the environment and our health, and suitability of these wastes for recycling as secondary raw materials.
In almost all regions specially the developing countries the way of living has changed thus the type of wastes have altered to more complex mixtures of safe and hazardous wastes beside the municipal organic wastes. The household wastes nowadays contain a lot of plastics, batteries which contain chemicals, unused drugs and medicines, and other chemicals. In addition, the waste resulting from small industries or from hospitals which may mix their hazard wastes with the normal garbage. After adopting all waste avoidance and resource recovery options for municipal garbage, wastes must be handled and disposed of in an environmentally friendly manner.

Since all waste treatment operations produce residues that cannot be reused or recovered, landfilling is one of the most extensively utilized municipal solid waste disposal strategies worldwide. Excess rainfall infiltrates the garbage layers of landfill, producing leachate. The most serious possible environmental impacts associated with landfill leachate include groundwater and surface water contamination. Because most landfills were constructed without engineered liners and leachate collecting systems in the past, this concern is likely to be the most devastating environmental impact from landfills.

2. Statement of the problem

The landfill's liner system is the primary line of defence against leachate and methane gas migration into the environment. A performance standard criterion is a feasible method for building a liner system. A performance standard specifies the expected performance of a lining system, and site-specific design requirements are created [1]. Landfill liners limits the leakage of leachate to the surrounding environment, when it comes to choose the right material for the liner, the engineering features of the compacted layer, such as low hydraulic conductivity, play a major role in preventing the penetration of rainwater while allowing landfill gases to seep out of trash [2]. Human health and the environment might be jeopardized if waste disposal facilities are not carefully selected and secured. Besides odor issues, the leaking of contaminated fluids from wastes into the ground, and then into the ground water, poses significant dangers to flora and wildlife, as well as human health. Mixing the leachate and subterranean reservoir generates a plume that extends hundreds of meters and damages the hydrogeological system. It is reported Leachate may be tracked using a number of indicators, such as a main hydrochemistry, stable isotopes and tracer elements, [3]. It is reported subsurface water around the Serum dump site contained significant levels of trace elements such as chlorine and ammoniacal nitrogen, [4]. Researchers at the Dhapa landfill site in Kolkata [5] investigated the health impacts of leachate plume and found that residents living near the Dhapa dump site had diarrhea, vomiting and stomach discomfort. In another study, similar findings in Chandigarh MSW employees, [6]. Clean water quality is closely connected to human health, it is a major concern for humanity. Achieving and maintaining long-term viability depends on preventing or reducing ground pollution.

This research aims to find a suitable alternative for clay to be used in landfill liners, since clay does not provide an economical alternative to be used in Egypt due to its geological scarcity. So, it is therefore essential to approach the most suitable alternative, by the following:

- Performing an experimental program using different types of bentonites available in the Egyptian market, to determine the suitability of these types.
- Determining the optimum sand bentonite mixture proportions using the bentonite available in the Egyptian market.
- Field application on landfill projects under-construction.

3. Clay and alternatives as landfill liners

Since they are almost impermeable, compacted clay is often utilized as a liner. Across most scenarios, compacted clays have a permeability of $1X10^{-7}$ cm/s or less. However, clay in some areas may be unobtainable due to the non-availability or the high cost of transportation, this where alternatives should be used [7]. Sand is a porous material, Where the permeability factor of sand varies from 1 to $1X10^{-5}$ cm/s). However, these results of sand can be significantly reduced when mixed with a relatively impermeable material such as bentonite. Because of their lower tendency for volumetric change response to climate fluctuations, compacted mixes of sand and bentonite are extensively utilized as a substitute for compacted clay in landfill barriers.

3.1. Sand bentonite mixtures

Sand-Bentonite mixtures are utilized as engineered liners to hold the leachate composed from waste. Since bentonite has a high swelling capacity, it serves as it ideal solution to produce low permeability barriers. Shrinkage happens in bentonite when the clay content is too high. Sand is added to bentonite to prevent shrinkage upon drying, giving mechanical stability to the combination. When exposed to water, elastic materials such as bentonite elevate a volume shift and swell. as bentonite absorbs water, swelling pressure develops under constant volume conditions.

Sand-Bentonite mixtures are an ideal type of soil to be used in engineering proposes such as hydraulic barriers such as landfill liners, sand play the role of the mixture's skeleton due to its strength and stability and the bentonite's fine particles reduces the permeability of the mixture by filling the gaps remaining between the sand grains [8]. To limit leachate flow through the barrier into the surrounding environment, low hydraulic conductivity, is the primary design and assessment concern for liners. However, owing to shipping costs or country-specific regulations, natural clay soil may not be readily available. A compacted sand-bentonite combination (8-15%) has been proposed as an alternative. It has been argued that samples containing 10% bentonite and 90% of sand had the lowest saturated hydraulic conductivity, with 99.9 % decrease, [9].

4. Experimental program

This Research aims to study the suitability of bentonite available in the Egyptian Market to control the hydraulic conductivity of sand-bentonite mixtures to be used in landfill liners. The study passed by two phases: laboratory experiments were used at first to determine the optimum bentonite percentage and the best type available in the market, then it was used in a full-scale application in 2 different landfills liners layers.

4.1. Materials and methods

Before any material can be used for the construction of a liner, number of laboratory tests are necessary. Initially, Proctor tests have to be carried out to establish the relation between dry soil density and moisture content. From the curve obtained the optimum moisture content for compaction in the field is determined. It was found that the lowest permeability is reached when the clay is compacted on the wet side of the Proctor curve.

4.1.1. Sand. The sand was (SP-SM) poorly graded silty sand based on the Unified Soil classification System [10]. sand obtained from a suggested site with an effective diameter (D10) equals 0.11, coefficient of uniformity (Cu) was 6.07 and Coefficient of curvature (Cc) was 1.33.

4.1.2. Bentonite. Three different types of bentonite from the Egyptian market were used to conduct this experiment, two types of bentonite were collected from two separate landfill construction sites, "Badr Site" in Behara Governorate and "Koum Oshem" in Fayoum Governate, while the third is a premium grade OCMA bentonite known as Nanogel-HY obtained from RP Minerals firm. Grain size distribution were obtained according to [11]. Atterberg limits were obtained according to [12] as shown in the following Table 1.

Table 1. Physical properties of soil constituents.

Constituent Property	ASTM test standard	Bentonite 1	Bentonite 2	Bentonite 3
Liquid limit, LL%	D 4318	67%	65%	426%
Plastic Limit, PL %	D 4318	27%	24%	42%
Plasticity index, PI %	D 4318	40%	41%	384%
% of Fines	D422	93%	88.20%	99%
Soil Classification	D 2487	CH	CH	CH

4.2. Measurement technique and equipment

Permeability tests using triaxial cell with back pressure system were conducted in this research with only one constant pressure system connected to the specimen if the outlet from the specimen is connected to an open burette as shown in figure 1. If the permeability is high enough to permit displacement of air from the voids in the sample by upward movement of water, the back-pressure system should be connected to the base, allowing drainage from the top figure 1 (b). If saturation is first achieved by application of back pressure increments, the direction of flow is immaterial, after disconnecting the pore pressure panel.

Flow is then downwards as indicated in figure 1 (c). The outlet pressure can be raised slightly above atmospheric by elevating the burette so that the water level is higher than the exit point from the specimen. For every meter increase in height the pressure is increased by 9.81 kPa. When hydraulic gradients exceeding unity are to be applied, upward flow can lead to instability and piping especially in non-cohesive soils. Downward flow gives stable conditions and is generally to be preferred. Bubbles of air or gas are likely to emerge from the specimen at these low pressures. Air can be removed by fitting an air trap, filled with water initially, as indicated in figure 1.

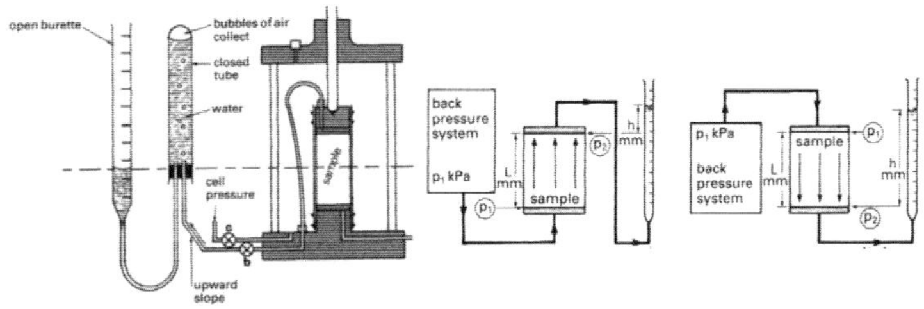

Figure 1. (a) Arrangement for triaxial permeability test using one back pressure system. (b, c) bubble trap for separating air from water expelled from a partially saturated specimen.

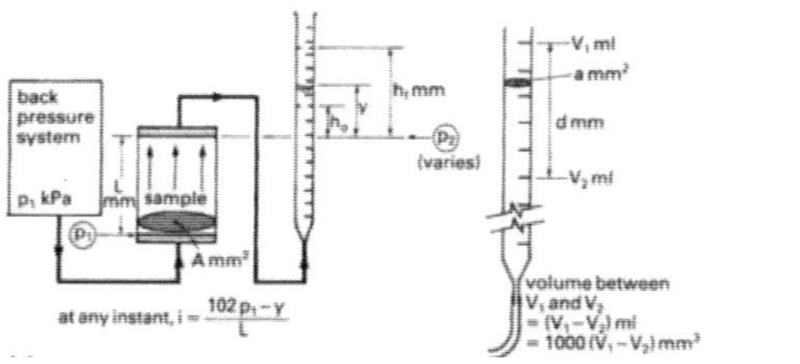

Figure 2. Triaxial permeability test under falling head condition (a) pressures and hydraulic gradient, (b) measurements for determining cross-sectional area of burette.

Figure 3. Specimens used in the triaxial setup

Since the inlet pressure is not much greater than the outlet pressure under the head of water in the burette, and a constant level overflow is not used, the variation in outlet pressure as the water level in the burette rises might be significant shown in figure 2. and the calculation of permeability modified as following.

$$K = 3.84\left[\frac{aL}{At}Log_{10}\left(\frac{102P_1 - h_0}{102_{P1} - h_f}\right)\right]X10^{-5}\frac{m}{s} \quad \ldots\ldots[1]$$

H_0= height of water level in burette above outlet end of specimen initially (mm)

h_f=corresponding height after time t (mm)

t= duration in minutes L= Height of sample

A= Area of sample p_1= back pressure

A triaxial permeability test under falling head conditions was used to evaluate each compacted specimen's hydraulic conductivity *k*. The specimens for hydraulic conductivity testing were mixed at a target water content that was two percentage wetter than w_{opt} for the specific mixture. In order for the permeant water to move upward through the specimen, each specimen was back pressured with 42 Kpa and a hydraulic gradient; *i* equal to 30. also, a confining pressure $б_3$ = 100 Kpa. Figure (3) shows the triaxial apparatus used in this study.

5. Results and Discussion

Drought causes cracks in waste containment materials, which creates preferential flow pathways for hydraulic leakage. The permeability of a sand-bentonite mixture may be met without shrinkage cracking as long because the bentonite content is not excessively high, which might result in shrinkage breaking during desiccation therefore contaminated leaking, bentonite is utilized as a sealing material in disposal systems. When wet, the clay fills the gaps in the sand, resulting in a combination with very low permeability. The sand component of the mixture reduces shrinkage during drying, and the sand particles are in contact below a specific bentonite level, giving mechanical stability and avoiding shrinkage. The granulated sand particles preserve the composite's strength and stability, while the tiny bentonite grains fill the gaps between them and minimize permeability. Hydraulic conductivity and swelling capacity depend on the confinement conditions, such as the temperature, content, and quantity of accessible water, the hydraulic conductivity of a landfill liner must be less than 1 x 10^{-9} m/s, and the system must be stable during the life of the landfill. Materials must be carefully selected, which is generally decided by local availability. Sand bentonite mixtures with 5 different bentonite contents, 6:14% by dry weight was used to conduct this experiment with 3 different types of bentonites. The 15 conducted specimens were used to measure the Hydraulic conductivity by falling head permeameter according to [13], also Grain size distribution, Atterberg limits and moisture-density relationship were conducted to identify the mixture classification, percentage of fines, liquid and plastic limits and the optimum moisture content for the specimen, as shown in table 2.

Table 2. Physical properties of sand-bentonite mixtures.

Specimen No and Bentonite Type	*Bentonite Content (%)*	*Measured Atterberg limits*		*Measured compaction properties ASTM D698*		*Bulk Density (g\cm³)*	*Hydraulic Conductivity (k)*
		Liquid Limit, LL (%)	*Plastic Limit, PL (%)*	*Maximum Dry Density (g/cm³)*	*Wop (%)*		
1.1	6	21	12	2.095	8.06%	2.26	$2.50X10^{-08}$
1.2	8	22	13	2.082	9.73%	2.296	$2.00X10^{-08}$
1.3	10	26	12	2.102	8.76%	2.1	$1.20X10^{-08}$
1.4	12	26	14	2.127	8.33%	2.316	$3.90X10^{-09}$
1.5	14	30	16	2.154	7.48%	2.3	$1.80X10^{-09}$
2.1	6	25	NA	1.977	8.04%	2.136	$1.30X10^{-08}$
2.2	8	25	12	2.005	6.37%	2.143	$9.70X10^{-09}$
2.3	10	25	14	2.098	4.67%	2.291	$5.60X10^{-09}$
2.4	12	28	13	2.133	7.58%	2.304	$1.90X10^{-09}$
2.5	14	33	14	2.155	6.62%	2.299	$4.70X10^{-10}$
3.1	6	25	NA	2.047	7.55%	2.207	$4.00X10^{-09}$
3.2	8	38	18	2.066	6.68%	2.21	$2.60X10^{-09}$
3.3	10	55	25	2.111	6.62%	2.218	$1.80X10^{-09}$
3.4	12	54	21	2.076	7.88%	2.224	$1.20X10^{-09}$
3.5	14	94	28	2.135	7.76%	2.291	$4.40X10^{-10}$

5.1. Effect of different mixtures ratios on Hydraulic Conductivity

5 different bentonite contents were used in this research, samples with bentonite contents 6% and 8% didn't reach the minimum hydraulic conductivity criteria needed for landfills liners, $1X10^{-9}$ m/sec.

5.1.1. Mixture of 90% sand to 10% Bentonite. As for the 10% bentonite to sand mixture the hydraulic conductivity results were $1.2X10^{-8}$ m/sec for Bentonite Type 1 and $5.6X10^{-9}$ m/sec for bentonite Type 2, and $1.8X10^{-9}$ m/sec for bentonite type as shown in figure 4.

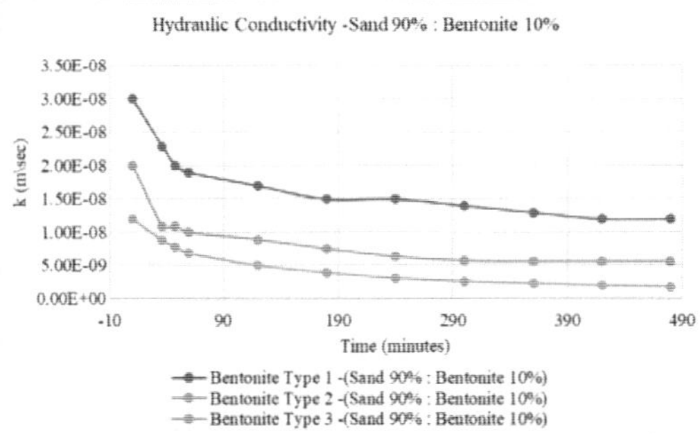

Figure 4. Hydraulic conductivity for 90S:10B for different bentonite types

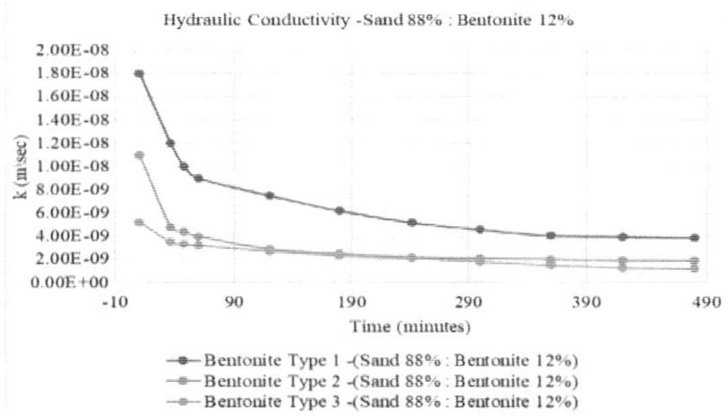

Figure 5. Hydraulic conductivity for 88S:12B for different bentonite types

5.1.2. Mixture of 88% sand to 12% Bentonite. By increasing the bentonite percentage to 12%, the Hydraulic conductivity decreased to 1.2X10^{-9} m/sec for Bentonite Type 3 and 3.9X10^{-9} and 1.9X10^{-9} m/sec for Type 1 and 2 respectively, As the amount of bentonite increased the gap between Bentonite Type 3 and the other 2 types decreased, Bentonite type 2 and type 3, k values were very close to 1X10^{-9} m/sec which could be obtained if more readings were taken after the 8 hours as shown in figure 5.

5.1.3. Mixture of 86% sand to 14% Bentonite. 14% bentonite gave the lowest hydraulic conductivity for all the 3 types of bentonite, showing a 1.8X10^{-9} m/sec and 4.7X10^{-10} m/sec for bentonite Type 1 and 2 respectively and 4.4X10^{-10} for bentonite type 3 with a 309% and only 6.8% lower than types 1 and 2. Both bentonite type 2 and 3 verified the hydraulic conductivity requirements, however bentonite type 1 didn't fulfil the requirements with a very close K value to 1X10^{-9} as shown in figure 6.

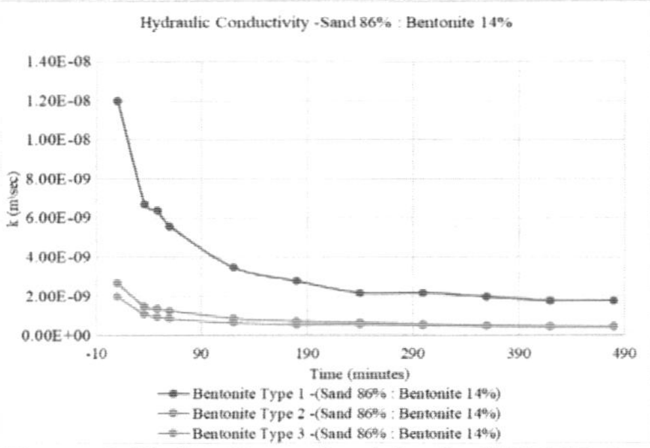

Figure 6. Hydraulic conductivity for 86S:14B for different bentonite types.

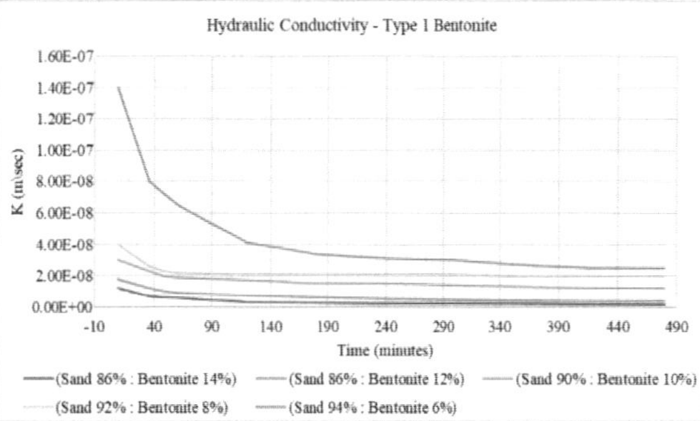

Figure 7. Hydraulic conductivity for type 1 bentonite.

5.2. Effect of bentonite Content on hydraulic conductivity of sand bentonite mixtures

5.2.1. Type 1 Bentonite. Figure 7 shows the k values for type 1 Bentonite for the 5 mixtures specimens varied between 2.50X10^{-8} m/sec for 6% Sand Bentonite mixture and 1.8X10^{-9} m/sec as the bentonite content increased to 14% showing a 1288.89% reduction in the hydraulic conductivity of the specimen compared with the 6% bentonite mixture. This Bentonite type did not fulfil the 1X10^{-9} m/sec hydraulic conductivity requirement for landfill liners.

5.2.2. Type 2 Bentonite. Type 2 bentonite showed better results compared with type 1 bentonite with an average 140% reduction in terms of Hydraulic conductivity, the K results varied between 1.3X10-8 m/sec for the 6% Bentonite Sample to 4.7X10-10 m/sec for the 14% bentonite specimen showing a tremendous reduction in hydraulic conductivity with 2665% compared with 6% bentonite specimen. Bentonite type 2 could fulfil the 1X10^{-9} m/sec requirements with 14% bentonite content shown in figure 8.

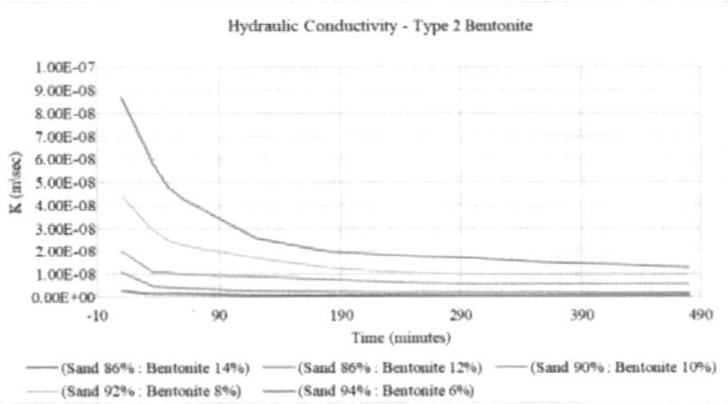

Figure 8. Chart Results for type 2 bentonite.

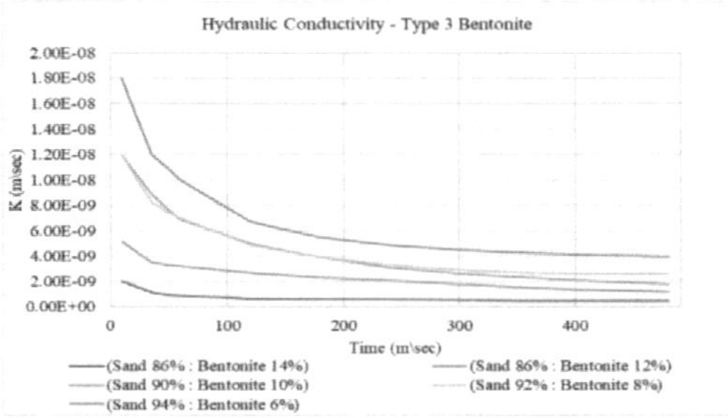

Figure 9. Chart Results for type 3 bentonite.

5.2.3. Type 3 Bentonite. Figure 9 shows the best results obtained in the experimental program, with an average reduction of 460% and 154% compared with bentonite type 1 and 2, respectively. The k value for 14% bentonite content specimen was $4.4X10^{-10}$ m/sec verifying the desired requirements, The 10% and 12% bentonite specimens showed a K value of $1.8X10^{-9}$ m/sec and $1.2X10^{-9}$ m/sec, respectively. Bentonite type 3 showed the closest values to $1X10^{-9}$m/sec for all specimens which may be achieved in longer time experiment.

5.2.4. Field Results for Different landfill sites. This laboratory experimental program was conducted to determine the lowest hydraulic conductivity of sand bentonite mixtures that can be obtained from the bentonite available in the Egyptian market to be used in a full-scale landfill project under construction in Egypt. The sand-bentonite mixtures were used in those projects due to its lower cost in both material and handling, as well as the regulation regarding razing the agriculture lands. After achieving the most applicable type of bentonite and sand-bentonite percentage 12% Bentonite mixtures were used in 2 different landfills in 2 different towns, as shown in figure 10. The results in both sites varied from $1X10^{-8}$ m/sec to $1X10^{-10}$ m/sec due to in-site environmental and human factors. Figures 11 and 12 provides the hydraulic conductivity measured over time for those two landfills.

Figure 10. Landfill Sites: (a) Beni-Suef Landfill. (b) Badr Landfill Site.

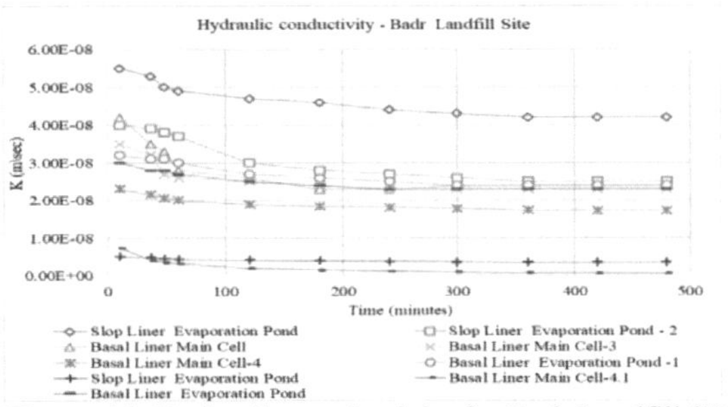

Figure 11. Hydraulic conductivity for Badr Landfill SB Samples.

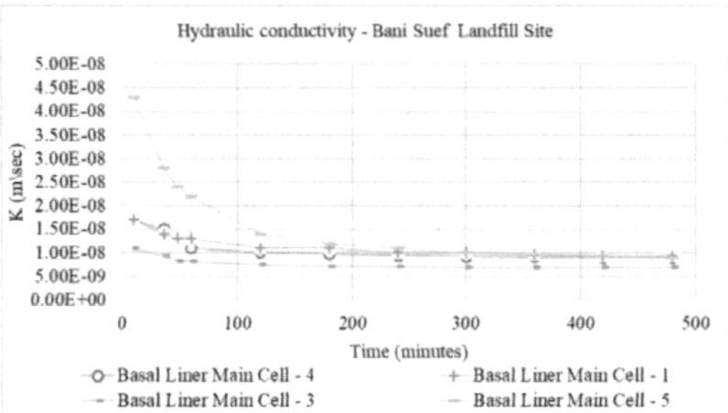

Figure 12. Hydraulic conductivity for Beni-Suef Landfill SB Samples.

6. Conclusions

The results of the experimental program show that bentonite type 3 can be safely used as an alternative for clay starting from 10% bentonite content which achieves the minimum requirement for landfill liners hydraulic conductivity equal to $1X10^{-9}$ m/sec. Also type 2 bentonite achieved the requirements but only for a bentonite content more than 14%. Type 1 showed the least satisfactory k values in all the five bentonite content values used in the experimental program without achieving $1X10^{-9}$ m/sec in any of the specimens. The conclusion can be summarized in the following few points:

- Bentonite in Egyptian market can be utilized as clay alternative in landfill liners under an adequate quality control program with bentonite content 12% and 14%.
- The presence of bentonite in all samples has significantly reduced the hydraulic conductivity of the sand-bentonite mixtures, by filling up the voids between soil particles.
- Bentonite Type 3 which was a nano-scale bentonite, had the best hydro-mechanical properties including the hydraulic conductivity reached $4.4X10^{-10}$ m/sec for the 14% bentonite mixture while Bentonite Type 2 showed $4.7X10^{-10}$ m/sec for the same bentonite percentage.
- The Specimens taken from the landfill had a high variance, so an adequate quality control program must be implemented to ensure the optimum water content is fulfilled so the hydraulic conductivity not exceed $1X10^{-9}$ m/sec.
- Additional bentonite added to all samples increased the maximum dry density. In addition to increasing compaction, bentonite particles function as lubricants between the sandy particles and act as an excellent medium, resulting in greater dry densities.

References

[1] J. Workman and R. Keeble, "Design and Construction of Liner Systems," *Academic Press,New York,* pp. 301-309, 1989.

[2] Blight and Fourie, "Experimental Landfill Caps for Semi-Arid and Arid Climates," *Waste Management and Research,* pp. 113-125, 2005.

[3] A. Tarazi, "Detecting leachate plumes and groundwater pollution at Ruseifa municipal landfill utilizing VLF-EM method," *Journal of Applied Geophysics,* pp. 121-131, 2008.

[4] Nagarajan and R., "Impact of leachate on groundwater pollution due to non-engineered municipal solid waste landfill sites of Erode City, Tamil Nadu, India," *Iranian Journal of Environmental Health Science & Engineering,* pp. 9-35, 2012.

[5] Maiti and S.K, "Characterization of leachate and its impact on surface and groundwater quality of a closed dumpsite," *Procedia Env. Sci.,* pp. 391-399, 2016.

[6] K. Ravindra and Kaur, "Occupational exposure to the municipal solid waste workers in Chandigarh, Indria," *Waste Management & Research,* pp. 1192-1195, 2016.

[7] S. M. Das, "A Study on the Effect of Chemicals on the Geotechnical Properties of Bentonite and Bentonite-Sand Mixtures as Clay Liners," *International Journal of Engineering Research & Technology,* pp. 660-666, 2016.

[8] V. Srikanth, "A Laboratory Study on the Geot. Characteristics of Sand Bentonite Mixtures and Role of Particle Size of Sand," *Int. J. of Geosynth. and Ground Eng.,* 2016.

[9] Ameta, "Effect of Bentonite on Permeability of Dune Sand," *Electronic Journal of Geotechnical Engineering,* 2008.

[10] ASTM D2487-11, "Standard Practice for Classification of Soils for Engineering Purposes (Unified Soil Classification System)," 2011.

[11] ASTM D422-63, "Standard Test Method for Particle-Size Analysis of Soils," 2007.

[12] ASTM D4318-17e1, "Standard Test Methods for Liquid Limit, Plastic Limit, and Plasticity Index of Soils," 2017.

[13] ASTM D5567- 94, "Standard Test Method for Hydraulic Conductivity Ratio (HCR) Testing of Soil/Geotextile Systems," 2018.

Testing Daylight availability for cliff dwellings by adding Clerestory: A case study in Taba, Egypt

Hala A Nabil[1,*] and Ahmed A Fikry[2]

[1] Part-time Lecturer, Future University in Egypt & Researcher at FECSFS in Egypt
[2] Architecture Engineering Department, Faculty of Engineering, Cairo University, Egypt

*Corresponding Author Email: hala.ali@fue.edu.eg

Abstract. The succession of day and night affects human beings' psychological and health status. Therefore it is essential to transfer the proper amount of daylight. The one-directional daylight source causes the most significant problem in dwellings: the deep gloom sectors. The Clerestory is an old strategy that provides daylight and air inside zones. We develop a strategy to solve daylight problems for earth-sheltered and excavated buildings. The main goal is to measure the enhancement of daylight penetration of excavated mountainous buildings settled in cliffs at various slopes after setting Clerestory. The research presents a simulation study to test Clerestory's capability to penetrate daylight by using the DIVA plug-in as a tool. The study model is a touristic chalet with a total built-up area of 100sq—meters in Taba, Egypt. The chalet overlooks the Red Sea with a south-oriented facade. The exposed roof's material and the Clerestory distance inward the facade control daylight penetration and cause a homogeneous daylight distribution.

Keywords: Daylight, Clerestory, Earth-shelters, DIVA.

1. Introduction

Daylight is an essential item that urges studying and testing various techniques, especially inside the excavated buildings, as daylight represents the primary source of natural light. Therefore, opening placement and dimension change depending on daylight availability [1]. The Clerestory role is to collect the maximum daylight throughout the day for the inner zones where the openings are unavailable. That deep understanding of the surrounding environment manifests under religious beliefs, traditions, and social customs.

The Clerestory function increases the daylight in deep gloom sectors, but the stunning view is unavailable [2, 3]. On the opposite side, the side-lit window offers an outstanding view but introduces insufficient daylight in deep gloom sectors and may cause the "*cave effect*". The "*cave effect*" appears in the spaces covered with side-lit windows at the perimeter because of the contrast between dark and over-lit zones [2, 3]. This effect also appears in the excavated cliff buildings with one side-lit window, where the daylight graduates from brightness at perimeters to darkness at deep gloom sectors.
The excavated cliff building's architectural concept, entirely composed of a cavity in rocks, acts as an entrance [4] which is usually the source for natural ventilation and daylight penetration.

2. Background

The excavated buildings differ from other building types since their features appear in the interior spaces with the earth mass features [5,6] as it is developed from the primitive man's shelter by drilling rock to simulate the natural caves [5]. At the same time, the cliff buildings' preferred locations in mountainous places are mountain terraces, slopes, and Gullies [4]. Therefore, it is connected with location and height, affording important features as view and privacy.

The Clerestory is the oldest daylight strategy that combines side lighting with top lighting to enhance daylight in deep gloom sectors, [7]. The Clerestory is an Ancient Egyptian strategy that allows daylight inside enclosed hypostyle halls [8]. The method applied by inserting the Clerestory at the higher middle level of the main hall creates a height difference. The giant stone slots settled along the wall's outer edge allow daylight and fresh air inside the temple [8]. Nowadays, the Clerestory, defined as a side-lit window higher than the visual plane, allows daylight inside the room [9].

An enormous development in daylight simulation has spread widely of daylight software tools. In 2006, Mardaljevic invented climate-based daylight modelling (CBDM) [9, 10]. This method predicts daylight using both sky condition and sun by using hourly weather data and its sensitivity to the daylight changes over time [3, 9, 10]. Since the late 1990 AD., the researchers have formulated many climate-based daylight metrics to determine daylight design distribution inside space; one of them is Daylight Autonomy (DA) used in this research [9, 10].

3. Method

3.1. Study model

Taba is a touristic coastal city in the South Sinai Peninsula, East Egypt. The town overlooks the Red Sea with high mountains formation, parallel to the Red Sea coast of the southeast [11]. The geographical location by Longitude is E° 34.895, and Latitude is N° 29.497 [12]; the climate zone classification is a hot climatic zone (1B), [13]. The excavated cliff building orientation depended on the mountain inclination direction. Thus, the schematic study model was East-oriented, with one exposed facade to the Red Sea. The Study model was a separated chalet with a total built-up area of 100square meters and an indoor height of 3meters. The facade and the Clerestory were double clear glass with a visible transmittance of 80%. The illuminance level required for dwelling function is 300 Lux [1]. Diva plug-in calculated the daylight performance by setting a reference grid of 0.5meters x 0.5meters at a work plane height of 0.8meters. Table 1 show indoor surfaces reflectance values for daylight calculations.

Table 1. Light Reflectance Values for the study model [1, 14].

Inner Surface	Ceiling	Wall	Floor
Finishing Material	White paint	Limestone	Limestone
Reflectance%	80	50	40

3.2. Methodology

The method aimed to measure the Clerestory availability to enhance daylight of excavated cliff buildings so that the study model passed through three primary phases. Phase one measured the daylight availability with a fully glazed facade. Phase two measured the Clerestory effect in enhancing daylight by adding 0.6meter height with 10meter double glazed Clerestory, while phase three measured the daylight amount by eliminating the fully glazed facade to test the exposed roof reflectance effect and distance width. Besides that, we enforced three variables to apply the methodology. Variable one was the cliff natural slope angle, while variable two was the percentage of the light reflectance material for the exposed roof. The third was the Clerestory fixation method, as shown in Figure 1.

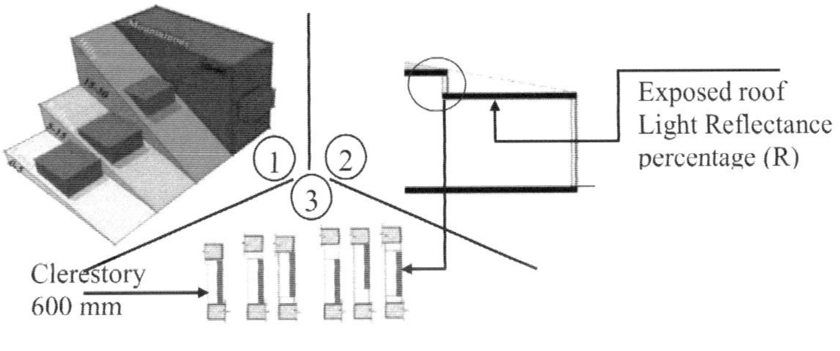

Figure 1. The three affecting variables on the study model by Author.

The first variable was the cliff natural slope angle, so the simulation implemented slope angles from 10 to 90. The inclination varied every 20 degrees. The variation in the cliff angle inclination created many alternatives for the excavated building when it intersected with the cliff mass. As much as the cliff slope's inclination angle increased, the cliff mass covered area decreased from the semi sheltering stage to the complete sheltering stage. The second variable was the light reflectance percentage for the study model exposed roof (R), as the simulation accomplished using four various finishing materials to show the material reflectance's effect on daylight penetration, as values shown in Table 2. Finally, the Clerestory fixation methods represented the third variable in the simulation process. The alternative types manifested the Clerestory impacts on daylight penetration through changing the lintel height and the sill height.

Table 2. Light Reflectance percentage for the exposed roof of the study model [14, 15].

Symbol	R1	R2	R3
Finishing Material	White Paint	Limestone	Granit
Reflectance %	80	40	20

Integrating the mentioned variables determined the study model designing types. The cliff slope angle with the Clerestory height influenced the study model's exposed roof distance (X) and inner height (Y), creating multiple study models, as shown in Table 3, 4, & Table 5.

Table 3. The study model types with Clerestory alternatives.

Model Type	**Model Design**	**Model Type**	**Model Design**
A	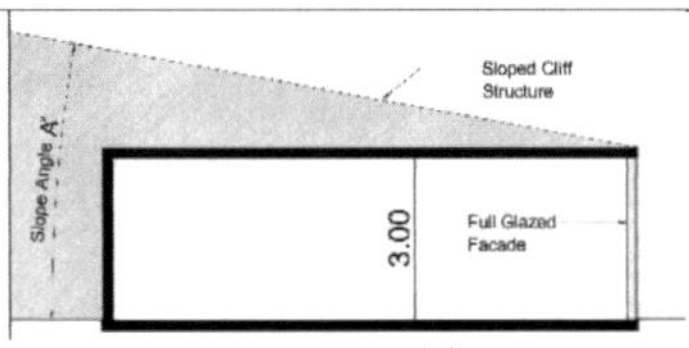 A1	**B**	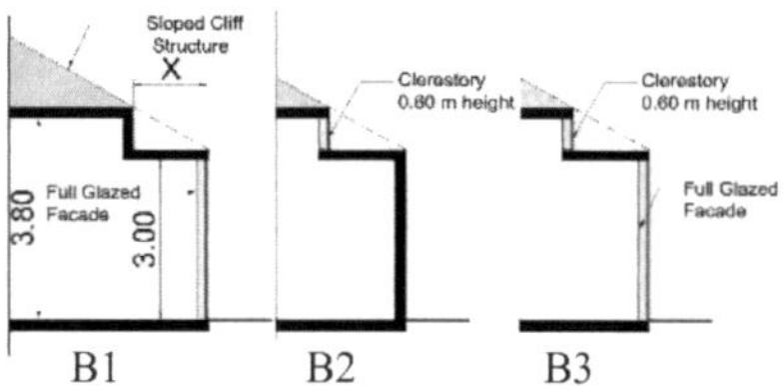 B1 B2 B3
C	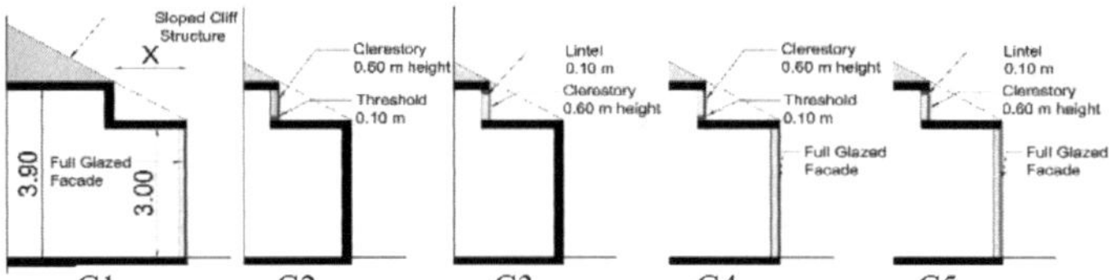 C1 C2 C3 C4 C5		
D	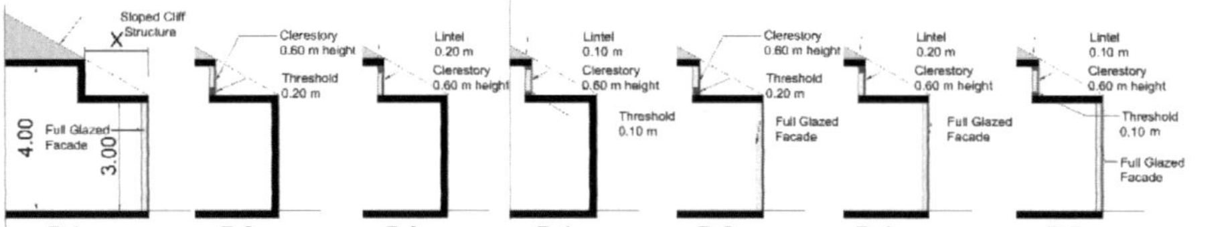 D1 D2 D3 D4 D5 D6 D7		

Table 4. Values of the exposed roof distance (X) for cliff slope angles.

Cliff Slope angle	**Building Type**		
	B	**C**	**D**
A°	**m**	**m**	**m**
10	4.55	5.10	5.70
30	1.40	1.55	1.70
50	0.70	0.75	0.80
70	0.25	0.30	0.35
90	0.00	0.00	0.00

Table 5. Description of the study model types with Clerestory alternatives.

Building Type	Height under Clerestory (Y)	Clear height	Facade WWR.	Clerestory WWR.	Lintel Height	Threshold Height
	m	m	%	%	m	m
A1	–	3.0	100	0	–	–
B1	3.0	3.8	100	0	–	–
C1	3.0	3.9	100	0	–	–
D1	3.0	4.0	100	0	–	–
B2	3.0	3.8	0	100	–	–
C2	3.0	3.9	0	85	–	0.1
C3	3.0	3.9	0	85	0.1	–
D2	3.0	4.0	0	75	–	0.2
D3	3.0	4.0	0	75	0.2	–
D4	3.0	4.0	0	75	0.1	0.1
B3	3.0	3.8	100	100	–	–
C4	3.0	3.9	100	85	–	0.1
C5	3.0	3.9	100	85	0.1	–
D5	3.0	4.0	100	75	–	0.2
D6	3.0	4.0	100	75	0.2	–
D7	3.0	4.0	100	75	0.1	0.1

4. Results and Discussion

4.1. Phase One: Fully glazed facade

Figure 2 illustrates a comparison study between four types (A1, B1, C1, & D1), showing ununiformed daylight distribution at illuminance level 300 Lux with a penetration percentage of about 71% for various cliff slopes. The fully glazed facade's effect concentrated in the front area with a dark zone in the back area; this contrast between brightness and darkness led to *"the cave effect"* with an uncomfortable illuminance level.

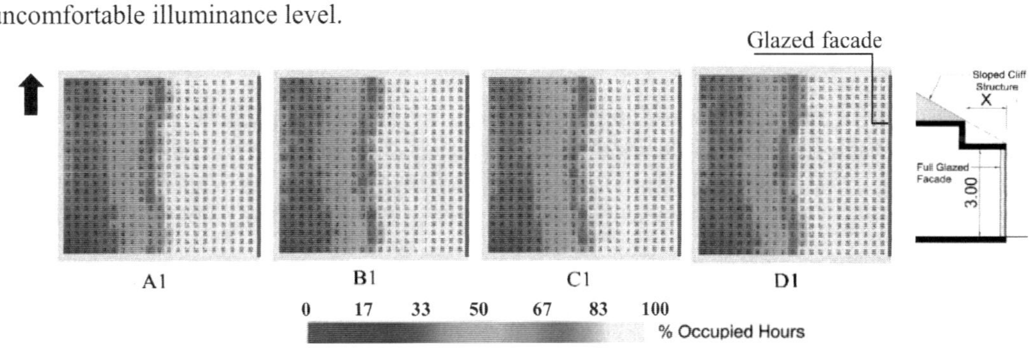

Figure 2. Daylight Autonomy maps show the effect of the fully glazed façade.

4.2. Phase Two: Fully glazed facade with Clerestory

In Phase 2, implementing the Clerestory in the study model caused a significant increase in daylight. Table 6 showed that the annual area coverage of the daylight at illuminance level 300 Lux ranged from 81% to 94%, which mean approximately 20% enhancement for the slight slopes, while only 10% enhancement for the steep slopes.

The daylight autonomy maps for the tested cliff slope angles in Figure3 illustrated the enhancement in daylight penetration by integrating the fully glazed facade and the Clerestory, besides the homogeneous distribution all over the area. According to the cliff slope angle, the daylight autonomy gradually increased in low tilt angles and decreased in sharp tilt angles. A fundamental change in the study models dark zones due to adding the Clerestory, as far the Clerestory moved away from the facade line as much as the dark zones lighted.

Table 6. Daylight autonomy percentages with the fully glazed facade and the Clerestory.

Slope angle	Daylight Autonomy %- (R1)						Daylight Autonomy %- (R2)						Daylight Autonomy %- (R3)					
	B3	C4	C5	D5	D6	D7	B3	C4	C5	D5	D6	D7	B3	C4	C5	D5	D6	D7
10	93.9	94.2	93.8	93.9	93.1	93.7	92.9	93	92.8	92.7	92.1	92.5	91.9	91.7	91.5	90.9	90.6	90.9
30	87	87.1	87.6	87.5	87.9	87.6	85.8	85.8	86.2	86.3	86.8	86.4	85.2	84.9	85.3	85.3	85.9	85.5
50	84.1	84.1	84.4	83.5	84.9	84.4	83.7	83.7	83.9	83.1	84.1	84	83.3	82.9	83.4	82.9	84.1	83.6
70	83.1	82.8	83.1	82.5	83.3	82.9	82.9	82.4	83.1	82.2	83.1	82.7	82.6	82.3	82.7	82.3	82.9	82.4
90	81.2	81.7	81.5	81.6	81.5	81.9	81.2	81.7	81.5	81.6	81.5	81.9	81.2	81.7	81.5	81.6	81.5	81.9

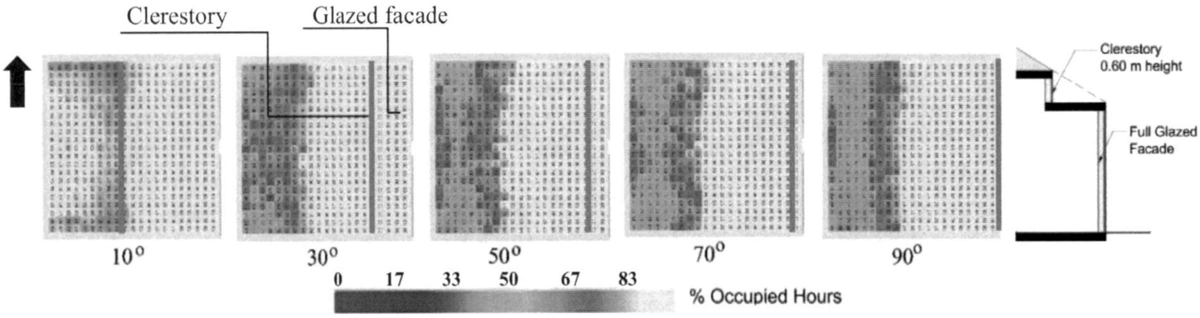

Figure 3. Daylight Autonomy maps for type (D7) with exposed roof material (R2).

4.3. Phase Three: Clerestory only

Phase 3 aimed to isolate the fully glazed facade effect to show Clerestory's impact on the study model. In Table 7, the Clerestory provided 45% of the area coverage for illuminance level at 300 Lux for high reflectance material (R1). While only 30% area coverage of illuminance for low reflectance material (R4). By looking at Table 7, the percentage of daylight area coverage at illuminance level 300 Lux increased by increasing the cliff slope angle, then decreased gradually.

Table 7. Daylight autonomy percentages for the study model with Clerestory only.

Slope angle	Daylight Autonomy %- (R1)						Daylight Autonomy %- (R2)						Daylight Autonomy %- (R3)					
	B2	C2	C3	D2	D3	D4	B2	C2	C3	D2	D3	D4	B2	C2	C3	D2	D3	D4
10	41.5	39.5	39.2	36.5	35.7	36.8	33.8	32.4	31.9	29.9	29.3	30.3	27.9	26.8	26.3	24.7	24.4	25.2
30	45.2	44.5	45.7	43.6	45.1	44.9	37.8	36.9	37.8	35.7	37.1	37.1	32.6	31.8	32.2	30.7	31.7	31.9
50	41.3	41.1	42.9	39.1	42.4	41.6	35.7	35.6	36.7	33.6	35.9	35.7	32.1	32.3	32.4	30.8	31.9	32.3
70	37.1	35.4	37.9	33.3	37.4	36.2	32.9	32.7	33.9	31.2	33.2	33.3	31.2	31.2	31.5	29.9	30.9	31.6
90	26.6	29.3	26.3	28.9	26.2	28.9	26.7	29.3	26.3	28.8	26.2	28.9	26.6	29.3	26.3	28.8	26.2	28.9

The comparison study for type (B2) in Figure 4 illustrated the daylight penetration inside the study model due to implementing the Clerestory and isolation of the fully glazed facade effect. It observed that the impact of Clerestory distance inward the facade and the variation in the cliff angle affects the daylight distribution. The daylight reached the maximum percentage in the middle angles of the cliff, where the Clerestory wall in the middle distance between the facade wall and the rear wall.

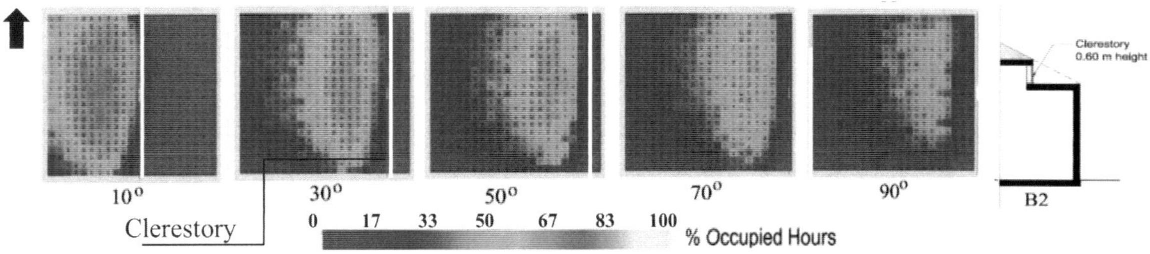

Figure 4. Daylight Autonomy maps for various cliff slope angles and light reflectance material value for the exposed roof (R3).

The previous results show that Clerestory enhances the daylight by 20% for illuminance at 300 Lux. It observed that many influences vary the area coverage percentage of daylight illuminance level in the study models. The influences separated, the first one relevant to the indirect daylight reflected from the study model's exposed roof. The other is that direct daylight penetrates through the Clerestory. Regarding the indirect daylight reflected from the study model's exposed roof, the distance (X) from the Clerestory and the Light reflectance material (R) affect it.

The mentioned point explains the variation in Table 7, as the daylight increases as much as the cliff slope angle decreases due to the exposed roof distance (X) increase. So the Clerestory moves inward the facade, which allows a uniform distribution for the daylight inside the study model by covering the space floor area's dark zones. However, in an instance, as the Clerestory becomes in-depth, the daylight covers the opposite wall much more than the dark zones of the study model's floor area.

By excluding the fully glazed facade from the simulation process, the results show the daylight's actual amount penetrates from the Clerestory, whatever it is direct or reflected from the exposed roof. In addition, it shows the daylight amounts in the previous Table 7, indicating that the exposed roof area and its finishing material affect the daylight distribution inside the study model.

The daylight percentage increases owing to increasing the distance (X), which increases due to the area reflecting the daylight. Even though the daylight percentage increases, the daylight percentages decrease despite the distance (X) increase. Thus, another internal factor reduces the daylight percentages, affecting distance (X) and the Clerestory daylight amount. The daylight autonomy also increases due to the closer distance between the Clerestory and the wall opposite it.

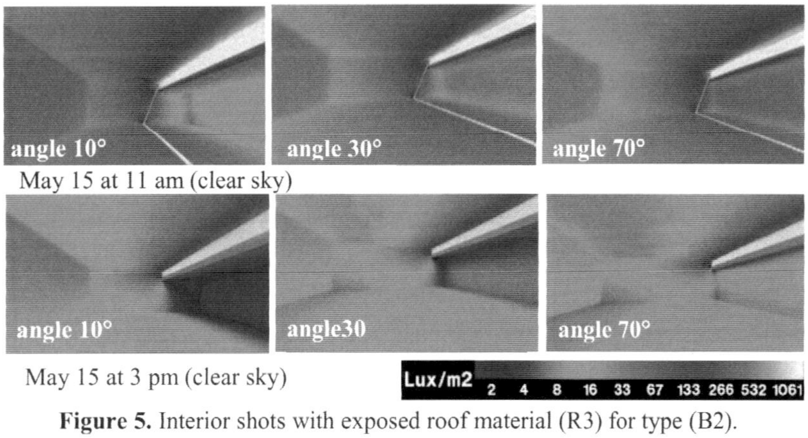

Figure 5. Interior shots with exposed roof material (R3) for type (B2).

Figure 5 illustrates interior shots for the daylight distribution inside the study model for slope cliff angles 10, 30 and 70. The simulation test the daylight distribution for a selected day, 15 May, at 11 am and 3 pm. From angle 70 to angle 10, the daylight gradually becomes more concentrated on the opposite wall and penetrates the dark part of the floor area. By looking at the angle 10 shots, the Clerestory distance from the inner opposite wall decrease, so the daylight penetrates the wall and the dark zone more than the inner side of the facade.

The exposed roof acts as a light shelf, reflecting the direct daylight, which simultaneously reflects and penetrates through the Clerestory reflects and penetrates through the Clerestory. Furthermore, as shown earlier, the roof's exposed parts depend on the cliff mass sloping angle. Therefore, an increasing change in the daylight area coverage occurs as the exposed roof Light reflectance material (R) ability to reflect daylight increases, as shown in Figure 6.

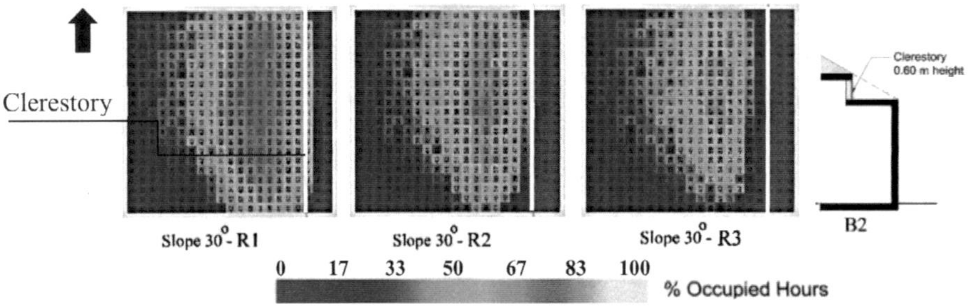

Figure 6. Daylight Autonomy maps for type (B2) at slope angle 30°.

As shown earlier, the suggested clerestory types have significantly enhanced the daylight amount inside the space area. Figure 7 illustrates a comparison study that shows the link between the changes in the daylight percentage related to changing the clerestory types. In the graphs (a, b & c), the variation in daylight percentage changes slightly between one type to another due to exchanging both the lintel height and sill height. The influence of Clerestory "D2" is less efficient than other Clerestory types for the daylight percentage. In contrast, the Clerestory "C3" is the most efficient Clerestory type for daylight percentage. The Clerestory "D2" efficiency decreases in the increase in sill height and indoor height.

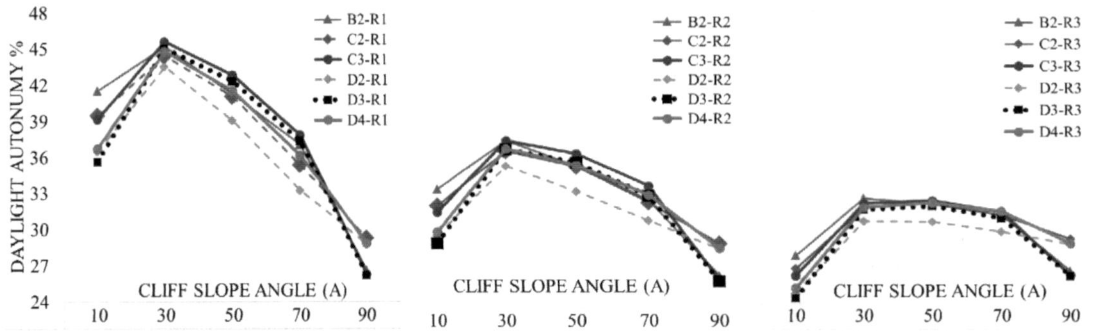

	10	30	50	70	90
B2-R1	41.54	45.22	41.32	37.10	26.63
C2-R1	39.50	44.46	41.08	35.38	29.27
C3-R1	39.22	45.66	42.87	37.90	26.27
D2-R1	36.54	43.57	39.08	33.28	28.82
D3-R1	35.69	45.11	42.35	37.40	26.20
D4-R1	36.78	44.85	41.56	36.19	28.90

(a) Light reflectance material R1

10	30	50	70	90
33.74	37.74	35.64	32.84	26.63
32.40	36.87	35.63	32.65	29.27
31.86	37.75	36.69	33.99	26.27
29.92	35.67	33.56	31.15	28.82
29.30	37.05	35.95	33.17	26.20
30.26	37.10	35.72	33.26	28.90

(b) Light reflectance material R2

10	30	50	70	90
27.91	32.61	32.12	31.22	26.63
26.82	31.84	32.28	31.23	29.27
26.26	32.20	32.39	31.49	26.27
24.76	30.73	30.67	29.85	28.82
24.42	31.67	31.95	30.98	26.20
25.22	31.87	32.26	31.56	28.90

(c) Light reflectance material R3

Figure 7. Comparison of the Clerestory alternatives with the reflectance material values of the exposed roof.

5. Conclusions

The research declares the enhancement in the daylight of excavated cliff buildings inside a touristic chalet by implementing Clerestory. Taba city characterizes by high topography and various cliff masses' slope angles. The study explains the relationship between the cliff slope angle, which affects the study model indoor design, resulting from that the daylight amount penetrates from the Clerestory. Furthermore, the Clerestory distance inward the facade edge, the exposed roof area, and the exposed roof material influence the daylight from the study model type to another one. The research manifests that the one-directional opening of the excavated cliff buildings does not penetrate the whole space area with daylight, even with a fully glazed facade. On the contrary, the Clerestory enhances the daylight penetration annually by an average of 10% for cliff masses with high cliff angle inclination and 20% for low and middle cliff angle inclination. Besides that, the Clerestory solves the cave effect inside the study model.

The first influence in results is the Clerestory distance inward the facade edge, which controls the daylight distribution. The floor area illuminance level by Clerestory decreases as much as the Clerestory becomes in depth towards the opposite wall. Conversely, as much as the Clerestory becomes away from the fully glazed facade, the floor area lighted by Clerestory increases, which causes an over-lit zone in the front part of the study model. Thus, the Clerestory and the opposite wall affect the daylight penetration. The Clerestory placement is significant, as the target is to penetrate deep gloom sectors where the main glazed facade is incapable of lighting them.

The other influences like the exposed roof area and finishing material indirectly reflect the daylight through the Clerestory. The white colour material with a reflectance of 80% shows high daylight penetration values. However, weather factors like dust, rain, and wind affect the white colour in real life, so the Chromatic ability to reflect the light decreases. In addition to that, the most suitable materials are compatible with mountain environments like limestone, sandstone, and granite, with a light reflectance value of less than 40%. Even so, with low reflectance material value, the Clerestory enhances daylight penetration.

The study focused on daylight only by eliminating the effect of sunlight and the thermal. So, the study model may require solutions to solve the discomfort and cooling loads that, visas versa, affect the amount of daylight penetrating from the one-directional opening.

Acknowledgements
Thanks to Meteonorm for their support and help with the availability of Taba city climatic data and weather files in EPW format.

References

[1] Grondzik WT, Kwok AG, Stein B, Reynolds JS 2010 *Mechanical and Electrical Equipment for Buildings* Eleventh E. (John Wiley & Sons.Inc.) The USA p 493, pp 588-590.
[2] Boubekri M 2014 *Daylighting Design: Planning Strategies and Best Practice Solutions*. (Birkhauser Verlag GmbH). Switzerland pp65-71.
[3] Meek C, Wymelenberg K 2015 *Daylighting and Integrated Lighting Design*. (Taylor & Francis Group) Great Britain pp 54-56, pp 62-63.
[4] Golany GS 1992 *Chinese Earth-Sheltered Dwellings: Indigenous Lessons for Modern Urban Design*. (University of Hawaii Press) The USA pp 85-100.
[5] Università degli studi di Firenze 2012 *Common Identity - New Perspective The Rupestrian Settlements in the Circum-Mediterranean area.* Crescenzi C, Caprara (DA*dsp*-UniFi) Firenze p 13, pp19-25, p 41.
[6] Correia, M., Dipasquale, L., & Mecca S 2010 *Heritage for Tomorrow - Vernacular Knowledge for Sustainable Architecture*. (Firenze University Press) Italy pp 115-120.
[7] Osso A, Gottfried D, Walsh T, Simon L 1996 *Sustainable Building Technical Manual*. (Public Technology Inc.) The USA pp 94-100.
[8] Baker N, Steemers K 2014 *Daylight Design of Buildings : A Handbook for Architects and Engineers* Second Edi. (Taylor & Francis Group) The USA pp 32-34.
[9] Meyers R 2012 *Encyclopedia of Sustainability Science and Technology* (Springer) New York pp 2810-2812,p 2817, pp 2832-2833, pp 2846-2848, pp 2858-2863.

[10] Konis K, Selkowitz S 2017 *Effective Daylighting with High-Performance Facades: Emerging Design Practices* (Springer International Publishing) Switzerland) pp 41-49.
[11] EGSMA, NARSS, UNDP, UNESCO 2004 *Geological Atlas of SINAI.* (EGSMA, NARSS, UNDP, UNESCO) Egypt pp 18-19.
[12] Meteonorm 2010 *Irradiation data for every place on Earth* (Meteonorm) http://www.meteonorm.com/en/
[13] ASHRAE 2007 *ANSI/ASHRAE/USGBC/IES Standard 90.1-2007: Energy Standard for Buildings Except Low-Rise Residential Buildings* (ASHRAE, Inc.)The USA pp 112-114.
[14] Materials Council 2012 *Whiter than White* (Creative Materials Consultants Limited) https://www.materialscouncil.com/wp-content/uploads/2012/10/Whiter_than_White_report_by_Materials_Council.pdf
[15] Thomas R 2006 *Environmental Design-An introduction for architects and engineers*, Third Edit. (Taylor & Francis Group) New York p 98.

A sustainable road safety approach for pedestrians in new cities of Egypt

Mona Azouz[1, *] **and Ahmed Fahim**[2]

[1] Architecture Institute, Housing and Building National Research Centre (HBRC), 87 El-Tahrir Street, El-Dokki, Giza, Egypt
[2] Housing and Building National Research Centre (HBRC), 87 El-Tahrir Street, El-Dokki, Giza, Egypt

*Corresponding Author Email: mona.azouz@yahoo.com

Abstract. Over the last decade, Egypt's growing population and dynamic economy have resulted in a need for greater infrastructure development; consequently, the risk of pedestrians crossing at grade has increased tremendously. For reducing fatalities caused by the conflict between pedestrians crossing at grade and vehicles, pedestrian bridges have been established. Noting that these bridges which are dedicated to human safety are highly expensive, therefore without the fulfilment of conditions of safety, convenience and ease of accessibility they might be a wasted investment. In this respect, site visits and a SWOT analysis were conducted on three pedestrian bridges in New Cairo for monitoring and evaluating the impact of these interventions. The assessment of strengths, weaknesses, opportunities and threats of these projects formed a base for upgrading and the development of new projects. Although, there are some attempts for using escalators for encouraging usability of pedestrian bridges, the research propose a national project using small solar PV systems for the operation of outdoor escalators in a more sustainable manner, which would in turn support the 2030 Agenda for Sustainable Development. This requires joining efforts of all relevant stakeholders in coordinated and rigorous action towards achieving road safety targets.

Keywords: Pedestrian bridges, Outdoor solar escalators, Small solar PV systems, Pedestrians' safety.

1. Introduction

In 2015, the United Nations adopted 17 Sustainable Development Goals (SDGs), as a universal call to action to end poverty, protect the planet and ensure that by 2030 all people enjoy peace and prosperity [1]. Among goal targets was target 3.6 that seeks to halve road traffic deaths and injuries by 2020 in addition to target 11.2 that focuses on providing safe, affordable, accessible and sustainable transport systems for all and improving road safety.

In 2018, the World Health Organization Global Status report on Road Safety showed that the Africa region had 40% Pedestrian deaths among road user categories—the highest amongst all regions [2]. In September 2020, the UN General Assembly adopted resolution A/RES/74/299 "Improving global road safety", proclaiming the Decade of Action for Road Safety 2021-2030, with the ambitious target of preventing at least 50% of road traffic deaths and injuries by 2030. WHO and the UN regional commissions, in cooperation with other partners in the UN Road Safety Collaboration, have developed a Global Plan for the Decade of Action, released in October 2021 [3].

According to Egypt's Road Safety Country Profile [4], the least safe for the road users goes for the pedestrians as 'figure 1', as 83% of roads are assessed for being the least safe for pedestrians according to the International Road Safety Assessment Programme (IRAP) that provides road star ratings which give a simple and objective measure on the level of safety which is 'built-in' to the road for the road users (5 Star roads are the safest while 1 star roads are the least safe).

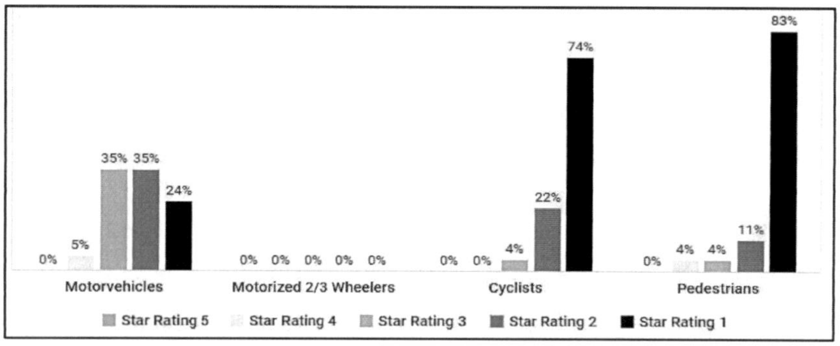

Figure 1. Fatalities by user comparison chart [4].

In 2014, Egypt has launched National Roads Project to support economic development, encourage investors, reduce traffic congestion and improve overall safety. Since 2018 the government has spearheaded significant investment in the road system to ensure that it will have the capacity to cope with long-term development plans. New Cairo has witnessed road development in land transport segment as being a link between the Capital and the New Administrative Capital. Road developments has encouraged vehicles to drive at high speed causing an increase in road accidents and pedestrians were exposed to deadly hazards from fast-moving vehicles and dramatic scenes of death accidents have been regularly appearing in the media. By recognizing that these roads became unsafe for pedestrian, the Egyptian government has established pedestrian bridges to secure the crossing of citizens and reduce traffic congestion. In order to ease and encourage the usability of these bridges, some of these bridges were being provided by outdoor electric escalators.

2. Research Objectives

The cost of establishing pedestrian bridges is very high and in order not to be a wasted investment, the aim of the research is to propose a national project for a new solar energy application; using PV solar systems for operating outdoor escalators to encourage the use of the established bridges for pedestrians' safety, comfort and convenience, thus, contributing in Egypt's progress towards road safety targets and the increase of renewable energy share in the context of the 2030 Agenda for Sustainable Development, in addition to decreasing the urban electricity utilization pressure.

3. Methodology

The research was based on collecting data from Egypt-PV project, international companies for outdoor escalators available in the local market and a structured observational analysis for 3 established pedestrian bridges in New Cairo to be used as a tool for recording human behaviour. Detailed data were collected from 22 site visits and 120 photos that have been taken as part of the overall field data collection effort. These data were collected during nearly a whole year starting from December 2020 and included variables such as time, behaviours and conflicts. In addition to personal one-to-one interviews were conducted to collect more data from a random sample of pedestrians in order to study the pedestrians' behaviour and attitude in each location. A SWOT analysis has been applied on the three projects that provide opportunities for upgrading of the projects or for up-scaling and the development of new projects based on their strengths, overthrown the weaknesses, considering opportunities and resisting possible threats.

4. Case Studies of Implemented Projects

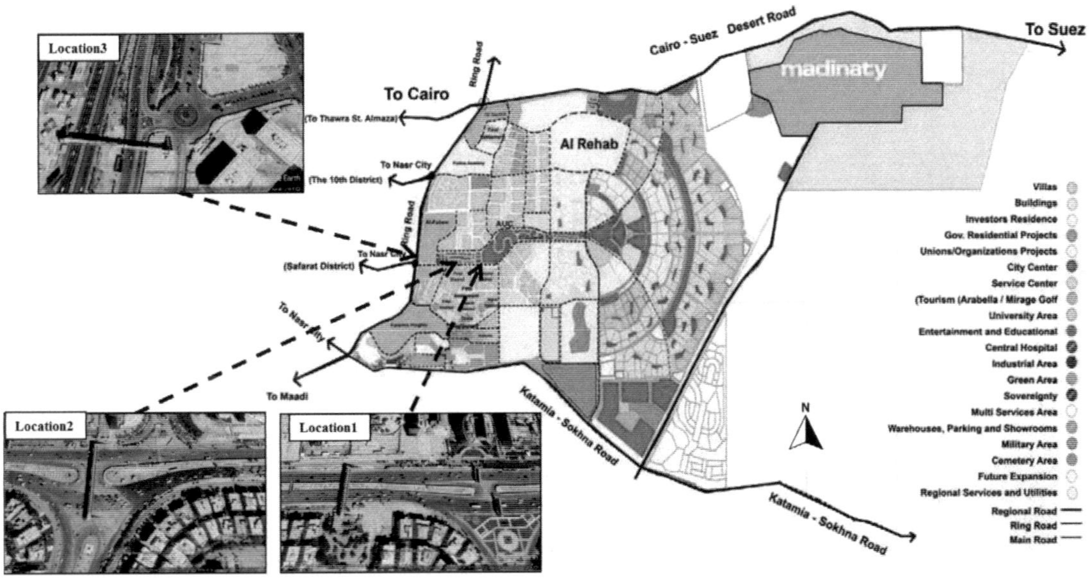

Figure 2. The location of the selected 3 pedestrian bridges serving the City Center of New Cairo [5], (Modified by the author).

4.1. Locations Selected for Study

Selection of case studies has been carried out based on initial pilot surveys and Pedestrian Risk behaviour. Three pedestrian bridges serving the City Centre of New Cairo city have been selected where both the density and the traffic volume are high. The City Centre was designed to be the core of services of New Cairo and representing a point of attraction for citizens that includes shopping malls, banks, office and administrative buildings in addition to a hotel and healthcare facilities, hence, these infrastructure bridges serve the entire vicinity. People come from different places using public and semipublic means of transportation. The three projects are representing three different locations with different conditions, restrictions and other parameters of pedestrian overcrossings as 'figure 2'. The locations of the first two bridges are on one of the main streets in New Cairo which is the Teseen Street, (street no. 90), that is considered a main link connecting the Capital with the New Administrative Capital. The third bridge is on the Ring Road and is mainly serving Cairo Festival Mall one of the biggest shopping malls in Egypt. Buses and microbuses are the main transportation mode by which pedestrians reach the three locations and their stops are available on both sides of each bridge.

4.2. Survey analysis

The analysis consist of two parts namely General Observations on sites and personal interview for monitoring and evaluating impact of interventions.

4.2.1. General Observations

From the overall observation it was found that most of the pedestrians in the first two locations are employees and workers but in the third location besides that was visitors of Cairo Festival shopping mall especially on weekends. Pedestrians typically seek out short, direct routes to their destinations [6] as 'figure 3'and if there is a chance or possibility they prefer to cross at grade. For the first two locations, two peaks are experienced for pedestrians, one from 6am till 10am and the other from 3pm till 5pm, but for location 3the use stretches longer till 1am for night shifts. The above stated locations were assessed based on location & length of the bridge, possibility of at-grade crossing and usability.

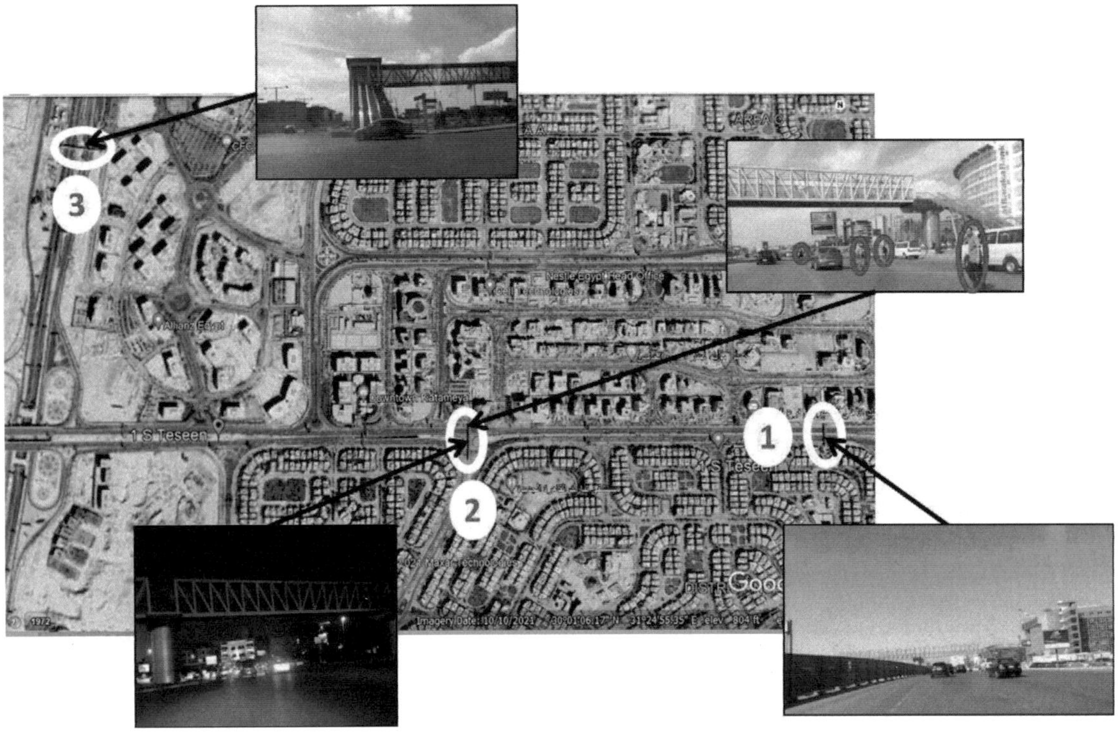

Figure 3. The selected 3 pedestrian bridges serving the City Center of New Cairo.

Location 1

Location & length of the bridge: This location is situated on South Teseen Street, (street no. 90), as stated before and its length is nearly 75m. *Possibility of at-grade crossing:* The first and obvious factor is the difficulty of crossing at-grade because of the barrier of the monorail construction sites that hinders people from crossing the street. *Usability:* Nearly 80% of the pedestrians use the bridge and 20% crossing at-grade beneath the openings in the barrier of the monorail construction site or at the U-Turn after or before the bridge depending on the direction of the street by a distance ranges between 170m to 285m.

Location 2

Location & length of the bridge: This location is situated also on South Teseen street and the length of the bridge is nearly 115 m. *Possibility of at-grade crossing:* As under the bridge is a U-Turn and there is no median obstacle, people tend to cross the street at grade level, which is more convenient, but at the same time more dangerous. And also, the presence of a hump before the bridge helps in decreasing the velocity of cars, thus, encouraging crossing the street at grade. *Usability:* Nearly 95% of the pedestrians are crossing

at-grade and only 5% are using the bridge. In addition, there is no adequate lighting and the darkness in night discourages the use of the bridge and might leads to antisocial activities.
Location 3. Cairo Festival City Ring Road Bridge
Location & length of the bridge: This location is situated on the Ring Road and serves one of the biggest shopping malls in Egypt. The length of the bridge is nearly 150 m. *Possibility of at-grade crossing:* Being a high way, it is very dangerous to cross at grade that's why in this location there is no possibility of crossing at grade and the provision of a barrier was very important and observed very useful to promote usage of the bridge. *Usability:* The presence of electric escalators and elevators promote the usage and ease accessibility of the bridge.100% of the pedestrians are using the pedestrian bridge as there is no other option even when the escalators are out of use.

4.2.2. Personal Interview:
Interviewing pedestrian at the three locations was to get an idea about their perspective on the ease of usability of these bridges and to point out the factors that have a direct impact on usability. A sample of 20 pedestrians (19 males and one female) was selected of age ranges from 18 to 50 years. 100% of the sample was for people working in the City Center. It was concluded that people's choices differ with age and ability. Almost 100% of the pedestrians would prefer at-grade crossing noting even if they were capable of using these facilities because they are always in a hurry going work or home. In addition 25% of the sample was carrying stuff, luggage and tools and it would be uncomfortable for them to use the stairs.

4.3. SWOT Analysis
Strengths. Using pedestrian bridges ensures pedestrian safety. For Location 1 and 3: Median barrier of the monorail construction sites and the median barrier on the Ring Road force people to use pedestrian bridges. The electric escalators in location 3 ease the usability and accessibility of the bridge.
Weaknesses. Pedestrians typically seek out short, direct routes to their destinations. The usability analysis of these bridges found that they are poorly designed for people with physical restrictions and lead to circuitous walking routes that typically increase travel distances and time. The bridge in location 2 typically represents a wasted investment as there is accessibility of pedestrians to cross at street level where there is a U-turn and no median barrier. The electric escalators in location 3 lead to increase in the consumption of energy from non-renewable sources associated with the increase of emissions that negatively impact the environment. No adequate lightning in location 1 and 2, the darkness in night discourages the use of pedestrian bridges and might lead to antisocial activities.
Opportunities. The use of outdoor solar escalators will increase the usability of the pedestrian bridges and decrease the harmful emissions resulted from the use non-renewable source of energy for operating the escalators, in addition to the presence of multiple competitors companies in the local market for outdoor infrastructure escalators and the availability of funds from the United Nations Road Safety Fund that finance high impact best practices to increase road safety and minimize and eventually eliminate road crash trauma for all road users [7] and [8].
Threats. A growing awareness of accessibility issues due to an aging population and people with physical restrictions.

4.4. Survey Findings
The study has resulted into a conclusion that pedestrians shall always prefer at grade crossing whenever possible, that's why the aim of the research is to ease usability and accessibility of these bridges in a sustainable approach even to those who have physical restrictions as pregnant women, aged pedestrians and kids as a step forward for the ease and accessibility for all.

5. Proposal of a National Project: Outdoor Solar Energy Infrastructure Escalators

Egypt is at the heart of the global solar belt and is blessed with abundant solar energy [9], which should provide an opportunity to use photovoltaic (PV) technologies for energy production instead of using traditional unrenewable sources of energy whenever possible. Solar energy is considered the most promising clean and renewable energy resource.

As road development should be always in line with the safe systems approach and coping with Egypt's great efforts to reduce greenhouse gas emissions for supporting the implementation of Egypt 2030 strategy for sustainable development, as well as the United Nations Sustainable Development Goals, especially the goal of ensuring access to affordable, reliable, sustainable and modern energy for all, in addition to the Paris agreement to reduce greenhouse gas emissions [10], the research propose using Photovoltaic Systems to utilize the available solar energy in Egypt in generating electricity for the operation of outdoor escalators to encourage the usability of pedestrians bridges. This proposal will be applied on the pedestrian bridge in location 2 where 95% of the pedestrians are crossing at-grade as mentioned before.

Photovoltaic panels systems are used for converting solar energy into direct current; a photovoltaic inverter is used for converting direct current (DC) into alternating current (AC) to be used for operating outdoor escalators in the selected location. Generally, the life span of PV modules lasts up to 25 years and the inverters life span ranges from 7-10 years. Most of PV panels available in the local market are guaranteed by the manufacturers with a performance warranty of 20-25 years. The PV system must be well exposed to the south orientation. There must be no obstruction to the access of the sun's rays from the buildings or trees or any other barriers. If panels are to be installed on a structure, the structure should support the weight of the panels and should be accessible for the panels' maintenance.

5.1. Operating hours

From the survey, two peaks are experienced, one from 6 am till 10am and the other from 3pm till 5pm, the use stretches longer into the evening for social meetings, overtime, etc. The duty cycle for the escalator will be chosen to be 18 hours/day from 6am to 12am.

5.2. Technical data requirements

Infrastructure escalators should comply with the EN115 Public Service requirements and to be suitably enhanced for longer duty cycles and heavier passenger load profiles. Step width is 1000mm to allow two adult passengers to stand on each step, inclination is 35 degrees which offers the most compact dimensions, vertical rise measured equals 7500 mm, step band speed is to be 0.50 m/s to provide rapid displacement while not disregarding comfort and safety [11], rated capacity is 9000 persons/hour, motor power is 15 KW [12]. Duty cycle: 18 hours/day.

5.3. Calculating energy required daily in kWh (kilowatt-hours)

Taking into consideration the percentage the power loss due to passing by electrical components such as the inverter as its efficiency is 95% (5%of electricity is lost) as well as the cables etc. And the loss factor ranges from 20 to 30% that must be added to the daily consumption. So the energy needed to be produced by the panels: daily consumption + energy lost =energy required daily in kWh [10].

Therefore energy required daily in kWh=15*18+20%of daily consumption=270+54= 324 KWh

And the energy required daily for two escalators going up on the two sides of the bridge=324 x 2= 648 KWh.

5.4. Calculating area required solar panel system

By using the calculator on Egypt PV project web site (assumed 5 peak hours for Cairo), **t**he area required for solar panels systems of a capacity 150 KW is 900 m² [10] , which means that the area required for two escalators going up on the two sides of the bridge approx.= (648 x 900)/150 =3888 m²

Area required for maintenance equals 20% of area required for panels= 3888 x 0.2 =777.6 m²

Therefore the total area required for the solar panel system=3888+777.6=4665.6m m^2 approx. 4670 m^2 According to the total area require the placement of photovoltaic panels could be as 'figure 4'.

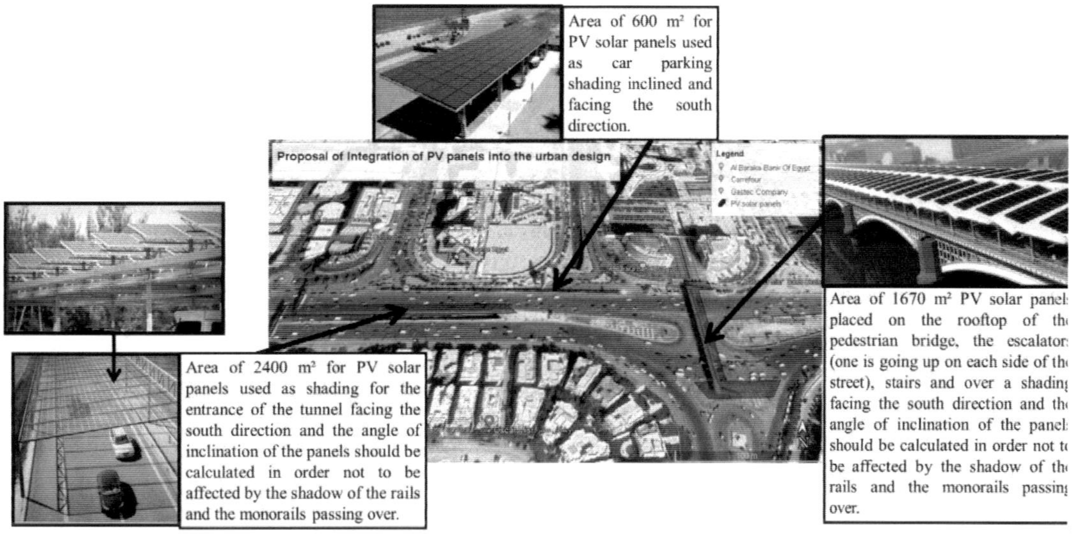

Figure 4. Proposal of integrating PV panels into the urban design with total area of 4670 m^2.

5.5. Connecting to the National Grid

The total energy required daily for the operation of the two outdoor escalators for 18 hours is calculated as shown before in KWh. This energy is used for the operation during sun peak hours (estimated 5 hours a day for Cairo) and the rest of the energy generated is sent to the grid as 'figure 5'which should be equivalent to the energy required for operating the escalator for the rest 13hours, thus, the net metering of the energy used from the grid should be zero. Noting, the continuous operation of the escalators during the sun peak hours even in the event of a power cut off. In this case a net metering connected with the National Grid is used for monitoring the efficiency of the system, at the end of each month the difference between the excess energy sent to the grid daily during sun peak hours and the electric energy from the grid that is consumed in operating the escalator during the rest of time should be zero.

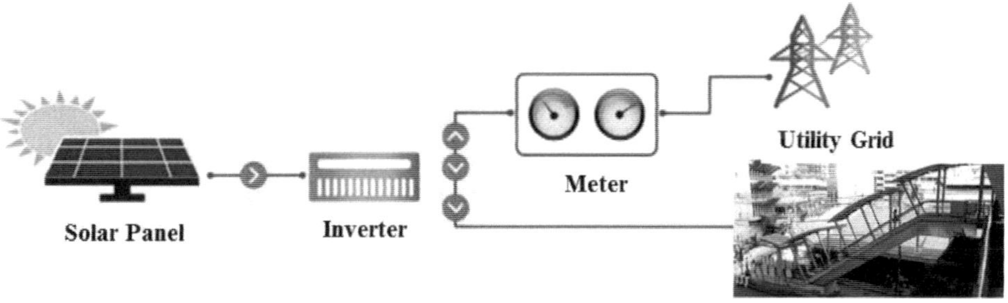

Figure 5. Connecting the system to the National Grid, [10], (Modified by the author).

5.6. Maintenance of PV system

The cleaning of the system is to be at least once every two weeks, in addition to visually inspection of the system's cabling for any cracks or breaks in the cabling throughout the system that could cause it to generate less electricity and checking the readings on the system's inverter because if the system is producing less energy, especially on clear sunny days, then it's likely that the system is in need of some maintenance.

5.7. Using energy-saving technology

As the innovative inverter control that increases motor efficiency under light loads and automatic operation that controls the travel speed depending on the passenger traffic volume detected by sensors and using long-lasting, low power consumption LED lights for all lighting options.

6. Conclusion

Planning safer roads in a sustainable approach is an investment that will ensure a better future for all. Moreover, future man-made structures may play an important role in reducing the overall impact of the built environment on the natural environment. The design of such smart structures will require a higher level of inspiration that designs the structure in a more comprehensive way in addition to the need of the cooperation of all relevant stakeholders.

7. Recommendations

For future projects in new cities, the use of renewable sources of energy should be incorporated into the design from the early stages of the urban planning; moreover, the integration of photovoltaic (PV) technologies for energy production should be applied whenever possible not only for the operation of outdoor solar escalators, elevators, lighting fixtures but also for any other innovative application.

References

[1] Desa UN 2016 *Transforming our world: The 2030 agenda for sustainable development* (United Nations).
[2] World Health Organization 2018 *Global status report on road safety 2018* (Geneva: World Health Organization) Licence: CC BYNC-SA 3.0 IGO.
[3] The United Nations Economic Commission for Europe, World Health Organization and United Nations Regional Commissions 2021 *Global Plan for the Decade of Action for Road Safety 2021-2030* (World Health Organization).
[4] World Bank 2020 *Guide for Road Safety Opportunities and Challenges: Low and Middle Income Country Profiles* (Washington, DC., USA: World Bank).
[5] Hafez R M 2017 New cities between sustainability and real estate investment: A case study of New Cairo city *HBRC journal* **13(1)** p 89-97.
[6] Kost C, Mwaura N, Jani A and Eyken C V 2018 *Streets for walking & cycling- Designing for safety, accessibility, and comfort in African cities* (The United Nations Human Settlements Programme and the Institute for Transportation and Development Policy).
[7] UN Road Safety Fund 2020 *A World where Roads are Safe for every Road User, everywhere* (UNECE Publications).
[8] UN Road Safety Fund 2021 *Build Back Better- A New Decade of Action for Road Safety- 2020 Anual Report* (UNECE Publications)
[9] Kosmopoulos P, Askary H E-and Kazadzis S 2020 *Solar Atlas of Egypt* (Geo-Cradle Project) Accessed 4 December 2021 URL: http://www.nrea.gov.eg/Content/files/SOLAR%20ATLAS%202018%20digital1.pdf.
[10] Egypt PV 2020 Accessed 11 October 2021 URL:https://egypt-pv.org/?lang=en.
[11] KONE Corporation 2001 *Planning Guide for Kone Escalator and Autowalk (EN115)* (KONE Inc.) Accessed 6 November 2021 URL:https://www.kone.us/Images/KONE-Escalator-AutoWalk-Planning-Guide_tcm25-18783.pdf.
[12] KÖHLER Elevator and Escalator 2013 *Passenger Conveyor Solutions* (KÖHLER Elevator and Escalator GmbH) Accessed 10 November 2021 URL:https://koehlerelevator.de/wp-content/uploads/2019/04/Harmony_Escalator_2016.pdf.

The role of energy conservation standards in raising quality of life: Low- Cost Housing as a Case Study

Ayman A. Amin[1], Noha M Ezz Eldin[2, *], Alshaimaa H. Mohammed[3]

[1] Department of Architecture, Faculty of Engineering, Banha University, Banha, Egypt

[2,3] MTI University, Department of Architecture, Cairo, Egypt

*Corresponding Author Email: Nohaezz1@gmail.com

Abstract. Low-cost housing is spread all over Egypt from the north to the upper, many issues appeared on the horizon, such as high maintenance costs, high levels of energy consumption, and the emergence of many practices for residents of these units in an attempt to adapt between their needs and the design of low buildings. One of the most important challenges impacting the quality of life is energy conservation. This paper aims to raise the quality of life in low-cost housing from the perspective of energy conservation. The research aims to ensure enough sustainable energy to maintain a quality of life. This study is based on questionnaires conducted by residents of low-cost houses projects in Egypt, namely; Mokattem and New Cairo. Trying to find the most important problem and put forward ways to improve and raise the quality of life, especially from the perspective of energy conservation. Studying the Living satisfaction in low- cost houses. These questionnaires were analysed on a proposal design and how to apply it.

Keywords: Low-cost housing, Energy conservation, Quality of life (QOL).

1. Introduction

Over the last few years the Egyptian government has pay attention to the approach of the low-cost housing .This is becoming an issue of great importance recently as a most suitable way to obtain decent housing Compatible with the economic conditions of Egyptian society [1]. Providing a suitable residential environment is considered one of the critical challenges government encounters especially since 52% of the population is with average income. The interest of the government is shown through the number of low-cost buildings in Fig no. (1) which represents that the number of low-cost units and economic housing alone is about 84,981 during the period 2019-2020. In light of this, Understanding quality of life is a key measurement for cities' liveability and habitability [14], especially for Low –cost housing.

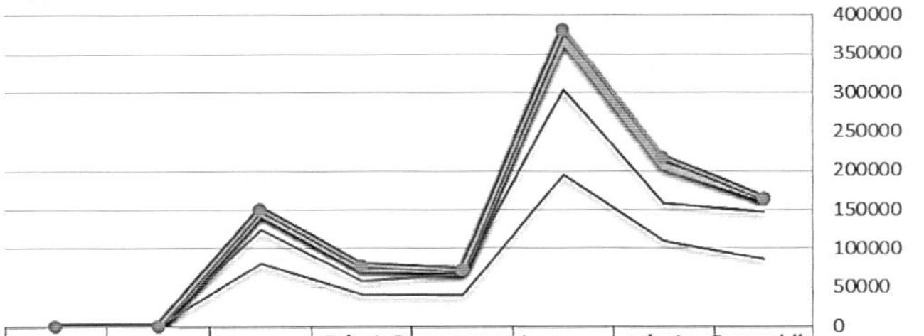

				Private	Governement		private	Gov public
			Total investement	Investement		Total	No of unites	
up-Avarage			4838.9	4562.2	276.7	12932	11916	1016
Average			22194.4	16858	5336.4	64393	46672	17721
Low cost			44086.6	16946.3	27140.3	109118	48418	60700
Total			78979.7	39994.6	38985.1	194249	109268	84981

Figure 1. Number of low-cost housing units built by Egyptian government according to Central Agency for Mobilization and Statistics Issue: July 2021.

2. Quality of Life

The expression "quality of life" (QOL) applies toward an assessment made regarding important components of a human's life or society [3]. It can be categorized as individual (physical and psychological health), interpersonal (social relationships), and -contextual (environment) aspects. It can also be could described by them or determined by 'physical indicators [4]. QOL has been determined as an interaction of social, health, economic and environmental conditions that have an impact on the development of the individual and society [5]. It is important to understand the impact of QOL on sustainable development [6].

2.1 Energy conservation QOL indicators for low -cost QOL housing

Several studies have explored the effects of energy consumption on **QOL** and human development (Index HDI) [7].Recent QOL indicators have started to focus on low-cost housing evaluation [9], the research adopted appropriate and important indicators in energy conservation. A suitable set of indicators to measure quality of low-cost housing should provide relevant data on both the physical indicators of a residence [8], as well as the inside environment which includes fundamental services. For instance well-designed housing has been identified as an important factor in promoting quality of life, orientation, internal spaces, electricity, water system, indoor running water, bathroom standards, and cooking facilities. These indicators were designed according to Sustainable Development Goals (SDGs) [10]. Neighborhood aspects are represented in the surrounding environment, level of noise, the quality of transportation services, etc. Monthly income is an essential signal of satisfaction in the household realm [11].The following table adapts several indicators of QOL according to the related items for energy conservation.

2.1.1 Parameters regarding energy saving:

More than 70% of the total energy consumed in residential buildings is attributed to lighting and HVAC (heating, ventilation and air conditioning) needs. Energy-saving parameters in Low-cost housing play a role in dictating the energy efficiencies and are liable for energy-saving programs.

2.1.2 Parameters regarding energy production:

Access to energy is a key pillar for human wellbeing [15]. Our ultimate goal is to ensure that everyone has access to enough sustainable energy to maintain a high standard of living. For energy production, the study has obtained several indictors related to energy conservations on a scale inside and outside the low-cost housing.

Table 1: QOL indictors for energy conservation.

		Indicator	**Determinant purpose**
Building Design	Building Mass	Building mass has an impact on day lighting, quality and hence energy consumption. Housing units are built close together on a large lot.	Preserving open space through housing clusters, housing units are built close together on a large lot, with the rest of the lot left as common open space. Each unit has a small open space.
	Building Orientation	It is important to consider the actual quantity of solar radiation on the facades of a residential unit ,as it affects the thermal load of the residential unit and controls the thermal behavior and the amount of thermal comfort of the space.	
	Optimal space standard and circulation	The prorate area of the house important for lower risk of respiratory infections. Optimal living space is also relevant to mental health. Poverty is improved access to key services and opportunities better life satisfaction.	
	Building Envelope	The building envelope protects the interior environment from less desirable exterior conditions. It is an important aspect in determining how much energy will be needed to heat or cool, light and ventilates the building.	
	Building Shape	The differences in shape between the buildings were found to have a large impact [2] and accounted for 10%-20% of their final energy demand [3].	

Openings	The most important factors regarding openings include the amount of sunshine that enters a room, the view, and human privacy. The best window room can vary. The favorite direction to the air movement path and the position of the openings completely permeates the vacuum.
Sustainable construction materials	The use of less carbon-intensive materials (e.g. cement, steel) or materials that store carbon (e.g. wood). These could [4] support de-carbonization of the industry sector, hence less need for energy-intensive resource extraction. Natural ecosystem protection limiting expansion of urban footprint.
conservation of natural resources	The Water management system can be achieved by energy recovery, and conserving and reusing water using low-flow faucets.
Green Space Factor.	The ratio of land consumption to the population growth. A well-insulated roof can negate the need for mechanical systems and reduce heating, cooling needs. Green roofs can reduce storm water volumes by up to 85% and the ceiling temperature, [5].
Spreading bike and car roads	Creating pedestrian- and bike-friendly spaces and walking/running paths, etc. to encourage exercise and discourage unnecessary driving.

3. Analytical Study

The study relies on two main factors the first is through a survey of visiting social low – cost housing models in Cairo (Mokattam- New Cairo) and observation the changes made by the residents to achieve their needs Fig (1,2,3), which the Egyptian government designed and implemented during the past ten years. For energy conservation we must study smart finishing to produce energy on landscape and public buildings as shown in table (3).

Figure 2. Changing the area and the shape of windows by closing or increasing the window space.
Ref. Take by Author

Figure 3. Using temporary curtain to cover balconies workplaces under buildings.
Ref. Take by Author

Figure 4. Narrow spaces between buildings to obtain maximum privacy.
Ref. Take by Author

4. Questionnaire results

The data used in the study was obtained from a questionnaire given to low - cost housing residents. Factors that may influence the study are quality of life and housing environment satisfaction (physical Aspects). The questionnaire was published on websites and distributed to low-income housing residents to study the problems of raising the quality of life through the concept of energy conservation.

The results were analysed by the SPSS program to obtain the percentage, mean, St Deviation of the results which is a essential elements in the design proposal.

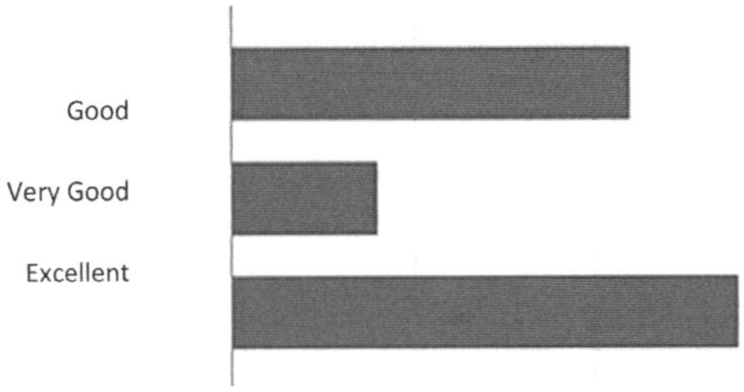

Figure 5. The value of the above-described approach is questionable for stratifying the residents on low- cost house location for public services found that 46% was closed and narrow to public services.

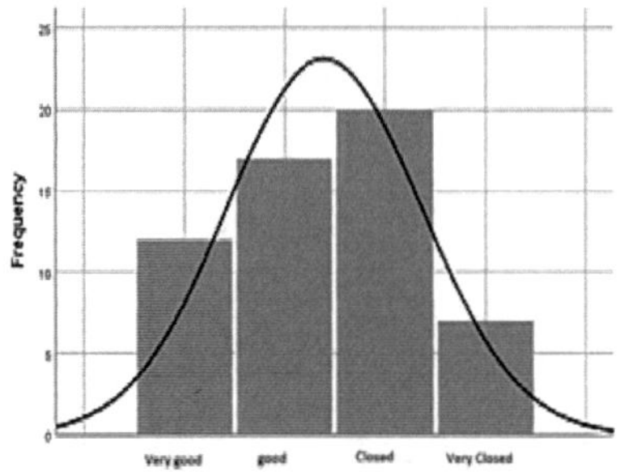

Figure 6. Housing units are built close together found that it was 43% for good spaces between the building and 21 % for much closed spaces between the building.

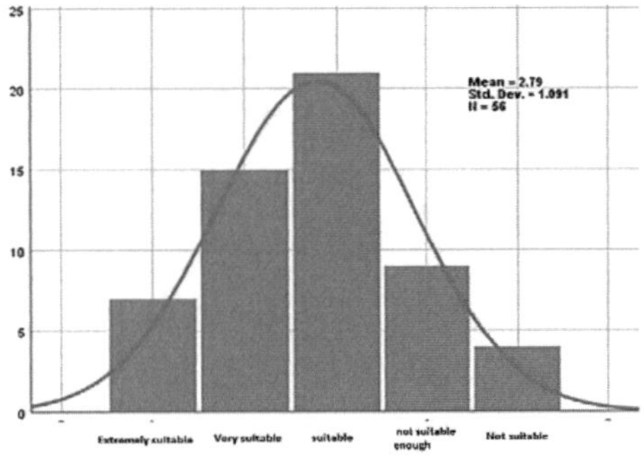

Figure 7. For the orientation of widows the results found it suitable for natural 34% was think it was not suitable and not suitable very enough and 56% was staffing to windows orientation.

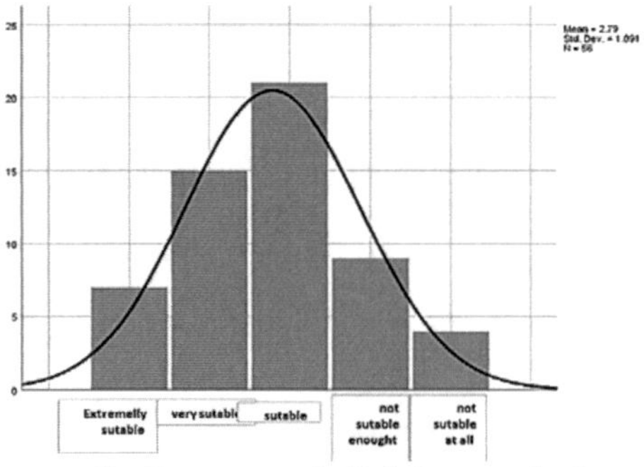

Figure 8. For natural lighting and the widows area found that 49% thought that it was very suitable and suitable for natural lighting 51% thought it was not suitable enough.

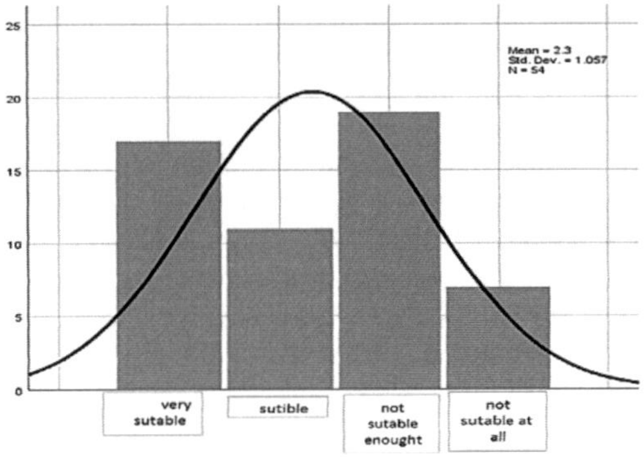

Figure 9. For The internal spaces inside the unit disposed of it was found that 53% thought it was suitable for them and 47% have thought it was not suitable.

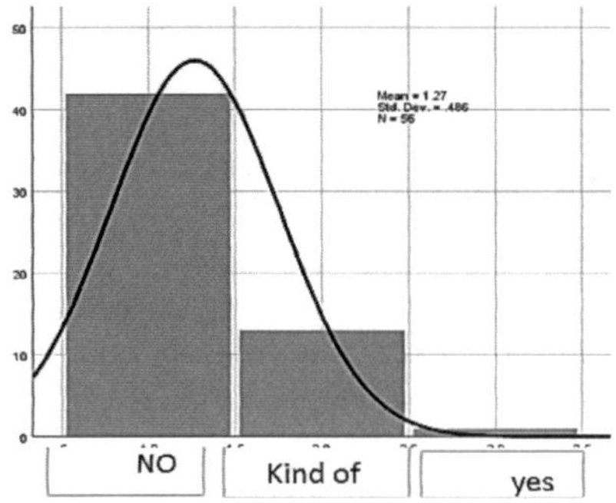

Figure 10. For the suitable methods to dispose of construction waste properly 75% of them are suitable.

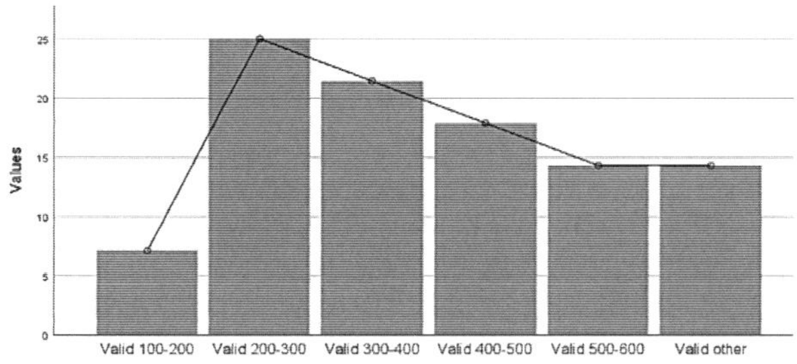

Figure 11. For the range of monthly electric bill 30% mentioned that it was 400-600 L.E. It is so high compared to their income .45% between 200-400 L.E and it is considered high according to low-cost housing.

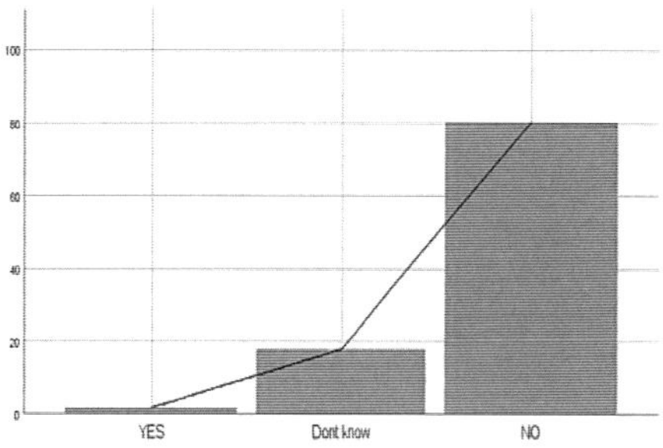

Figure 12. For conserving and reusing rainwater 80% represents that there is no method for conserving rain.

5. Result and Discussion:

The analytical study was based on result in the previous questionnaire, which showed varying rates in satisfaction on the design and rates of energy saving and comfort within spaces, Based on the results, we make the following observations:

5.1. Building design

- **Building Mass :**Separating Housing units Preserving open space through housing clusters
- **Building Orientation:** The unit should take 3 different directions at less In order to improve the rates of lighting and thermal comfort.
- **Building Shape:** The use of the Square shape in spaces design Preserving open space through housing clusters.
- **Optimal space standard and circulation:** Taking into account the proportions in the design spaces significantly contributed to the reducing waste area in spaces and be more accommodate future needs.
- **Openings**: There must be windows in every space regard the favorite direction and air movement path, Fig. 13.

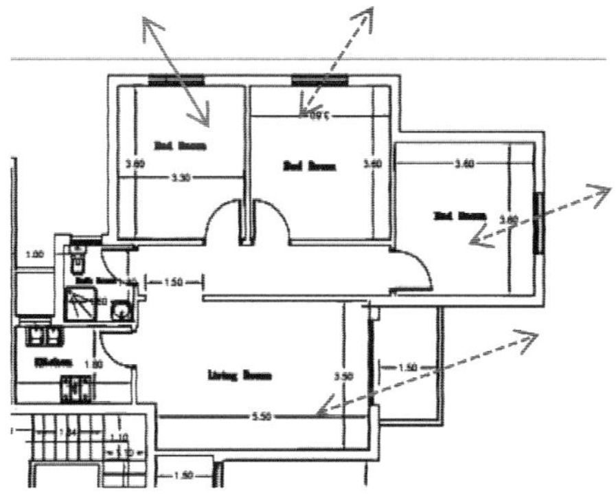

Figure 13. Designing the windows in different orientation allowing good air movement path.

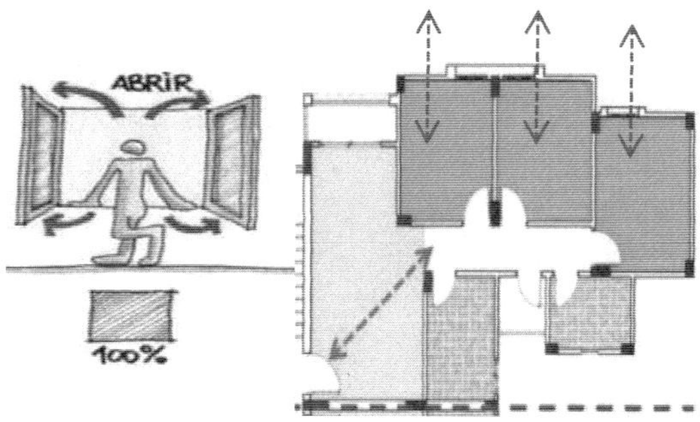

Figure 14. The current situation in low cost housing windows was designed in one orientation.

5.2. Urban and energy production

Regarding energy production in QOL indicators for Low -cost housing some important strategies are presented to achieve the production of energy:

- **Conservation of natural resources:** Water management system can be achieved by energy recovery, and conserving and reusing water. Using low-flow faucets.
- **Green Space Factor**: Raising the rate of green elements in Law cost housing urban spaces, encourage people to use Green roofs techniques which can reduce storm water volumes by up to 85% and the ceiling temperature.
- Spreading bike and car roads Creating pedestrian- and bike-friendly spaces and walking/running paths.
- **Sustainable construction materials:**
  - Roads with Solar Panels: It is a new technique that combines environmental corridors and lighting together with lighting path instead of traditional lighting poles.
  - Artificial wind tree: Wind tree is used for the generation of energy from wind the leaves act as mini wind turbines which are capable of providing power.
  - Solar PV Parking Structure: putting Parking Lot Solar PV ,By power shading structure we are not only able to make more use of the land Parking lot solar
  - Wind and solar - powered hybrid street lamp: It can transfer solar and wind power to electric power and incorporates Stable operation, non-cable working, No conventional energy consumption, long life span.
  - Tilfs for energy production: It converts the kinetic energy into electric energy and is stored in the units under the title to be used in various activities inside the building. One step produces 5-7 watts.
  - Photovoltaic cells: The building's facades and surfaces are supported by photovoltaic units to transform solar energy into electrical energy and can be placed on the hospital buildings in a decent manner providing aesthetic areas.

6. Applied Study

Designing the units may have been the main factor explaining the rising of consuming energy its related to some more factors (Building Mass- Building Orientation- Building Envelope- Building Shape) From that the research suggest an Architectural proposal design plan of a low-cost housing regarding energy and economic issues.

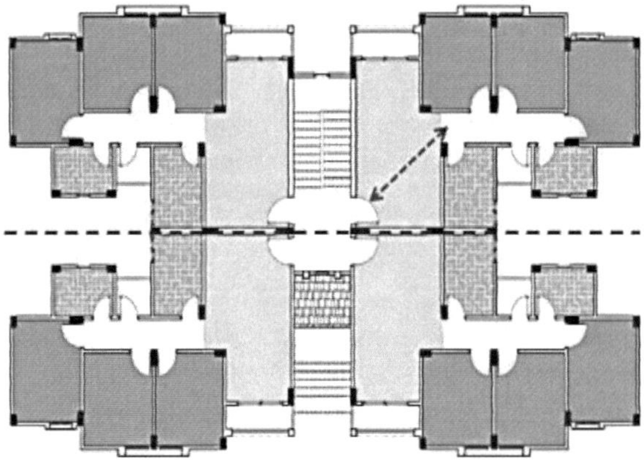

Figure 15. The current area is 90 m^2 and orientation is 2 different directions at maximum.

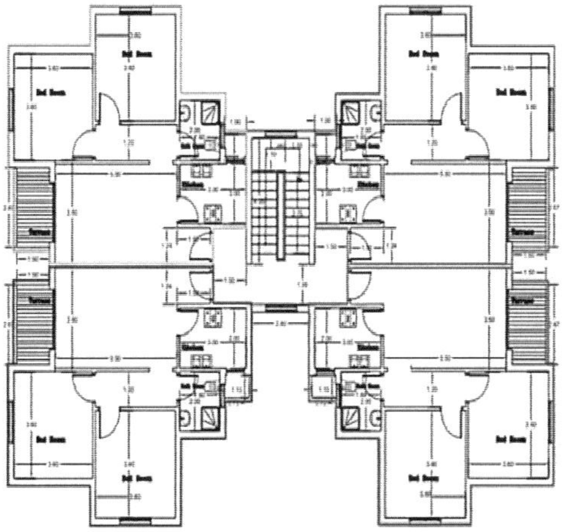

Figure 16. Design proposal for the $70m^2$ separating the building mass –regarding the orientation for every unit – avoiding the cross-sections in circulation.

Figure 17. Design proposal for the $90m^2$ separating the building mass –regarding the orientation for every unit – avoiding the cross-sections in internal circulation.

7. Conclusions

The analytical study focused on assessing and analysing the questionnaire and suggesting the necessary .modifications to solve the previously mentioned problems through the evaluation of rationalization and production of energy in terms of different items as introduced in a table.
The quality-of-life indicators were studied in terms of (energy rationalization and production). Conducting a questionnaire that measures the extent to which these criteria are applied, as low housing costs by the resident population. These indicators were analysed to make a redesign of the current situation of low-cost housing. The research proposed a housing unit of minimum surface area for one room not less than $3.3m^2$. Attention to the direction of the openings of the doors, and considering the shape of the square-shaped unit. Orientation per unit to 3 directions instead of 2 directions, and considering privacy.

8. Recommendations

1) Applying the indicators of energy conservation by taking into account the design standards, both internally and externally in Low -cost housing.
2) The smart technique should be focused on the start of planning instead of treatment, and it is not considered expensive if it is compared to with the executed quantity and the time element.
3) The possibility of transforming existing Low -cost housing to energy producers building through the cultivation of solar cell surfaces, the use of wind and solar-powered hybrid street lamps, solar-powered glass, energy-producing floors and landscape can be easily achieved from the preliminary stages of design as shown in the proposed design.
4) The researchers recommend the application of energy-producing technologies in there various forms inside public buildings attached to low-cost housing to security aspects of maintenance work and to produce energy to reduce expenditures on the citizen's priest.

References

[1] Al-Qawasmi, J. (2020). "Measuring Quality of Life in Urban Areas: Toward an Integrated Approach." Int. J. Environ. Sci. Nat. Resour 25: 67-74.
[2] Fassio, O., C. Rollero, and N. De Piccoli, *Health, quality of life and population density: A preliminary study on "contextualized" quality of life.* Social indicators research, 2013. **110**(2): p. 479-488

[3] Co-operation, O. f. E. and Development (2019). Accelerating Climate Action: Refocusing Policies through a Well-being Lens, OECD Publishing.
[4] Cutter, S. L. (1985). Rating places: A geographer's view on quality of life, Assn of Amer Geographers.
[5] Danielski, I., M. Fröling and A. Joelsson (2012). The impact of the shape factor on final energy demand in residential buildings in nordic climates. World Renewable Energy Forum, WREF 2012, Including World Renewable Energy Congress XII and Colorado Renewable Energy Society (CRES) Annual Conference; Denver, CO; 13 May 2012through17 May 2012; Code94564.
[6] Del Pero, A. S., W. Adema, V. Ferraro and V. Frey (2016). "Policies to promote access to good-quality affordable housing in OECD countries."
[7] Gasper, D. (2010). "Understanding the diversity of conceptions of well-being and quality of life." The Journal of Socioeconomics 39(3): 351-360.
[8] IEA, G. E. (2019). "CO2 Status Report 2018." International Energy Agency, Paris 562.
[9] Karim, H. A. (2012). "Low cost housing environment: compromising quality of life?" Procedia-Social and Behavioural Sciences 35: 44-53.
[10] Muhammad, S. and A. Sabo, (2021) "The Impact of Energy and Electricity Consumption on Quality of Life in Africa." Research Journal of Business Management 4(1):104-109
[11] Robinson-Smith, G., M. V. Johnston and J. Allen (2000). "Self-care self-efficacy, quality of life, and depression after stroke." Archives of physical medicine and rehabilitation 81(4): 460.
[12] Sapkota, J. B. (2014). "Access to infrastructure and human development: Cross-country evidence." Perspectives on the Post-2015 Development Agenda. Tokyo: JICA Research Institute.
[13] Stamenković, M. G., M. J. Miletić, S. M. Kosanović, G. D. Vučković and S. M. Glišović (2018). "Impact of a building shape factor on space cooling energy performance in the green roof concept implementation." Thermal Science 22(1 Part B): 687-698.
[14] Group, T.W., *The World Health Organization quality of life assessment (WHOQOL): development and general psychometric properties.* Social science & medicine, 1998. 46(12): p. 1569-1585.
[15] Vijayaraghavan, K. (2016). "Green roofs: A critical review on the role of components, benefits, limitations and trends." Renewable and sustainable energy reviews 57: 740-752.
[16] Mohamed, Noha , Criteria to Evaluate Energy Saving and Production in Hospitals "Nursing Units", PHD thesis, Architectural Engineering, Faculty of Engineering, Cairo University, Giza, Egypt, 2019.

IOP Conf. Series: Earth and Environmental Science 1056 (2022) 012033 doi:10.1088/1755-1315/1056/1/012033

VPL-based Code Compliance Checking for Building Envelope Design Using OTTV Calculation

A S Eid[1, *], M M Aboulnaga[1], A H Mahmoud[1]

[1] Department of Architecture, Faculty of Engineering, Cairo University

[*]Corresponding Author Email: asma2sale7@cu.edu.eg

Abstract. Building Envelope plays a vital role in reducing thermal loads, minimizing energy consumption, and ensuring sustainability. The Overall Thermal Transfer Value (OTTV) is a recognized method to evaluate the thermal performance of buildings. Building Information Modelling (BIM) has added great value to the AEC industry as an efficient tool to facilitate the building design process. With the development of BIM, more properties and material characteristics are assigned to the project design and building model that are getting more complex. Meanwhile, performing OTTV assessment and checking compliance with the assigned values involving time-consuming calculations with the possibility of ambiguity. Hence, automation of the code compliance checking process has become on-demand within the BIM environment. The objective is to develop a Visual Programming Language (VPL)-based script for automating the OTTV rule-based code compliance checking process. The applied method is to integrate BIM, VPL, and OTTV requirements; and develop a visual script that is responsible for extracting parameters from the BIM model, performing OTTV calculation, comparing output with assigned standard values of OTTV aiming answering the research question whether automated code compliance checking process can contribute to code compliance facilitation effectively. The results indicate that integrating VPL into the code compliance checking process has promising signs in facilitating the compliance checking process, increasing the accuracy and reliability of outcomes.

Keywords: Building envelope, overall thermal transfer value (OTTV), code compliance checking (CCC), Building information modelling (BIM), visual programming language (VPL).

1. Introduction

Building energy codes are becoming more vital in energy efficiency policies. They play an important role in raising awareness of building energy conservation, promote energy efficient designs in buildings, and formation of a basis for building energy performance assessment and developing energy efficiency programs [1]. According to the IEA global status report for buildings and construction 2019; the buildings and construction sector accounted for 36% of final energy use and 39% of energy and process-related carbon dioxide (CO_2) emissions in 2018 [2].

Renewable energy, coupled with energy efficiency gains, can contribute to 90% of CO_2 reductions needed to be achieved by 2050 and sustainable cities can contribute to achieving Climate Change adaptation (CCA) and resilience while eliminating Climate Change (CC) risks [3], [4]. Meanwhile, the application of BIM has shown exponential growth in facilitating building performance design, simulation, and optimization with attempts to dwindle the gap between simulated and actual building performance [5], [6] that aims at minimizing energy consumption and mitigating CO_2 emissions from buildings, reducing thermal loads, and ensuring sustainability. This may contribute to achieving environmental rating systems criteria efficiently in shorter times. Nevertheless, the building envelope is acting as a climatic barrier between the outdoor and indoor environment. Thus, the building performance is affected, mainly by facades' design and local climatic characteristics.

Thermal building performance can be measured by OTTV via determining the amount of heat transfer into the building through its outer envelope assuming it is completely enclosed [7]. With CC manifestation, it is important to enforce the use of thermal insulation materials that provide the least heat impacts on buildings [8]. Hence automating CCC for building envelope design using BIM-based OTTV calculations for mitigating CO_2 emissions might ease the process of complying with such needed criteria.

2. Building Information Modelling BIM

BIM was described as "a digital representation of a building, an object-oriented three-dimensional model, or a repository of project information to facilitate interoperability and exchange of information with related software applications" [9]. BIM has developed a new technological paradigm shift for many disciplines [10]. It eliminates the possibility of ambiguity, inconsistency, and inaccuracy in design and raises the percentage of efficiency and productivity as well as allows professional users to integrate external applications and features by add-ins or plugins development that can assist in automating repetitive or complicated functions [4], [11], [12]. Accordingly, integrating BIM-based performance design support is widely applied in the early design stages to assure the most efficient and optimal design solutions for projects in terms of the required assessment criteria [13].

2.1. Embedded BIM Applications

Embedded BIM applications can support both Building Design Assessment and Automated code compliance checking to be more efficient and accurate outputs.

2.1.1. Building Design Assessment. Multiple measures of integrating and finding solutions for Embedded BIM applications are being performed [9]. These applications have been mainly focusing on building design assessment whether through green building rating systems assessment or CCC. For instance, many researchers focused on the embedment of BIM for automating the LEED rating system assessment process [11]. Han et al., Nguyen et al., and Zhang and Chen have performed multiple measures to evaluate the building design at an early stage, calculate the potential acquired points of the LEED rating system [12], [14], [15]. LEPOST, a web-based tool created by Zhao et al. to eliminate LEED application submission costs and facilitate the green building design process [16].

Not only LEED but also other green building rating systems such as the Malaysian Carbon Reduction and Environmental Sustainability automation Tool (My-CREST) developed by Ohueri et al. with the assistance of BIM [17]. All the previously mentioned research focused on sustainable building rating systems assessment. Accordingly, embedding BIM into building design assessment has a significant contribution to facilitating the complicated evaluation design process. As for CCC, it will be explained in the following section.

2.1.2. Automated Code Compliance Checking CCC. CCC automation is defined by Ismail et al. as "a process or system which solely assess the design through its objects, attributes, and relations without modifying the design" [18]. Eastman et al. divided the automated code compliance checking process into four steps including: "i) rule interpretation and logical structuring of rules for their application; ii) building model preparation, where the necessary information required for checking is prepared; iii) rule execution phase, which carries out the checking; and iv) rule reporting of the checking results" [19]. Many researchers tried to document and classify all the previous measures of CCC automation with different techniques [4]. For instance, Ismail et al. have classified the previous research based on the adopted approach [18]. Dimyadi and Amor divided the previous measures according to a timeline based on the emergence of the IFC open data model. Also, Choi and Kim arranged the implemented previous research focusing on the type of the adopted rule-based code to check compliance with the assistance of

BIM [20]. While Eid et al. have collected most of the individual measures that have been performed till mid-year 2019 [4].

It is clear from the above literature that most of the previously implemented applications for CCC with the assistance of BIM were dependent on hard-coded techniques related to Application Programming Interface (API) until the emerge of Visual Programming Languages for BIM. Hence, VPL is an imperatively essential and significant method to support architects and designers with limited knowledge about the coding process to produce efficient buildings.

3. Visual Programming Language VPL

VPL was described as "a flow-based language, that represents complex structures as a flow of information with visual syntax and semantics to be more readily understandable for non-professional programmers" [21]. The VPL in Architecture is mainly used for generating geometry, automating repetitive tasks, or checking compliance with specific rules [4]. The visual code checking language (VCCL) has been used recently by limited researchers as an alternative for hard-coded methods [4]. In the recent few years, multiple research studies have been performed using VCCL or VPL for automated CCC. The most used VPL platforms in the recently conducted research are Dynamo for Revit, Grasshopper for RhinoCeros, and Marionette for Vectorworks.

3.1. VCCL and Building Energy Codes. Many researchers have developed measures to support VCCL. The potentiality of building codes translation with visual programming language was introduced by Preidel and Borremann where the VCCL on the German fire safety code was implemented [22]–[24]. Zhou et al. have carried out work on VPL code translation of the CCC automation for building envelope energy efficiency by focusing on the OTTV calculations [25]. Natephra et al. have also developed the BOTTVc system that worked on the optimization of the building envelope design assessment based on the OTTV requirements [13]. In addition, Kim et al. have also used the VPL approach to represent KBimCode for automated CCC based on the Korean building code [26].

3.2. Overall Thermal Transfer Value OTTV. OTTV is a measure of heat transfer into the building through its envelope that was primarily proposed by The American Society of Heating, Refrigerating and Air-Conditioning Engineers (ASHRAE) [39], [40]. Multiple countries have also developed OTTV standards about ASHRAE Standards with modifications to comply with local climatic characteristics and construction methods [9] for instance, Singapore [41] and Hong Kong [42], [43]. Moreover, Wong and Chan have explained in detail the method to calculate OTTV based on the Hong Kong code of practice [47]. While Hui has figured out the limitations of OTTV calculations and how it can be improved [1].

The OTTV formulas were developed to calculate OTTV for shaded and unshaded external envelopes (walls and windows). The OTTV calculation is the sum of the total rate of heat transfer through individual walls or roofs divided by the total gross area of individual walls or roofs. The OTTV standard values in ASHRAE and the Egyptian Energy Efficiency Code are mentioned in table 1. The (OTTVi) for each façade/roof can be calculated by the following equations [45], [46] that are valid when the walls and windows are not shaded.

$$OTTV_i = ((A_w \,.\, U_w \,.\, \alpha \,.\, TD_{eq}) + (A_f \,.\, U_f \,.\, \Delta T) + (A_f \,.\, SF \,.\, SC)) \,/\, A_i \quad (1)$$
$$= ((1-WWR) \,.\, U_w \,.\, TD_{eq}) + (WWR \,.\, U_f \,.\, \Delta T) + (WWR \,.\, SF \,.\, SC) \quad (2)$$

Another formula applies for walls and windows with shaded parts as follows [27]:

$$OTTVi = (Q_w + Q_{gs} + Q_{sol}) \,/\, A_i \quad (3)$$
$$= ((A_w \,.\, U_w \,.\, TD_{eq}) + (A_f \,.\, U_f \,.\, \Delta T) + (A_f \,.\, SC \,.\, SSF \,.\, SF))/(A_w + A_f) \quad (4)$$

Where; *$OTTV_i$ = Overall thermal transfer value for individual façade/roof (W/m^2), A_w = Area of opaque walls (m^2), U_w = Thermal transmittance of opaque wall (W/m^2 °C), α = Absorptivity of opaque wall, TD_{eq} = Equivalent temperature difference (°C), A_f = Area of fenestration (m^2), U_f = Thermal transmittance of fenestration (W/m^2 °C), ΔT = Temperature difference (°C), SF = Solar factor of fenestration (W/m^2), SC = Shading coefficients of fenestration, , SSF = Solar shading factor of external shading devices, A_i = Total area of opaque walls and fenestration, and WWR = Window-to-wall ratio.*

Table 1. OTTV Standard Values [7], [28].

Code	Energy Efficiency Code, Egypt				ASHRAE, USA				
Location/ Latitude	Reg.1	Reg. 2,3,4,5	Reg. 6	Reg. 7,8	10° & 20° N	30° N	40° N	50° N	60° & 70° N
$OTTV_{Wall}$	90W/m²	95W/m²	--	100W/m²	92W/m²	100W/m²	107W/m²	115W/m²	123W/m²
$OTTV_{Roof}$	26W/m²				26W/m²				

4. Methods and Scope of Work

Following the light of correspondent research measures; Ghannad et al. (2019), Natephra et al. (2018), and Seghier et al. (2017) developed an integrated framework to establish a relationship between a BIM model and needed values that are available in the BIM database to automate the workflow of the OTTV assessment and rule-based compliance checking process, reduce the needed time for the checking process and minimize the possibility of ambiguity and inconsistency[13], [29], [30]. The main aim is to develop a VPL based script for automating the OTTV rule-based code compliance checking process. The authors adopted the same method developed by Natephra et al. 2018.

The Methodology design consists of two core phases that can be used for establishing a VPL-based design compliance checking process [15]. The first stage is the integration of BIM, VCCL, and OTTV requirements assigned in the Energy Efficiency Egyptian Code (EEEC). This stage consists of two main steps: i) to determine the OTTV equation data requirements, parameters, and the assigned value; and ii) to check whether the needed calculation data and parameters are identified in the BIM model and can be easily extracted or have to be added as shared parameters. The next stage focuses mainly on generating the visual programming script that is responsible for extracting data and information from the BIM model; calculating the required coefficients for the OTTV; accommodating these coefficients into the OTTV equation and comparing the calculated output with the assigned values of OTTV. The following method incorporates all developed visual codes and such method has been tackled however few researchers explain the developed codes in detail.

4.1. BIM Modelling – 3D Building Design. Autodesk Revit software has been selected for developing the BIM model as it is one of the most used software programs for BIM. Dynamo (an embedded plugin-in Autodesk Revit) has been also selected for VPL scripting as it is more integrated with Revit. A new Revit template has been created and customized for OTTV calculations with full detailed geometry and material properties to facilitate the BIM modelling process. Missing parameters for objects and materials characteristics have also been added as shared parameters to be easily extracted for OTTV calculation.

4.2. Acquisition of required coefficients for OTTV calculation. Even though much research has adopted the same parameters, few of the applications described the detailed visual code. Hence, the next section presents the eight parameters application in the code OTTV calculation.

4.2.1. Area of Wall and Fenestration (A_W, A_f). A dynamo script has been developed to calculate area of wall and fenestration (A_W, A_f). It depends on selecting target groups which are walls and openings in each façade from the BIM model and calculating their areas.

4.2.2. Thermal Transmittance (U-value). The U-value is described as the amount of heat transfer of a structure divided by the temperature difference across the structure under certain climate conditions (temperature gradient of 24 °C, at 50% humidity with no wind). It can be calculated by the following equation [31]:

$$U = Q / (A \,.\, TD_{eq}) \tag{5}$$

Where *; Q is the amount of heat transfer through material, A is the area of material and* TD_{eq} *is the equivalent temperature difference.*

The U-value coefficient for walls and roof have been added as shared parameters and exists in the BIM model in material properties and can also be extracted with the VPL developed script.

4.2.3. Equivalent Temperature Difference (TD_{eq}). The equivalent temperature can be simplified to be measured by the difference between outside and inside the building. The TD_{eq} value is based on the solar radiation absorption period, solar radiation absorption coefficient, material density, angle, and wall orientation [32]. The TD_{eq} coefficient has been added as shared parameters with consistent value based on the building material characteristics in the BIM model including density, thickness, and orientation in the BIM model and can be extracted from a spreadsheet file and read with the VPL developed script.

4.2.4. Solar Factor (SF). The solar factor is the hourly radiation for horizontal and vertical surfaces in a one-square-meter area [33], [31]. SF is provided as a constant coefficient that users can choose the values for horizontal and vertical surfaces in each orientation. It can be extracted from a spreadsheet file and read with the VPL-developed script.

4.2.5. Solar Shading Factor (SSF). The solar shading factor (SSF) for external shading devices can be determined based on the orientation, type of projection, distance of projection, and dimensions of openings. The value of SSF ranges from zero to one. In case of no shading on openings, the solar shading factor is considered as one [27].

4.2.6. Shading Coefficient of glazing (SC). The shading coefficient (SC) is defined as the ratio of the solar heat transmittance through glass under a specific set of conditions to the solar heat gain relative to that double-strength sheet clear glass (3 mm clear float glass) under the same conditions [34]. SC has a maximum value of 1. The lower the SC, the lower the amount of solar heat transmitted and the higher the shading performance of the glass. The SC value is always provided in the technical data sheet of the glass type. Otherwise, it can be calculated based on the type of shading device and extracted from the spreadsheet [27]. The SC value has been added as shared parameters and exists in the BIM model in glass properties and can also be extracted with the VPL-developed script.

4.3. Extracted coefficients compensation for OTTV calculation and assessment. The extracting coefficients' components for calculating and assessing the OTTV are very significant. Following to extracting all needed parameters and coefficients for OTTV calculation from the BIM model such as U-value, TDeq, and α. Other obtained data BIM database or spreadsheet has been included in previously developed VPL script to get the required coefficients to be accommodated in the OTTV calculation, e.g., WWR, SC, and SF.

4.4. Information output for the OTTV Calculation. All these coefficients are compensated in the OTTV equation VPL script as input variables. The calculation output is compared to the assigned value of OTTV in the Egyptian energy efficiency code. If the calculated value is not compatible with the minimum assigned OTTV value, then the BIM model would fail in the OTTV test with an error message. In this case, the user is supposed to recheck the properties of the existing materials and try other alternatives with better thermal performance characteristics. The developed code for OTTV calculation and assessment is indicated in Figure 1. Also, the developed visual script for illustrating the OTTV assessment results on the BIM model via highlighting the "pass" and "fail" components with RG colours is shown in Figure 2. This has been applied using equation (3) and equation (4).

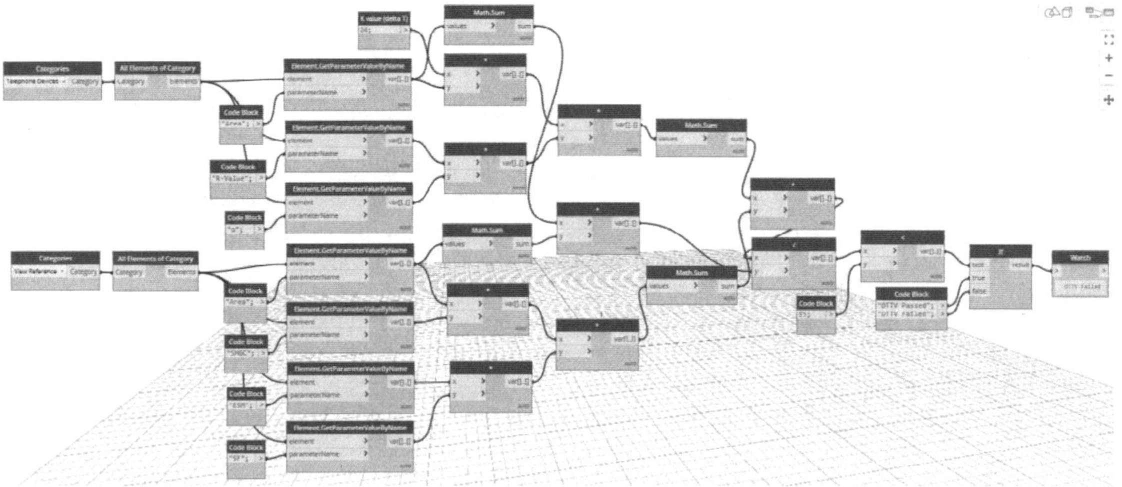

Figure 1. Developed VPL script for OTTV calculation and assessment. Developed by Authors using Dynamo for Revit 2.1 version.

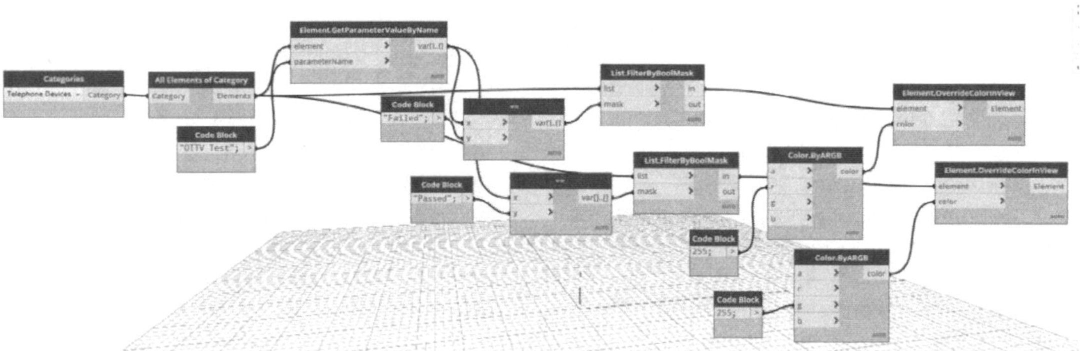

Figure 2. Developed VPL script for illustrating OTTV assessment results. Developed by Authors using Dynamo for Revit 2.1 version.

5. Proof of Concept

The developed VPL code for checking OTTV values has been applied and tested on five different commercial buildings (a, b, c, d and e) in five different climatic regions located in Egypt assigned in the Energy Efficiency Egyptian Code (EEEC). These buildings have been selected for availability of data of the building, the ease of accessibility of the building on the daily basis. 3D BIM model was created for each building by Autodesk Revit Architecture 2020 student version based on the original drawings and/or onsite visits. The VPL code was developed by Dynamo for Revit 2.1 version. It was noticed that the drawings were slightly different from the existing building after recently applied modifications to the building design.

5.1. Testing and Validation of the OTTV Calculation Output. The developed visual code has been tested and implanted on the five buildings 3D BIM model. The required parameters have been extracted in this calculation. Both façade areas and windows areas have been calculated for all elevations for each building by the developed visual code. The resulted output is indicated in Figure 3. The OTTV for building (2) and building (5) has exceeded the assigned value of OTTV in the EEEC and the test result is FAIL. On the other hand, the OTTV test result for building (1), (3) and (4) is PASS as the OTTV value does not exceed **90** or **95** Watt /m^2 for each. Validation was conducted to make sure that the outcome of the OTTV calculation is significant and real to be applied and generalized. The validation includes OTTV outputs of the baseline building where each parameter is calculated mathematically and compared the outcome results with its significance and to ensure coherent results. It was found that there is a minimal error ratio in comparing the resulted output ranges from 0.002 to 0.056 that could be acceptable.

6. Results and Discussions

This research-based on using visual programming to develop assigned rules for extracting data or information exchange to ensure reliability. The results indicate that VPL rule-based code compliance checking has promising signs in contributing to facilitating compliance checking process effectively, increasing the accuracy and reliability of outcomes. However, and based on previously performed research measures, exchanging data using the Visual Programming tool is yet fully stable so far which may affect the stability of the final output [35].

The visualization of system outputs needs more developing efforts to be more indicative and appealing for users. Gerrish et al. have discussed a similar issue regarding building energy performance visualization in BIM applications [5]. The possibility to integrate the visualized results not only into the BIM model but also into the project's Architectural sheets and documents need to be furtherly studied. It is also vital to calculate the accurate percentage of cost and time reduction of processing this tool can contribute to by applying it to multiple case studies keeping in mind that the decrease of calculation time might vary based on multiple factors. Also, the developed system can only deal with regular-shaped buildings; more efforts need to be done to deal with more complex and freeform-shaped buildings.

7. Conclusions

This research has presented a computational VPL BIM-based workflow for OTTV automated code compliance-checking. The developed tool can assess OTTV automatically of a building in a short time accurately compared to the ordinary method performed mathematically. The developed system is meant to extract the thermal and physical properties of the BIM model components for calculating the Overall Thermal Transfer Value and checking its compatibility with the OTTV assigned values. The OTTV assessment system is based on both OTTV calculation output and BIM-VPL functionalities compliance. The Dynamo scripts are expected to save project time and cost by speeding up the design process and avoiding building performance assessment rework.

The prototype developed tool is supposed to assist users in decisions making regarding the optimum selection of envelope materials with respect to OTTV requirements. The applied workflow in this research can be reused, modified, and implemented for automating different tasks related to checking compliance with assigned regulations and evaluating its impacts through the building design process. Furthermore, this research is considered as a proof of concept that architects and designers with the assistance of BIM and VPL scripting can develop their customized tools that fit their criteria and requirements rather than commercial tools.

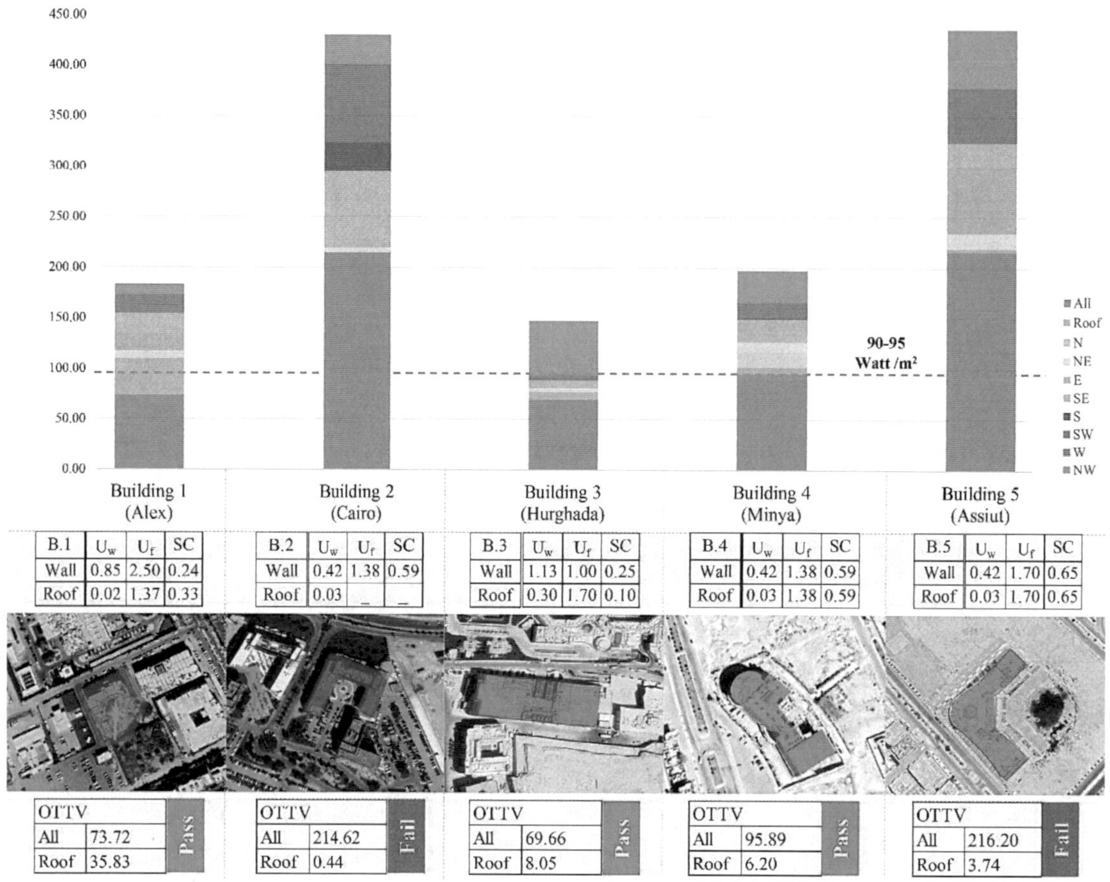

Figure 3. Summary of OTTV Output for Building 1, 2, 3, 4 and 5. Data and Charts developed by Authors. All maps produced by Google Earth [36 – 40].

References

[1] Hui Sam CM, "Overall thermal transfer value (OTTV): how to improve its control in Hong Kong. In Proc. of the One-day Symposium on Building, Energy and Environment," *Proc. One-day Symp. Build. Energy Environ.*, vol. 16, no. October 1997, pp. 12–1, 1997.

[2] IEA, "Global Status Report for Buildings and Construction 2019," 2019.

[3] IRENA, "Untapped potential for climate action: Renewable energy in Nationally Determined Contributions," 2017.

[4] A. S. Eid, M. M. Aboulnaga, and A. H. Mahmoud, "Future Cities for Climate Action : Automated Code Compliance Checking in Reference to Energy Efficiency Building Regulations," no. Cc, pp. 3–5, 2019.

[5] T. Gerrish, K. Ruikar, M. Cook, M. Johnson, M. Phillip, and C. Lowry, "BIM application to building energy performance visualisation and managementChallenges and potential," *Energy Build.*, vol. 144, pp. 218–228, Jun. 2017.

[6] A. C. Menezes, A. Cripps, D. Bouchlaghem, and R. Buswell, "Predicted vs. actual energy performance of non-domestic buildings: Using post-occupancy evaluation data to reduce the performance gap," *Appl. Energy*, vol. 97, pp. 355–364, Sep. 2012.

[7] J. Vijayalaxmi, "Concept of Overall Thermal Transfer Value (OTTV) in Design of Building Envelope to Achieve Energy Efficiency," *Int. J. Therm. Environ. Eng.*, vol. 1, no. 2, pp. 75–80, 2010.

[8] S. Gadallah and M. Aboulnaga, "CLIMATE ACTION AND SDGS' ATTAINMENT: INSULATION MATERIALS' IMPACTSASSESSMENT,"*sustainablemediterraneanconstruction*, vol. 11, pp. 76–80, 2020.

[9] R. Miettinen and S. Paavola, "Beyond the BIM utopia: Approaches to the development and implementation of building information modeling," *Autom. Constr.*, vol. 43, pp. 84–91, 2014.

[10] B. Succar, "Building information modelling framework: A research and delivery foundation for industry

stakeholders," *Autom. Constr.*, vol. 18, no. 3, pp. 357–375, May 2009.

[11] A. S. Eid and M. GamalEddin, "An Automated BIM-Embedded Approach for Rule Based Checking for Green Building Design," *Int. Conf. Adv. Struct. Geotech. Eng. ICASGE'19*, Mar. 2019.

[12] T. H. Nguyen, S. H. Toroghi, and F. Jacobs, "Automated Green Building Rating System for Building Designs," *J. Archit. Eng.*, vol. 22, no. 4, p. A4015001, 2016.

[13] W. Natephra, Y. Nobuyoshi, and T. Fukuda, "Optimizing Evaluation of Building Envelope Design for Thermal Performance using a BIM- based Overall Thermal Transfer Value Calculation." 2018.

[14] Y. Han, A. Motamedi, Y. Nobuyoshi, and T. Fukuda, "Green Building Design Support System Based on BIM and LEED," 2017, no. May, pp. 41–44.

[15] C. Zhang and J. Chen, "LEED Embedded Building Information Modeling System," in *AEI 2015*, 2015, no. September, p. 2015.

[16] J. Zhao, K. P. Lam, T. Biswas, and H. Wang, "An online platform to automate LEED energy performance evaluation and submission process," *Constr. Innov.*, vol. 15, no. 3, pp. 313–332, 2015.

[17] C. C. Ohueri, W. I. Enegbuma, K. K. Kuok, and N. M. Wong, "Preliminary Evaluation of Synergizing BIM and Malaysian Carbon Reduction and Environmental Sustainability Tool," vol. 1, pp. 218–227, 2019.

[18] A. S. Ismail, K. N. Ali, and N. A. Iahad, "A Review on BIM-based automated code compliance checking system," *Int. Conf. Res. Innov. Inf. Syst. ICRIIS*, 2017.

[19] C. Eastman, J. min Lee, Y. suk Jeong, and J. kook Lee, "Automatic rule-based checking of building designs," *Autom. Constr.*, vol. 18, no. 8, pp. 1011–1033, 2009.

[20] J. Choi and I. Kim, "Development of Rule-based Building Code Compliance Checking System for BIM Based Quality Improvement," *Information*, vol. 20, no. 4(B), pp. 2929–2936, 2017.

[21] Dynamo BIM, "What is Visual Programming? | The Dynamo Primer," *Autodesk - Dynamo Project*, 2019.

[22] C. Preidel and A. Borrmann, "Refinement of the visual code checking language for an automated checking of building information models regarding applicable regulations," *2017 ASCE Int. Work. Comput. Civ. Eng. IWCCE 2017*, no. 1, pp. 157–165, 2017.

[23] C. Preidel and A. Borrmann, "Automated Code Compliance Checking Based on a Visual Language and Building Information Modeling," *Proc. 32nd Int. Symp. Autom. Robot. Constr. 15-18 June*, pp. 256–263, 2015.

[24] C. Preidel and A. Borrmann, "Integrating Relational Algebra into a Visual Code Checking Language for Information Retrieval from Building Information Models," *Icccbe*, no. June, 2015.

[25] H. Zhou, S. Lee, and H. Ying, "VPL-Based Code Translation for Automated Compliance Checking of Building Envelope Energy Efficiency," in *Construction Research Congress 2018*, pp. 1–12.

[26] H. Kim, J. Lee, J. Shin, and J. Choi, "Visual language approach to representing KBimCode-based Korea building code sentences for automated rule checking," *J. Comput. Des. Eng.*, vol. 6, no. 2, pp. 143–148, 2019.

[27] Danish Energy Management A/S, "Algorithms for Calculation of Energy Consumption and OTTV - Compliance Method Alternative 1 of the Energy Efficiency Building Code," UNDP, 2011.

[28] S. C. M. Hui and J. C. Lam, "Overall thermal transfer value (OTTV)- a review," *Hong Kong Eng.*, vol. 19(9), no. September 1991, pp. 26–32, 1991.

[29] P. Ghannada, Y.-C. Leea, J. Dimyadib, and W. Solihin, "Automated BIM data validation integrating open-standard schema with visual programming language," *Adv. Eng. Informatics*, vol. 40, pp. 14–28, 2019.

[30] T. E. Seghier, L. Y. Wah, M. H. Ahmad, and W. O. Samuel, "Building Envelope Thermal Performance Assessment Using Visual Programming and BIM , based on ETTV requirement of Green Mark and GreenRE," vol. 4, no. 3, pp. 227–235, 2017.

[31] W. S. Wong and H. W. Chan, "Energy and Use of Energy: Calculation and Application of OTTV and U-value," *Hong Kong Inst. Archit.*, 2012.

[32] C. Singhpoo, N. Punnucharoenwong, and C. Benjapiyaporn, "Study of the Effect of Temperature Differences on the Overall Thermal Transfer Value of buildings," in *Energy Procedia*, Nov. 2015, vol. 79, pp. 348–353.

[33] Building Authority H.K., "Code of Practice for Overall Thermal Transfer Value in Buildings." Building Authority Hong kong, 1995.

[34] Saint-Gobain, "Saint-Gobain glossary of terms and meanings," *Saint Gobain Glass*, 2020. https://uk.saint-gobain-building-glass.com/en-gb/saint-gobain-glossary.

[35] K. Kensek, "Visual programming for building information modeling: Energy and shading analysis case studies," *J. Green Build.*, vol. 10, no. 4, pp. 28–43, 2015.

[36] Google Earth, 2022. *N 31° 10' 34.7268 E 29° 55' 50.412*, 1:1.500. Google Earth [online] Available through: < https://earth.google.com/web/@31.17320955,29.93089155>.

[37] Google Earth, 2022. *N 30° 1' 20.7372 E 31° 4' 10.7436*, 1:1.500. Google Earth [online] Available through: < https://earth.google.com/web/@30.02087775,31.07149565>.

[38] Google Earth, 2022. *N 27° 15′ 28.9836 E 33° 49′ 51.0096*, 1:1.500. Google Earth [online] Available through: < https://earth.google.com/web/@27.25616009,33.82882701>.

[39] Google Earth, 2022. *N 28° 5′ 8.43 E 30° 48′ 10.8468*, 1:1.500. Google Earth [online] Available through: < https://earth.google.com/web/@28.08557333,30.80322976>.

[40] Google Earth, 2022. *N 27° 16′ 26.706 E 31° 17′ 17.7072*,, 1:1.500. Google Earth [online] Available through: < https://earth.google.com/web/ @27.27419958,31.28809468,>.

Soil-Pile Interaction for the Foundations of Bridge Abutments in Soft Clay Formations

O M Emara[1, *], A L Fayed[2], H Arafat[3] and Y M ElMossallamy[2]

1 Structural Engineering department, Horus University, Damietta, Egypt.
2 Structural Engineering Department, Ain Shams University, Cairo, Egypt.
3 Deputy Head of Board of Trustees, Future University, Egypt, Ex-Minister of Transport.

*Corresponding Author Email: oemara@horus.edu.eg.

Abstract. The success of bridge construction and operation on soft grounds requires proper planning and execution of subsurface investigation. Embankments and/or approaches constructed on weak and compressible soils suffer from time settlement induced by the embankment loads and result in down-drag and lateral forces acting on the piles supporting the bridge abutments. The lateral loads acting on piles from the movement of soil are of significant concern in the design of pile foundations since they may result in additional straining actions and deflection in the piles. Furthermore, distress of the piles and accordingly adverse effect on the stability of the supported structures might occur due to such phenomenon. This problem is studied in this paper using the 3-D numerical modeling approach utilizing the finite element method. Verification of the developed numerical model was carried out by simulating a case study of a centrifuge test and comparing the predicted and measured results. The numerical model is used to study the effect of the developed lateral forces due to the consolidation of clay on the lateral displacement and abending moment of the adjacent piles.

Keywords: soil-pile interaction, passive lateral loads, soft clay, abutment, approach, embankment.

1. Introduction

Construction of bridge abutment and approach-embankment on soft soil is a challenging topic for geotechnical engineers due to the high compressibility of the clay soil (time-settlement process due to consolidation and creep) and low shear strength. The settlement of embankments can cause significant vertical and horizontal movements of the soil adjacent to the embankment. Therefore, piles supporting the bridge abutments might be subjected to axial and lateral loads induced by such soil movements (passive loads). Many methods are applied by geotechnical engineers to deal with this problem. The soil-pile interaction problem has been investigated over the last three decades using field or centrifuge tests accompanied by finite element analysis, e.g.,[1-3], and [5-7].

2. Case study

Ellis,(1997),[4], investigated the behavior of pile bridge abutment approach embankment over soft clay soil rested on medium dense sand. A centrifuge test model was performed to model the effect of lateral thrust acting on the abutment. A series of two plane strain tests (EAE3 and EAE4) were carried out using the 10m beam centrifuge at the Geotechnical Centrifuge Centre in Cambridge. In EAE3, the full pile bridge abutment and embankment were modeled as a prototype with slow embankment construction in 30min (210 days in full scale) and investigated its effect on soft soil. In EAE4, using the same model of EAE3 but with faster embankment construction in 3 min (21 days in full scale). The soil profile consisted of a clay layer (10 m depth) that was built in the centrifuge test creating a normally consolidated clay using kaolinite subjected to overburden pressure of 20KPa. The undrained shear strength of the clay was 25 KN/m^2. The substratum stiff soil (10m depth) was built using fine Leighton Buzzard sand with particle sizes of 0.06 - 0.2mm. A 1.0m thick sand layer over the clay was constructed to accelerate the groundwater excess from the clay layer.

The abutment and pile structure were constructed using an aluminum tube with a young's modulus of 70*10^6 MPa and the external diameter of the prototype piles was 1.27m with 0.122m wall thickness that is corresponding to the prototype of 19 mm length and 1.2 m pile diameter. The pile spacing/diameter ratio was 5.25 along the rows and 3.94 between the rows.

To construct the embankment behind the abutment, a sand hopper was mounted on the top of the centrifuge model. The embankment was constructed in four-stage for 3 minutes or 30 min according to the type of tests and the four equal lifts to reach the final stage of the embankment.

Pore pressure transducers (PPTs) were applied for measuring the excess pore water pressure from the clay layer during the stage of embankment construction and afterwards. The results of PPTs were including the excess pore water pressure in 21 days (embankment construction) and when achieving the full dissipation of excess pore pressure after 1000 days (porotype scale). To measure, the lateral displacement of the soil, pile head displacement and bending moment of piles, strain gauges, and LVDT (linear variable differential transformers) were fixed in the piles and abutment walls as shown in 'figure 1'.

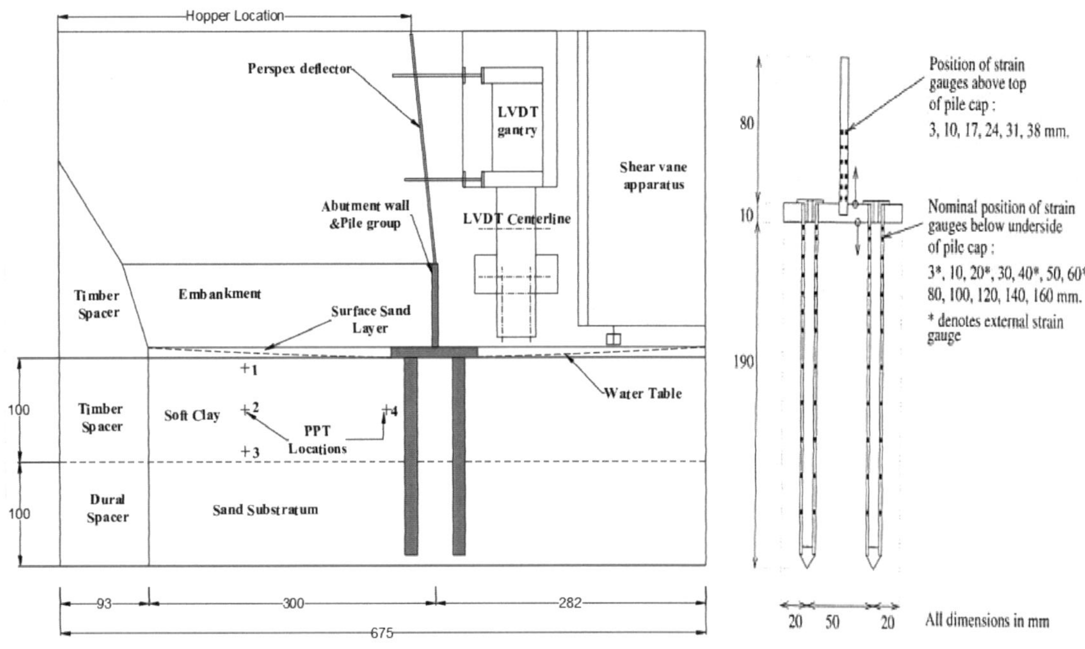

Figure 1. The general arrangement of the centrifuge model test.

3. Background analysis of the case study

3.1. Simulation of the centrifuge test in Finite Element Model

3.1.1 Geometry and Properties of subsoils

The centrifuge tests have been modeled using a finite element model (Plaxis 3D, V2020) to compare their results with the centrifuge test results. The geometry of the finite element model was 19.8m, and 50m in x and y directions respectively, and the Z-direction is defined according to embankment height (upper vertical boundary) and the bottom of the sand substratum (lower vertical boundary). Flow boundary conditions were required for consolidation analysis so the hydraulic boundary condition in vertical and horizontal boundaries are free to permit the groundwater dissipation under embankment load.

Hardening soil model was used for fill, filter sand, and the stiff sand as a constitutive model for more accurate results in deformation and displacement while the soft soil with creep model is applied for soft clay soil. For drained conditions, drained behavior was assigned for all sand layers due to the high permeability of sand material relative to the clay layer that was modeled as undrained material considering both consolidation and creep parameters. Table 1 shows the parameters of the model that is used in the numerical model.

Table 1. Soil parameters adopted in the F.E. model.

Type of soil	Clay	Sand filter, Sand stratum	Fill
Depth	*0-10m*	*10m-20m*	*Height =8m*
Soil model	*SSC*	*HS*	*HS*
γ *(kN/m³)*	*16.6*	*19.5*	*17.5*
K_x=K_y *(m/days)*	$2.298\text{x}10^{-4}$	*0.86*	*0.86*
K_z *(m/days)*	$1.149\text{x}10^{-4}$	*0.86*	*0.86*
C_c	*0.43*	-	-
C_s	*0.07*	-	-
C_α	*0.006*	-	-
C' (KPa)	*1*	*1*	*1*
φ*' (angle)*	*23*	*35*	*35*
Ψ *(angle)*	*0*	*5*	*5*
υ *(Poisson ratio)*	*0.15*	*0.2*	*0.3*
E_{50}^{ref} (KN/m^2)	-	$25\text{x}10^3$	$11\text{x}10^3$
E_{oed}^{ref} (KN/m^2)	-	$25\text{x}10^3$	$11\text{x}10^3$
E_{ur}^{ref} (KN/m^2)		$75\text{x}10^3$	$33\text{x}10^3$
OCR	*1*	-	-
POP	*20*	-	-

Where c_c: compression index, c_s: swelling index, c_α: creep index, OCR: over consolidation ratio, POP: overburden pressure, $E_{50}{}^{ref}$: 50% triaxial modulus of deformation at reference normal stress of 100 kPa,

E_{ur}^{ref}: Unloading–reloading modulus of deformation, E_{oed}^{ref}: modulus of deformation obtained from oedometer test at a reference pressure of 100 kPa.

3.1.2. Simulation of Structural Elements

To get a good comparison between the centrifuge test results and the calculated results from the finite-element method, the volume pile element was used. The properties of the volume pile which the piles were assigned with a material data set representing the concrete properties like unit weight and young's modulus according to the material used in the centrifuge test. Therefore, the unit weight of piles and its young's modulus is taken 25 kN/m^3 and 70x10^6 KN/m^2, respectively. Linear elastic-non porous material was chosen as a constitutive model for all structure materials. The property of the interface around the volume pile is related to the surrounding soil properties by the strength reduction factor R_{int} = 1.0. Figure 2 demonstrates the soil and pile models.

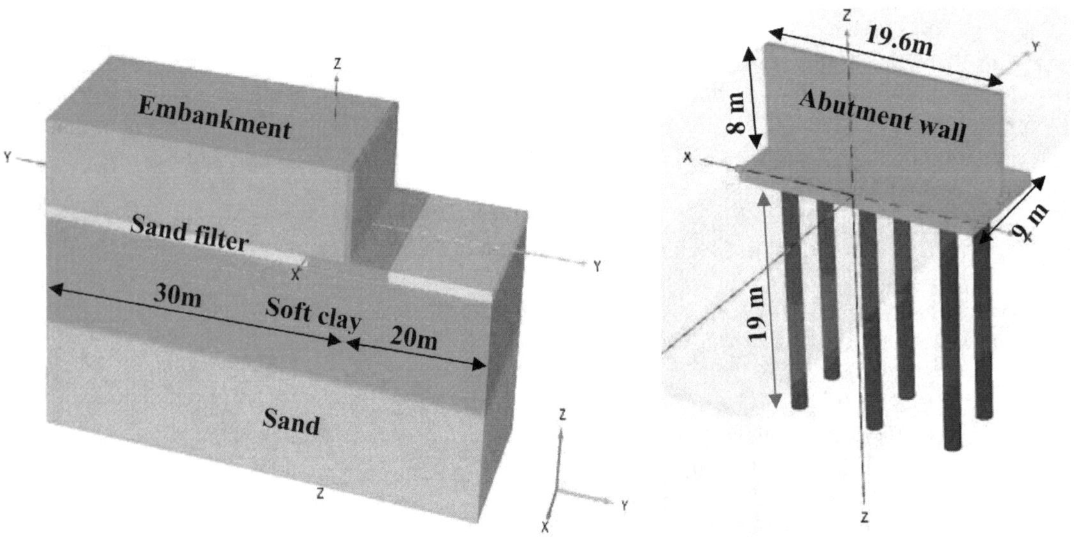

Figure 2. Simulation of structural elements as a volume.

3.1.3. Initial stage and stage of construction

The calculation stage has five stages, first stage was the initial stress calculation. In the initial stress, the K_o (earth pressure at rest) is 0.69 for clay and 0.42 for sand. The at-rest earth pressure coefficient "k_o" depends on the value of the over-consolidation ratio which is equal to 1. The second phase included the simulation of piles with a time interval of 1 day. The third phase involved a simulation of the pile cap and abutment wall and then placing the sand filter above clay soil (to accelerate the dissipation of water from the clay layer). Four stages including the construction of the embankment. The final phase is the consolidation analysis.

4. Discussion and Analysis

4.1. Lateral displacement of Rear Pile

The back analysis results and centrifuge results are represented in 'figure 3'. The maximum lateral pile displacement at the pile head is 17 cm at 21 days (short term) and for 1000 days is 25cm in centrifuge test. For numerical analysis results (FEM), the maximum pile displacement is 15cm and 21 cm for 21 days and 1000 days, respectively. Accordingly, lateral displacements in measured results are higher than the corresponding calculated values by about 11.7% at 21 days (short term) and 16% at 1000 days (long term). Hence, good agreement between calculated and measured performance can be recognized.

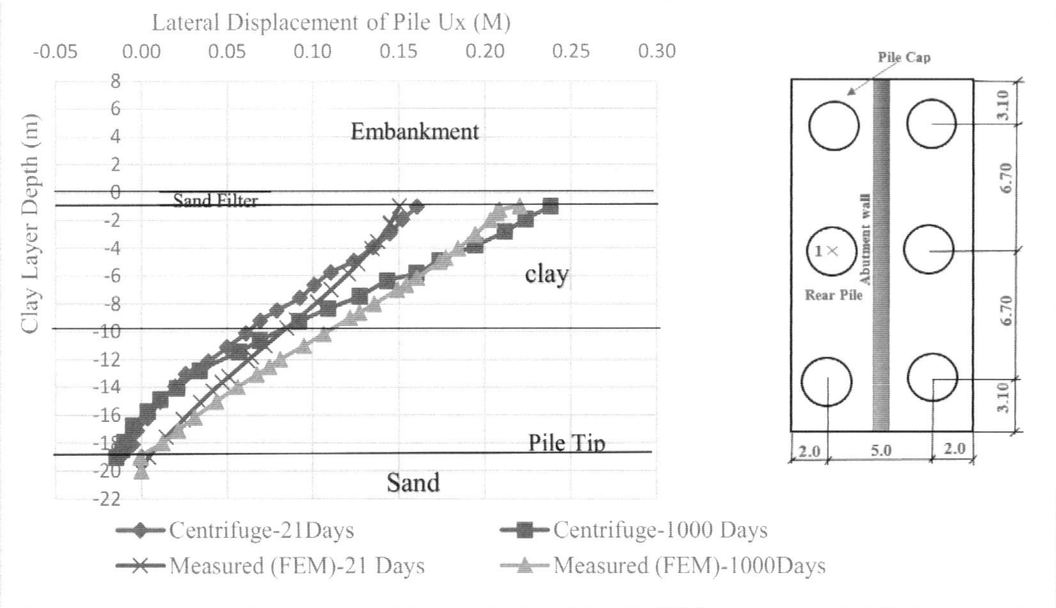

Figure 3. Comparison between Measured and Calculated Results for Lateral Pile Displacement at 21 days (construction time) and 1000 days (Consolidation time).

4.2. Bending Moment of Rear and Front Piles

For rear piles (piles near the embankment), the bending moment was measured along with the depth of the pile with the maximum bending moment at the pile head and then decrease in value to be zero at the pile tip (stiff substratum). Comparing the bending moment between measured and calculated results, it can be recognized that for the centrifuge test, the maximum bending moment at 21 days is 8.8 KN.m and 12.76 KN.m at 1000 days (consolidation) and it is lower than the value of bending moment of numerical analysis (FEM) with a difference of 18.6 % and 15.5 % for 21 days (construction) and 1000 days (consolidation), respectively.

For front Piles (piles on the other side of the embankment), it is found that the same performance and trend of bending moment in centrifuge test and numerical results. The maximum value of bending moment at the pile head in numerical analysis was 8 KN.m at 21 days and 11.45 KN.m at 1000 days while the value of maximum bending moment in centrifuge test results is 5kN.m (at 21 days) and 8 KN.m at (1000 days (consolidation). Relating the centrifuge test results with numerical analysis, the difference in construction time (21 days) is -37.5% and -30.13% at 1000 days (consolidation). According to these results, the value of bending moment for the rear pile is higher than that of the front pile, the rear piles are closer to the embankment and are directly affected by lateral loads generated from soil movement around the pile. This is a different phenomenon compared with pile group under external applied horizontal loads where front piles take higher loads than rear piles. 'figures 4 and 5' show the comparison between numerical analysis and centrifuge test results in bending moment of the rear pile and front pile with pile depth.

The difference between the measured and calculated bending moments is higher than that of horizontal displacements. This is due to either the connection between the pile heads and the plate simulated the foundation in the lab model or due to the accuracy of back estimating the bending moments from strain gauge measurements.

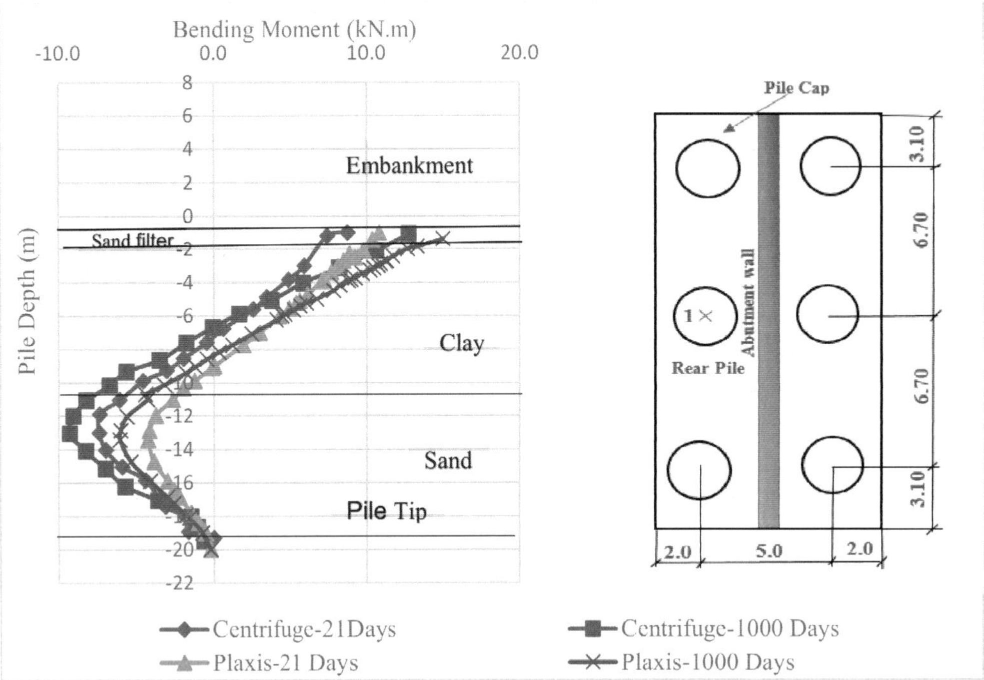

Figure 4. Comparison of Bending Moment of Rear Pile with Depth for Numerical Analysis (3D-FEM) and Centrifuge Test Results.

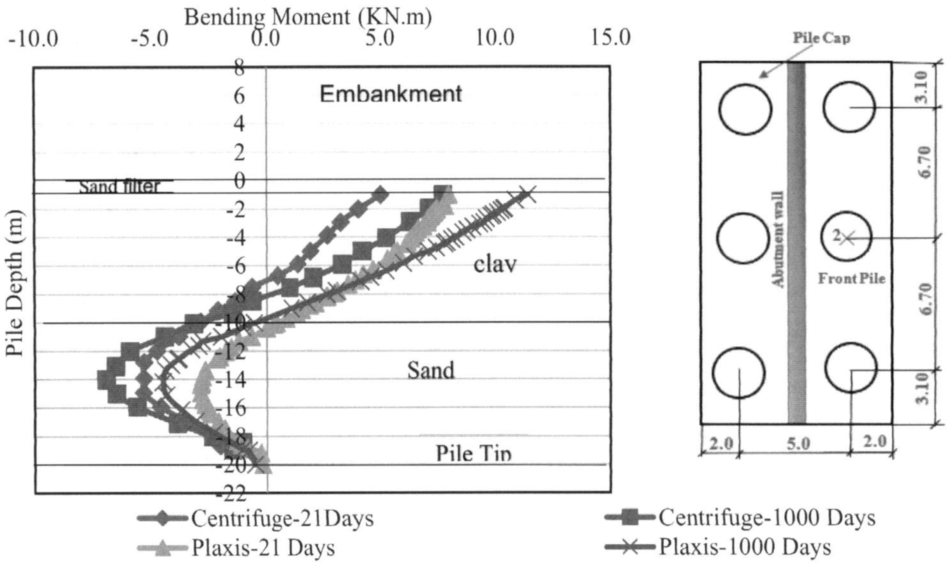

Figure 5. Comparison Of Bending Moment of front Pile with Depth for Numerical Analysis (3D-FEM) and Centrifuge Test Results.

4.3 Lateral Soil Displacement

'Figure 6' shows the variation of the measured and calculated results for lateral soil movement in the clay layer. According to this figure, it is evident that the maximum value of lateral soil displacement "U_x "is recorded at a depth of 2.5m measured from the top of the soft clay layer in numerical analysis. The measured results show the maximum value of U_x of soil at the ground surface of the clay layer. The trend of lateral soil displacement with depth in measured and calculated results was different. The discrepancy between the measured and calculated horizontal soil displacement is most probably due to the anisotropy of the soft clay fabric developed during the building of the lab model. Nevertheless, comparing the maximum U_x of the soil at a depth of 2.5m of clay layer, it can be detected that the calculated values of horizontal displacement from the numerical analysis at 21 days and 1000 days are 19 cm and 33cm respectively were for the corresponding measured values are20 cm and 38 cm. The difference between measured and calculated results was 5% (21 days) and 13.15% (1000 days). Therefore, neglecting the upper 2.0m, there is a generally good agreement between measured and back-calculated horizontal displacements.

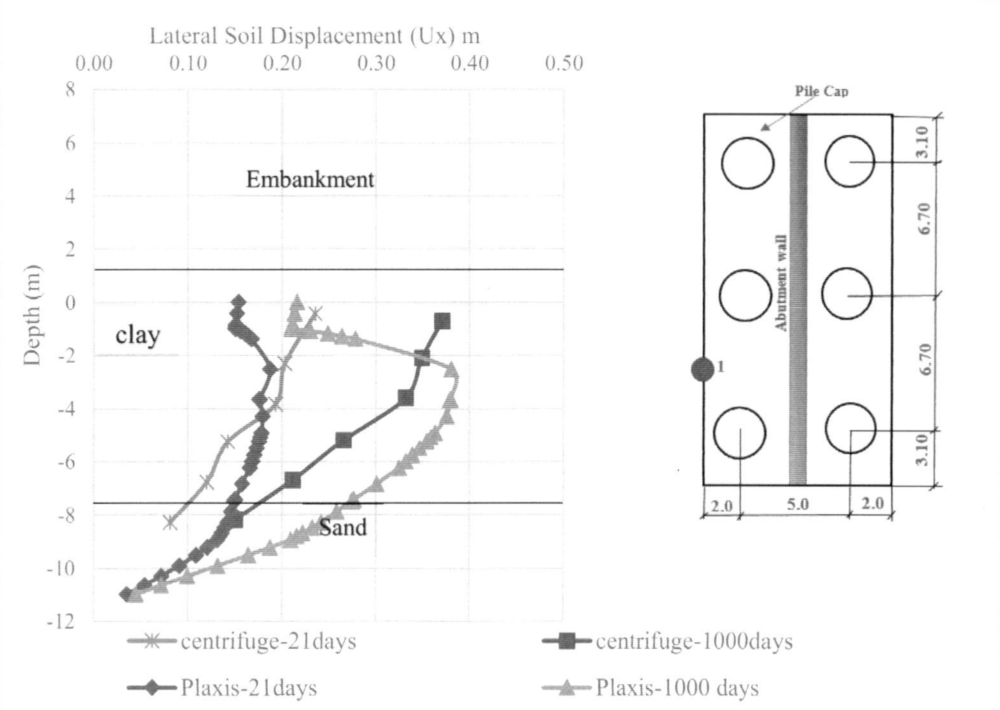

Figure 6. Comparison of Lateral Soil Movement between Measured and Calculated Results between Two Rear Piles.

5. Conclusions

In this research, the three-dimensional finite element program is adopted to simulate the interaction between the piles supporting the bridge's abutment and the approach embankment in soft soil formations, where the performance of piles subjected to soil movement due to the embankment load is the main concern. In the back analysis, the soft soil creep model is used for the clay soil while the hardening soil model is used for the embankment fill. The main conclusions can be summarized as follows:

1. The three-dimensional finite element model can simulate the performance of piles that are subjected to passive lateral loads and study the lateral soil movement in short-term as well as long-term conditions.
2. There is a good match between the measured and calculated lateral pile displacement and bending moment for the piles during construction and consolidation time.

3. The discrepancy between the measured and calculated horizontal soil displacement is most probably due to the anisotropy of the soft clay fabric developed during the building of the lab mode.
4. The performance of front and rear piles in the case of passive lateral loads is completely different from the performance of the pile group under external horizontal loads.

References

[1] Appel G C, Desai C S,1976, "3-d Analysis of Laterally Loaded Structures". In Proceedings of the *II International Conference on Numerical Methods in Geomechanics,* Blacksburg, Virginia.
[2] Brown D A, and Shie C F,1990. "Some numerical experiments with three-dimensional finite element model of a laterally loaded pile", *Computers and Geotechnics*, Vol. 12, pp. 149-162.
[3] Carter J P 1982, A numerical method for pile deformations due to nearby surface loadings, In proc. 4th int. *conf. on numerical methods in geomechanics*, Edmonton, **Vol. 2**, pp.811–817.
[4] Ellis E A 1997, "Soil-Structure Interaction for Full-Height Piled Bridge Abutments Constructed on Soft Clay." Unpublished doctoral dissertation, University of Cambridge, London.
[5] Ellis E A and Springman S M 2001, "Modeling of soil-structure interaction for a piled bridge abutment in plain strain FEM analyses", *Computers and Geotechnics*, **Vol. 28**, No. 2, pp.79-98.
[6] Ellis E A and Springman S M 2001, "Full-height bridge abutments constructed on soft clay," Geotechnique, p. 51:3–14.
[7] Kahyaoglu M R, Imancli G, Ozturk A U, and Kayalar A S, 2009. "Computational 3D finite element analyses of model passive piles." *Computational Materials Science*, **Vol. 46**, No. 1, pp. 193-202.
[8] Kahyaoglu M R, Imancli G, Ozturk A U, and Kayalar A S, 2011, "Numerical Analyses of Piles Subjected to Lateral Soil Movement." *KSCE Journal of Civil Engineering* 16(4):562-570.
[9] Palmeira E M, Fahel A R S, and Araújo G L S, 2013 "Behavior of geogrid reinforced abutments on soft soil," *geotechnical Engineering Journal of the SEAGS& AGSSEA*, **Vol.44**, No 4 Dec.
[10] Saride S, Puppala A J and Archeewa E 2009, "Bridge Approach Settlements-An Issue Due to the design of construction practices,", IGC, Guntur, INDIA.

Behavior of a vertical pile in loose sand under the effect of an inclined load

F M Abdrabbo[1], K E Gaaver[2] and M M Eldooma[3, *]

[1] Faculty of Engineering, Alexandria University, Egypt.
[2] Civil Engineering Department, Faculty of Engineering Scince, Nyala University, Sudan.

[*]Corresponding Author Email: musabeldooma@gmail.com

Abstract. The paper investigates the behavior of single vertical steel pipe pile embedded in sand subjected to compressive inclined load through MIDAS GTS/NX 2019 FEM package. The effect of the loading inclination angle and pile length and diameter on the performance of the pile were investigated. Inclined load carrying capacity and pile stiffness were presented. The pile head displacement and pile stiffness are significantly affected by load inclination angle. It was observed that, p-y curves of the pile-soil system are independent of loading inclination angle. In addition, the pile length and diameter are markedly affecting the behavior of the pile.

Key Words: Piles, Inclined load, Pile stiffness, Displacement, Ultimate load.

1. Introduction

Many structures supported on piles are subjected to an inclined compressive loads such transmission towers, wind energy converter towers, and offshore structures. The complex cases of the applied loading condition in such structures pile foundation should satisfy the safety and serviceability conditions as well. Normally the axial loading of the pile is predominant. But in special cases like piles supporting offshore wind energy converter towers, the vertical load is minor compared by lateral load [1]. The study of vertical pile subjected to inclined loading was investigated during 70s of the last century on the basis of experimental results from loading tests performed on small-scale piles, Meyerhof and his co-researches [2–4]. Theoretical analyses for assessing the ultimate bearing capacity of a vertical pile subjected to an inclined load were carried out by Meyerhof and Ranjan [5].

In case of inclined loaded pile, current design practice involves independent analysis of vertical and lateral pile capacities and does not consider the interactions of vertical and horizontal load components. Therefore, to predict the load-displacement response of a vertical pile under inclined load, it is assumed that the lateral displacement of the pile head is independent of the vertical component of the inclined load. Similarly, while estimating the ultimate lateral resistance, it is considered that the vertical load component of the inclined load does not influence the ultimate lateral resistance of the pile [6]. The inclination angle (α) markedly influence the bearing capacity of the pile, in a way that the horizontal load improves the axial behavior of the pile [7], this result is in contrast with Meyerhof and his co-workers [3,8]. Experimental study was carried out to investigate the capacity of an inclined vertically loaded pile near ground slope. The authors conducted small scale tests on steel pipe piles, the study revealed that both ultimate vertical and lateral loads were decreased with the increase of inclination angle with vertical. And the pile length and the inclination angle have significant effects on the pile ultimate load [9]. Furthermore, the increases of inclination angle the pile lateral deflection along the pile increased. The stiffness and bearing capacity of the pile under inclined load increase with the increase of the load inclination angle measured from the horizontal [10]. For a preliminary assessment of the bearing capacity of piles embedded in cohesionless and cohesive soils under inclined loads, a simple-to-use approach is proposed [11].

Limited experimental works on the vertical pile under inclined loading yielded a semi-empirical relationship relating the components of the applied inclined load to the ultimate vertical and ultimate horizontal loads of the pile as follows [12]:

$$\left(\frac{V}{V_u}\right)^2 + \left(\frac{H}{H_u}\right)^2 = 1 \tag{1}$$

Where, V and H are vertical and horizontal load components, and V_u and H_u are the ultimate pile capacities under vertical loading and horizontal loading, respectively.

With the advance of numerical modeling finite element method was utilized to simulate vertical pile subjected to inclined load. A more comprehensive study on the behavior of vertical hollow steel pipe pile subjected to inclined loads was preformed and presented. The effect of Load inclination angle (α), and pile length (L) and diameter (D) have been investigated and discussed.

2. 3D Numerical modelling

An extensive numerical investigation using finite element software MIDAS GTS/NX, 2019 has been performed. The study aimed to a better understanding of the behaviour of vertical pile under the effect of an inclined load (R). Steel pipe piles of different lengths (L) and different diameters (D) with wall thickness (t) of 50.0 mm embedded in loose sand have been analysed. The pile was modelled as linear, elastic, and non-porous material with a unit weight of $\gamma = 78$ KN/m^3, modulus of elasticity E = 210 GPa, and Poisson's ratio ν=0.2. While the soil has been modelled as loose sand simulated under undrained condition by using Modified Mohr-Coulomb constitutive model (MMC) built in software, Table 1. A typical mesh discretization used in the study is shown in the Figure 1. The boundaries of the model were simulated using the automatic constraint built in the software in which the vertical boundary was constraint against any lateral movement, only vertical movement was allowed. The bottom boundary was constrained against the movement in all directions.

The vertical boundaries were located at a distance equal to 32 times the pile diameter from center of the pile. While the bottom boundary was located at about 16 times the pile diameter measured from the ground surface which is corresponding to 2.6L, as illustrated in Figure 1.

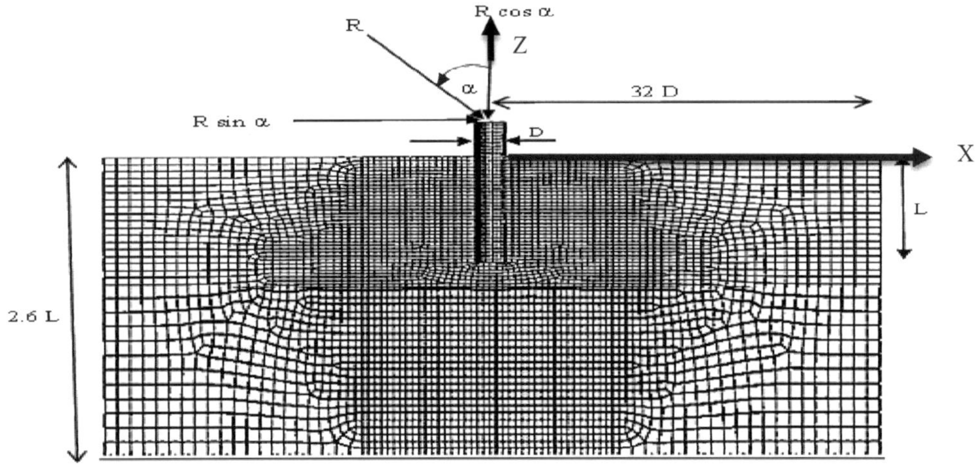

Figure 1. Model Mesh and sign conventions.

The mesh around the pile was refined to avoid stress concentration zones effects. The interface between the pile and the soil was modelled using the strength reduction factor built in software. The recommended value of Strength Reduction Factor SRF was 0.6.

Piles of diameters 3.0, 5.0 and 7.0 m and of lengths 20.0, 30.0 and 40.0 m were analysed. An extensive numerical investigation using finite element software MIDAS GTS/NX, 2019 has been performed. The study aimed to a better understanding of the behaviour of vertical pile under the effect of an inclined load (R). The parameters used in this simulation are shown in Table 1.

3. Simulation Process

The initial step in simulation was to simulate the soil body force using K_0-state, K_0 was taken as equal to (1- sinΦ). Followed by construction stage that the steel pile was activated as well as the interface between the pile and the soil, then the displacements which were resulted from these two steps were cleared to zero. In loading stage, each inclination loading angle (α) has a corresponding loading

Table 1. undrained soil Parameter after [14].

Parameter	Value	unit
Unit weight (ɤ)	16.5	KN/m3
Young's modulus (E_{oed}^{ref}) (Tangential Stiffness in oedometer test loading)	25000	KN/m2
Secant stiffness in standard drained triaxial test (E_{50}^{ref})	25000	KN/m2
Elastic modulus at unloading (E_{ur}^{ref})	75000	KN/m2
Poisson's ratio (μ)	0.3	-
Undrained Cohesion (Cu)	0.1	KN/m2
Friction angle (ϕ)	27	[°]
dilatancy angle (ψ)	0	[°]
Lateral earth pressure coefficient (K0)	0.54	-

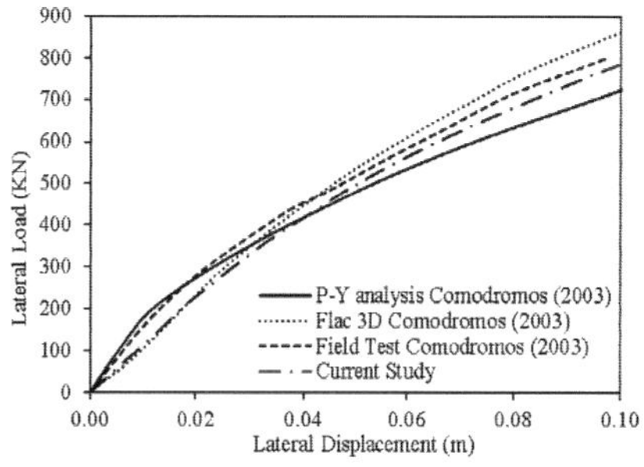

Figure 2. Validation of the current model with Comodromos [13].

situation consist of a couple of lateral (H) and vertical (V) load components which are applied simultaneously to provide a resultant loading force (R) with a constant inclination angle (α) with the vertical. Each loading stage consists of 100 incremental loading steps for each α. The loads were applied at different α, by using load control condition.

4. Sensitivity Analysis

To reduce computational effort without a lack of accuracy, preliminary analyses were carried out to examine the mesh fineness and the geometry of the soil domain. Figure 3. presents the effect of the lateral boundary from the pile center line (X) and the depth of the soil domain (Y) and the mesh fineness around the pile of a pile loaded at α=45°. The figure reveals that a domain of (X=32D), (Y=2.6L) and medium fine mesh is suitable to avoid boundary effects and save computational time.

5. Model Validation

To ensure reasonable precision of the developed numerical model, the obtained results were compared by filed, laboratory and numerical results documented in the literature, [13]. The field load test conducted by Comodromos [13] was simulated using MIDAS GTS/NX 2019. The field tests were conducted on a large diameter concrete pile of 1.0 m diameter and 52 m length embedded in stratified

Table 2. properties of soil and pile after [13].

Thickness of the layer (m)	Material type	Constitute model	Density (KN/m³)	Elastic modulus (E) (MPa)	Poisson ratio (ν)	Friction angle (deg)	Dilation angle (deg)	Cohesion (kPa)
0-36	Soft clay	Mohr Coulomb	20	2.7	0.45	0	0	27
36-48	Medium-stiff clay	Mohr Coulomb	20	11.0	0.45	0	0	110
48-70	V. Dense sand	Mohr Coulomb	22	100.0	0.3	40	10	0
0-52	pile	Elastic	25	25000	0.2	--	--	--

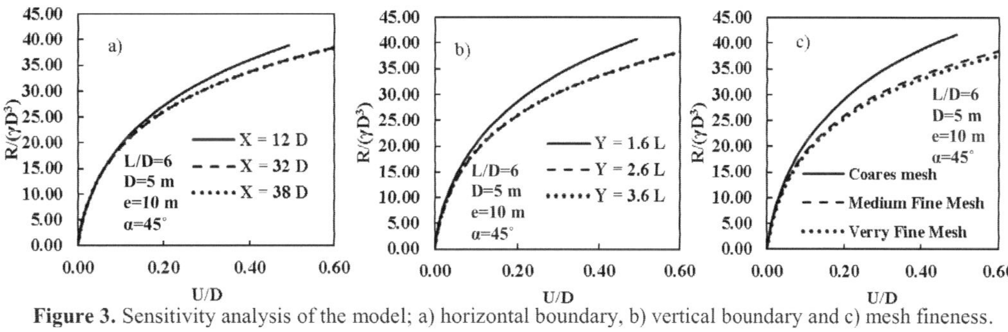

Figure 3. Sensitivity analysis of the model; a) horizontal boundary, b) vertical boundary and c) mesh fineness.

soil layers shown in Table 2. The ground water table was located at the ground surface. The pile was simulated as an elastic material, its properties shown in Table 2. While the interface between the pile and the soil was simulated using the interface friction coefficient ($\mu = 0.67$ Tan(φ)), where; φ represent the friction angle of the soil. The horizontal boundary was taken as 32 times the pile diameter in which the lateral movement was constrained. Whereas the vertical boundary was set at the bottom of the last soil layer depth that is 70.0 m from the ground surface by constrain the movement at all direction. A lateral load increments were applied to the pile head until the lateral displacement reached 10% of the pile diameter at which the maximum load capacity of the pile was considered. The obtained results from current study were compared by field observations and by analytical results caried by [13] using p-y curves and FLAC 3D software. The lateral load-displacement relationship of the current study was compared with the field test results, analytical and numerical analysis results Figure (2). The Figure revealed that the p-y analysis overestimates the lateral displacement of the pile at high load level, while the numerical analysis carried by Comodromos [13] using FLAC 3D software slightly underestimate the lateral displacement. While the achieved results agree reasonably with the filed test result.

6. Parametric Study

An extensive 3D numerical analysis simulating the pile soil system has been made using MIDAS GTS/NX, 2019 Package. The piles extend above ground surface a distance (e) equal to 10.0 m and embedded in sandy soil to a depth (L). The total length of the pile is (L+e). Therefore, the pile at the ground surface is subjected to H, V and M, where H is the lateral force component, V is the vertical force component and M is the moment. The pile was subjected to loading control conditions.

7. Results

7.1. *Pile Load-Displacement relationship*

An inclined load acting on the pile (R) produces horizontal displacement (U_h) and vertical displacement (U_v) at the pile head and rotation (θ). The resultant displacement is (U). Figure 4. indicates that the initial stiffness of the pile (K_g) and the ultimate load (R_u), at displacement (U) equal to 10 % of the pile diameter, increased with the decrease of the inclination load angle (α). This is attributed to the increase of vertical component of the inclined load with the decrease of α. Figure 4. and (4) revealed that for α equal to 45° and more, the displacement vector (U) is resulted mainly from horizontal displacement (U_h).

Comparison of Figure 5. with Figure 6., the lateral displacement is the control of failure load of the pile, since a lateral displacement of 10% (D) is achieved at smaller load compared by vertical displacement, except at inclination angle equal to zero with vertical.

Investigation of Figure 4. through Figure 7. revealed that at small, inclined load ($R/\gamma D^3$) equal to and less than 20.0, the vertical component of the load (R) is small, whatever the inclination angle (α), to mobilized appreciable vertical displacement component, the vertical displacement component is less than 2% of the pile diameter. While the lateral displacement varies from 2% to more than 10% of the pile diameter depending upon the load inclination angle (α). The corresponding rotation increased from 0.3° to more than 1.0° in accordance with α.

In case of load inclination angle (α) equal to and bigger than 15°, the failure of the pile is due to excessive lateral displacement and rotation. Figure 7. indicates to that the failure load (R_u) and the serviceability load (R_s) are highly dependent on the load inclination angle (α). The serviceability load (R_s), at rotation angle 0.2 produces negligible vertical displacement at pile head, but with horizontal displacement (U_h/D) varies from 0.15% D to about 10% according to load inclination angle (α).

Figure 8. indicates that the pile ultimate load increases with the increase of the embedment length (L) as well as the initial tangent stiffness, but it seems that the initial tangent stiffness of a pile having length bigger than 30.0 m is not appreciably affected by pile embedment length (L).

The increase of pile ultimate load is attributed to the increase of the mobilized friction stress along the pile shaft and the increase of bearing stress at the pile base. For long piles the pile vertical stiffness is mainly due to friction developed along the pile shaft and little contributed by bearing stress at pile base. Therefore, the increase of pile length from 30.0 m to 40.0 m contribute very little on pile

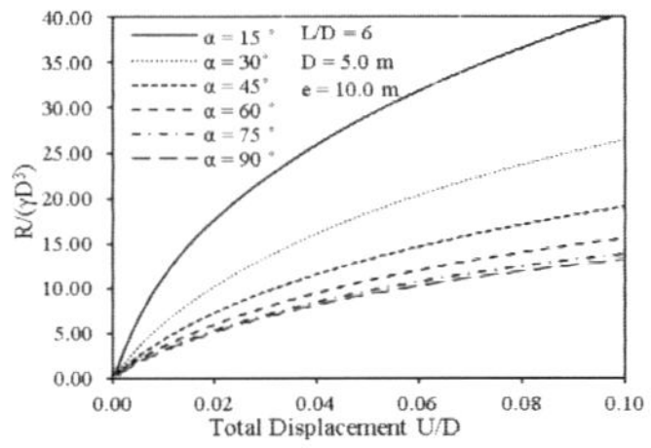

Figure 4. Total displacement-load curve for different values of α.

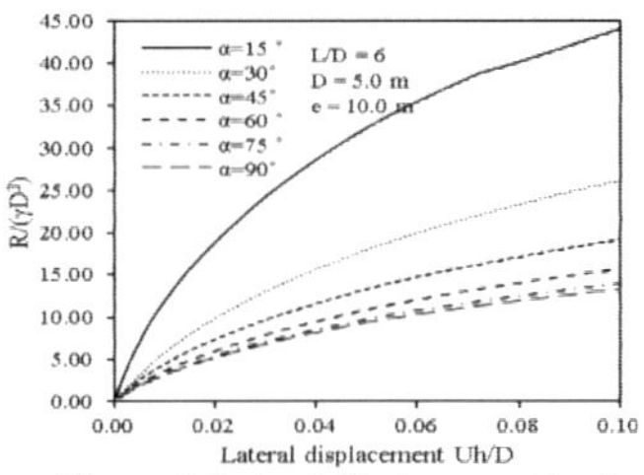

Figure 5. Lateral displacement-load curve for different values of α.

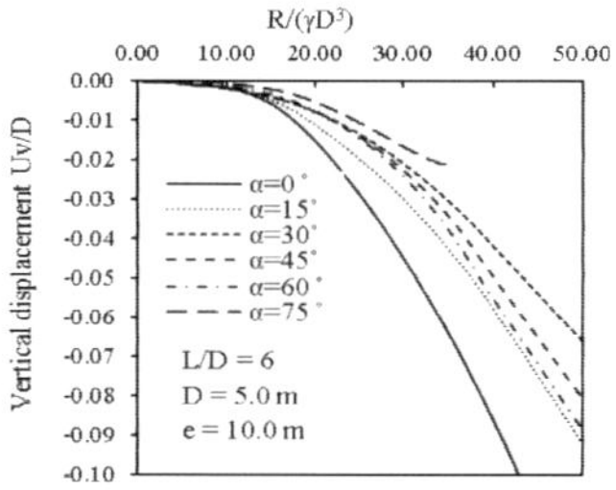

Figure 6. Vertical displacement-load curve for different values of α.

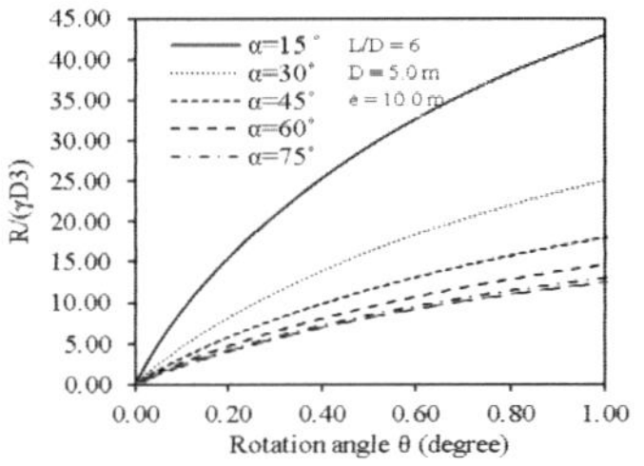

Figure 7. Pile head rotation angle-load relationship for different values of α.

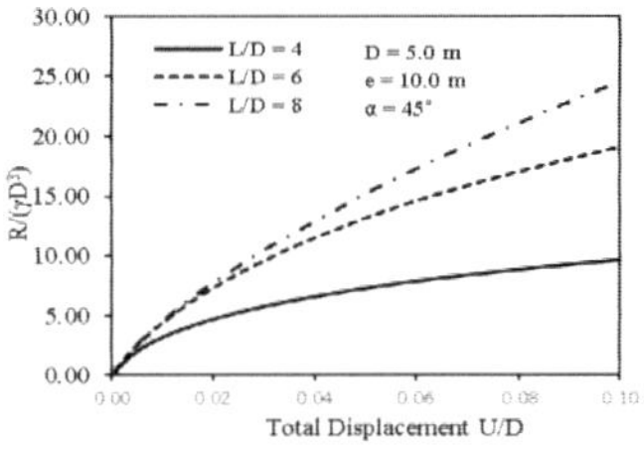

Figure 8. Total displacement-load curve for different values of α and different values of pile lengths.

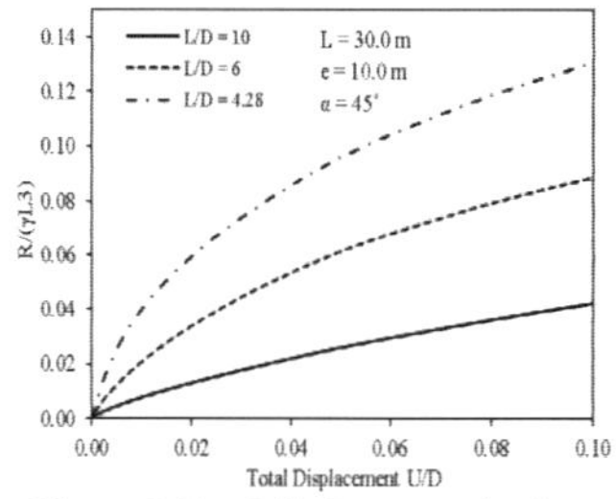

Figure 9. Total displacement-load curve for different values of pile diameters.

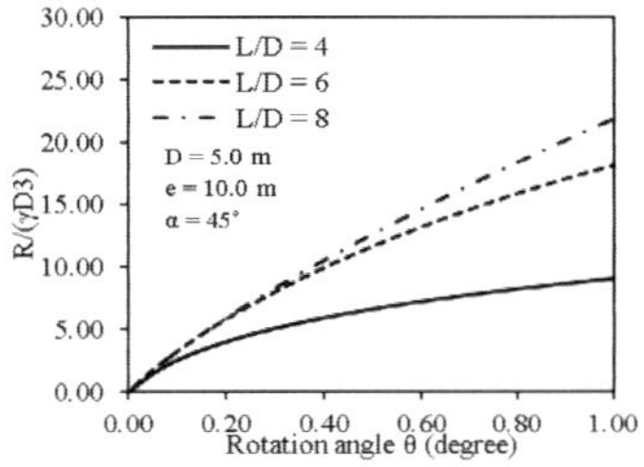

Figure 10. Pile head rotation angle-load relationship for different values of pile lengths.

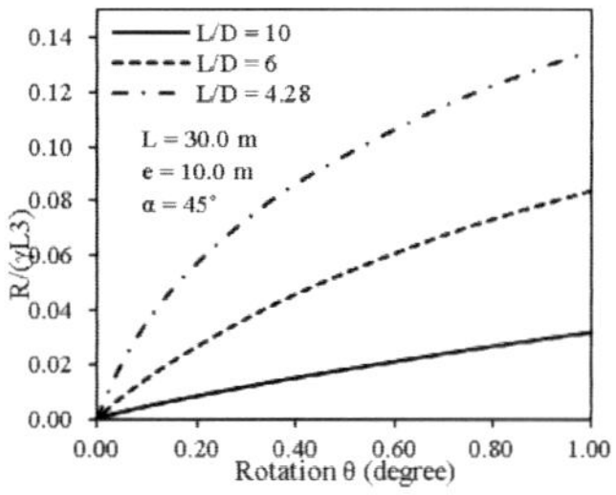

Figure 11. Pile head rotation angle-load relationship for different values of pile diameters.

vertical stiffness (K_v). At the mean time the increase of pile length beyond the pile critical length Lc has not appreciably affected pile lateral stiffness, (K_h).

Figure 9. presents the effect of the pile diameter on the response of pile. The initial tangent stiffness and the pile ultimate load increased with the increase of pile diameter (D).

Figure 10. and Figure 11. show the effect of pile length (L) and pile diameter (D) on pile head rotation (θ) of a pile loaded by inclined load at α equal to 45°. The lateral displacement of the pile decreases with the increase of pile diameter and with the increase of pile length, due to the increase of pile lateral stiffness, therefore the pile rotation decreased. At small inclined load, the rotation of the pile with length bigger than 30.0 m is not affected by the pile length (L). This can be attributed to that the pile critical length (L_c) is equal to about 30.0 m. the pile critical length (L_c) is not pile and soil parameters only but affected also by the load level.

Figure 12. indicates that the ratio (U_h/U_v) varies from about 5.0 to about 70.0 as the load inclination angle increased from 15° to 75°. The ratio slightly decreases with the increase of ($R/\gamma D^3$) and becomes independent of ($R/\gamma D^3$) when equal and greater than 20.0. It is worth to note that (U_h/U_v) is the inclination angle (β) of pile head displacement vector with vertical. The angle β is bigger than α.

Figure 13. presents the normalized failure loads versus the load inclination angles (α), different ultimate loads criteria were considered. The figure indicates that the total displacement (U), the horizontal displacement (U_h) and rotation angle (θ=1.0°) criteria predict almost the same ultimate load (Ru). While the vertical displacement (U_v) criterion predicts ultimate load (R_u) about three times that predicted by (U), (U_h) and (θ) criteria.

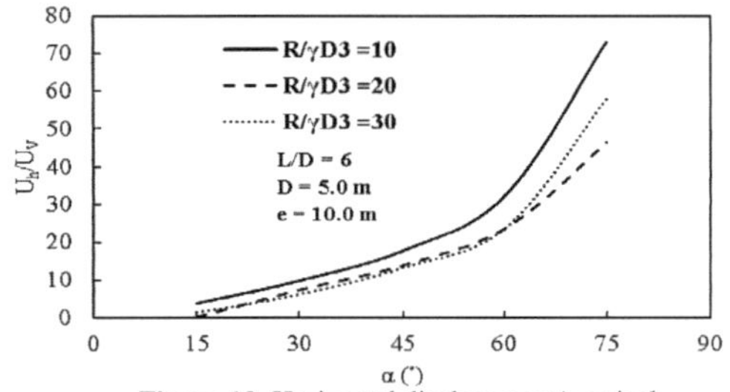

Figure 12. Horizontal displacement/ vertical displacement ratio at different α

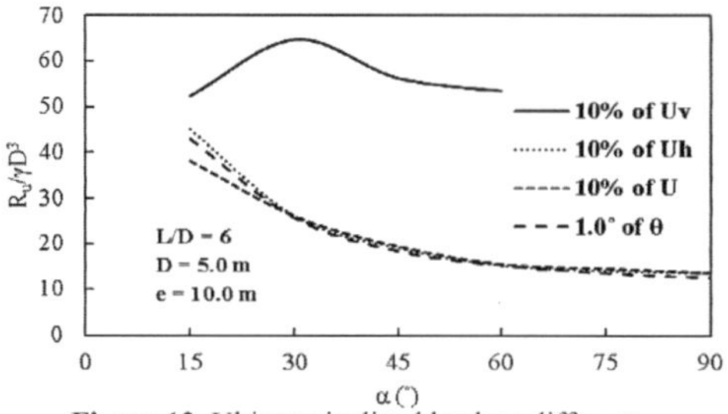

Figure 13. Ultimate inclined loads at different criteria.

7.2. Total stiffness

The initial tangent stiffness of the pile (K_g) was calculated as the slope of the tangent line of displacement-load relationship (R-U) for various values of inclination angles (α) as illustrated in Figure 14-a. as well as for various pile lengths (L) and pile diameters (D), Figure 14-b. and 14-c. respectively. The initial tangent stiffness was normalized as $R/(U\gamma D^3)$ in case of pile diameter (D) is constant and as $R/(U\gamma L^3)$ in case of constant pile length (L). It is noted that the initial tangent stiffness decreases with

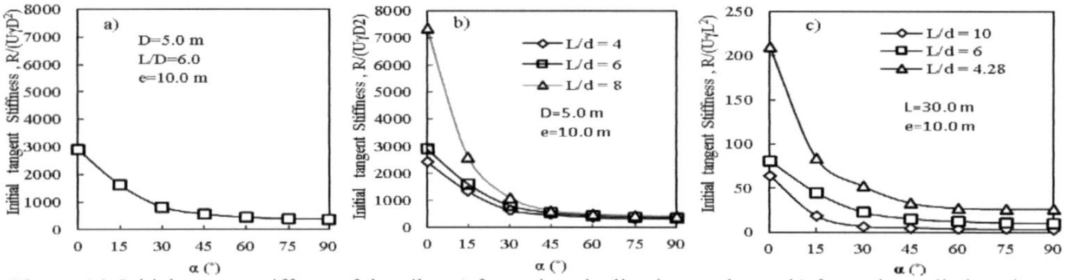

Figure 14. Initial tangent stiffness of the pile: a) for various inclination angles α., b) for various pile lengths and c) for various pile diameters.

the increase of α due to the increase of the horizontal component of the applied load (R) which tend to increase the failure potential of the pile laterally. While the increase of the pile length (L) or pile diameter (D) increases the stiffness of the pile-soil system.

Figure 14. indicates that the initial tangent stiffness (K_g) is not appreciably affected by α greater than 45°. This may be attributed to that, for α greater than 45°, the dominant contribution of the total stiffness is the lateral pile-soil stiffness. Also pile length bigger than 30.0 m has not appreciably affected the stiffness (K_g), while the initial tangent stiffness (K_g) increases with the increase of pile diameter. While, for α greater than 45°, the pile length and the pile diameter have not appreciable effect of the initial tangent stiffness (K_g) of pile-soil system.

7.3. Lateral stiffness

The initial tangent lateral stiffness of the pile-soil system (K_h) was calculated as the slope of the tangent line of (R_h-U_h) relationship and presented in Figure 15-a.

The initial tangent lateral stiffness (K_h) decreased with the increase of load inclination angle (α), decreasing of pile length(L), Figure 15-b. and decreasing of pile diameter (D), Figure 15-c. due to the decrease of lateral stress afforded by the soil. The initial total stiffness (K_g) is tenfold the lateral stiffness

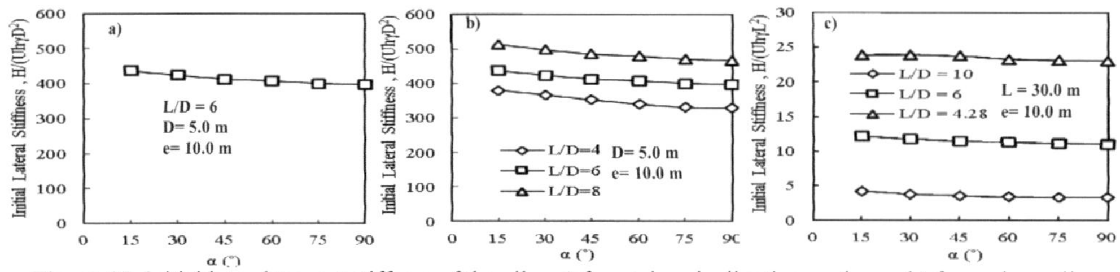

Figure 15. Initial lateral tangent stiffness of the pile: a) for various inclination angles α., b) for various pile lengths, and c) for various pile diameters.

(K_h). Therefore, the main contribution of the initial total tangent stiffness (K_g) of a pile subjected to inclined load is the pile vertical stiffness.

7.4. Rotation stiffnesses

The rotation stiffness of the pile was calculated from the moment resulted from the horizontal component of the applied load R at ground surface versus rotation angle of the pile head relationship. Figure 16-a. presents the initial tangent rotation stiffness of the pile normalized as ($M/(\theta\gamma D^4)$) or ($M/(\theta\gamma L^4)$) versus α.The initial tangent rotation stiffnesses of the pile (K_θ) decreases with the increase of α. It was noted that the trend of the rotation stiffness (K_θ) is similar to that trend of (K_h) with α. Initial tangent rotation stiffness increase with the increase of pile length and pile diameter Figure 16-b. and Figure 16-c.

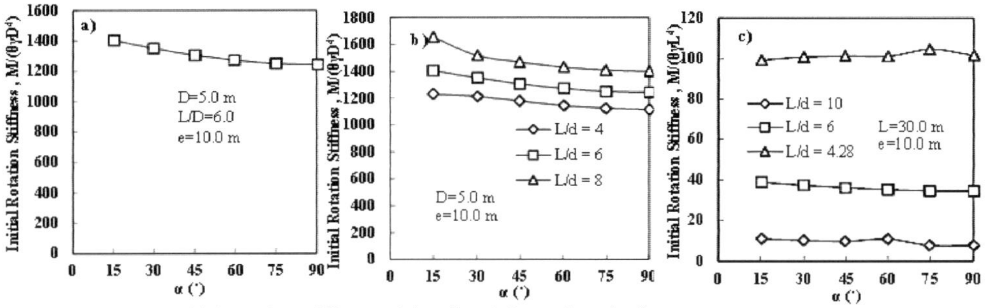

Figure 16. Initial rotation stiffness of the pile: a) for various inclination angles α., b) for various pile lengths, and c) for various pile diameters.

7.5. p-y curves

The following is an attempt to explore the effect of load inclination angle (α) on p-y curves of piles embedded in loose sand. The p-y curves at different pile depths resulted from inclined load R acting at various inclination angles (α) are presented in Figure 17. Where p is the horizontal stress at certain pile depth (z) (KN/m) and y is the corresponding horizontal displacement. The figures revealed that the p-y relationships are independent of α.

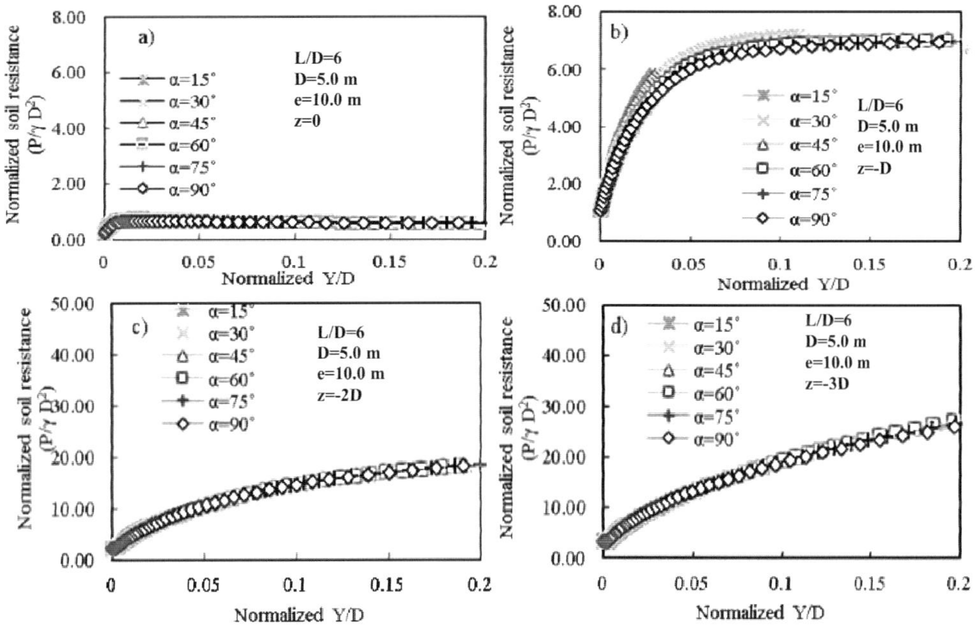

Figure 17. p-y relationships along the pile at various inclination angle (α).

8. Conclusion

The paper presents an attempt for understanding the response of vertical piles subjected to inclined load. The pile is embedded in loose sand. Piles with various pile lengths and pile diameters were analysed. The paper revealed the following conclusions:

1- The load inclination angle (α) has appreciable effect on pile displacement (U), in a way that U increases as α increased.

2- The initial tangent total stiffness (K_g), initial lateral stiffness (K_h), and initial rotation stiffness (K_θ) of the pile decreased with the increase of α, while (K_h and K_θ) increased with the increase of pile length (L) and pile diameter (D).

3- The initial total stiffness (K_g) of a pile subjected to inclined load embedded in loose sand is tenfold the initial lateral stiffness (K_h).

4- Piles subjected to inclination load of 45° and more attained their ultimate load (R_u) due to excessive lateral displacement (U_h). The ultimate inclined loads (R_u) predicted by lateral displacement (U_h) and total displacement (U) equal to 10% of the pile diameter (D) revealed almost the same values, that agree with ultimate load (R_u) at pile head rotation equal to 1.0°.

5- The horizontal and vertical displacements of inclined loaded pile revealed a displacement vector with inclination angle (β) with vertical bigger than the load inclination angle (α).

6- The p-y relationship of the pile at different relative depth z/L of the pile is independent of load inclination angle (α).

References

[1] Zhang L, Silva F and Grismala R 2005 Ultimate Lateral Resistance to Piles in Cohesionless Soils *J. Geotech. Geoenvironmental Eng.* **131** 78–83

[2] Chari T R and Meyerhof G G 1983 ULTIMATE CAPACITY OF RIGID SINGLE PILES UNDER INCLINED LOADS IN SAND. *Can. Geotech. J.* **20** 849–54

[3] Meyerhof G G and Sastry V V R N 1985 BEARING CAPACITY OF RIGID PILES UNDER ECCENTRIC AND INCLINED LOADS. *Can. Geotech. J.* **22** 267–76

[4] Meyerhof G G, Yalcin A S and Mathur S K 1983 Ultimate pile capacity for eccentric inclined load *J. Geotech. Eng.* **109** 408–23

[5] Meyerhof G G and Ranjan G 1972 The Bearing Capacity of Rigid Piles Under Inclined Loads in Sand. I: Vertical Piles *Can. Geotech. J.* **9** 430–46

[6] Roy S, Chattapadhyay B C and Sahu R B 2013 Pile behavior under inclined compressive loads-A model study *Electron. J. Geotech. Eng.* **18 K** 2187–205

[7] Conte E, Troncone A and Vena M 2015 Behaviour of flexible piles subjected to inclined loads *Comput. Geotech.* **69** 199–209

[8] Sastry V V R N, Meyerhof G G and Koumoto T 1986 BEHAVIOUR OF RIGID PILES IN LAYERED SOILS UNDER ECCENTRIC AND INCLINED LOADS. *Can. Geotech. J.* **23** 451–7

[9] Sakr M A and Nasr A M 2013 Behavior of vertical piles embedded in sand under inclined loads near ground slope *18th International Conference on Soil Mechanics and Geotechnical Engineering: Challenges and Innovations in Geotechnics, ICSMGE 2013* vol 4 (IOS Press) pp 2851–4

[10] Conte E, Troncone A and Vena M 2016 On the Pile Response to Inclined Loads *Procedia Eng.* **158** 206–11

[11] Conte E, Pugliese L, Troncone A and Vena M 2021 A Simple Approach for Evaluating the Bearing Capacity of Piles Subjected to Inclined Loads *Int. J. Geomech.* **21** 04021224

[12] Meyerhof G G and Ranjan G 1972 The Bearing Capacity of Rigid Piles Under Inclined Loads in Sand. I: Vertical Piles *Can. Geotech. J.* **9** 430–46

[13] Comodromos E 2003 Response prediction of horizontally loaded pile groups *Geotech. Eng.* **34** 123–33

[14] Eid M, Hefny A, Sorour T, Zaghloul Y and Ezzat M 2018 Numerical Analysis of Large Diameter Bored Pile Installed in Multi Layered Soil: A Case Study of Damietta Port New Grain Silos Project *Int. J. Curr. Eng. Technol.* **8.**

IOP Conf. Series: Earth and Environmental Science 1056 (2022) 012036 doi:10.1088/1755-1315/1056/1/012036

Experimental Investigation of the Axial Behaviour for Battened Tubular Columns

Essam A Amoush[1, *]

[1] Civil Engineering Department, Higher Technological institute, Tenth of Ramadan City, Cairo, Egypt

[*]Corresponding Author Email: essamamoush2008@yahoo.com

Abstract. This paper describes a series of experiments on batten tubular specimens with various slenderness ratios. The experimental study was employed to satisfy the expectations of behaviour and axial load capability of battened tubular columns. Twelve specimens composed of four tubes having different global column slenderness and various ratios of the unbraced chord slenderness-to-the global column slenderness were chosen to cover stub, short and medium columns. The tubes were assembled with welded batten plates. A numerical model with a thin shell element considered both geometric and material nonlinearities was used to verify the experimental results. Tests result showed different failure modes, such as local, local-flexural and flexural buckling. By comparing, the finite element and tested results have a good agreement. The results revealed the strength of the battened columns, which was related to the effect of global slenderness and spacers on the buckling behaviour. Additionally, the stub, short and medium column strengths were enhanced by 8%, 25%, and 78%; respectively, when the number of spacers increased to four batten plates. Eventually, predicted strengths of the battened column by AISC and EC3 specifications were slightly un-conservative.

Keywords: Battened columns, Tubular columns, Built-up columns, Axial Buckling, Experimental axial strength.

1. Introduction

The research on steel members has completely shifted in the recent decade, with a great reliance on built-up sections. When individual components do not meet the structure strength or stiffness criteria, built-up elements are often used. Such sections aid the economical use, which has limited reserves. The study of battened tubular columns is rarely reported. The use of batten tubular columns here is thought to be novel since they nearly have not been used previously for built-up batten columns.

This section presents a review of recent literature on built-up batten columns. Yanlin Guo and Jie Wang [1], investigated the shear stiffness and behavior of multi-tube latticed steel columns. El Aghoury et al. [2], evaluated experimentally the capacity of batten columns comprising four equally slender angles and explored the influence of local and global slenderness ratios on the capacities of those columns. Anbarasu et al. [3], Dabaon et al. [4], Roy et al. [5], and Zhang and Young [6], studied the built-up columns which addressed on the sections combining of back-to-back two channel sections. As a result, open built-up sections form, which is weaker in torsion if compared versus closed sections. Although the usage of closed built-up sections has been indicated better torsional resistance, there has been limited researches on this subject, such as Reyes and Guzman [7], Li et al. [8], Zhang and Young

[9], and Kherbouche and Megnounif [10] which has mainly focused on singly symmetric closed built-up sections. However, Vijayanand S. and Anbarasu M. [11], carried out a finite element analysis to satisfy the expectations of behavior and ultimate axial load capability of batten columns. Additionally, they found that the column strength was influenced by global column slenderness and plate slenderness.

Additional studies to understand more completely the key tenets of the behavior of the batten tubular columns are required. Therefore, this study is carried out. The study describes an experimental investigation implemented on pin-ended built-up sections comprising four-tube sections. A related finite element modeling was conducted by using ABAQUS software package [12]. The effect of interaction between the global column slenderness and the ratio of unbraced chord slenderness-to-the global column slenderness on batten columns behavior and capacities was discussed. Local, local-flexural and flexural buckling failure modes were the most observed modes. The experiments and the FE results were compared with the predicted strengths by American and European specifications.

2. Experimental study

2.1. Section Geometry

Twelve specimens of battened tubular columns were fabricated and experimentally tested under axial compression loads. The specimens composed of four tubes (each tube has D x t = 25 mm x 1.0 mm) and assembled by welded spacer plates (750 mm x 50 mm x 1.5 mm). were selected. Since, the specimens were classified into three groups 1, 2 & 3 related to their overall slenderness ratios. Furthermore, the ends of the specimens were assembled using welded box-shape with dimensions 100 mm x 75 mm x 1.5 mm. To ensure equal load distribution over the specimen cross-section, end thick plates with dimensions 100 mm x 100 mm x 20 mm were used. The dimensions of the specimens and the study parameters such as the global column slenderness and the ratio between the unbraced chord slenderness-to-the global column slenderness are shown in figure 1 and listed in Table 1. Eventually, the specimens labeled as follows: the number of intermediate batten plates, the global column slenderness, and the ratio between unbraced chord slenderness-to-the global column slenderness. For example, the label "B1-20-1.9" defines the following specimen: the first two symbols (B1) shows that the number of intermediate batten plate is 1, the "20" shows that the global column slenderness equals 20 and "1.9" shows that the ratio of unbraced chord slenderness-to-the global column slenderness ratio is 1.9.

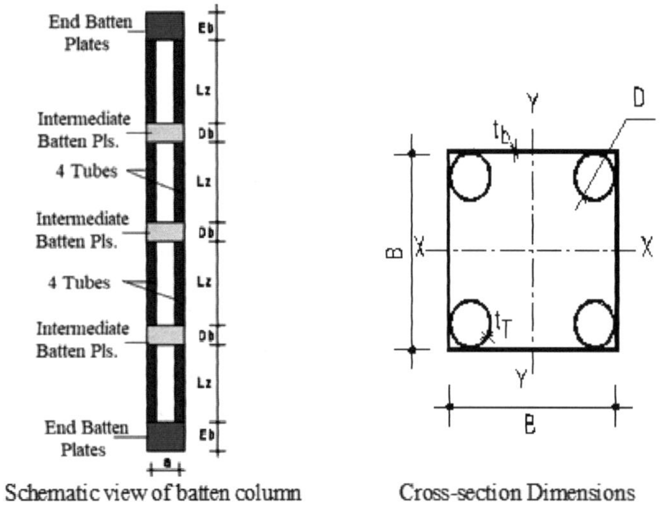

Figure 1. Typical details of the tested specimens.

Table 1. Specimens Dimensions.

	Specimens B x B 100 mm x 100 mm	L_c mm	L_z mm	Global column slenderness $\lambda_c=L_c/r$	Unbraced chord slenderness $\lambda_z=L_z/r_i$	Critical length @ $\lambda_c=\lambda_z$	Ratio λ_z/λ_c
Group 1	B0-20-3.5	750	600	20	70.65	170 mm	3.5
	B1-20-1.9		325		38.27		1.9
	B2-20-1.2		200		23.55		1.2
	B3-20-0.8		137.5		16.19		0.8
Group 2	B1-40-1.9	1500	650	40	76.54	340 mm	1.9
	B2-40-1.3		450		52.99		1.3
	B3-40-1		325		38.27		1
	B4-40-0.7		250		29.44		0.7
Group 3	B1-60-2	2250	1025	60	120.69	510 mm	2
	B2-60-1.4		700		82.42		1.4
	B3-60-1		512.5		60.35		1
	B4-60-0.8		400		47.10		0.8

2.2. Test setup and procedures

Figure 2 shows the test setup and instrumentation for pin ended battened specimens. The specimens were axially loaded with load control using a 300 kN hydraulic jack. To prevent the upper loaded end from the horizontal movements in x-direction and y-direction, an x-bracing attached with the reaction frame was used. The verticality of the specimen was achieved using a theodolite apparatus. Despite of all previous precaution, a small eccentricity (ranged from 1 mm to 2 mm) was occurred at the specimen ends, which was very difficult to avoid. The load- axial shortening and load-lateral displacement were recorded using a data logger system. A horizontal transducer (LVDT) was used at the mid-height of the specimen to record the lateral displacement. In addition, the axial-shortening was recorded by using a vertical (LVDT) at the upper end of the specimen. To make a full contact, an initial small axial load equals about (1/20) from the predicted ultimate load was applied, and then the load increased by regular load steps up to the failure load.

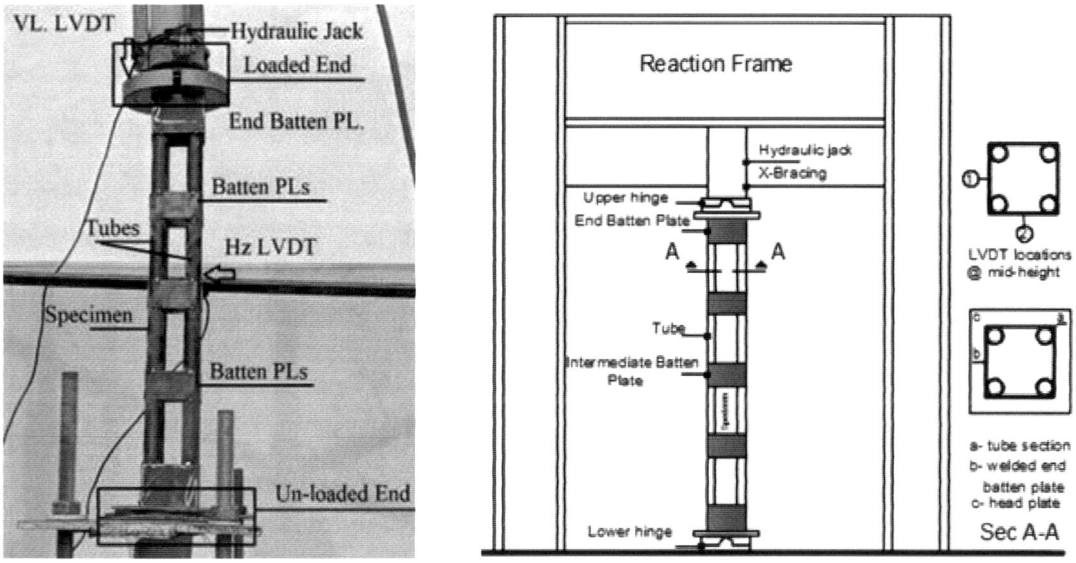

Figure 2. Test setup.

2.3. Material Properties

For small tube, ASTME8/E8M-13a [13] recommended that the specimen used in the tensile coupon test is a full-size tubular section. Therefore, four steel tubes specimens were prepared and tested to find out the uni-axial yield stress and ultimate tensile strength. The tests result revealed that the steel material properties were closely matched with the steel grade-37, where average yield and ultimate tensile stresses equal to 240 MPa and 370 MPa; respectively, as shown in Fig. 3. However, their nominal values of shear and Young's modulus, as well as Poisson's ratio, equal to 81 GPa, 210 GPa, and 0.30; respectively.

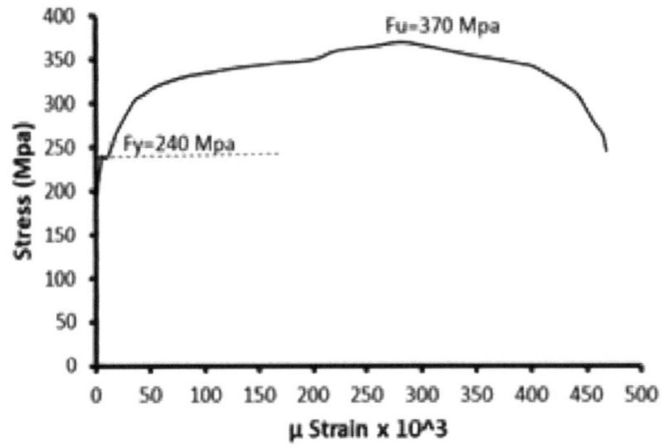

Figure 3. Stress–strain relationship.

3. Discussion of test results

3.1. Failure modes and deformed shapes

A series of local buckling waves were observed at buckling load with a half-wavelength nearly equal to the distance between the intermediate batten plates, while at failure load the failure mode was changed. In general, the buckling behavior of the batten tubular column is controlled by the column constraint and its effective length. First, for stub specimens (Group 1), it is observed that specimen B3-20-0.8 having global slenderness, λ_c=20 and unbraced chord slenderness, λ_z=16, was failed by crippling and local buckling. However, the other specimens, B0-20-3.5, B1-20-1.9, and B2-20-1.2 having λ_z, equal 70, 38, 23; respectively, were failed with local buckling of the tube between batten plates as shown in figure 4.

Figure 4. Failure modes of group specimens with λ_c=20.

Second, for short specimens (Group 2), B1-40-1.9, there was a small eccentricity ranged from 1 mm to 2 mm occurred during the test setup, so that local buckling and flexural buckling were noticed. Whereas, the local-flexural buckling of the tube was clearly observed in specimens B2-40-1.3, having λ_c=40, and local-flexural buckling & failure of the welding of batten plate were noticed in specimens B3-40-1 and B4-40-0.7 as illustrated in Fig. 5. Finally, for intermediate specimens (Group3), B1-60-2, B2-60-1.4, B3-60-1, and B4-60-0.8 having λ_c equals 60 and λ_z equal to 120, 82, 60 & 47; respectively, the local flexural-torsional buckling failure shape was observed as shown in Fig. 6.

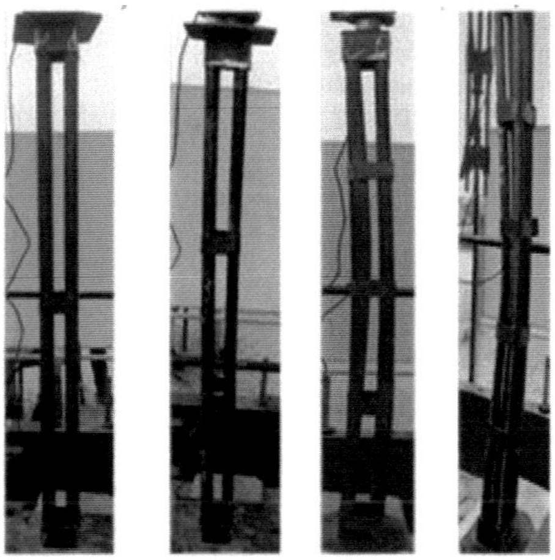

Figure 5. Failure modes of group 2 specimens with λ_c=40.

Figure 6. Failure modes of group 3 specimens with λ_c=60.

3.2 Load-carrying capacities

In general, the strength characteristics of batten columns were governed by the ratio of unbraced chord slenderness-to-global column slenderness. In effect of lower unbraced chord slenderness, a better performance was observed than ones with higher. In one hand, for stub specimens, the axial load capacity was slightly increased by about 1.8% when the ratio (λ_z/λ_c), decreased from 3.5 to 1.9. However, the capacities were increased by 3.1% and 2.4% when the ratio (λ_z/λ_c), decreased from 1.9-to-1.2 & 1.2-to-0.8; respectively as plotted in figure 7-a. In second hand, for short specimens, the effect of decreasing the ratio (λ_z/λ_c), was clearly observed, when the (λ_z/λ_c) ratio decreased from 1.9-to-1.3, 1.3-to-1.0 & 1.0-to- 0.7, the columns capacities increased by about 12.4%, 4.4% and 6.2%; respectively, as shown in figure 7-b. Eventually, for intermediate specimens, the effect of decreasing the (λ_z/λ_c) from 2.0 to 1.4, the column capacity increased by about 43.5 % , which considered as the high enhancement. However, the enhancement of the strength is ranged from 1.7 %-to-21.8 % when the (λ_z/λ_c) ratio decreased from 1.0-to-0.80 & 1.4-to-1.0; respectively, as shown in figure 7-c. Accordingly, at the same global column slenderness the axial compression strengths of batten columns enhanced by about 8%, 25%, and 78% when the ratio of unbraced chord slenderness-to-global column slenderness, (λ_z/λ_c), changed from 3.5-to-0.8 for G1, 1.9-to-0.7 for G2, and 2-to-0.8 for G3, respectively.

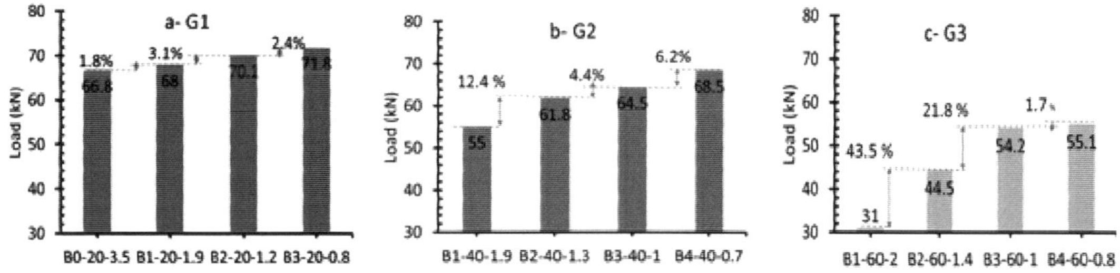

Figure 7. Effect of decreasing the ratio of unbraced chord slenderness-to-global column slenderness (λ_z/λ_c).

4. Finite Element Analysis

4.1. Model Discerption

A Finite Element Modeling using the ABAQUS software package [12], was used to simulate the models of this experimental study. The column cross-section components were modeled using thin shell (S4R) elements with four-nodes and six degrees of freedom at each node. In the other side, a full contact concept was used to simulate the welding between the column components. A typical finite element with an aspect ratio of meshing elements close to unity was considered.

Those models were built by adopting the measured centerline cross-sectional dimensions and the material properties obtained from the tensile coupon tests. Since, the modulus of elasticity and yield stress for the used steel material were taken as 210,000 and 240 MPa; respectively. The geometric and material nonlinearities were considered in the FE modeling.

The critical buckling loads and relevant buckling modes were estimated by elastic buckling analysis, and the technique of the lowest buckling mode was used. Consequently, the obtained buckling modes were used as input for the geometric imperfections in the nonlinear analysis. To initiate the nonlinear analysis, the global imperfection, Δ, with L/1000 was included into the FE model.

A master node was created at the centered of the column cross-section and connected to the slave nodes using the rigid body tie constraint to simulate the pin end conditions. Via reference points, the end conditions were applied through that master nodes. Also, via successive load steps control, the load was accumulated. Since, the nonlinear static RIKS method was used in analyzing the model. In addition, thick end plates were used to prevent the warping at the ends. Both rotations about the z-axis and translations in the X and Y directions were avoided at the upper loaded end. However, the lower unloaded end was prevented from translation in directions of X, Y, and Z, and rotation about the z-axis as shown in figure 8.

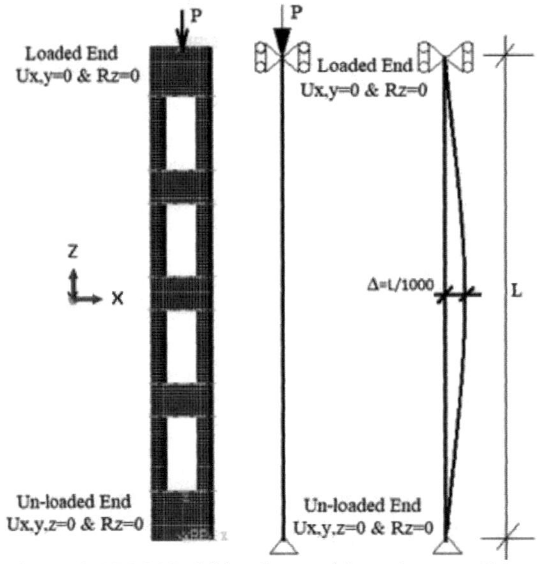

Figure 8. FEM Model loading and boundary conditions.

4.2. The Numerical Model Verification

The results of the FEM are comparable to those obtained through laboratory tests, as presented in Table 2. Moreover, the comparison revealed that the ratios of the experimental to finite element ultimate capacities (P_{EXP}/ P_{FEM}) are extremely close to unity. In addition, the results demonstrate two things. First, the failure modes where the FEM are closely matched with the experimental results as shown in figures. 9-11. Second, the load and displacements (axial-shortening and lateral displacement) relationships where the experimental and FEA results are very close and approximately identical as shown in figure 12.

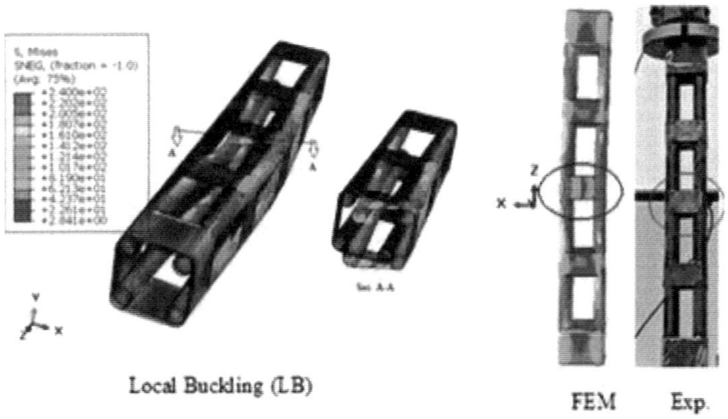

Figure 9. Comparison between FEM and Experimental failure mode for specimen B3-20-0.8.

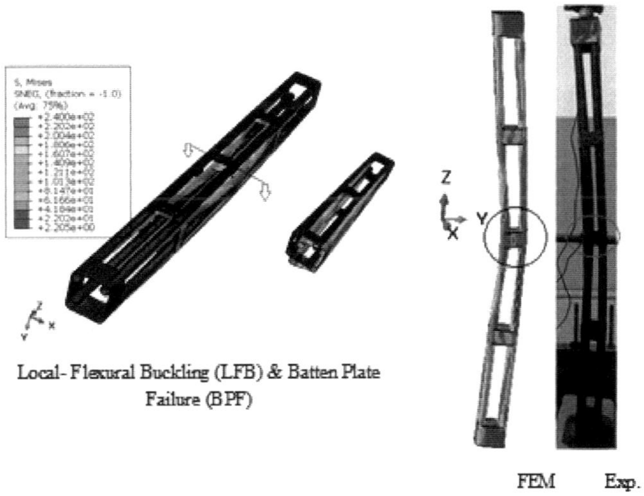

Figure 10. Comparison between FEM and Experimental failure mode for specimen B3-40-1.0.

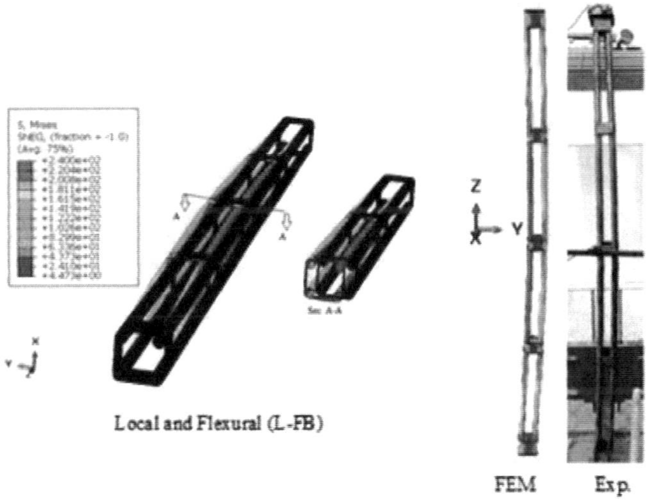

Figure 11. Comparison between FEM and Experimental failure mode for specimen B3-60-1.

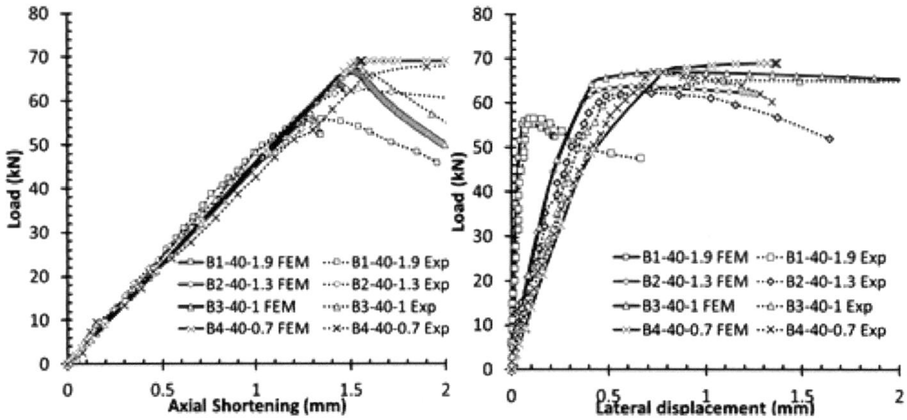

Figure 12. Load-displacement relationships for short specimens having λ_c=40.

5. Comparison Between Experimental Axial Capacities and Design Rules

The experimental axial capacities (P_{EXP}) are compared with the un-factored design compression resistance predicted by AISI [14] and EC3 [15] for built-up battened columns as listed in Table 2. By comparing the experimental results to the design rules, it should be pointed out that the strengths which predicted by the AISI is slightly conservative while the strengths predicted by EC3 are slightly un-conservative.

Table 2. Comparison between experimental strengths and design strengths.

Specimen	P_{EXP} (kN)	Failure Mode	P_{FEM} (kN)	Failure Mode	P_{EXP}/P_{FEM}	AISI P_{AISI} (kN)	EC3 P_{EC3} (kN)	P_{EXP}/P_{AISI}	P_{EXP}/P_{EC3}
B0-20-3.5	66.8	LB	67.07	LB	0.996	56.3	61.64	1.187	1.084
B1-20-1.9	68.0	LB	70.06	LB	0.971	65.4	69.4	1.040	0.980
B2-20-1.2	70.1	LB	71.81	LB	0.976	67.98	71.8	1.031	0.976
B3-20-0.8	71.8	LB	72.91	LB	0.985	68.8	72.9	1.044	0.985
B1-40-1.9	55.0	L-FB	56.08	L-FB	0.981	51.7	59.6	1.064	0.923
B2-40-1.3	61.8	L-FB	63.70	L-FB	0.970	58.86	66.4	1.050	0.931
B3-40-1	64.5	L-FB & BPF	67.00	L-FB	0.963	62.31	69.3	1.035	0.931
B4-40-0.7	68.5	L-FB & BPF	69.03	L-FB	0.992	63.91	70.8	1.072	0.968
B1-60-2	31.0	L-FTB	31.88	L-FTB	0.972	32.91	39.3	0.942	0.789
B2-60-1.4	44.5	L-FTB	46.63	L-FTB	0.954	45.84	57.2	0.971	0.778
B3-60-1	54.2	L-FB	55.49	L-FB	0.977	52.4	64.5	1.034	0.840
B4-60-0.8	55.1	L-FB	56.00	L-FB	0.984	55.67	67.6	0.990	0.815
Average					0.977			1.038	0.917

6. Conclusions

Twelve experimental tests for pin-ended batten specimens are successfully constructed and executed from steel grade 37. The main parameters governed the strength and behavior of batten column specimens, are unbraced chord slenderness and global column slenderness. The geometric load eccentricity, which produced from the column response of imperfection, is kept constant and taken as L/1000. The FE strengths are higher than the experimental test strengths, with an average percentage of less than 5%. In general, the ultimate stress in the column decreases when the global column slenderness increases.

- Built-up columns with a lower unbraced chord slenderness performed better than ones with a higher. At the same global slenderness, the axial compression strengths were enhanced by 8%, 25% and 78

% for the stub, short and intermediate columns when the ratio of the unbraced chord slenderness-to-the global slenderness were altered from 3.5-to-0.8, 1.9-to-0.8, 2-to-0.8; respectively.

- The local, local-flexural, and flexural buckling were the governed failure modes per the current study.
- The experimental and FE results were very close to the predicted strengthens by design rules AISI and EC3 for the short specimens.
- The batten-columns strengths which predicted by the AISI was slightly conservative while their strengths predicted by EC3 was slightly un-conservative.
- Further (experimental, numerical, and analytical) studies on this topic are recommended for future research so that appropriate limits and sufficient recommendations can be included in the design rules.

References

[1] Yanlin Guo and Jie Wang 2009 *Instability behavior and application of prismatic multi-tube latticed steel column* (Journal of Constructional Steel Research 65) pp12–22.

[2] El Aghoury M A, Salem A H, Hanna M T and Amoush E A 2010 *Experimental investigation for the behavior of batten beam-columns composed of four equal slender angles* (Thin-Walled Structures 48) pp 669–683.

[3] Anbarasu M, Bharath K P and Sukumar S 2014 *Study on the capacity of cold-formed steel built-up battened columns under axial compression* (Lat. Am. J. Solid. Struct. 11) pp 2271-1375.

[4] Dabaon M, Ellobody E and Ramzy K 2015 *Experimental investigation of built-up cold formed steel section battened columns* (Thin Walled Struct. 92) pp 137-145.

[5] Roy K, Ting T C H, Lau H H and Lim J B P 2018 *Nonlinear behavior of back-to-back gapped built-up cold-formed steel channel sections under compression* (J. Constr. Steel Res. 147) pp 257-276.

[6] Zhang J and Young B 2015 *Numerical investigation and design of cold-formed steel built-up open section columns with longitudinal stiffeners* (Thin-Walled Struct. 89) pp178-191.

[7] Reyes W and Guzman A 2011 *Evaluation of the slenderness ratio in built-up cold-formed box sections* (J. Constr. Steel Res. 67 No. 6) pp 929-935.

[8] Li Y Q, Li Y L, Wang S K and Shen Z Y 2014 *Ultimate load-carrying capacity of cold-formed thin-walled columns with built-up box and I section under axial compression* (Thin-Walled Struct. 79) pp 202-217

[9] Zhang J and Young B 2018 *Experimental investigation of cold-formed steel built-up closed section columns with web stiffeners* (J. Constr. Steel Res. 147) pp 380-392.

[10] Kherbouche S, Megnounif A 2019 *Numerical study and design of thin-walled cold formed steel built-up open and closed section columns*, Eng. Struct. 179 No. 15) 670-682.

[11] Vijayanand S and Anbarasu M 2019 *Strength and behavior of cold-formed steel built-up batten columns: tests and numerical validation* (Journal of Structural Engineering Vol. 46, No. 2) pp 154-164.

[12] ABAQUS User's Manual 2004 *Version 6.14* Hibbit, Karlsson, and Sorenson (Inc., Providence, RI).

[13] ASTM E8/E8M-13a *Standard Test Methods for Tension Testing of Metallic Materials.*

[14] American Iron and Steel Institute 2016 *North American Specification for the Design of Cold-Formed Steel Structural Members* (Washington DC-USA: American Iron and Steel Institute AISI).

[15] Eurocode3 BS EN1993-1-3 2006 *Design of Steel Structures. Part1-3: General Rules e Supplementary Rules for Cold-Formed Members and Sheeting* (European Committee for Standardization, Brussels).

Numerical modelling of apex connections in cold-formed steel gable frames

S A Mourad[1,*], A M Naguib[1], H H Alanwar[1]

[1] Structural Engineering Department, Faculty of Engineering, Cairo University, Giza, Egypt

[*]Corresponding Author Email: smourad@eng.cu.edu.eg

Abstract. Cold-formed steel (CFS) sections have gained acceptance as primary elements in steel frames due to their exceptionally high strength to weight ratio. The main purpose of this research is to study the behavior of the apex joint in portals frame when subjected to monotonic vertical load. Numerical models for apex connections are developed using a finite element program to study the behavior of the connection. The numerical model is first verified by comparing its results with available experimental and numerical results. After the verification, the effect of several parameters on the connection performance was investigated. The parameters included the effect of the connecting gusset thickness and length as well as the effect of the distance between bolts. Moment-rotation relationships for different CFS apex connection numerical models were analyzed. The numerical models showed that the most significant modes of failure were gusset plate or beam web buckling. Increasing the distance between the two connected sections decreased the connection moment capacity, whereas increasing the thickness of the connecting plate enhanced the moment capacity.

Keywords: Cold-formed steel connections, Finite element modeling, Apex connection, Moment-rotation curves, Ultimate capacity, Failure modes for connections.

1. Introduction

Cold-formed steel (CFS) systems are becoming popular in building construction as both load-bearing and non-load-bearing elements. This is due to their high strength-to-weight ratio and ease of construction compared to alternative materials such as concrete, timber, and heavy steel systems. Most codes of practice for designing cold-formed steel structures are concerned with evaluating capacities of sections and member resistance against both local and overall buckling. The design of connections is still limited and specified only for some cases due to the limited number of studies on the behavior of cold-formed portal frame connections [8]. Design of these connections relies on the engineering experience and knowledge of the designer due to the absence of detailed design standards.

Several resources, such as Kulak et al. [7], provide general information about the behavior and design of structural steelwork connections. However, these resources deal exclusively with hot-rolled gauge construction connections. Design resources primarily oriented towards constructing cold-formed steel, such as Yu [19], provide little details on connection design. Although the literature includes several references to joint studies in constructing cold-formed steel, these are mainly limited to working on particular types of connections, usually linked to joint strength determination.

Chung et al. [4] developed a finite element nonlinear model to represent two groups of flexural tests on industry-standard cold-formed steel C-and Z-section beams. The experimental tests investigate beams laterally supported with detailing of the compression flange so that local buckling failures occurred in the first group and distortional buckling failures in the second group. Haidarali et al. [5] developed a finite element model to present local buckling as well as the combination of local-distortional buckling of lateral restrained CFS beams. Two series of FE models for buckling behavior of laterally restrained CFS Z-section beams were generated. Ye et al. [18] developed a design methodology to optimize the cold-formed steel folded-flange section subsequently using it. The efficiency of the folded-flange beam section was demonstrated in conjunction with nine standard channel prototypes to study the effects of the inclined lips, return lips and intermediate web stiffeners. Anbarasu [2] performed studies about the post-buckling behavior, ultimate strength and design of CFS lipped channel beams that was subjected to uniform bending at the main axis and affected by the interaction of the local-distortional buckling mode. Rinchen et al. [15] provided a detailed investigation into the numerical modeling of portal frames using single C-section in CFS using shell elements. Numerical models simulated the behavior of six tested portal frames using a single CFS C-section at the University of Sydney. Phan et al. [14] developed a different analytical design method using Eurocode 3 effective width method to estimate the maximum flexural strength of CFS bolted moment connections by taking into account the effects of bimoment. The expected analytical capacities were contrasted with the experimental findings of the CFS apex connection tests. Mojtabaei et al. [12] investigated the local buckling failure in bolted moment connections in cold formed steel back-to-back channel beams that were connected through the webs using a gusset plate.

This study aims at contributing to the development of standards on the connection design of CFS gable frames. Nonlinear finite element analysis (FEA) is performed using a software program [1] and validated by previous experimental results. The effect of various components of the apex joint behavior was investigated.

2. Numerical Modelling

A finite element model is developed for the proposed apex connection. This modeling protocol will not only enable engineers to predict the apex connection's maximum moment capacity, but will also provide an insight into the failure mechanism of the proposed apex connection subjected to concentrated loads. Furthermore, the existence of a verified finite-element model allows a parametric study to be conducted for a wide range of parameters, which would neither be feasible nor economical to perform experimentally.

2.1. General model details

ABAQUS software was used in this research for the numerical model. The apex connection geometry consisted of brackets bolted to the webs of the C sections back-to-back using eighteen M16 8.8/TF bolts. Figure 1 shows the assembly of the apex connection in the ABAQUS software. The specimen geometry and practical details.

In this research, two types of elements were employed in this connection; S4R, which is a shell element and C3D8R, which is a solid element.

S4R is defined as a four-node doubly curved general-purpose, first-order reduced integration finite membrane strain shell element with hourglass control. The hourglass technique allows mesh distortion at zero strain because of using a reduced-integration element type to reduce the running time. This element was used to model the brackets and channel sections. It was discretized into an element size of 15x15 mm near bolt holes and 15x50 mm in other parts to decrease computational cost and running time. This type of element allows two-sided contact with changes in the thickness so it is suitable for bolted moment connections [13].

C3D8R is defined as a three-dimensional hexahedral eight-node linear brick, first-order interpolation reduced integration solid element with hourglass control. This element is more appropriate to simulate the ultimate behavior of bolted connection for the thin-walled steel [6]. It was discretized into an element size of 10x10x10 mm. The solid elements in ABAQUS can be used for linear and nonlinear analyses, including large deformations, plasticity, and contact.

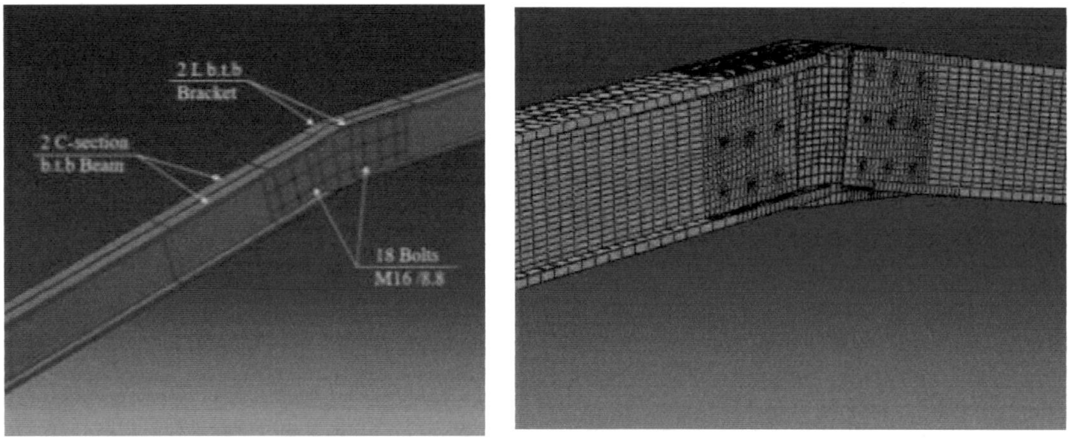

Figure 1. Apex connection modeling.

Schafer et al. [17] investigated the sensitivity to element choice and meshing in numerical modeling of cold-formed steel. It was investigated that mesh density has a clear effect on the reaction of cold formed steel members in finite element analysis. A coarse mesh cannot accurately capture local buckling modes but can be suitable for producing the global and distortional buckling modes. On the other hand, a medium or fine mesh can simulate all buckling modes, including distortional, local and global, with reasonable accuracy. Once a reasonable meshing is used, the difference in response between different types of elements becomes minor. For these reasons, as depicted in Fig. 1, a relatively fine mesh is used in this modeling effort. Also, it was found that using a mesh size for the CFS elements of around 10x10 mm provides a balance between accuracy and computational time [18].

2.2. Material properties

The material modeled in the numerical analyses is supposed to be linearly elastic with proportionality limit stress of 350 MPa. The stress-strain curve outside the proportionality limit is outlined by piecewise linear sections using the true plastic stress-strain relation [16]. Further details may be found in Naguib [10]. The residual stress and the increase in yield strength at the corner due to the cold-forming are not taking into account in the numerical model. In the high-grade (~550MPa) CFS section, the longitudinal residual stresses are always focused at the rounded corners compared to the flat segments of the sections [11]. Compared to its overall cross-sectional area, the total corner area of the C section is negligibly small. Since it is commonly assumed that the unfavorable effect of residual stress roughly negates the favorable effect of corner strength enhancement, their effects have been entirely ignored [17].

2.3 Boundary conditions and Constraints

All nodes within the region bounded by the outermost bolts clamping the brackets were subject to beam style multi-point constraint to simulate the clamping motion supplied to the web by the load brackets and the support brackets. To simulate the boundary conditions at supports; the x-axis translational degree of freedom (DOF) and z-axis rotation were released while other DOFs were restrained. The horizontal translational DOF in the x-direction at the midpoint of the apex bracket was restrained to prevent rigid body motion.

Contact between brackets and channel sections was modeled using surface-to-surface contact pair including a pure "master-slave" algorithm. This formula identifies the contact surfaces and executes constraints on the nodes of the slave surface, letting only the master surface to penetrate the slave surface. The friction coefficient used between each contact surface is 0.3. Contact between bolt's shank and bolt's holes of the brackets and channel sections was modeled using shell-to-solid coupling formulation. The left support was restrained from displacement and permitted to rotate about x-axis only. The right support was allowed to move in the z-direction and to rotate about x-axis only.

2.4 Loading

In this model, two concentrated vertical loads were applied at the top of the channel sections with displacement control. The loads were located at a distance equal to 1.5 m from the apex point. In the static nonlinear analysis, the displacement magnitude increases along increments with defined time and total increments number. The subjected displacement increases during the increase of the increments, the subjected displacement increases. So, the poor elements and the failure modes are collected. The model considers the material and geometrically nonlinearity. for the solution of nonlinear equations, the complete Newton–Raphson method was introduced. The convergence was accomplished by the automatic stabilization factor using the 2.0E-4 damping coefficient [3].

3. Numerical Model Verification

Lim et al. [9] conducted experimental tests on four different configurations of the apex connection. The apex joints were tested under pure bending. They were tested horizontally on the laboratory floor. The maximum distance between the lateral restraint points along the apex joint was 1036 mm. Figure 2 compares the load-deflection behavior of the numerical model with the results obtained by Lim et al. [9]. The ultimate moment capacity of the experimental specimen was 89.4 kN.m whereas the numerical value was 98.5 kN.m. The error between the experimental and nonlinear analysis is about 10%. Comparing the deflections from the numerical model and the experimental results, it was observed that the error between the numerical and the experimental deflection at ultimate load is less than 2%. Based on these results, it may be concluded that the numerical model may be used to simulate the experimental model with acceptable accuracy. The mode of failure were early web buckling of the channel-sections. No buckling was observed at the apex brackets. A good agreement was observed between the experimental and numerical failure modes of the channel section, as shown in Figure 3.

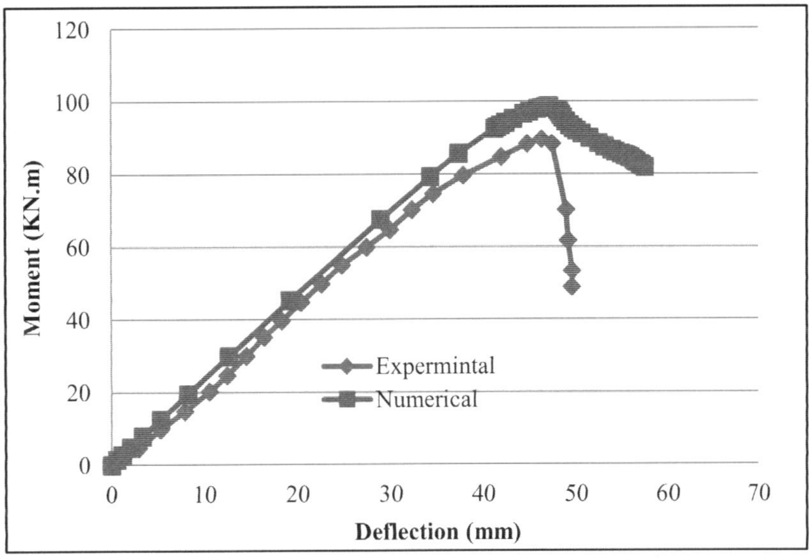

Figure 2. Comparison between the numerical and experimental and results.

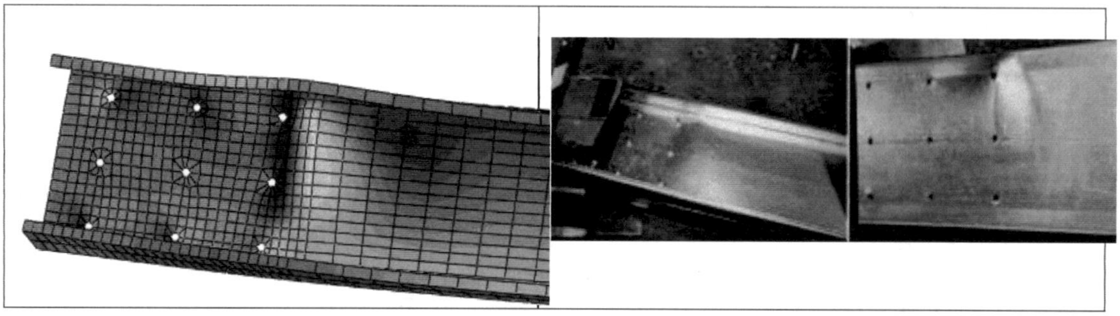

Figure 3. Failure mode comparison between numerical and experimental model.

4. Parametric study

The failure modes of the apex connection included yield of bracket, yield of the channel section, local buckling of the bracket, and local buckling of the channel section. The parameters studied included bracket length, bolt spacing, and bracket plate thickness. Each sample is donated by (Tt – L – S), where t is the bracket thickness, L is the bracket length, and S is the bolt spacing (pitch). The models may be divided into three groups:

1. **Group 1**: Six different length cases were studied: the length of bracket equals 440, 525, 600, 675, 750, 825, mm, the pitch distance (S) between the bolts equals 115, 157.5, 195, 232.5, 270, 307.5 mm, respectively. In this case, the gap (G) between the two beams at the joint was fixed. All six cases of the group were studied in two different thicknesses of brackets; for Group 1-A, 3mm, and Group 1-B, 4 mm.
2. **Group 2**: Six different distances between bolts were changed; the pitches of the bolts were 115, 157.5, 195, 232.5, 270, 307.5 mm. The length of the bracket was fixed to be equal to 825 mm. The six cases of the group were studied in two different thicknesses of brackets; for Group 2-A, 3mm, and Group 2-B, 4 mm.
3. **Group 3**: Four cases were studied to demonstrate the effect of thickness of the bracket. The different thicknesses of the bracket were considered t= 1, 2, 3, and 4mm for the case of bracket length equal 525and pitch distance 157.5mm

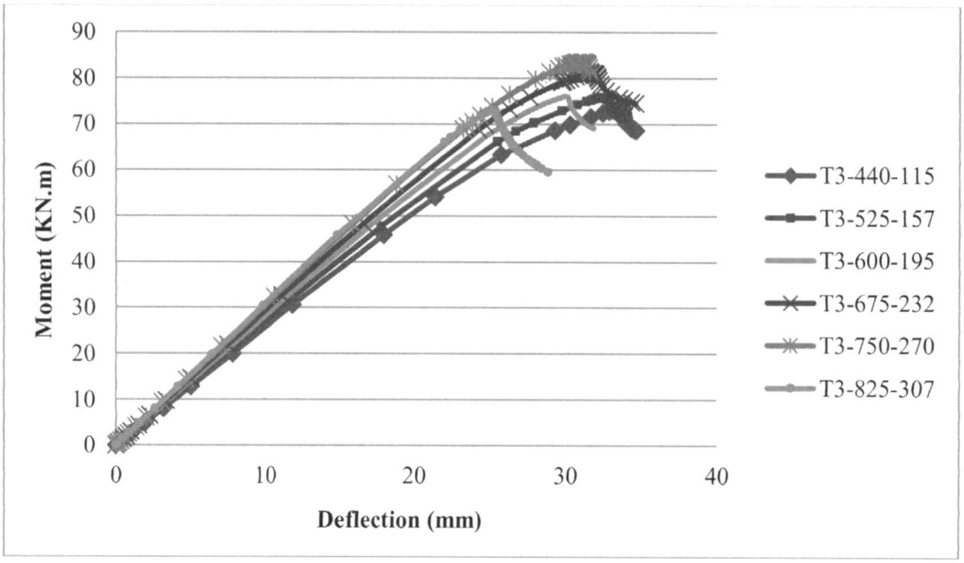

Figure 4. Moment–Deflection relationship for group 1-A, using two brackets with t = 3 mm.

Table 1. Results of group 1 – Variation in bracket length.

Group -1 Fixed Gap:			Mult(t=3)= 98.9 kN.m			Mult(t=4) = 98.9 kN.m		
Sample Group A	Mmax (KN.m)	Effeciency of joint	Increase %	Sample Group B	Mmax (KN.m)	Effeciency of joint	Increase %	Mmax (t4/t3)%
T3-440-115	73.34	74.16	–	T4-440-115	81.47	82.38	–	11.09
T3-525-157	75.82	76.66	3.38	T4-525-157	81.94	82.85	0.58	8.07
T3-600-195	76	76.85	3.63	T4-600-195	88.3	89.28	8.38	16.18
T3-675-232	81.13	82.03	10.62	T4-675-232	88.839	89.83	9.05	9.50
T3-750-270	83.416	84.34	13.74	T4-750-270	93.272	94.31	14.49	11.82
T3-825-307	73.1	73.91	-0.33	T4-825-307	95.232	96.29	16.89	30.28

Note: Efficiency of joint = M_{max} / M_{ult}.

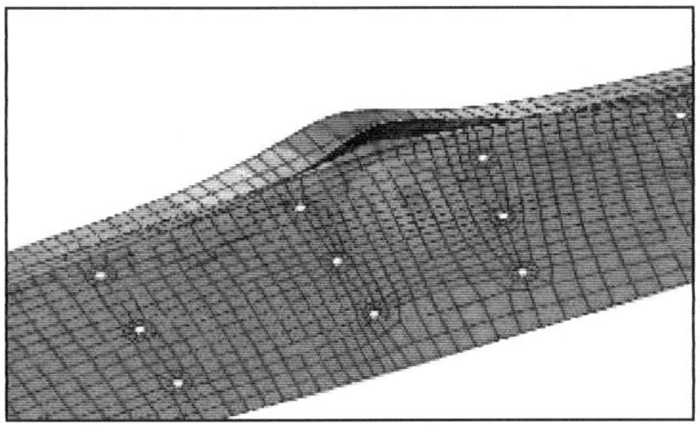
(a) Sample T3-750-270

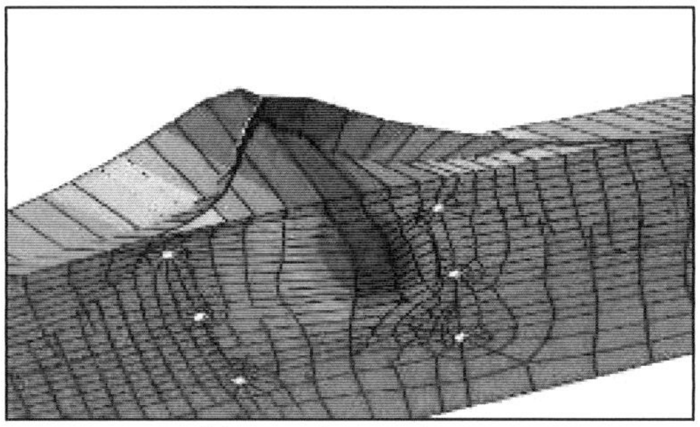
(b) Sample T3-825-307

Figure 5. Mode of failure for the cases (T3-750-270) and (T3-825-307).

From the investigation of Figure 4 and Table 1, it can be observed that:
The increase in the bracket length has a positive impact on the ultimate moment of the considered connections. For example, the connection moment capacity increased in (T3-750-270) with 13.74% more than the base control sample (T3-440-115). The efficiency of the joint increased to provide moment resistance equal to 84.34% of the ultimate moment capacity of the beam section. The results show the relation between the increase in length/depth ratio to the increase in ultimate connection capacity. As the bracket length increase, the inertia of the bracket section increased. Besides, the distance between bolts (S) increases to fix the gap distance and bolts edge distance dimensions, which increases the shear resistance of the connection bolt group as the polar inertia of the bolt group increased. However, in the case of (T3-825-307), the increase in bracket length, L/d exceeds 2.2, shows a converse effect. Sudden reduction in moment resistance by 14.11% occurred at L/d exceeded 2.2 with S/t exceeded 90. It can be explained by comparing the modes of failure shown in all cases with the mode of failure appeared in case (T3-825-307). The main mode of failure is buckling in bracket flange, as shown in Fig. 5(a), while in case (T3-825-307), bracket web buckling significantly appears as shown in Fig. 5(b). As the connection moment resistance increases, due to the increase in bracket length, conjugate compression force is induced on the upper portion of the bracket web. In the case of (T3-825-307), the compression force exceeded the buckling critical force for the bracket unstiffened web (Gap region), which is a function in gap length and bracket height. As a support to our interpretation, this sudden reduction did not appear in group B (4 mm thickness) as will be discussed in the next section.
For Group 2, six different distances between bolts were changed to study the effect of the distance between the bolts (pitches) on the behavior of the apex connection. The selected pitches of the bolts were 115, 157.5, 195, 232.5, 270, 307.5 mm. The gaps between the two beams connected were 318.44, 313.98, 239.46, 164.95, 90.42, 15.9 mm respectively. Figure 6 indicates the dimensions of the brackets.

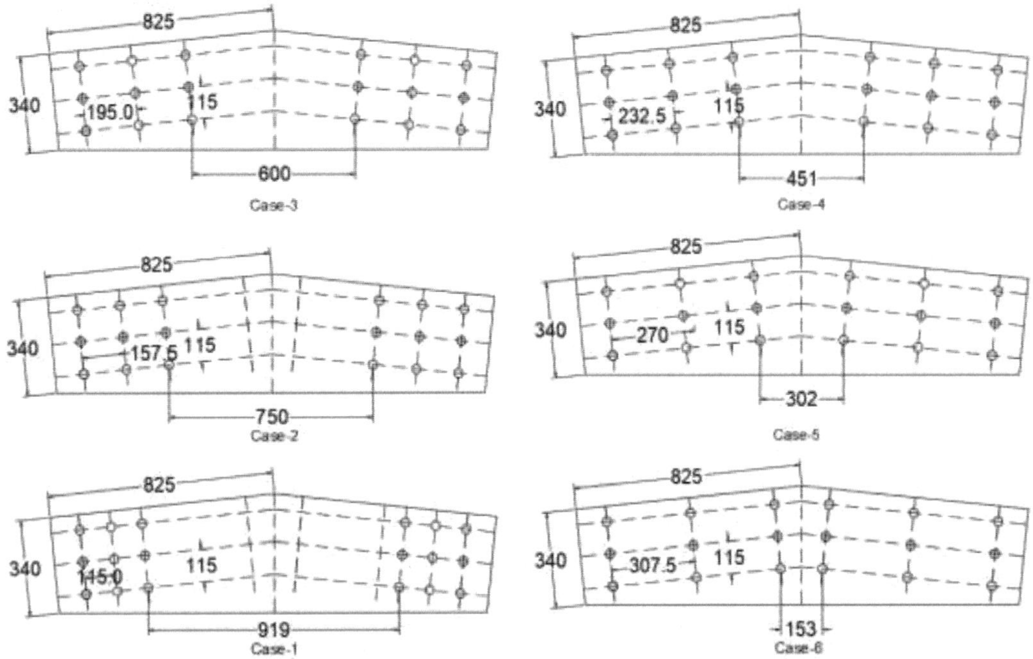

Figure 6. Details of brackets for group 2.

Table 2. Results of group 2 – Variation in bolt spacing.

Group -2 Fixed bracket length Mult(t=3)= 98.9 KN.m Mult(t=4)= 98.9 KN.m								
Sample Group A	Mmax KN.m	Effeciency of joint	Increase %	Sample Group B	Mmax KN.m	Effeciency of joint	Increase %	Mmax (t4/t3)%
T3-825-115	71.87	74.67	–	T4-825-115	81.47	82.38	–	1.13
T3-825-157	72.81	73.62	1.31	T4-825-157	87.66	88.63	7.60	1.20
T3-825-195	74.614	75.44	3.82	T4-825-195	90	91.00	10.47	1.21
T3-825-232	64.396	65.11	-10.40	T4-825-232	91.45	92.47	12.25	1.42
T3-825-270	68.98	69.75	-4.02	T4-825-270	93.57	94.61	14.85	1.36
T3-825-307	73.1	73.91	1.71	T4-825-307	95.23	96.29	16.89	1.30

Figure 7 illustrates the moment-rotation relationship for the six models used to study the effect of the distance between bolts on the rotational stiffness of the CFS moment connection, where the rotational stiffness is the slope of the straight line of the shown curves ($K_\theta=\Delta M/\Delta\theta$). We used constant bracket thickness t= 3 mm and constant bracket length L= 825 mm with variable bolt distances S= 115, 157, 195, 232, 270, and 307 mm.

From the investigation of Fig. 7 and Table 2, it can be observed that:
The rotational stiffness ($K_\theta=M/\theta$) increases with increasing the distance between bolts (S). For example, in sample (T3-825-195) the rotational stiffness increased by 11.66 % more than the controlled sample (T3-825-115). As shown in figure 4-11', the top three samples have different Moment-rotation relationship than the last three samples with S/t exceeded 77.3. For example, sample (T3-825-232) rotational stiffness increased by 19.7% than sample (T3-825-195). The last three samples show a larger rotational stiffness that shows a more rigidity behavior that can be explained by investigating the modes of failures.

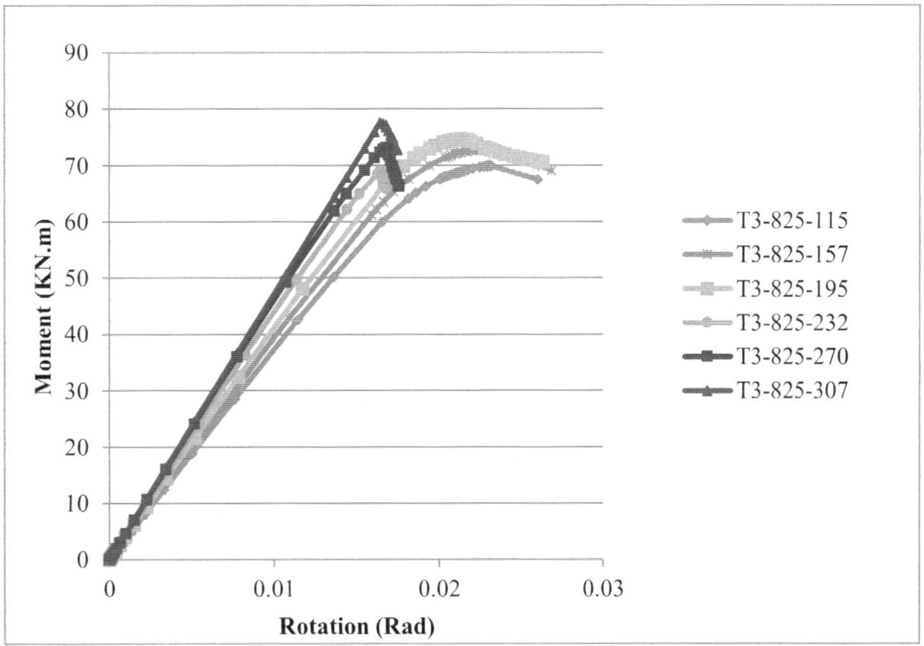

Figure 7. Moment–Rotation relationship of group 2-A, using brackets with t = 3 mm.

Figure 8 illustrates the moment-rotation relationship of the studied cases to investigate the effect of the bracket thickness on the rotational stiffness of the CFS moment connection. From the investigation of Fig. 8 and Table 3, it can be observed that:

The increase in bracket thicknesses significantly improves the connection's ultimate moment. The results show that the improvement significantly appears in the smaller thicknesses, such as 2 mm increased by 162% more than 1 mm, and 3 mm increased by 93% more than 2 mm. In comparison, the increase in connection moment is only 7.8% from 3 to 4mm. Bracket buckling failure is the main mode failure in case the thickness of the bracket is less than the thickness of the beam section, which explains the significant increase in connection moment capacity in case of smaller thicknesses. On the other hand, when the bracket section is greater than the beam section thickness, the failure occurred in the beam section. This explains the lower increase in moment capacity in the larger thickness values.

Table 3. Results of group 3 – Variation in bracket thickness.

Group 3	Fixed length – Fixed pitch		
Mult=	98.9	kN.m	
Sample Group A	Mmax kN.m	Effeciency of joint	Increase %
T1-525-157	14.988	15.15	–
T2-525-157	39.28	39.72	162.076328
T3-525-157	75.821	76.66	93.0269857
T4-525-157	81.772	82.68	7.84874903

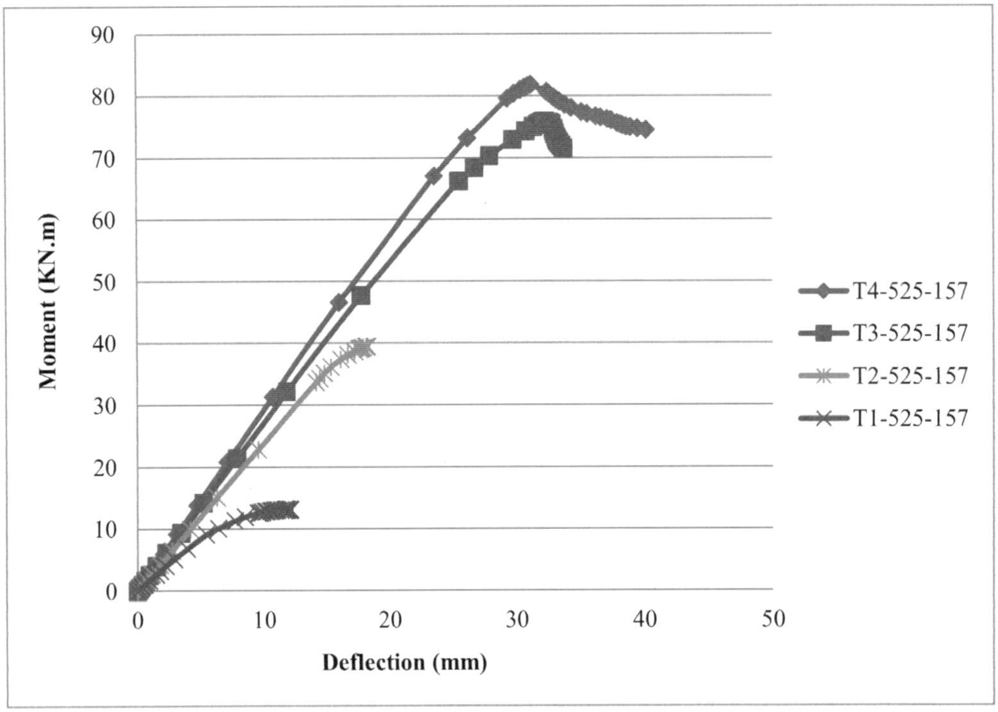

Figure 8. Moment–Deflection relationship of Tt-525-157, with t=1, 2, 3, and 4 mm.

5. Conclusions

Based on the numerical investigation, a set of graphs have been presented to show the effect of various parameters on the load capacity of the CFS bolted moment connection. The main conclusions derived based on the parametric study presented are:

The increase in bracket length improves the connection moment resistance. The effect of increasing the bracket length is more effective when the thickness of the bracket is larger than the thickness of the beam section. For small thicknesses, $L/d < 2.2$ is recommended for maximum enhancement for connection moment.

Increasing the bolt spacing improved the connection moment capacity. The study shows that the reduction change in connection behavior for S/t exceeds 70 in large bracket length ($L/d > 2.5$). The results show that the bolt spacing is more effective in higher bracket thicknesses ($L/t < 250$) than lower bracket thicknesses ($L/t < 250$).

The results show that the thickness of the bracket is significant for connection behavior at small thicknesses (Bracket thickness smaller than beam section thicknesses). As the bracket thickness increases, the dominant mode of failure changes from bracket buckling failure to beam section failure. The study recommends bracket length to be larger or equal to the beam section thicknesses, especially with connection of $L/d > 2.2$, and not to be smaller than 75% of the beam section thicknesses for apex connections.

References

[1] ABAQUS, ABAQUS/Standard User's Manual, Version 6.14-4 EF1. Pawtucket, RI, USA, (2009).

[2] Anbarasu, M., "Local-distortional buckling interaction on cold-formed steel lipped channel beams", Thin-Walled Structures, 98, Part B, (2016), 351-35.

[3] Badr, Ahmed R., Elanwar, Hazem H., Mourad, Sherif A., "Numerical and experimental investigation on cold-formed walls sheathed by fiber cement board", Journal of Constructional Steel Research, 158, (2019), 366-380.

[4] Chung, K.F., Ip, K.H., "Finite element investigation on the structural behaviour of cold-formed steel bolted connections", Engineering Structures, 23, (2001), Issue 9, 1115-1125.

[5] Haidarali, Mohammad R., Nethercot, D.A., "Finite element modelling of cold-formed steel beams under local buckling or combined local/distortional buckling", Thin-Walled Structures, 49, (2011), 1554–1562.

[6] Kim, T.S., Kuwamura, H., "Finite element modeling of bolted connections in thin-walled stainless steel plates under static shear", Thin-Walled Structures, 45, (2007), Issue 4, 407-421.

[7] Kulak, G.C., Fisher, J.W., Struik, J.H.A., "Guide to design criteria for bolted and riveted joints", New York, Wiley (1987).

[8] Lee, Y.H., Tan, C.S., Mohammad, S., Tahir M. Md, Shek, P. N., "Review on cold-formed steel connections.", The Scientific World Journal, Vol. 14, (2014).

[9] Lim, J.B.P., Nethercot, D.A., "Ultimate Strength of Bolted Moment-Connections Between Cold-Formed Steel Members", Thin-Walled Structures, 41, (2003), 1019–1039.

[10] Naguib, A. M., "Numerical modeling of apex connections in cold-formed steel gable frames", M.Sc. Thesis, Struct. Eng. Dept., Faculty of Engineering, Cairo University, Egypt (2021).

[11] Moen, C.D., Igusa, T., Schafer, B.W., "Prediction of Residual Stresses and Strains in Cold-Formed Steel Members", Thin-Walled Struct, 46, (11), (2008), 1274–1289.

[12] Mojtabaei, S. M., Becque, J., Hajirasouliha, I., "Local Buckling in Cold-Formed Steel Moment-Resisting Bolted Connections: Behavior, Capacity, and Design", Journal of Structural Engineering, 146, (9), (2020).

[13] Öztürk, F., Pul, S., "Experimental and Numerical Study on a Full-Scale Apex Connection of Cold-Formed Steel Portal Frames", Thin-Walled Structures, 94, (2015), 79–88.

[14] Phan, Duoc T., Mojtabaeib, S. Mohammad, Hajirasoulihab, I, Lau, T.L., Lim, J.B.P., "Design and Optimization of Cold-Formed Steel Sections in Bolted Moment Connections Considering Bimoment", Journal of Structural Engineering, 146, 8, (2020).

[15] Rinchen, Rasmussen, K.J.R., "Numerical Modelling of Cold-Formed Steel Single C-Section Portal Frames", Journal of Constructional Steel Research, 158, (2019a), 143-155.

[16] Rinchen, Rasmussen, K.J.R., "Behavior and Modelling of Connections in Cold-Formed Steel Single C-Section Portal Frames", Thin-Walled Structures, 143, (2019b), 106233.

[17] Schafer, B.W., Li, Z., Moen, C.D., "Computational Modeling of Cold-Formed Steel", Thin-Walled Structures, 48, (2010), 752–762.

[18] Ye, J., Mojtabaei, S., Hajirasouliha, I, Shepherd, P., Pilakoutas, K., "Strength and Deflection Behaviour of Cold-Formed Steel Back-to-Back Channels", Engineering Structures, 177, (2018), 641–654.

[19] Yu, W.W., "Cold-Formed Steel Design", 2nd Ed., New York, Wiley (1991).

Project Manager Skills affecting Construction Projects in Egypt

R M Abdelmasseh[1, *], **H A Bassioni**[2] and **E F Gaid**[1]

[1] Orascom Construction Egypt, Cairo, Egypt
[2] Construction and Building Engineering Department, College of Engineering and Technology, Arab Academy for Science, Technology and Maritime Transport, Alexandria, Egypt

*Corresponding Author Email: ramy.eng@gmail.com

Abstract The research was to explore the relationships between Construction Project manager skills and the performance of projects in Egypt. Specifically, this study investigated the relationship between the project manager skills (Technical Skills, Human skills, Conceptual Skills, Political and Social Skills) and the dimensions of project performance (Project Cost and Project Time). Hard skills include technological abilities, background, industry knowledge, and project experience. That define the tangible portion of a project manager's resume as the 21st century recognition of project management. Data has been collected from 104 expert engineers who have a good realistic background in the field of construction industry; questionnaire survey is the used method. The results of analysis propose that risk management abilities have a significant association with project cost and time performance. The Egyptian construction sector may use the findings as a guideline to choosing project managers with the 'suitable' skills profile or focusing training and development programs on skills crucial for successful projects.

Keywords: Construction Management, Project Manager Skills, Project performance.

1. Introduction

Construction projects consider frequent challenges and sometimes Lack of clarity that tests the efficiency over credentials of project managers [1] in every part of the project life cycle there are multiple different characteristics restricted by specific requirements were frequently updated for each project. As a result, project managers must communicate with a vast and diverse group of individuals, however he does not have direct authority over many of them [2]. In this unstable and frequently changing environment project, managers' skills are critical in operating as the hub that integrates all project components and pushes them together toward a successful project handing over. Furthermore, a combination of skills necessarily fully or sometimes a certain needed percentage for project managers to pass the challenging task, but it is a significant duty. The capacity to understand people in the workplace and use that knowledge to influence others to act in ways that further one's personal or organizational goals [3]. The construction business is an important partner to Egypt's economy and national income, as well as one of the country's fastest-growing sectors [4] so searching factors affecting the performance is very important.

2. Project Performance

Project performance evaluation is the process of evaluating performance relative to success in basic criteria of project success in terms of time and cost, [5] observed that several prior studies had studied the performance of construction projects, [6]. Identified Financial Stability, Project Performance, Quality, Health and Safety Standard, Assets, Customer Relationships, Consultant Relationships, Management Capabilities, Claims and Contract Disputes and Subcontractor Relationships, Mention as the key success parameters for construction projects. It is argued that the increasing importance of construction time and cost, which considered serving as a crucial benchmark for evaluating project performance, [7].

3. Project Manager Skills:

3.1 Technical Skills
Technical skill is the ability to do a given type of operation, particularly one requiring systems, processes, procedures and techniques. Effective project managers should have the necessary technical expertise or knowledge needed by the project they manage. Technical skills provide advanced expertise and analytical abilities in the use of the relevant methods and techniques [2].

3.1.1 Scheduling
The project manager must consider the timeline while executing different activities, consider the activities prioritization, and decide when the activities are due. In addition, he can manage the available resources to the proper activities, Evaluate the available time and any shortages, taking into consideration the issues that might force the project to be delayed [8].

3.1.2 Budgeting and Cost Management
One of the most important factors in ensuring the completion of the project according to the specified budget. Define the types and volume of resources required to complete project tasks in the first step. The next step by updating all resources forecast costs and reallocating the budget to the project activities, As the last step all project changes should be managed by the project manager [9].

3.1.3 Quality Management
Construction quality exceeds or meets standards according to design documents [9]. Following procedures that should guarantee the meeting of the project standard required based on the periodic project performance evaluation, observation the results compared with the standards and getting methods to maintain the displeasing performance [8].

3.1.4 Administration of Document and Contract
The perspective of the processes would include or exclude construction contracts in line with construction industry requirements and rules. In addition, he must handle many administrative parts of the projects and establish the contractual ties between all involved parties. Manage and regulate payments, as well as variation orders, claims, data requests, closeout reports, periodical records system and management reporting activities [10].

3.1.5 Risk Management
Project managers must be proactive in controlling project risks in order to achieve the project objectives. Risk management consists of the five phases, creating the context, identifying risks, analysing risks, evaluating risks, and treating risks. Communication with project stakeholders is required at every stage to ensure that all risks were addressed. Each stage should be tracked and evaluated on a regular basis in order to keep track of current progress and revise the risk management plan.

3.1.6 Procurement Management
Procurement management is the procedure necessary to get materials and services from sources other than the project team. The procurement management procedure identifies procurement strategy and create procurement forms, establish procurement policy, award a contract to chosen external parties after receiving bids and selecting an external party. The step, which entails monitoring the performance of the contract and making, required modifications. Finally, close the completed project.

3.2 Human Skills

Human skills is the project manager ability to generate a cooperative mode within the team he leads and works effectively as a member of the group. Human skill is caring with working with people whereas technical skill is caring with working with "things" (processes or physical objects) [9].

3.2.1 Self Confidence

Self-confidence is considered a worthy asset in most societies. One of the key reasons for success, according to psychology is believing in oneself. Psychologists established that self-esteem, as well as the functional ones on which we will concentrate related to their effectiveness [11].

3.2.2 Leadership and Power in Project Management

Due to strong global competition, leadership is considered the most important aspect and a key part of a good project. complicated negotiations, integration issues, and a dynamic move toward decentralization of project management operations its relevance are growing through guiding the project team and other stakeholders toward the project's goals within the specific constraints.

3.2.3 Communication

The project process requires formal and informal contact at various scales within the internal and external project stakeholders. The basic management skills of efficient communication are fundamental to the project's success. The project manager will provide orders, attend meetings, transmit suggestions and knowledge from and to members of the project team, supervisors, stakeholders, vendors, functional administrators, other project managers, and even outside staff. Listening's importance, challenges to good listening, and principles for successful listening [12].

3.2.4 Motivation

Overall project success mainly depends on the performance and motivation of project team members and other stakeholders, who do their best and produce better quality results. It's also vital to consider that motivation is an internal phenomenon that makes a team makes an extra effort [12].

3.2.5 Managing and Resolving Conflict

Project managers have to develop appropriate skills to manage conflict effectively which must be considered as a project key activity, which is related to both top management and project manager's intervention to control the level of conflict between the involved parties. Using several interpersonal attitudes and identifying an appropriate conflict to manage the conflict [12].

3.2.6 Negotiation

Negotiation is a life fact. Everyone negotiates something every day. Benefits and salaries are negotiated with employers, purchase prices are negotiated with vendors, and schedules are negotiated with customers, a negotiation is a basic approach for reaching an agreement with others. By which is reached an agreement when both parties have a combination of shared and opposed interests.

3.2.7 Managing Stress

Project management is no exception from the fact that there is a certain amount of stress associated with each occupation. Stress is one of the results of disputes is stress and changes, which are unavoidable in a project environment and sometimes necessary to improve project performance. As a result, it has become an integral part of project life, and stress management has to be regarded as a critical component of project manager skills, [13].

3.3 Conceptual

The skill to perceive the project as a whole, understand how the organization's many duties interact with one another and how modifications in any element may have an influence on all of them. Extending to see the relationship between individual business, the industry, the society, the governmental and economics of the project.

3.3.1 Visualizing

The ability to see the relationship between the project and the organization to perceive the project as a whole. Project managers will have a clear understanding of what needs to be completed at each stage and adapt the plan when available new knowledge [14].

3.3.2 Scoping

Is the ability to decide and regulate which must and must not be in the project, [15]? It is essential for the project to verify that all work requires the project manager's scoping skill and that just the work required to finish the project effectively has been considered. To avoid future disputes, it is also critical to draft a contractual agreement in which the scope of the project must be clear and specified.

3.3.3 Integration

Scoping skills are used by project managers to guarantee that all project activities are clearly recognized, harmonized, and coordinated [16]. During the planning stage, for example is to integrate all project activities, as well as the cycle and scheduling of tasks and resources necessary to achieve the project. The project manager should coordinate all design details, such as structural, architectural, electromechanical and interior, throughout the design stage and ensure that each aspect functions as required. Also during the construction process to decide resources and effort focusing, coordinate multiple activities, and determine trade-offs between conflicting targets and alternatives [17].

3.4 Political and Social

The ability fruitfully recognize others at the project, using that expertise to affect others to deal in strategy that value one's personalized and/or structural aims [18]. Many human beings do now no longer apprehend the crucial distinction between human and political skills. Human skill is defined as the comfort and ease with which all stakeholders communicate. Some project managers, on the other hand, are well preferred but their projects fail. Political skills are essential in this circumstance.

3.4.1 Social intelligence

It is stated that persons with strong political skills are attentive to others and are sensitive to a wide range of social conditions [15]. These people understand social relations, can analysis their own and others' behaviour in social situations. They are highly sensitive to others therefore; they are seen to be astute in interacting with people. In summary, those with high social awareness comprehend social contexts and interactions interpersonal that occurred.

3.4.2 Interpersonal influence

One of the characteristics of persons with excellent political skills is their engaging private style, which has a great impact on the Surroundings. They are adaptable and may alter their attitude to each set due to revealing specific reactions from others. In the human skill model, this component separates from interpersonal skill. Interpersonal skill refers to the capacity to form relationships and get along with people, whereas interpersonal influence is related to the ability to affect positive relations with others in order to assure the greatest achievement of a project or organization.

3.4.3 Networking ability

The power to establish and use this network of humans is a key part of political talent. People in the networking conceded have values that are important [15], also required for reaching project and personal objectives. Project managers with strong networking skills are frequently dealmakers, great negotiators, and resolving conflicts experts.

3.4.4 Apparent sincerity

Others see people with good political skills have a high level of honesty and sincerity [15]. This element is important in affecting others because it focuses on the perceived intentions of specific attitude displays. When there are no hidden objectives behind the attitude presented, the influence efforts will be successful. Humans are high in seeming sincerity apparent sincerely and confidence since they do not appear to be aggressive or forceful.

4. Research Methodology

Surveys are performed based on systematic sampling. It is exceedingly unusual for complete population surveys to be performed. The sampling by principles of statistical, which ensures a sufficient sample, is used for speed and efficiency. Questionnaires were used for sampling.

4.1 Questionnaire Design:

The questionnaire was separated into two parts in order to collect such information. After respondent's general information about the, name, company and job, the first part 10 queried for response demographics, time and cost performance impact for the considered projects. The second part was designed to contain the application of nine project manager skills in the considered Project. These skills were taken from the discussed literature studies that reported the project manager skills. Each skill was rated on a five-point Likert scale rating from one to five, with 1 entity 'No application', 5 indicating 'Rigorous Application'.

4.2 Survey and Sampling Approaches:

Random sampling was undertaken in the first instance of the survey and due to weak response, A non-random snowballing technique was then followed, in which the researcher obtains prospective subjects from existing participants, thus building up the sample in a snowball manner [19]. The study is a specimen or a subset of the overall population that is taken to illustrate what the rest of the population is like. [20]. Just a smaller proportion of the population is used as a norm through scientific researchers, referred to as a survey. The following formula (1) has been utilized to assess the sample size of an infinite population formula (2) Creative Testing Method [21].

$$SS = \frac{Z^2 \times P \times (1-P)}{C^2} = \frac{1.645^2 \times 0.5 \times (1-0.5)}{0.07^2} = 138 \quad (1)$$

Where,

SS = Size of Sample.

Z = Z Value (e.g. 1.645 for 90% confidence level).

P = Percentage of people who make a choice, given as a decimal (0.50 for sample size needed).

C = Confidence interval (0.07)

Correction for finite population, when population is 324, then

$$SS_{new} = \frac{SS}{1+\frac{SS-1}{pop}} = \frac{138}{1+\frac{138-1}{324}} = 97.1 = 100 \quad (2)$$

5. Analysis

The focus of this analysis was specifically construction engineers working in Egyptian construction with more than 10 years of experience, the questionnaires were directly circulated to Top management, Project managers, Functions managers, construction team, technical office and project control to clarify the goals of this study. The delivery of questionnaires involves experience engineers in public agencies, municipalities, non-governmental organizations, private sector, Owners, Consultants, Contractors and subcontractors. The researcher distributed 124 questionnaires to the intended audience. The researcher obtained 104 questionnaires from these targets. This study included demographic-multi data Figure 1.

58.6% form 10-15 years of experience, 23.1% from 16-20 years of experience, 9.6% from 21-25 and 8.7% more than 25 years of experience.

Job title of the respondent Demographics 36.5% form project control team, 28.8% form construction team, 25% form Project manager/deputy and 9.6% form others.

Project Type Demographics 58.6% form commercial and residential, infrastructure and Roads 25% and industrial 16.3%.

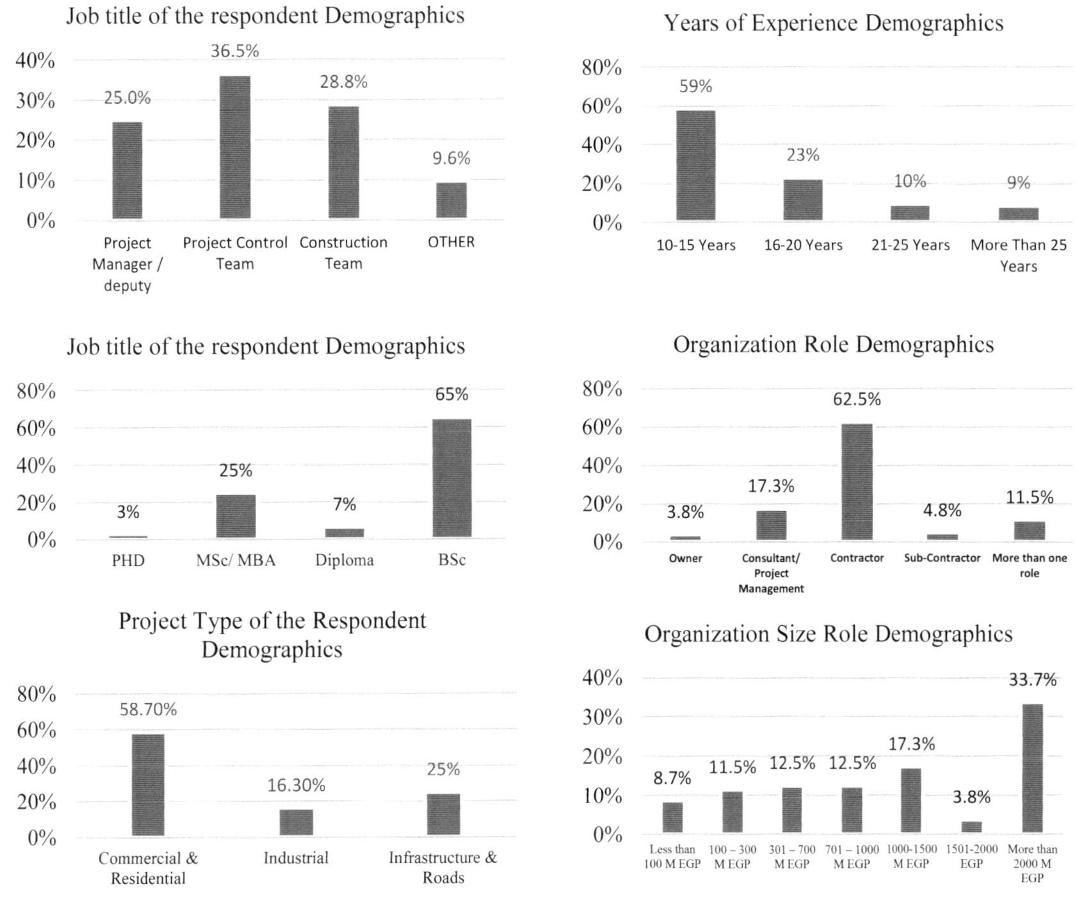

Figure 1. Demographics Data.

6. Factors effectiveness the Construction Projects Performance in Egypt

The study's descriptive statistics show the relative relevance index and rank of factors affecting the construction projects performance in Egypt. The following steps show a summary of factors ranking according to each type of target group:

1- Consultant/project management ranked scheduling skills in top, with a RII of 0.68 while contractor, sub-contractor and More than one role ranked Human skills in the first position with RII equal 0.78, 0.89 and 0.80.

2- Human skills have been ranked by Commercial & Residential and Infrastructure & Roads Projects in the first position with RII equal 0.78 and 0.76 while Industrial projects ranked documentation and Contract Administration skills in the first position with RII equal 0.78 Table 1.

3- Human skills has been ranked in the first position for all projects volume over than 2000MEGP except project volume from 1500-2000M EGP which ranked Risk Management skills in the first position with RII equal 0.80.

Table 1. The relative effectiveness index of factors affecting the construction projects performance in Egypt according to project type.

Commercial & Residential Projects		Industrial projects		Infrastructure & Roads projects	
Skill	RII	Skill	RII	Skill	RII
Human	0.78	D.C and Contract Admin.	0.78	Human	0.76
Political	0.74	Human	0.77	Procurement Management	0.72
Conceptual	0.71	Scheduling	0.76	Political	0.71
Quality Management	0.70	Conceptual	0.74	Budgeting	0.69
D.C and Contract Admin.	0.70	Budgeting	0.72	Scheduling	0.69
Scheduling	0.70	Political	0.71	D.C and Contract Admin.	0.68
Budgeting	0.69	Risk Management	0.71	Conceptual	0.67
Procurement Management	0.68	Procurement Management	0.69	Quality Management	0.66
Risk Management	0.62	Quality Management	0.68	Risk Management	0.59

7. Results:

The correlation coefficients indicate that ٢٣ items along with ٦ components of technical skills should be retained, the 6 components of technical skills named Scheduling, Budgeting and Cost Management, Quality Management, Document and Contract Administration, Risk Management and Procurement Management skills. All 7 items of human skill were retained. The three items of conceptual skill were retained named Visualizing, Scoping and Integration. In addition, all 4 items of Political and social skills items were retained named Social intelligence, Interpersonal influence, Networking ability and Apparent sincerity. Pearson Correlation nonparametric analysis was performed to correlate between the skills components and project time and cost performance. Pearson Correlation was chosen because the data has a normal distribution. The correlation analysis findings are shown in Table 2.

Table 2. Correlation analysis between skill components and project performance.

	Cost Performance		Time Performance	
	Pearson Correlation	Sig. (2-tailed)	Pearson Correlation	Sig. (2-tailed)
Scheduling Skills	0.168	0.088	0.220	0.025
Budgeting Skills	0.244	0.012	0.232	0.018
Quality Management Skills	0.114	0.248	0.159	0.018
D.C and Contract Administration Skills	0.114	0.248	0.061	0.541
Risk Management Skills	0.245	0.012	0.235	0.017
Procurement Management Skills	0.100	0.312	0.081	0.411
Human Skills	0.054	0.588	0.125	0.205
Conceptual Skills	0.086	0.387	0.201	0.041
Political Skills	0.054	0.588	0.090	0.364

8. Conclusion and Recommendations

This study has tested nine skills of projects managers to determine their effect on construction projects performance in Egypt in terms of project time and cost performance. The following Conclusion:

- Budgeting and risk skills have a significant influence on project cost performance.
- Scheduling, Budgeting, Risk Management and Conceptual skills have a positive impact.
- The necessity for a project manager to have an applicable strategy of risk management.
- Construction organizations can use the results as a guide in assigning the project managers according to their skills. They might also focus the development strategy on the important skills that improve the construction project's performance.

References

[1] R. Y. Sunindijo, "Project manager skills for improving project performance," *International Journal of Business Performance Management,* 2015.

[2] S. El-Sabaa, "The skills and career path of an effective project manager," *International journal of project management,* 2001.

[3] G. R. Ferris, D. C. Treadway, R. W. Kolodinsky, W. A. Hochwarter, C. J. Kacmar, C. Douglas and D. D. Frink, "Development and validation of the political skill inventory," *Development and validation of the political skill inventory. Journal of management,* pp. 31(1), 126-152, 2005.

[4] M. M. Marzouk and F. G. Emad , "Assessing Egyptian construction projects performance using principal component analysis," *International Journal of Productivity and Performance Management,* 2018.

[5] S. . S. A. Shaban, "Factors affecting the performance of construction projects in the Gaza strip," *Journal of Civil engineering and Management,* pp. 15(3), 269-280, 2008.

[6] S. M. Dissanayaka and M. K. Mohan, "Evaluation of factors affecting time and cost performance in Hong Kong building projects," *Engineering Construction and Architectura Management,* pp. 6(3), 287-298, 1999.

[7] D. W. W. Chan and M. K. Mohan, "Compressing construction durations: lessons learned from Hong Kong building projects," *International journal of project management,* pp. 20(1), 23-35, 2002.

[8] R. U. Farooqui, M. Saqib and S. M. Ahmed, "Assessment of critical skills for project managers in Pakistani construction industry," in *In Proceeding of Conference on Construction in Developing Countries*, 2008.

[9] P. M. Institute, A guide to the Project Management Body of Knowledge (PMBOK guide), Newton Square, PA: Project Management Institute, 2017.

[10] R. L. Katz, Skills of an effective administrator, Harvard Business Review Press, 1974.

[11] E. R. Fisk, Construction Project Administration Prentice Hall, New Jersey, USA, 1997.

[12] R. Bénabou and J. Tirole, "Self-confidence and personal motivation," *The quarterly journal of economics,* vol. **117(3)**, pp. 871-915, 2002.

[13] V. K. Verma, Human Resource skills for Project Manager, Newtown Square, Pennsylvaina 19073 USA: Project Management Institute Four Campus Boulevard, 1995, February.

[14] P. Sundström and Z.-V. Annika , "Organizing for innovation in a product development project: combining innovative and result oriented ways of working–a case study," *International Journal of Project Management,,* pp. 27(8), 745-753, 2009.

[15] R. Y. Sunindijo and X. Z. Patrick, "CHPT construct: essential skills for construction project managers," *International Journal of Project Organisation and Management,* 2011.

[16] L. C. Stuckenbruck, Integration: The essential function of project management, 2, 56-81 ed., Project management handbook, 1998.

[17] R. S. Goodwin, "Skills Required of Effective Project Managers," *Journal of Management in Engineering,* pp. 9. 10.1061/(ASCE)9742-597X(1993)9:3(217), 1993.

[18] K. K. Ahearn, G. R. Ferris, W. A. Hochwarter, C. Douglas and A. P. Ammeter, "Leader political skill and team performance," *Journal of management,* pp. 30(3), 309-327, 2004.

[19] R. E. Glor, T. M. Townsend and M. F. Bernard, "Sampling reptile diversity in the West Indies with mouse glue-traps," *Herpetological Review,* vol. **31**, no. 12, p. 88, 2000.

[20] S. G. Naoum, Dissertation research and writing for construction students, Routledge, 2007.

[21] S. K. Thompson, Simple random sampling, Sampling, 2012.

Human Resources Management influence on projects performance: A Study of Contractors Professionals Working on Construction Projects in Egypt

M S Ghattas[1,*] H A Bassioni[2] and E F Gaid[1]
[1] Orascom Construction Egypt, Cairo, Egypt.
[2] Construction and Building Engineering Department, College of Engineering and Technology, Arab Academy for Science, Technology and Maritime Transport, Alexandria, Egypt.

[*]Corresponding Author Email: Merhan.saad@orascom.com

Abstract. The Research is about the human resources management influence on Project Performance, A study related to contractors professionals that are working on construction projects in Egypt. Ten major HRM Factors (objectives and goals, job description , staffing and recruitment, training, communication, leadership, team relationship, retaining people, trust and values and finally appraisal and motivation) were assessed for their relation and influence on both cost and Time project performance, A sample of 105 respondents replied with how HRM factors were implemented and how were their team effectiveness on their Egyptian construction projects, also giving their actual project measuring of performance ,analysis had been done including all collected data, descriptive analysis and also correlation tests were implemented in this study, mainly 8 HR factors out of 10 were significantly correlated with cost performance, also 4 out of 10 HR factors were significantly correlated with time performance. Results indicating that HRM Factors have an impact and relation to project performance, but not a governing factor.

Keywords: Construction Management, Human Resources Management, Project performance.

1. Introduction.

Construction Industry is considered as one of the most labor-intensive industries, human resource management (HRM) is given inappropriate caring. Furthermore, the center of interest regards to HRM is mainly at centralized head-office, but most of problems and operational challenges arise on projects. [1]. It is worthy to notice that studies testing trends related to project management in publishing generally showed a steady but rather low interest in HRM/personnel management issues (e.g. [2] and [3]), and it is argued that based on extended research that the project could be considered as a temporary organization added to a permanent structure, to have HRM system is considered to be rare [4]. The aim of this study research is for measuring the implementation of human resources management in Egyptian construction contractors and investigate its effectiveness on the project's performance of in terms of cost and time.

2. HRM definitions.

Human Resource Management (HRM) has been defined as a coherent, and a strategic tactic to the management of an organization's most valuable assets [5]; that is, working people who individually and collectively contribute to the achievement of its goals and targets. In this era of globalization, the success of business organizations highly related to an efficient and effective use of its infrequent resources. Among the required resource, human resource is the most powerful. Also, HRM can be defined as a process of acquiring, motivating, developing, and maintaining human resources to reach specific required outcomes in an organization [5].

3. Developed theories and Models linking HRM & organizational performance.

3.1 Harvard Model: -

The Harvard Model makes an important link between HRM decisions, the business environment and an organization's performance. Capability to consider both the circumstances and the stakeholders' interests is important in construction projects, Here the mutual interdependence of those involved, those affected and those who can affect the project's results must be taken into consideration through appropriate HRM decisions and policies, [1] as shown in Figure 1.

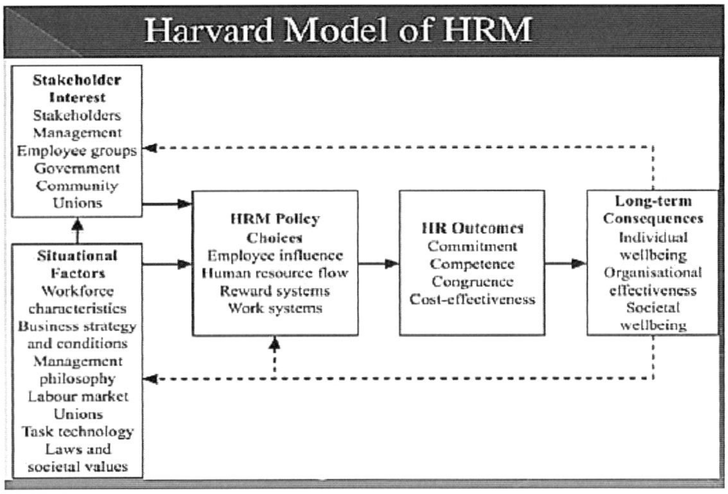

Figure 1. The Harvard Model.

3.2 The Michigan Model

The Michigan realizes the essential important area for motivating, promoting and awarding personnel, finds that when you try to manage the assets of human and utilize them in optimum case, you can reach all your goals. But experimental proofs have not produced any significant link for practical systematic and consistent with hard HRM. [6] , as shown in Figure 2.

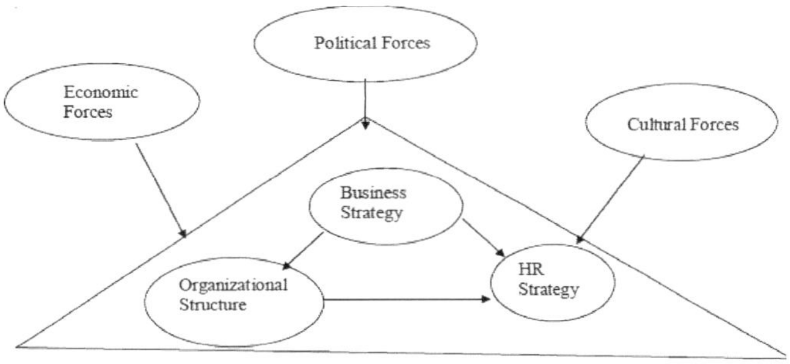

Figure 2. The Michigan Model.

3.3 The Guest Model: -

Planning of project Manpower as an important side of the functional plans and must have the power to reach the achievement for all organizational goals' situation. Whereas, Management can't adapt easily or flexible. The Guest Model trusts on total positive change in quality by preparing and delivering quality products leading to quality treatment to its manpower as valuable assets. [6], as shown in Figure 3.

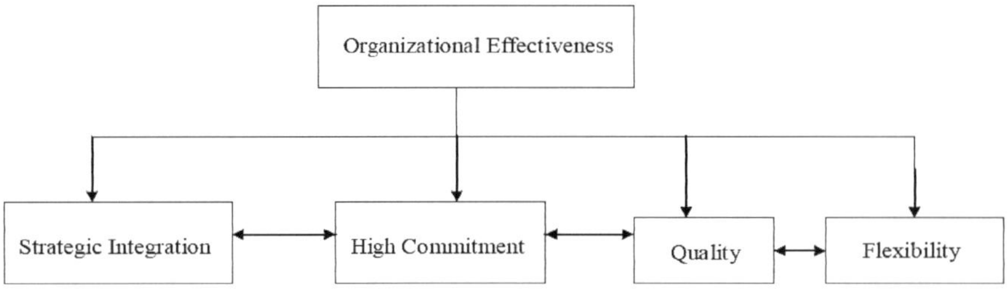

Figure 3. The Guest Model.

4. HRM in a project-oriented company.

Over recent decades, Notion of 'Projectification' has been arise in major sectors of economics. It was obvious that a transformation happened in firms of project base. Where most of tasks are performed in projects as well as administrative support is provided by permanent structures [7]. A new organizational type needs to be project-based, as the main effort goes for delivering products and services to customers. The core in the companies which are project-oriented are Human resource management (HRM), affecting the way the organization develops and utilizes human resources, and how employees experience the employment relationship. Many researchers and theorists produced knowledge about HRM through publishing their work in journals and books, such method for building up knowledge, make a frame for the shape of debates raised in the researches. [8]

To conclude, the mixing of research or theorizing on project-based companies and HRM in the project management and general management literature is rather rare and often focuses on the temporary employment.

5. Researches and Studies examining HRM influence on PM.

At present, HRM-performance repetitive studies are mainly giving an attention on when and how HRM influences performance rather than testing its effect on performance, [9]; On the other hand surprisingly, 35 HRM/innovation empirical studies have been reviewed in the last 25 years [10], showed that most of past studies have focused on studying direct HRM performance links. So-called HRM-organizational performance is divided into four levels; employee outcomes (attitudinal and behavioral), organizational outcomes (productivity and efficiency), financial outcomes (sales and profit) and market-based outcomes [11] On the Other hand, [12] found no clear supportive evidence that declared the existence of this mediation effect. Instead, he discovered that HR outcomes do not directly add value to innovation, but rather, we found that providing an innovative and supportive environment is what fosters higher levels of innovation. He concluded in his context he surveyed that innovation is a reached when organizations allow an innovative climate that supports creativity, risk-taking and flexibility to new ideas rather than when individual trails for positive work culture (job satisfaction, involvement and Organizational Behavior).

6. Researches Identifying Factors affecting HRM and their impact on project performance.

It has been argued that the influence of various HRM activities on business result of construction companies in Croatia considering 18 independent variables arranged in 4 groups and 7 variables for control in their model [13]. The 4 groups were: Selection of employee candidates variables, Lifelong learning and on-job training variables, Motivation variables, transfer of knowledge and organizational culture management variables. They proposed other human resources activities, the most reliable significant impact on business results was material motivation activities. Finally, they recommended to construction employers not to neglect the HRM activities influence, especially for material motivation, on the overall performance of the company.

Many researches tried to measure HRM factors and its practices, which are reflected in the project team effectiveness. Ten Main HR factors have been chosen to address the degree of HRM factors implementation on project, after merging some factors. Also, the project performance will be measured in terms of Time and Cost. Finally, it will be examined either there is a direct influence between applying good HRM practice on projects and project performance or not? And if it is available to which extent it will show an impact.

7. Measurements of HRM Factors and project performance main indictors, and historical related studies.

7.1 Measurements of HRM Factors: Many HRM factors could have an impact on construction projects, as mentioned before, finally 10 major factors have been chosen to be measured by using a Likert scale from 1 strongly disagree High to 5 strongly agree, revealing the degree of implementation of the factor in the chosen project, according to collected data from a questionnaire.

7.1.1Objectives and Goals: Every personnel on the team of project has to be informed with a well-defined, clear and consistent common goals, and always strives to meet team goals and objectives [14] and [15].

7.1.2 Job description & career path: To have many potential positions to be promoted for desired employees, job description and career path have to be considered, [16] and [17].

7.1.3 Staffing and Recruitment: Matching people in a Proper way with job vacancies and the goal for such selection process is the organization, [18]. Overqualified or underqualified personnel, or people who aren't fitting for any reason either the job or the organization's culture, they will not be effective and most probably will leave their job, voluntarily or in any other way. Staffing and selection process and insure their importance in HRM practices have been also studied [18].

7.1.4 Training: Training, personal development and creation of knowledge is considered to reach a motivated, efficient and creative leads to innovative industry [1]. Training has been referred as an important factor for development of an effective team [13].

7.1.5 Communication: Individuals to share an understood information, [20], has been discussed as one of the most important team effectiveness factors [19].

7.1.6 Leadership: It is the role of convincing people to change their attitude, motivates them to work together to achieve project objectives. It has been revealed that the importance of Team leadership within the project members [17].

7.1.7 Team relationship: Reflecting the ability of a group of people to stick together to work on reaching job target, including team structure, type and bonds, also managing raised conflicts, [1] and [19].

7.1.8 Retaining people (Turn over & retention): Retention is defined as the long-term employment relationship that result between employees and their organizations due to work-life balance practices, growth opportunities, and training programs, they are both about retaining qualified staff. Importance of retaining people as they are the main organization assets have been outlined, [21] and [22].

7.1.9 Trust and Values: Trust is described as a psychological situation, with all feelings and expectations [19], resulting ideas about others are trustable or not, positive culture comes mainly from team values, the importance of trust and values have been also discussed, [12] and [23].

7.1.10 Motivation, Appraisals and rewards: Motivation is the reasons and pushing forces that inspire employees to do their work assignments [14], whereas a reward is a benefit or bonus designed to help in motivating staff to produce even beyond their usual performance expectation [24]. Importance of motivation and rewards have been discussed [13].

7.2 Measurements of project performance main indictors Factors: Project data should be collected, showing HRM impact on construction management performance indicators, according to collected data from a questionnaire, two main indictors revealing the project performance, from Cost and Time point of view, description as follows:

7.2.1 Project Cost: It has been argued that the project cost aspect part in the project performance survey can coupled with the way of how cost had been managed rather than its final performance, so it was dealt mainly with the question referring to performance concern, at Project Efficiency part at Project Success Assessment Questionnaire, [19] and [25].

7.2.2 Project Time: It has been argued that the project schedule aspect part in the project performance survey was dealing with the way of how time schedule had been managed rather than its final performance, so it was dealt mainly with the question referring to performance concern at Project Efficiency part at Project Success Assessment Questionnaire, [19] and [25].

8. Research Methodology

The investigation of HRM Factors and their effectiveness on the project performance, also prioritization of the factors according to their impact is a complex task, which requires a systematic approach to data collection from projects, also analysis of the results to give a meaningful answer to research questions. This section outlines the methodology and techniques used to fulfill the aim of the research. And presents research design and selection of appropriate research strategy including perspectives, research different types, research methodology, data collection, sample size and ways of selection, collection tools, including how was it to manage during every single stage during mentioned processes. Quantitative method is applied in this study were selected to show these goals:

- To measure and evaluate HRM Factors of team effectiveness for Contractors Professionals working on Construction Projects in Egypt.
- To show and determine how effective do HRM Factors in construction projects are affecting construction project performance.
- To check the relation between HRM Factors and project performance in construction projects

9. Sampling and Population

The population of the study is Contractors Professionals personnel working on Construction Projects in Egypt completed within the last six years, where HRM Factors could be applied in an effective manner. The respondents should be mainly: (Project Manager-Construction Manager-Site engineer-Project Control-Technical Office-Contracts-Admin and HR), these contacts are asked for voluntary participation via phone or direct contact interaction, questionnaires to be sent via emails. The sample size is a significant feature of any empirical study in which the goal is to make inferences about a population from a sample. This study presents a summary of how to calculate the survey sample size in social research and information system research [26], [27], check Equation 1. and Equation 2.

$$n = \frac{(P\ X\ (100-P)\ Z^2\ )}{E^2} \qquad \text{Equation 1.}$$

n is the sample size, P is the percentage occurrence of a state or condition (50), E is the percentage maximum error (8), Z is the value corresponding to level of confidence (1.645 for 90% confidence level).

$$n = \frac{50\,X\,(100-50)1.645^2}{8.0^2} = \frac{6{,}765}{64} = 105\,samples \qquad \text{Equation 2.}$$

10. Data Analysis:

A Total of 105 Contractors Professionals in Egypt replied to the questionnaire, on Google forms, they are all working on Construction Projects in Egypt. Everyone had chosen only one project, in Egypt finished within the last 6 years. The collected data had been quantitatively analyzed, the researcher used descriptive and relations correlation analysis, and the following analysis were performed: First, Descriptive analysis to demonstrate sample characteristics based on demographics analysis. Second, Correlation test to find out the existing correlations between independent variables and dependent variable.

10.1 Descriptive analysis:

- *Categorizing by project category:* 33 projects were residential Building (representing 31.4% from total sample), 29 projects were non- residential (representing 27.6% from sample), 25 projects were Industrial one (representing 23.8%), while 18 projects were heavy Civil (about 17.1% from sample size), which shows that all types of projects were represented inside the sample (Figure 4).
- *Categorizing by project contract amount:* Figure 4 shows that 27 projects were more than 2.5 Billion (25.7% from sample), 23 projects were from 1 to 2.5 Billion (21.9% from sample).

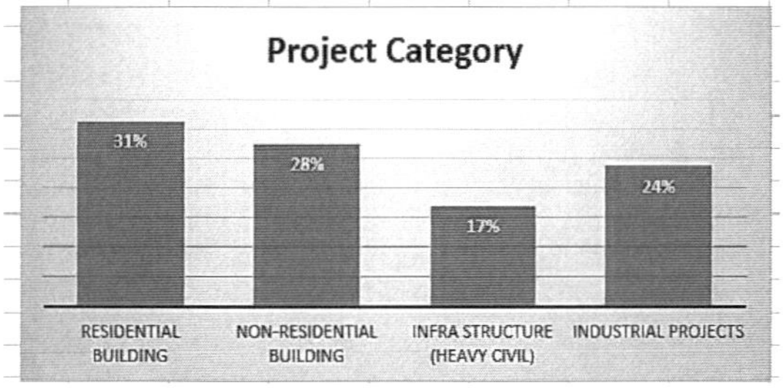

Figure 4. The project Category

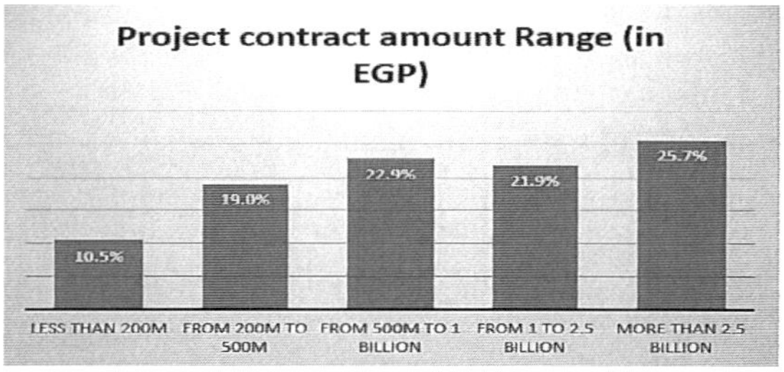

Figure 5. The Project contract amount range

10.2 Correlation test:

The correlation test had been done twice , for all HR independent factors versus the two dependent performance factors, First , it had been done with actual cost performance, results were indicating some significant values for 8 HR factors out of 10, where Sig. (2-tailed) < 0.05, it means that there is a relation , between every independent HR factor and the cost performance as a dependable one, the most related one was the Motivation & Appraisal Factor resulted 0.0000 , Cost correlation with Leadership 0.001,with Job Description 0.008 , with Team relationship 0.010,with Retaining people 0.017, with Training 0.019, with Staffing recruitment 0.02, Finally with Communication 0.033.Only 2 Factors were Sig. (2-tailed) > 0.05, these factors are Goals and objectives and Trust and values (Table 1.)

Table 1. Correlation between HRM Factors and Project Cost Performance.

Cost performance	Pearson Correlation	Sig. (2-tailed)
Goals objectives	0.091	0.357
Job description	0.257	0.008
Staffing Recruitment	0.227	0.020
Training	0.228	0.019
Communication	0.208	0.033
Leadership	0.306	0.001
Team relationship	0.250	0.010
Retaining people	0.232	0.017
Trust Values	0.048	0.630
Motivation Appraisals	0.346	0.000

Second , it had been done with actual time performance, results were indicating some significant values for 4 HR factors out of 10, where Sig. (2-tailed) < 0.05, it means that there is a relation , between every independent HR factor and the time performance as a dependable one, the most related one was the Motivation & Appraisal Factor resulted 0.0000 , Cost correlation with Retaining people 0.01,with Job Description 0.011, Finally with Communication 0.044. Only 2 Factors were Sig. (2-tailed) > 0.05, these factors are Goals and objectives, Staffing recruitment, Training, Leadership, Team relationship and Trust and values (Table 2).

Table 2. Correlation between HRM Factors and Project Time Performance.

Time performance	Pearson Correlation	Sig. (2-tailed)
Goals objectives	0.078	0.432
Job description	0.246	0.011
Staffing Recruitment	0.176	0.072
Training	0.080	0.416
Communication	0.197	0.044
Leadership	0.165	0.092
Team relationship	0.141	0.150
Retaining people	0.251	0.010
Trust Values	0.062	0.531
Motivation Appraisals	0.339	0.000

11. Conclusion.

This Research was about testing the human resources management influence on Project Performance, A study of contractors professionals working on construction projects in Egypt. Ten major HRM Factors were assessed for their relation and influence on project performance, two main project performance indicators, actual cost and actual time performance were received, Results were as follows:

- Correlation results giving a relation between HRM Factors and project actual cost performance, mainly 8 Factors significant with cost performance, namely Motivation and Appraisal factor, Leadership, Job Description, Team relationship, retaining people, Training, Staffing recruitment and Communication.
- Correlation results giving a relation between HRM Factors and project actual time performance, namely 4 Factors were significant, Motivation and Appraisal Factor, Retaining people, Job Description, and Communication.
- Finally, it had been found through the quantitative analysis that there is a relation and impact on actual project performance, but it is not a governing relation.

References

[1] A. Dainty, M. Loosemore and H. Lingard, Human Resource Management in Construction Projects Strategic and operational approaches, Taylor & Francis Group, 2003.

[2] L. Crawford, J. Pollack and D. England, "Uncovering the trends in project management: journal emphases over the last 10 years.," *Int. J. Proj.Manag.,* vol. **24**, no. 2, p. 175–184, 2006.

[3] K. Bredin and J. Söderlund, "The HR quadriad: a framework for the analysis of HRM in project-based organizations.," *Int. J. Hum. Resour.Manag.,* vol. **22**, no. 10, p. 2202–2221., 2011.

[4] M. Huemann, Human Resource Management in the Project-Oriented Organization: Towards a Viable System for Project Personnel., Gower: Aderlshot., 2015.

[5] M. Armsrtong, A Handbook of Human Resource Management Practice, 10th ed., London: Kogan Page, 2010.

[6] A. Chatterjee and S. Guha, "Manpower Planning - Theoretical Perspectives of Utility Theory and Models," *Global Journal of Management and Business Research: A Administration and Management,* p. Volume

19 Issue 4 Version 1.0, 2019.

[7] J. Packendorff and M. Lindgren, "PROJECTIFICATION AND ITS CONSEQUENCES: NARROW," *South African Journal of Economic & Management,* pp. Special Issue **17** (2014) : 7-21, December 2013.

[8] M. Huemann, A. Keegan and J. R. Turner, "Human resource management in the project-oriented company:," *International Journal of Project Management,* p. 315–323, 2007.

[9] D. E. Guest, "Human resource management and performance: still searching for some answers," *Human Resource Management Journal,* vol. **21**, no. 1, p. 3–13, 2011.

[10] H. Seeck and M.-R. Diehl, "A literature review on HRM and innovation – taking stock and future directions," *The International Journal of Human Resource Management,* 2016.

[11] A. Savaneviciene and Z. Stankeviciute, "HumanResource Management Practices linkage with organizational commitment anf job satisfaction," *Economics amd Management*, 2011.

[12] A. M. Noor Egal, *Determining the Relationship between Human Resources Management Practices and Organizational Innovation: The Case of Somali Telecom and Banking Firms,* Eskisehir: Anadolu University, 2019.

[13] I. Š. Nukić and N. Šuvak, "Impact of Human Resources Management on Business Result of Croatian construction companies," *Organization, technology and management in construction: an international journal,* pp. 663 - 675, 2013.

[14] M. Nadeem, "Role of training in determining the employee corporate behavior with respect to organizational productivity: Developing and proposing a conceptual model," *International Journal of Business and Management,* vol. **5**, no. 12, 2010.

[15] M. Chand and A. A. Katou, "The impact of HRM practices on organisational performance in the Indian hotel industry," vol. **29**, no. 6, pp. 576-594, 2007.

[16] L. Sun, S. Aryee and K. Law, "High-performance human resource practices, citizenship behaviors and organizational performance: a relational perspective," *Academy of Management Journal,* p. 558–577, 2007.

[17] K. Bredin and J. Söderlund, "Project managers and career models: An exploratory comparative study," *International journal of project managment*, p. 89–902, 2013.

[18] G. Dessler, Human Resource Management thirteen edition, USA: Pearson, 2013.

[19] N. Azmy, *The role of team effectiveness in construction project teams and project performance,* Ames , Iowa: Iowa State University, 2012.

[20] D. Moore, Dainty and A.R.J., "Intra-team boundaries as an inhibitor to performance improvement in the UK construction industry," *Construction Management and Economics,* pp. 559-62, 2001.

[21] f. Benest , "The Role of Benefits in Winning the War for Talent," *International Foundation of Employee Benefit Plans,* 2008.

[22] S. Pai, S. Dheeraj and R. S. Kristam , "Human Resource Management Techniques Dealing With Deficit & Retention of Skilled Labor in Construction Industry," *International Journal of Scientific and Research Publications,* pp. 1-6, 2013.

[23] M. Armstrong, A Handbook of Human Resource management practice, London: Kogan Page, 2006.

[24] A. A. Katou and P. S. Budhwar, "The Effect of Human Resource Management Policies on Organizational Performance in Greek Manufacturing Firms," *Thunderbird International Business Review,* vol. 49, no. 1, 2007.

[25] H. M. a. H. Mughal, *Effectiveness of project teams and thier impacts on the performance of Saudi construction projects,* 2013.

[26] H. Taherdoost, "Determining Sample Size; How to Calculate Survey Sample Size," *International Journal of Economics and Management Systems,* vol. **2**, pp. 237 - 239, 2017.

[27] A. Bryman and E. Bell, Bussiness Research Methods, 3rd ed., Oxford univeristy press, 2011.

Evaluating the properties of hot asphalt mixtures modified by Metakaolin

Ahmed W. Oda[1,*], Ahmed El-Desouky[2], Hassan Mahdy[3], Osama M. Mousa[1]
[1] Department of Civil Engineering, Military Technical College, Cairo, Egypt.
[2] Dean of Belbeis Higher Institute for Engineering, Belbeis, Egypt.
[3] Ain Shams University, Cairo, Egypt.

[*]Corresponding author email: Ahooba122@gmail.com

Abstract. The mineral filler has an important role in filling the voids between aggregates inside the asphalt matrix. This study investigates the effect of using Metakaolin (MK) as a modifier on the mechanical properties of hot mix asphalt. MK was added to the mixtures as a part of the filler used. As known, the hot mix asphalt (HMA) consists of coarse aggregate, fine aggregate, filler material, and asphalt binder with or without existing modifiers. The components of the HMA were subjected to validation tests to check the validity of the materials for producing HMA The asphalt binder used was bitumen 60/70. Tests performed on bitumen were penetration test, softening point test, flash and firing point test. The optimum asphalt content based on the designed control mix was 5.2%. the control mix has no content of MK. MK was added as a partial replacement of filler used (limestone powder). Mk was added to the HMA by replacing 20%, 40%, 60%, 80%, and 100% of the filler. A comparative study was performed to evaluate the effects of HMA modification by MK. Marshall quotient (MQ) increases by 10.37% for the mix modified by 40% MK. MK has significant effects on the indirect tensile strength (ITS).

Keywords: Hot mix asphalt; Metakaolin; Mineral filler; modified HMA.

1. Introduction

In recent days, because of the growing traffic loads, it becomes very necessary to give more concern to enhancing asphalt pavement properties. Besides, the filler has an important role to fill the air voids inside the asphalt mixtures. This reduction improves the transfer of loads to provide stiffening and causes a reduction in temperature susceptibility, [1-5]. There are many types of modifiers that are used to enhance the performance of asphalt pavements. These modifiers may be added to the bitumen to enhance the binder properties as nanomaterials, crumb rubber, ….etc. Modifiers may be added to the aggregates or the filler. These modifiers are for example marble dust, fly ash, Metakaolin, and other modifiers.

A. W. Oda et al used Nanosilica (NS) and Nanoclay (NC) as binder modifiers. Both nanomaterials showed significant improvements in bitumen properties. Modified HMA by NS and NC also possessed higher stability, ITS, and better performance against the rutting effect under heavy traffic loads [6]. This research focuses on studying the effects of HMA modification by using different fillers. In this study, MK was added as a replacement for the filler used, limestone powder (LSP).

The use of MK in the construction industry as a partial replacement of cement started in the 1960s and the interest in this material increased in recent years. Kaolin's chemical composition is aluminous silicates hydrates associated with Manganese (Mn), Ferrite (Fe), Calcium (Ca), potassium (K), and Sodium (Na). MK is produced by thermal treatment of Kaolin at a temperature of about 650-800°C and a heating time of 90 minutes is found to have active Pozzolanic content which can be substituted or used as a filler material. MK has significant effects in enhancing the mechanical properties of cement concrete. Some researchers use MK as a cement replacement for producing high-strength concretes. The modified concrete shows better workability, higher strength, and more durability with reduced permeability compared to traditional concrete, [7].

Murana et al used MK as a partial replacement for PC to reduce the harmful effect of emission of CO_2 and to improve the performance of asphalt, [8].

Kareem et al. used nano-metakaolin(NMK) as a modifier for the asphalt binder. NMK was added by 1, 2, 4, 8, 16, and 20% by weight of the binder. The optimum NMK content was determined by the conventional bitumen tests. The optimum content was 16% NMK. It was found that the HMA modified by 16% NMK provides significant enhancements in the HMA properties with higher durability, [9].

2. Materials

Materials used in this research are:

a) Bitumen 60/70.
b) Coarse aggregates: (Aggregate A, Aggregate B, Aggregate C).
c) Fine aggregates: crushed sand.
d) Limestone powder: as a mineral filler.
e) Metakaolin: as a replacement of the filler.

Figures 1 and 2 present samples of bitumen and aggregates.

Figure 1. Bitumen samples.

Figure 2. Aggregate sampling.

3. Methodology and experimental work

An experimental work program was established to produce control mixtures (have no content of MK) and modified Mixes (contain different contents of MK). Bitumen, aggregates, filler, and MK were subjected to validation tests. The tests were performed according to AASHTO and ASTM. The results of the validation tests match the limits of ECP, 2008. Metakaolin was added by percentages 20%, 40%, 60%, 80%, and 100% by weight of filler. A comparative study was conducted to investigate the effects of modification HMA by different amounts of MK. Figure 3 illustrates the flow chart of the experimental work.

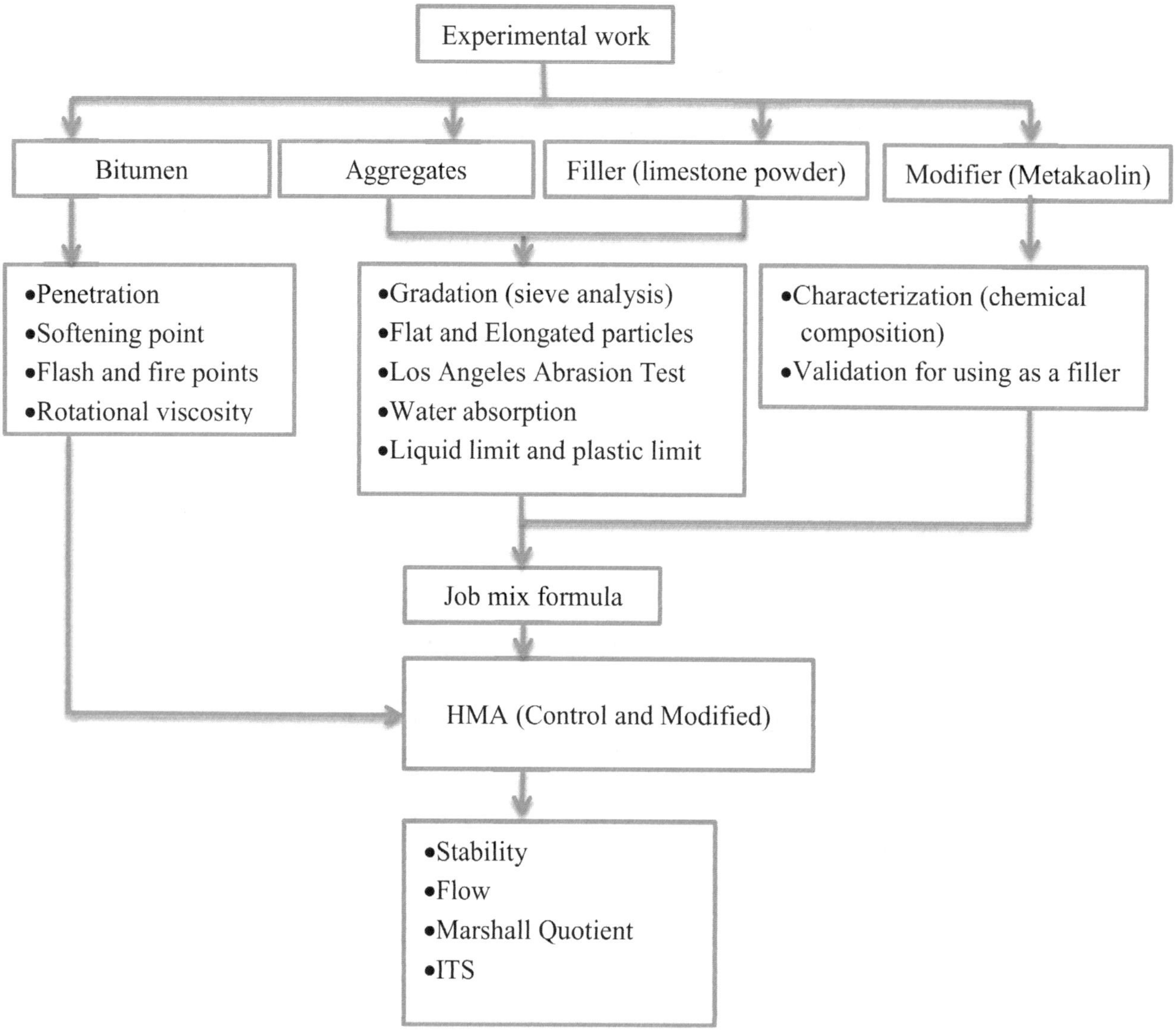

Figure 3. Flow chart of the experimental work.

4. Results and discussion

4.1. Bitumen testing

Bitumen used is bitumen 60/70 produced by Nasr Petroleum Company in Suez, Egypt. Bitumen was tested according to ASTM and AASHTOspecifications. The tests performed were penetration test, softening point test, flash and firing point test, and Brookfield viscosity test. Figure 4 presents photos of some conventional bitumen tests. Table 1 list the properties of bitumen characteristics and the ECP, 2008 limits. The results match the code limits.

Figure 4. Bitumen Characterizing.

Table 1. Specifications of bitumen.

	Standard	value	Unit	ECP, 2008 limits
Penetration @ 25°C for 5 sec.	AASHTO T 49	61	0.1 mm	60-70
Softening Point	ASTM D 36	48.5	°C	45-55
Flash point	ASTM D92	305	°C	>250 °C
Fire point	ASTM D92	336	°C	>250 °C
Rotational viscosity @135°C	AASHTO T201	345	Cst	>320

4.2. Aggregate testing

The aggregates used were divided into three coarse aggregates: Aggregate A, Aggregate B, and Aggregate C based on their particles size. In addition, crushed san was used as a fine aggregate. The mineral filler was limestone powder. Tests were performed on coarse and fine aggregate that were completely resulting from crusher millers. Table 2 illustrates the results of the shown tests listed below.

Table 2. Specifications of Aggregates.

Test	Standard	Agg. A	Agg. B	Agg. C	Crushed Sand	ECP, 2008 limits
LA abrasion after 500 rev.	ASTM C131	13.7	19.7	20	--	Max 40 %
Flat particles	ASTM D 693	3	3	2	--	Max 10 %
Elongated particles	ASTM D693	2	4	2	--	Max 10%
Water absorption	ASTM C 127	1.5	2.3	2.1	1.1	Max 5 %
LL & PI	ASTM D 4318	--	--	--	0	0
Bulk Specific Gravity	ASTM C29	2.67	2.53	2.5	2.58	

4.3. Metakaolin

The used MK in this research were produced locally and brought from the local market. Chemical and physical investigations were performed to detect the properties of MK, and the validation tests respectively.

Table 3. MK chemical composition.

Ingredients	% by weight
SiO_2	48.30
Al_2O_3	38.24
Fe_2O_3	3.22
TiO_2	3.27
MgO	2.13
CaO	2.01
Na_2O	1.80
K_2O	1.03

Table 4. MK physical properties.

Property	Value
Color	Off-white
Physical form	Powder

Table 5. MK and LMS validation tests.

	Standard	Values for MK	Values for LSP	Unit	ECP, 2008 limits
Sieve analysis	ASTM C117	70.3	68.1	%	Passing from sieve #200 > 65%
		90.4	86.7		Passing from sieve #100 > 85%
LL and PI	ASTM D4318	0	0	%	Not plastic

4.4. Job mix formula

The gradation used in preparing the desired mix is the gradation (4C) for use as a surface layer in the asphalt pavements. The job mix formula was shown in table 6, and figure 5.

Table 6. Job mix Formula.

Sieve size		JMF	Specifications	
Inch	mm		L.L	U.L
1"	25.4	100	100	100
¾ "	19	98.7	80	100
3/8"	9.51	74.6	60	80
#4	4.76	52.0	48	65
#8	2.38	41.6	35	50
#30	0.595	22.6	19	30
#50	0.297	16.5	13	23
#100	0.149	9.7	7	15
#200	0.074	5.2	3	8

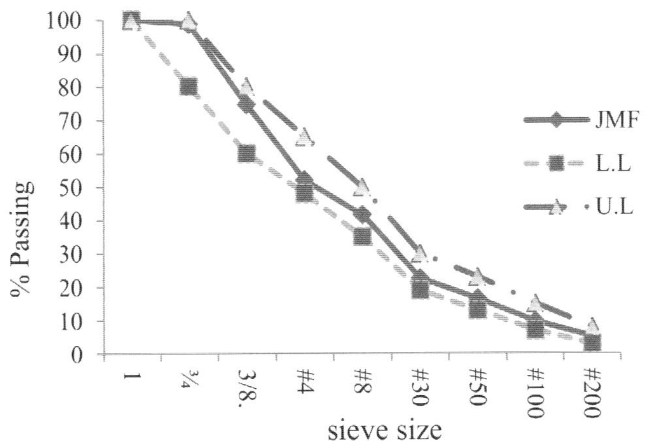

Figure 5. job mix formula

4.5. HMA Design

The control HMA (0% MK) was designed according to the Marshall mix design method. The optimum asphalt content was 5.2%. This asphalt content was used to produce the modified mixes by MK. As aforementioned MK was added to the asphalt mixture as a partial replacement for filler by the mentioned contents. The comparative study was based on the change in the HMA performance. Control and modified mixes were tested by Marshall Stability, flow, Marshall quotient, and Indirect Tensile Strength. Figures 6 to 8 present specimens and tests of HMA.

Figure 6. HMA specimens.

Figure 7. stability and flow measuring.

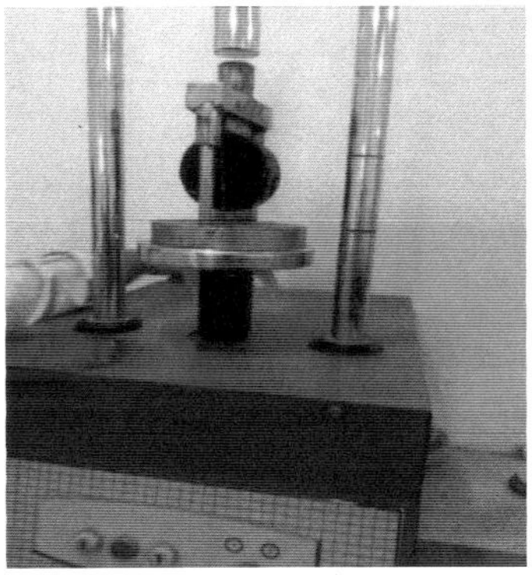

Figure 8. ITS measuring.

The mixes are named as mentioned in table 7. The results of the control and modified specimens are shown in figures from 9 to 16.

Table 7. Mixes numbering.

MK conten (%)	0	20	40	60	80	100
Mix Number	Mix 1	Mix 2	Mix 3	Mix 4	Mix 5	Mix 6

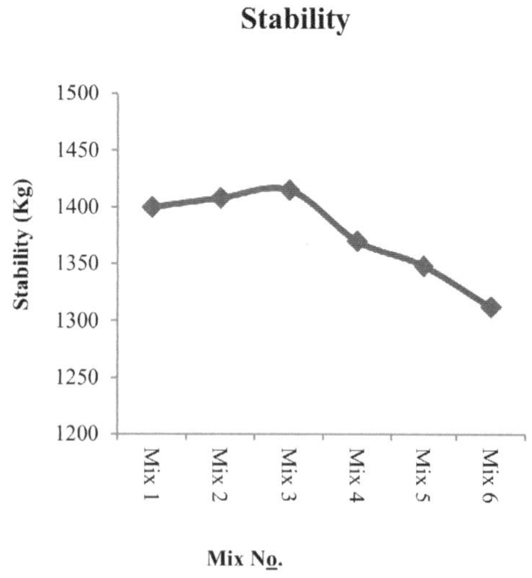

Figure 9. Stability.

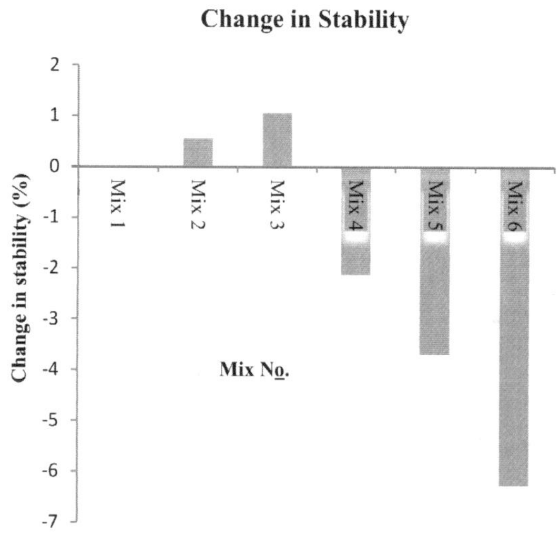

Figure 10. Changes in Stability values.

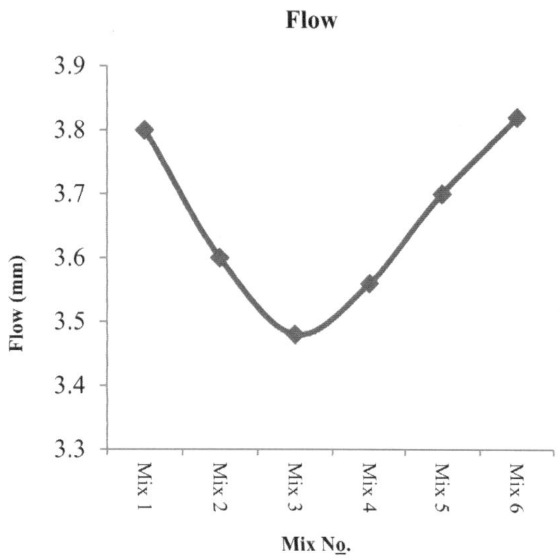

Figure 11. Flow.

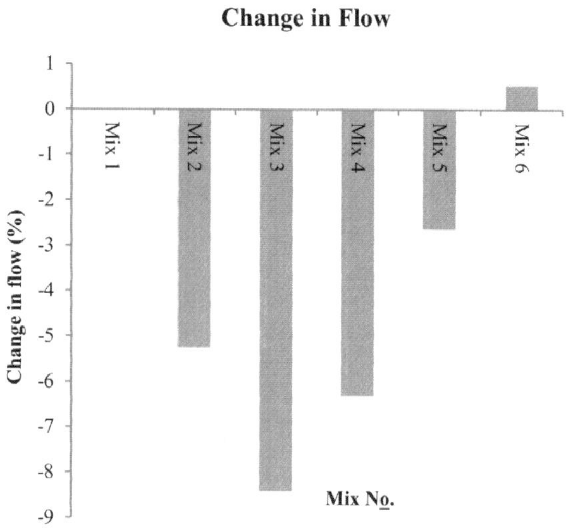

Figure 12. Changes in Flow.

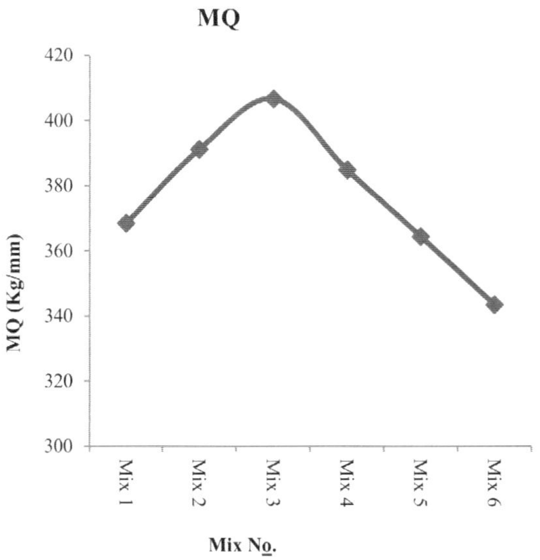

Figure 13. Marshall Quotient (MQ).

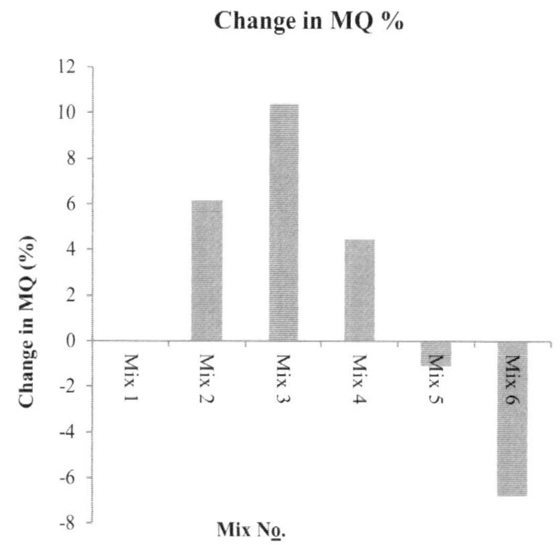

Figure 14. Changes in MQ.

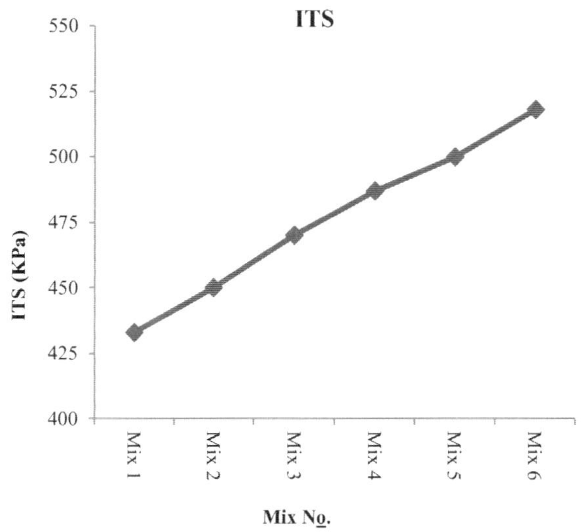

Figure 15. Indirect Tensile Strength.

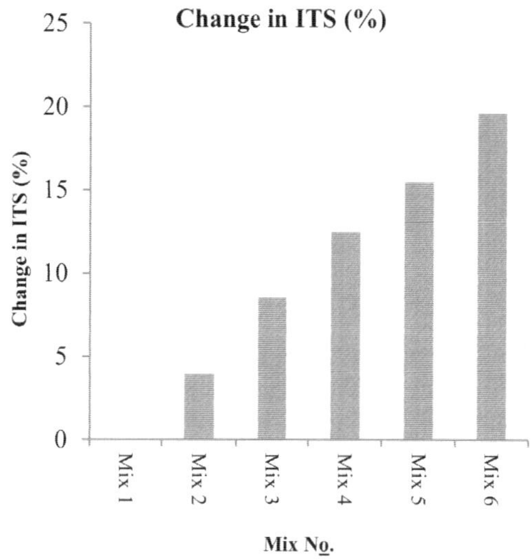

Figure 16. Changes in ITS.

From figures 13 and figure 14, stability increases slightly to reach 1408 Kg for mix 2 by a ratio of 0.57%. Then continues to increase by 1.07% at mix 3 to become 1415 Kg. For mixes 4, 5 and 6 stability decreases gradually with increasing the MK content.

Flow decreases with increasing the MK content to reach the lowest value at mix 3. At mix 3 the flow decreases by 8.42%. after that value the flow returns to increase at mixes 4, 5, and 6. The previous values of stability and flow affect directly the MQ values. MQ reaches 368.42 Kg/mm for mix 1then increases dramatically to reach 406.61 for mix 3 by 10.37 ratio of increase. MQ decreases gradually for mixes 4, 5, and 6 to reach 343.46 Kg/ mm.

The ITS shows significant improvements with increasing the MK content. The highest value of ITS is for mix 6 that has the highest content of MK. From the previous results, MK provides significant improvements in the HMA performance.

5. Conclusion

MK is a Pozzolanic material that has significant effects on the properties of hot asphalt mixtures. Stability increases slightly with the increase of the MK content to achieve the highest stability value at 40% MK content (mix 3). After that stability value, with increasing the MK content stability begins to decrease gradually and also slightly.
MQ values increase with the increase of MK content. The increase of MQ indicates higher mix stiffness. The best stiffness is achieved at 40% MK content. From the previous Mix 3 is considered the optimum mix that achieves better stability and MQ. That means higher resistance to traffic loads.
The ITS increases dramatically with the increase of MK content. This behavior indicates significant improvement to crack resistance and a high enhancement of resistance to damage due to moisture.

References

[1] A. M. Rashad and G. M. F. Essa, "Optimum Metakaolin Content in Cement Pastes," vol. 6, no. 11, pp. 1–7, 2018.
[2] N. A. A.-S. 3 and L. F. H. Mustafa A Kareem, Abbas Al-Hdabi, "Hot Asphalt Mixture Characteristics with Nano-Metakaolin Materials," in *International Conference on Civil and Environmental Engineering Technologies*, 2019.
[3] B. Xing, W. Fan, X. Lv, and C. Qian, "Study the effects of filler particle size and filler/asphalt ratio on the high and low temperature performance of asphalt mastics," *IOP Conf. Ser. Mater. Sci. Eng.*, vol. 493, no. 1, 2019.
[4] and M. F. A.-R. H. H. Jony, I. Y. Jahad, "The effect of using glass powder filler on hot asphalt concrete mixtures properties," *Eng. Technol. J.*, vol. 29, pp. 44–57, 2011.
[5] Y. Cheng *et al.*, "Influence of the properties of filler on high and medium temperature performances of asphalt mastic," *Constr. Build. Mater.*, vol. 118, pp. 268–275, 2016.
[6] A. W. Oda, A. El-Desouky, H. Mahdy, and O. M. Moussa, "Effects of asphalt modification by Nanosilica and Nanoclay on asphalt binder and hot mix asphalt properties," *IOP Conf. Ser. Mater. Sci. Eng.*, vol. 974, no. 1, 2020.
[7] A. M. Rashad, "Metakaolin as cementitious material: History, scours, production and composition-A comprehensive overview," *Constr. Build. Mater.*, vol. 41, pp. 303–318, 2013.
[8] A. Murana, A. Olowosulu, and S. Ahiwa, "Performance of Metakaolin as Partial Replacement of Cement in Hot Mix Asphalt," *Niger. J. Technol.*, vol. 33, no. 3, p. 387, 2014.
[9] W. Wang, X. Liu, L. Guo, and P. Duan, "Evaluation of properties and microstructure of cement paste blended with metakaolin subjected to high temperatures," *Materials (Basel).*, vol. 16, no. 6, 2019.

Improving the Performance Grade and Traffic Loading of Egyptian Asphalt Binders by Recycled Polyethylene Modification

Ahmed Eraky[1], H. S. Ayoub[2, 3,*], Omar Osman[4], Mohamed E. Ouf[1], Abdel Zaher E.A. Mostafa[1]

[1] Civil Engineering Department, Faculty of Engineering, Helwan University, Cairo, Egypt
[2] Department of Physics, Faculty of Science, Cairo University, Cairo, Egypt
[3] Scientific Center for Polymer Technology, 11585, El Mokatam, Cairo, Egypt
[4] Civil Engineering Department, Faculty of Engineering, Cairo University, Cairo, Egypt

*Corresponding Author Email: Dr.hsayoub@gmail.com

Abstract: Recent SUPERPAVE technology allows assessment and performance grading (PG) of modified asphalts according to AASHTO T 350–14 and ASTM D7405-15 standards. These testing protocols use modified - multiple stress creep recovery (MSCR) test, to yield the high temperature PG and the traffic level of the asphalt binder. Latest AASHTO T 350–14 test protocol is used to assess two major Egyptian bitumen products, Alex and Suez (pen 60/70) before and after polymer modification. These standard products have different fractional composition since they are obtained from different local origins. Low density polyethylene modification (LDPE) of such products leads to a major enhancement in their PGs and decreases their high temperature rutting potential. A characteristic comparison between virgin a and LDPE modified bitumen samples is provided to achieve the desired PG and traffic level for high reliability projects deployed in Egyptian hot climate regions. it was found that a 2% wt recycled LDPE modification, enhances the high temperature performance grade and the traffic grade from PG 58S-XX to PG 64S-XX for the bitumen produced by Alexandria Oil Company, and from PG 58S-XX to PG 70S-XX for the bitumen produced by Suez Oil Company.

Keywords: Asphalt binder performance grading, AASHTO T 350–14 Modified MSCR test, LDPE modified Bitumen.

1. Introduction

For many years, standard Egyptian 60/70 penetration grade bitumen, provided a satisfactory AC pavement performance [1-3]. Lately, a significant shortage of bitumen supply has been observed in Egyptian market, due to the tremendous road construction works, associated with the current national mega projects. As a result, the government attempted to solve the problem by increasing the local bitumen production [4, 5], through the main two production companies (Suez and Alexandria) as well as importing more crude oil from different countries, to cover this shortage in production [6]. However, A serious problem has been observed on the most of the newly constructed Egyptian AC pavements [7-10], which is rutting distress (as shown in Figure 1), due to the major increasing in the total amount of traffic volume via Egyptian roads network, stripping issues, horizontal deformation and early aging. These problems show that crude oil quality has reduced [11], oil refining processes have become more efficient and bitumen is extended by the addition of what would usually have been waste. As a result, AC pavement designer are forced to consider using modified binders especially, those modified by recycled polymers [12]. One of the best modifiers in this concern is the LDPE, that enhances asphalt binder high PG, with good mixing properties [13-15] and relatively low cost. Moreover, the difference in bitumen fractional composition [16, 17] for the same penetration grade, leads to a deep thinking of "Diagnosis before Treatment" concept [18], in which compatibility between bitumen and modifiers is matched, through binder testing.

Figure 1. Example of AC pavement rutting observed on some newly constructed highway.

Knowing that the viscoelastic characteristics of bitumen as an asphalt binder depends on temperature, load, time rate of loading and aging [19,20], SUPERPAVE performance grading protocol (NCHRP project 9–10), was the first test procedure, that directly relates those parameters to the field performance of the asphalt binders [21], through a number of tests. These tests include [22] rotational viscometer (RV) test, dynamic shear rheometer (DSR) test, bending beam rheometer (BBR) test, the direct tension (DT) test, rolling thin-film oven (RTFO) short time aging test and pressurized vessel aging (PVA) test. Latter, standard MSCR test (AASHTO TP 70–07 and ASTM D7405–08) replaced the DSR test used in the original SUPERPAVE PG system. Recently, the standard MSCR test was enhanced to modified-standard MSCR test (AASHTO T 350–14 and ASTM D7405-15) [23, 24]. Generally, the MSCR test has the ability to distinguish between different elastomeric characteristics of various modifiers, such as rubber and polymer. In this test, the asphalt binder specimen is forced beyond the linear behavior limit, to characterize the stress dependence and special rutting parameter called "non-recoverable creep compliance" [25] , defined as the ratio between the residual strain at the end of the recovery portion, to the applied stress at the creep portion and given by the equation [26]:

$$J_{nr}\,(\sigma, N) = \frac{\varepsilon_r\ - \varepsilon_0}{\sigma} \tag{1}$$

Where σ is the applied stress, N is the number of creep and recovery cycles, ε_r is the shear strain measured at the end of a recovery portion, ε_0 is the shear strain measured at the start of a creep portion. This parameter allows the assessment of asphalt binder into four traffic level grades according to SUPERPAVE criterion as shown in table 1.

Table 1. Traffic grade according to AASHTO M 332–14 criteria [23].

Traffic grade		Non-recoverable Creep Compliance @ 3.2 kPa (kPa^{-1})	Traffic level (million equivalent single axle load)	Traffic speed (km/h)
S	Standard	4.5	< 10	> 70
H	Heavy	2	10-30	20-70
V	Very heavy	1	> 30	> 20
E	Extremely heavy	0.5	> 30	< 20

A second important parameter given by MSCR test called "percent recovery" [26] and represents the ability of an asphalt binder to restore deformation after removal of the creep load, and given by the equation [26]:

$$R(\sigma, N) = \frac{\varepsilon_c\ - \varepsilon_r}{\varepsilon_c\ - \varepsilon_0} \times 100 \tag{2}$$

Where ε_c is the shear strain measured at the end of a creep portion. a third proposed parameter that is obtained by same test, called "the change in non-recoverable creep compliance" due to an incremental change in the applied stress [26]. It helps in quantifying the stress sensitivity of an asphalt binder, as a better replacement of the "percent difference". Also "non-recoverable strain rate" and "average recovered strain" were proposed as an alternative representation for the results [26]. It has been found that most of the unmodified Egyptian bitumen products fail according to conventional grading systems (penetration and viscosity grading), and according to standard SUPERPAVE performance grading system where high conservative level of analysis for high reliability projects are required [27]. This work, shed light on improving both, high temperature performance and traffic grade of standard 60/70 Egyptian bitumen, by LDPE modification, to meet the requirement of high reliability projects, assessed by modified-standard MSCR SUPERPAVE grading system.

2. Materials characterization

Two normal (virgin) bitumen samples of 60/70 penetration grade, produced by major Egyptian oil companies (Alexandria and Suez Oil Companies), were selected as test samples. These samples were subject to SARA fraction analysis test (ASTM 2007-11) at the central labs of the Egyptian Petroleum Research Institute (EPRI) . Because of the difference in bitumen local source, tests reveals that the asphaltene content in Suez sample was higher than that of Alex sample, while the malltene content of Alex sample, was higher than that of Suez sample, as shown in figure 2. According to experimental studies, high quality bitumen contains [28] asphaltens 5–25%, resins 20– 35%, aromatics 30–45% and saturates 5–15%. Therefore Egyptian bitumen samples cannot be considered of high quality because of their high aromatic and saturate content and low resin fraction. The presence of high saturate content, is the main reason behind AC pavement rutting in Egyptian highways. This could be owed to the sudden loss viscosity when saturates meltdown inside the colloidal network of bitumen as the temperature increases.

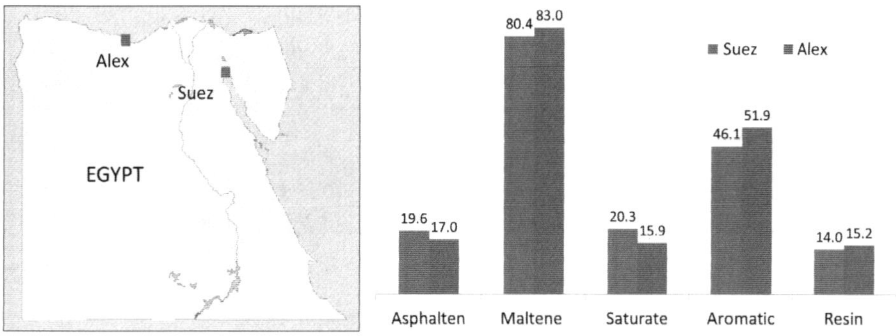

Figure 2. Local origin and SARA fraction percentages of both Alexandria and Suez oil co. grade 60/70 bitumen.

To overcome this problem, polymer modification is a must to enhance viscoelastic characteristics of Egyptian bitumen, by creating an elastomeric network inside the fluid for boosting its elastic properties over its viscous properties [12]. Since cost and environmental sustainability are principal issues in civil engineering, recycled LDPE were selected as an optimum modifier [13-15]. It has a very suitable melting point and density for mixing [14] with bituminous materials. Table 2 shows the specifications of recycled LDPE brand that was selected to modify our experimental virgin samples.

Table 2. Recycled LDPE modifier specifications

Brand	origin	density*	Melting point*	Melt flow rate*
Sabic LDPE1922 25kg bag Transparent granules	Saudi Arabia	919 kg/m^3	107 ^{o}C	22 deg/min at 190°C and 2,5kg

* Certified by the Scientific Center for Polymer Technology, Prof. Dr. / H.H. Hassan and partners, Cairo, Egypt.

3. Samples preparation

LDPE granules are crashed and mixed with un-aged virgin bitumen samples for 15min at 135 °C and 4000 rpm during RV test. Samples of 2, 4 and 6 LDPE wt% were prepared for both Alex and Suez bitumen. As shown in figure 3a, the viscosity of both samples increased exponentially as the concentration of the modifier increases. A 2% wt. LDPE was found to be the optimum modification level that causes a two fold increase in bitumen rotational viscosity, relative to its initial value at 135 °C, within the design workability criteria for Egyptian highways pavement construction. Therefore, only the 2% LDPE modified samples (Modified Alex and Modified Suez) will be subject of testing and comparison with unmodified samples (Alex and Suez).

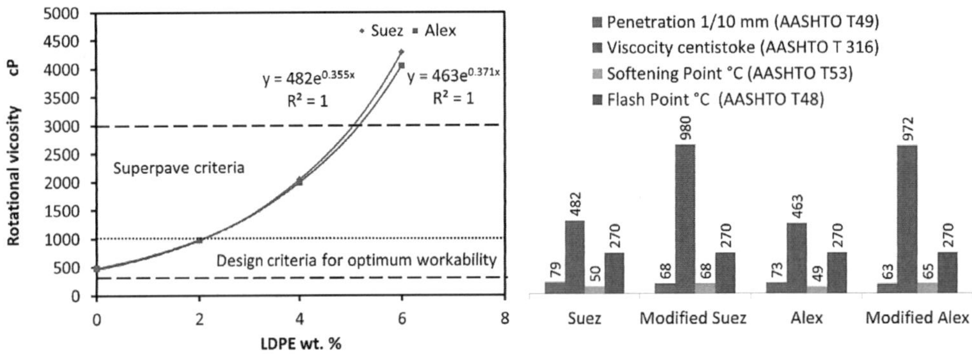

Figure 3. Alexandria and Suez bitumen samples (a) Selection of optimum modification percentage (b) virgin and modified sample routine testing results.

As shown in figure 3b, routine tests AASHTO T48, T49, T53 and T316 applied on test samples, revealed that a 2% wt. modification by LDPE has 2 effects. The first was an increase in viscosity and softening point for Suez sample by 103% and 36% successively, versus an increase of 109.9% and 32.6% successively for Alex samples. The second effect was a 16% and 15.9% decrease in penetration for Suez and Alex samples. However, the flash points remained the same for all samples. (Samples preparation and routine testing were performed through the laboratory facilities of the General Authority for Roads, Bridges and Land Transport (GARBLT), Ministry of Transportation (MOT), Cairo, Egypt)

4. Test procedure

RTFO aged binder samples were subject to DSR test according to AASHTO M 332-14 at standard bumping temperatures (58, 64, 70 and 76 °C) in order to obtain the rutting resistance $|G^*|/\sin\delta$ [29-33], and to fulfill the Superpave criterion for PG (where G^* and δ are the complex shear modulus and the phase angle between the applied shear stress and the resulting shear strain). Then, and according to the latest modified-standard MSCR test protocol, binder samples were tested using the standard 25 mm disc with gap setting of 1mm, at the passed maximum PG temperatures for 20 creep and recovery cycles and applied stress levels of 0.1 kPa as a first stage. The average non-recoverable creep compliance and the average percent recovery are given then by the equations [26]:

$$J_{nr0.1} = \frac{\sum_{N=11}^{20} J_{nr}(0.1,N)}{10} \tag{3}$$

$$R_{0.1} = \frac{\sum_{N=11}^{20} R(0.1,N)}{10} \tag{4}$$

Then, 10 creep and recovery cycles at stress levels of 3.2 kPa were applied at the second stage. The corresponding average non-recoverable-creep compliance and the average percent recovery are given then by the equations [26]:

$$J_{nr3.2} = \frac{\sum_{N=1}^{10} J_{nr}(3.2,N)}{10} \quad (5)$$

$$R_{3.2} = \frac{\sum_{N=1}^{10} R(3.2,N)}{10} \quad (6)$$

The creep time and the recovery time were 1 and 9 s successively. The idea behind this procedure is that the induced viscoelastic strain in the creep period can be recovered once shear stress disappears, differentiating between the permanent strain and the total strain, as good indication of rutting potential under high traffic level [26,29-31]. The percent difference in non-recoverable creep compliance and the percent difference recovery are indicators of stress-sensitivity or non-linear response of asphalt binders and given by the equations [26]:

$$J_{nrdiff} = \frac{J_{nr3.2} - J_{nr0.1}}{J_{nr0.1}} \times 100 \quad (7)$$

$$R_{diff} = \frac{R_{3.2} - R_{0.1}}{R_{0.1}} \times 100 \quad (8)$$

This test was performed by a Modular Compact Rheometer, model "Anton Paar MCR 302", one of the test facilities of the EPRI.

5. Experimental results

5.1 DSR rutting parameter

As shown in figure 4a, the complex shear modulus of the modified samples over the whole MSCR temperature range shows a higher values than for virgin samples, as an indication of high stiffness, with a decrease in their phase angles, as an indication of low energy loss, due to the presence of elastomeric network in the modified binders. The rutting resistance of modified samples was superior to that of the virgin samples especially for modified Alex sample just above 68 °C, as shown in figure 4b. Beyond this temperature, the modified Suez sample shows a better rutting resistance.

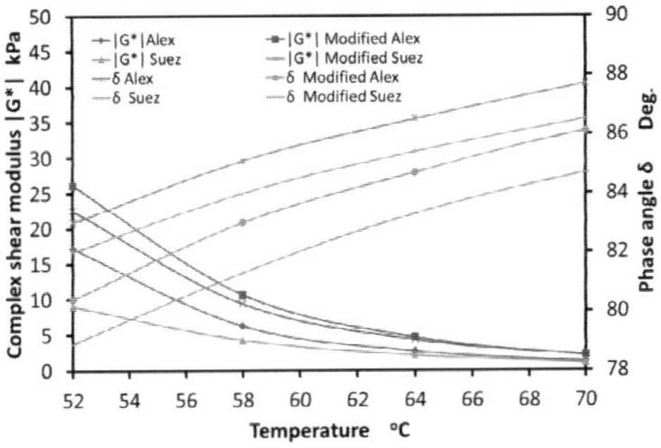

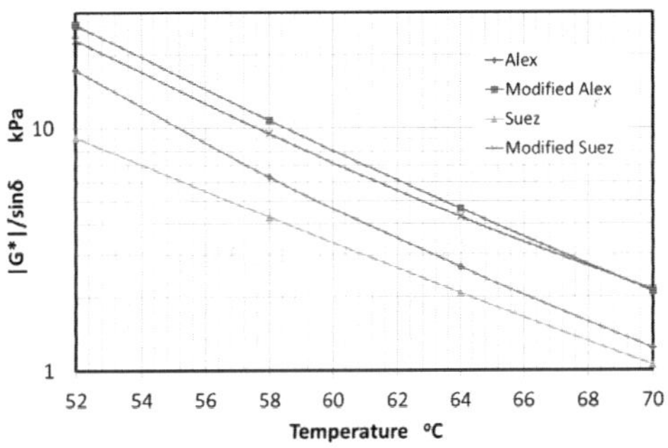

(a) (b)

Figure 4. Temperature behavior for virgin and 2% LDPE modified bitumen samples and its effect (a) on complex shear modulus and phase angle at periodic shear strain of 12 % magnitude and 10 rad/s frequency (b) on rutting resistance parameter.

According to AASHTO M 332-14, both Alex and Suez samples failed to fullfil the SUPERPAVE PG criterion at 64.01 °C, but modified Alex sample passed the same temprature, while modified Suez sample passed at 70.01°C. Therefore modified Suez sample was of higher PG than of modified Alex sample.

5.2 MSCR rutting potential

The shear strain response curves represented by figure 5a and 5b, shows that virgin bitumen samples at 64°C demonstrated larger shear strain with minor creep recovery during relaxation periods, while modified samples revealed oposit behavior at 70°C as an indication of better elastic response. Modified Suez sample showed a superior creep recovery behavior over modified Alex at both 0.1 kPa and 3.2 kPa stress levels.

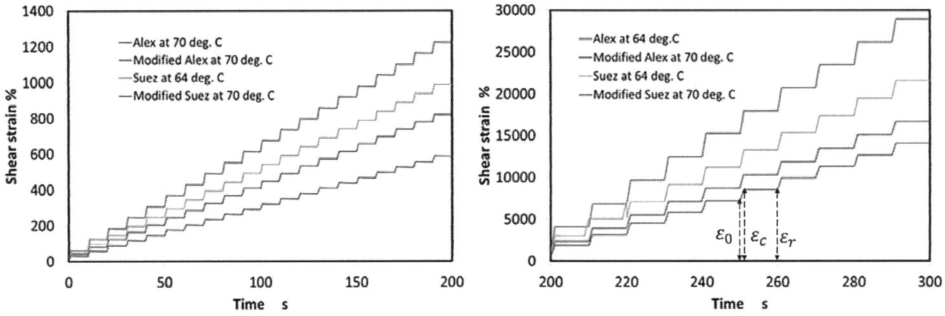

(a) (b)

Figure 5. Shear strain response curve for virgin and 2% LDPE modified bitumen samples (a) at stress levels of 0.1 kPa over 20 creep and recovery cycles (b) at stress levels of 3.2 kPa over 10 creep and recovery cycles.

By obtaining the values of ε_0 , ε_c and ε_r for each creep and recovery cycle and by solving the set of equations from 1 to 8, the MSCR behavior of the test samples can be illustrated as shown in figure 6a and 6b. Where the first bar-graph shows that modified Suez sample demonstrated the least average percent recovery and average non-recoverable creep compliance at stress levels of 0.1 and 3.2 kPa, among other samples, flowed by modified Alex sample, which indicates that they developed an elastomeric network at their molecular level. This network decreases the rutting potential of bitumen at high temperatures.

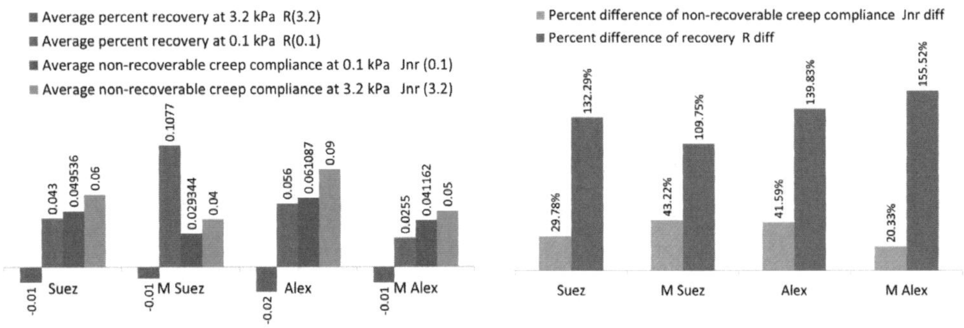

(a) (b)

Figure 6. MSCR parameters for virgin and 2% LDPE modified bitumen samples (a) Average percent recovery and average non-recoverable creep compliance at 0.1 and 3.2 kPa (b) Percent difference of non- recoverable creep compliance and percent difference recovery.

The second bar-graph shows that the percent difference of non- recoverable creep compliance for modified Alex sample, was the minimum among other samples including, modified Suez, which indicates lower stress sensitivity. On the other hand, the percent difference recovery for modified Suez was the maximum among other samples, which indicates a greater non-linear stress recovery dependence and more stable elastomeric behavior.

5.3 Traffic grade bumping

According to AASHTO M 332–14 incorporated with AASHTO T 350–14 test procedure, traffic level grade bumping can be achieved by plotting the percent difference recovery versus non-recoverable creep compliance for each sample, in order to determine the sample that plots above bumping lines indicates the corresponding traffic grade [23]. As shown in figure 7, allmost all samples failed to match the traffic level H. and therefore, the corresponding traffic level is S.

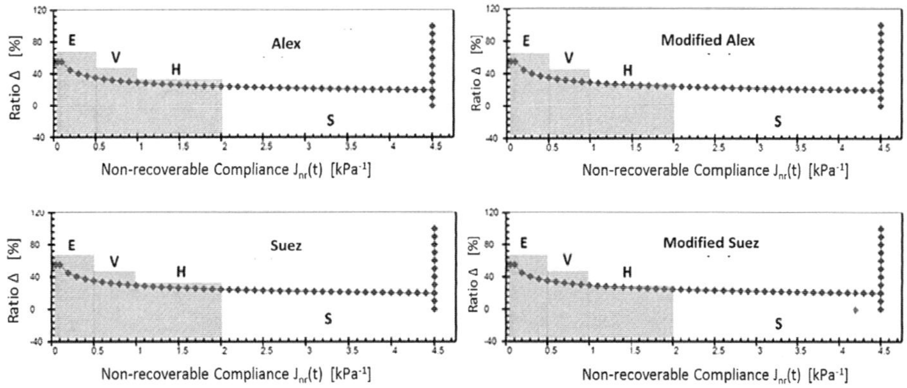

Figure 7. Traffic grade bumping through percent difference recovery versus non-recoverable creep compliance curves for virgin and modified samples.

6. Discussion

The MSCR rutting potential represented by the non-recoverable creep compliance, is more reliable than the DSR rutting parameter, since it was able to assess bitumen samples according to their stress dependent behavior, revealing the ability of asphalt binder to recover its deformation under extreme loading condition [32,33]. For example, the rutting resistance of modified Suez sample was smaller than that of modified Alexandria as previously shown in figure 4b, while its MSCR rutting potential was actually inferior to the same sample as shown in figure 6a. The reason behind this result is the fact that G* and δ test, failed to predict the non-linear behavior of the binder as function of stress level, since this test is performed at standard frequency, without taking into account the relaxation time of the viscoelastic response of the asphalt binder. And hence the value of G* cannot be considered as an indication of elasticity, even if it is combined to the phase angle, as a measure of energy loss during the stress cycling. In other word, DSR rutting parameter may not be able to differentiate between binder elasticity and stress sensitivity. Finally, table 3 summarizes all of this work findings.

Table 3. Conventional grading and SUPERPAVE performance grading of virgin and 2% LDPE modified Egyptian bitumen products of different sources

Bitumen	Alex	Modified Alex	Suez	Modified Suez
Nominal Penetration Grade	60/70	-	60/70	-
Nominal Viscosity Grade	AC-20	-	AC-20	-
	Low Quality		Low Quality	
Minimum required SUPERPAVE PG for Egyptian highways[34]		PG 64 -10		
Actual PG and traffic bumping assessed by AASHTO T 350–14	PG 58S-xx	PG 64S-xx	PG 58S-xx	PG 70S-xx

7. Conclusions

Most of the Egyptian normal 60/70 penetration grade bitumen products, require polymeric modification due to their high saturates content, that causes rutting problems in AC pavements, especially, when deployed in hot weather conditions and under high traffic loads. Modified-standard MSCR SUPERPAVE performance grading protocol is a suitable assessment test that meet the requirement of high reliability projects. By applying the AASHTO T 350–14 test procedure, it was found that a 2% wt recycled LDPE modification, enhances the high temperature performance grade and the traffic grade from PG 58S-XX to PG 64S-XX for the bitumen produced by Alexandria Oil Company, and from PG 58S-XX to PG 70S-XX for the bitumen produced by Suez Oil Company. This enhancement fulfills the Egyptian highways design criteria that requires a minimum high temperature performance grade of PG 64 – XX. This results promotes modified Suez over modified Alexandria bitumen for hot climate deployment.

References

[1] Abdallah, Ahmed M., Ahmed E. Abu Elmaaty, and Mohamed G. Alzoghby. "Assessment of Surface Friction Characteristics for Egyptian Highways." Assessment 15, no. 1 (2019): 41-54.
[2] Mahdy, Hassan AH, Khaled AA Kandil, and Mazen M. Elassem. "Skid Resistance of Egyptian Asphalt Roads."University Civil Engineering Research Magazine (CERM) Vol. (39) No. (2) April, 2017.
[3] Khedr, Safwan, and Maram Saudy. "SUPERPAVE Versus Conventional Techniques For Designing Normal And Modified Asphalt Mixes." (2016).
[4] Al-Taher, M. G., A. Mohamady, M. I. Attia, and M. A. Shalaby. "Technical Evaluation of Using Grouting For Producing Semi-Flexible Asphalt Concrete Mixtures." Bulletin of the Faculty of Engineering. Mansoura University 40, no. 4 (2020): 22-38.
[5] Abdel-Motaleb, M. E. "Impact of high-pressure truck tires on pavement design in Egypt." Emirates Journal for Engineering Research 12, no. 2 (2007): 65-73.
[6] Misr Petroleum Company, Arab Republic of Egypt, 1991- 2015.
[7] Ibrahim, E. M., S. M. El-Badawy, M. H. Ibrahim, and Emad Elbeltagi. "A modified pavement condition rating index for flexible pavement evaluation in Egypt." Innovative Infrastructure Solutions 5 (2020): 1-17.
[8] Basiouny, M. E., M. S. Eisa, A. M. Abdallah, and A. A. Abd Algahny. "The Effect of Diffraction in the Degree of Bitumen Penetration on Asphalt Mixture Used for Surface Layer." Journal of Scientific Research and Reports (2019): 1-10.
[9] Elshaeb, Maha A., Sherif M. El-Badawy, and El-Sayed A. Shawaly. "Development and impact of the Egyptian climatic conditions on flexible pavement performance." American Journal of Civil Engineering and Architecture 2, no. 3 (2014): 115-121.
[10] El-Maaty, Ahmed Ebrahim Abu. "Temperature change implications for flexible pavement performance and life." International Journal of Transportation Engineering and Technology 3, no. 1 (2017): 1-11.
[11] Saudy, SK Maram, and SAFWAN KHEDR. "Evaluation of normal and modified asphalt according to conventional and modern grading systems." DOI: 10.14455/ISEC. res. 2016.58, Proceedings of the First European and Mediterranean Structural Engineering and Construction, At Istanbul, Turkey (2016).
[12] F. L.Roberts, P.S.Kandahal, D.Y.Lee, T.W.Kennedy, Hot mix asphalt materials, mixture design, and construction, Second Edition, NAPA Research and Education Foundation, Lanham, Maryland,1996.
[13] Roja, K. Lakshmi, Amara Rehman, Eyad Masad, and Ahmed Abdala. "Low density polyethylene for asphalt binder modification." In Advances in Materials and Pavement Performance Prediction II, pp. 324-328. CRC Press, 2020.
[14] Ghuzlan, Khalid A., Ghazi G. Al-Khateeb, and Yazeed Qasem. "Rheological properties of polyethylene-modified asphalt binder." Athens Journal of Technology and Engineering 10 (2013): 1-14.
[15] Masad, Eyad, K. Lakshmi Roja, Amara Rehman, and Ahmed Abdala. "A review of asphalt modification using plastics: a focus on polyethylene." Texas A&M University: Qatar, Doha (2020).
[16] Rochlani, Mrinali, Sabine Leischner, Gustavo Canon Falla, Puneet Goudar, and Frohmut Wellner. "Influence of source and ageing on the rheological properties and fatigue and rutting resistance of Bitumen using a DSR." In Proceedings of the 9th International Conference on Maintenance and Rehabilitation of Pavements—Mairepav9, pp. 481-491. Springer, Cham, 2020.
[17] Remišová, Eva, and Michal Holý. "Changes of properties of bitumen binders by additives application." In IOP conference series: Materials science and engineering, vol. 245, no. 3, p. 032003. IOP Publishing, 2017.

[18] Adams, Jeramie J., Michael D. Elwardany, Jean-Pascal Planche, Ryan B. Boysen, and Joseph F. Rovani. "Diagnostic techniques for various asphalt refining and modification methods." Energy & Fuels 33, no. 4 (2019): 2680-2698.
[19] Wen, Haifang, Shihui Shen, Zhanpeng Ma, and Jingan Wang. "Modeling the effects of temperature and loading rate on fatigue property of asphalt binder." Journal of Testing and Evaluation 38, no. 6 (2010): 647-652.
[20] Aliha, M. R. M., and S. Shaker. "Effect of bitumen type, temperature and aging on mixed I/II fracture toughness of asphalt binders-experimental and theoretical assessment." Theoretical and Applied Fracture Mechanics 110 (2020): 102801.
[21] Bahia, Hussain U., Walter P. Hislop, Huachun Zhai, and Andres Rangel. "Classification of asphalt binders into simple and complex binders." Journal of the Association of Asphalt Paving Technologists 67 (1998).
[22] Bahia, Hussain U., D. I. Hanson, M. Zeng, H. Zhai, M. A. Khatri, and R. M. Anderson. Characterization of modified asphalt binders in superpave mix design. No. Project 9-10 FY'96. 2001.
[23] AASHTO (2014) AASHTO T350 – Standard Method of Test for Multiple Stress Creep Recovery (MSCR) Test of Asphalt Binder Using a Dynamic Shear Rheometer (DSR). American Association of State Highway and Transportation Officials, Washington, DC.
[24] ASTM (2010) ASTM D7405 – Standard Test Method for Multiple Stress Creep and Recovery (MSCR) of Asphalt Binder Using a Dynamic Shear Rheometer. American Society for Testing and Materials, West Conshohocken. DOI: 10.1520/D7405-10A.
[25] Hafeez, Imran, and Mumtaz Ahmed Kamal. "Creep compliance: a parameter to predict rut performance of asphalt binders and mixtures." Arabian Journal for Science and Engineering 39, no. 8 (2014): 5971-5978.
[26] Liu, Hanqi, Waleed Zeiada, Ghazi G. Al-Khateeb, Abdallah Shanableh, and Mufid Samarai. "Use of the multiple stress creep recovery (MSCR) test to characterize the rutting potential of asphalt binders: A literature review." Construction and Building Materials (2020): 121320.
[27] Aguib, Ashraf Ayman, and Safwan Khedr. "The mechanistic-empirical pavement design: An Egyptian perspective." In Functional Pavement Design, pp. 933-942. CRC Press, 2016.
[28] Paliukaite, Migle, Audrius Vaitkus, and Adam Zofka. "Evaluation of bitumen fractional composition depending on the crude oil type and production technology." In Environmental engineering. Proceedings of the international conference on environmental engineering. ICEE, vol. 9, p. 1. Vilnius Gediminas Technical University, Department of Construction Economics & Property, 2014.
[29] Hafeez, I, Mian, I, Zaidi, B, Kamal, MA, Ahmed, N, Ullah, S, Hussain, J, Riaz, K & Hussain, S 2013, 'To study the viscoelastic behavior of asphalt binders using under multi stress creep recovery test', *Life Sciences Journal*, vol. 10, no. 12s, pp. 921–5.
[30] White, Greg. "The multiple stress creep recovery test for airport asphalt binders." Road & Transport Research: A Journal of Australian and New Zealand Research and Practice 24, no. 3 (2015): 24-34.
[31] D'Angelo, J.; R. Kluttz; R. Dongré; K. Stephens e L. Zanzotto (2007) Revision of the Superpave High Temperature Binder Specification: The Multiple Stress Creep Recovery Test. Journal of the Association of Asphalt Paving Technologists, v. 76, p. 123-162.
[32] Behnood, Ali, and Jan Olek. "Stress-dependent behavior and rutting resistance of modified asphalt binders: An MSCR approach." Construction and Building Materials 157 (2017): 635-646.
[33] R. Delgadillo, K. Nam, H. Bahia, Why do we need to change G*/sin d and how?, Road Mater. Pave. Des. 7 (1) (2006) 7–27.
[34] Khedr, S. A. F. W. A. N., M. A. R. A. M. Saudy, and M. O. N. A. Khafagy. "Development of asphalt binder Performance Grades." In Proceedings of the 2 nd Australasia and South East Asia Structural Engineering and Construction Conference, Bangkok, Thailand, pp. 393-398. 2014.

Modelling for integrated sustainable urban transformation

Mohamad Farrag Fayad [1, *]

[1] Engineering faculty, Al-Azhar University, Cairo, Egypt

[*]Corresponding Author Email: engmohamad.farag@gmail.com

Abstract. The researcher proposes at this time an integrated framework for developing models of transition to sustainable ecological cities, and the goal is for cities to be able to adapt and be sustainable with different conditions. The contribution of this proposed model is to propose an effective scalable framework for sustainable urban transformation of communities that addresses the gaps and limitations of existing models. It takes into account the fundamental issues of urban societies including urban, environmental, social, economic and governance perspectives. The core of this framework consists of five dimensions that must be combined in order to achieve the objective of the proposed framework. It is the urban, environmental, social, economic and governance framework. Each of them has a number of main categories in addition to a number of criteria. The model contains an additional dimension, which is the dimension of information and communication technology that will be presented as an implicit dimension that will be included in all the five main dimensions, and these six frameworks have been collected in one equation in order to build the simulation model and then obtain the results of indicators and future expectation.

Keywords: Modelling, Integrated Sustainability, Urban transformation, Resource and cost Indicators

1. Literature Review

The researcher will discuss the concepts of sustainable urban transformation and modeling and simulation systems.

1.1. Sustainable urban transformation (SUT)

SUT defined as a subset of urban sustainability transitions, consisting of 1) sustainable places and their management and usage, 2) the sustainability transition of the urban development regime and 3) sustainability transitions in related societal sectors [1].

This definition is based on an understanding of urban transformation and urban sustainability, and thus the concept of sustainable urban transformation highlights and strengthens the sustainable process of urban transformation so that it aims to develop it. He defined sustainable transition instead of sustainability as a key to solving existing environmental and economic problems in addition to development problems, or in other words, it reconciles the environmental and economic necessities and desires to achieve sustainability [2]. Sustainable urban transformation places greater emphasis on structural transformation processes, whether radical or multidimensional, that can directly drive urban development in effective ways towards sustainability. Simply put, sustainable urban development is primarily aimed at urban development while urban transformation The sustainable is anchored towards development or change in urban areas, and therefore it is a dynamic process resulting from the interaction between the economic, environmental and social processes of the urban system through the capacity and scope of the urban area to perform its function and at the same degree of quality of life required by the population without specifying the options available for current and future generations [3].

1.2. Modelling System

Bratley defined modeling as a process of driving a model of a system with suitable inputs and observing the correspondingly outputs [4] but **Shannon** agrued that modeling as the most common way of planning a model of an applied framework and utilizing it to lead tests to comprehend the presentation of the framework as well as assessing elective administration methodologies and dynamic cycles utilizing reenactment results. The motivation behind modeling incorporates execution evaluation, verification, forecast, disclosure, preparing, amusement and training [5].

Studying the model instead of the system enables experimenting with several dialogues about the system by returning the model to the original state when conducting each dialogue, unlike the original system, which if changed, cannot be returned to its original state. For example, to study the urban development management system by changing policies and planning decisions may lead to The results cannot be reversed, and the model can be studied in hypothetical times. For example, a simulation of the system can be performed using the model and knowing some of the system's behavior for periods of several months or years in a few minutes. Also, through the model, the system can be studied before its creation and existence to determine which option is the best option. A model for each option and we simulate the factory's behavior under these options [6], [7]. Modeling and simulation have only recently become an academic discipline of its own. Formerly, those working in the field usually had a background in engineering [8].

2. Methodology

In order for cities to be able to adapt and be sustainable with different conditions, the researcher proposes a sustainable framework for developing models of transformation into sustainable eco-cities as in figure 1.

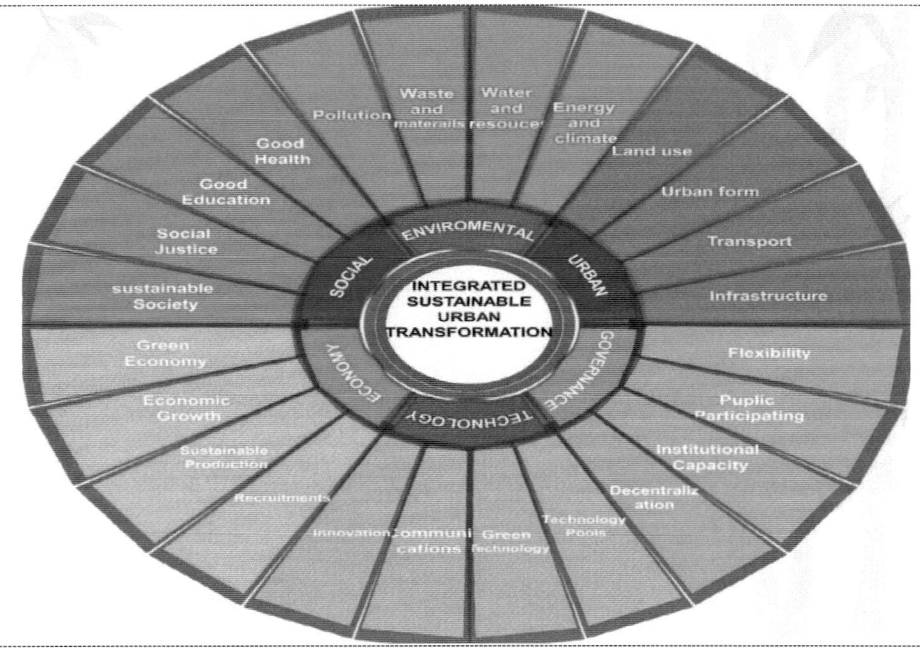

Figure 1.The proposed model for integrated sustainable urban transformation [9].

The contribution of this proposed model is to propose an effective scalable framework for sustainable urban transformation of societies that addresses the gaps and limitations of existing models, taking into account the fundamental issues of urban societies including urban, environmental, and socio-economic and governance perspectives. This framework consists of five dimensions that must be combined in order to achieve the objective of the proposed framework. It is the urban, environmental, social, economic and governance framework. Each of them has a number of main categories in addition to a number of criteria. The model has an additional dimension, which is the Information and Communication Technology (ICT) dimension that will be presented as an implicit dimension that will be included in all five main dimensions. At hand: It is of the type of maximization, where we work to make the return and benefit from the project as high as possible. 1/ Decision variables: The following matrix shows the decision variables and the mechanisms for applying the foundations of sustainable transformation. The use of dynamic programming techniques as a simulation model to solve the issues of managing the sustainable transformation of Egyptian cities needs first to identify the decision variables of the issue, then formulate goals and seek to minimize or maximize it according to its nature, and then know the constraints on the issue. Each transformation project can be considered a problem and since the transformation project consists of stages, each stage can be considered a partial problem, and the search for a partial conditional optimal solution to it.

- The objective function of the question at hand: It is of the type of maximization, where we work to make the return and benefit from the project as high as possible.
- Decision variables: The following matrix shows the decision variables and the mechanisms for applying the foundations of sustainable transformation.

Table 1. The main and sub-mechanisms of sustainable transformation with the proposed proportions for each mechanism [9].

Category	Subcategory	*General weight*	Criteria	*Detailed weight*
Urban framework	land use	0.25	optimal land use	0.10
			public services	0.45
			green areas	0.15
			housing	0.30
	urban form	0.25	urban fabric	0.15
			heritage areas	0.15
			urban character	0.15
			roads	0.55
	transport	0.20	sustainable transport	0.12
			transport facilities	0.64
			transport policy	0.11
			active transport	0.13
	Infrastructure	0.30	green Infrastructure	0.15
			infrastructure efficiency	0.75
			infrastructure management	0.10
Environmental Frame work	Energy and climate	0.35	energy supply technology	0.30
			sustainable energy action plans	0.20
			greenhouse	0.20
			carbon emission	0.30
	Water and resouces	0.25	water Safety Plans	0.15
			wastewater recycling	0.15
			resource use efficiency	0.15
			domestic resource use	0.55
	Waste and materails	0.20	integrated solid waste management plan	0.25
			waste reuse and recycling	0.20

			management of waste incinerators	0.25
			sustainable local materials	0.30
	Pollution	0.20	air quality management	0.20
			reducing water pollution	0.30
			reducing soil pollution	0.20
			pollution prevention measurements	0.30
Social framework	Good Health	0.45	health and safety issues	0.10
			Providing medical facilities	0.45
			access to health services	0.15
			public awareness	0.30
	Good education	0.35	educational development	0.15
			access to educational services	0.15
			educational facilities	0.15
			Educational Administration	0.55
	Social justice	0.10	Fair distribution of services	0.12
			Fair Distribution of Income	0.64
			Public Participation	0.11
			Employment Opportunities	0.13
	Sustainable Society	0.10	demographics	0.15
			thriving and sustainable societies	0.65
			community communication	0.10
			Community participation	0.10
Economy framework	green economy	0.25	Boosting the local economy	0.10
			Diversity of economic activities	0.45
			Effective use of resources	0.15
			Balance between income and spending	0.30
	Economic Growth	0.25	Economic capacity	0.55
			Promote domestic industry	0.25
			Develop new investments	0.10
			Provide a healthy economic environment	0.10
	Sustainable production	0.25	delivery services	0.34
			Product quality	0.35
			cost effectiveness	0.15
			Effective Pricing	0.16
	Recruitments	0.25	green jobs	0.15
			Employment Opportunities	0.75
			Jobs chances	0.10
			work environment	
Technical framework	Technology Innovation	0.30	manufacturing	0.30
			research and development	0.15
			marketing	0.55
	communications	0.25	electronic activities	0.40
			communication systems	0.30
			digital networks	0.30
	green technology	0.20	Electronic waste recycling	0.30
			Sustainable production	0.40
			NATO green technology	0.30
	technology pools	0.25	technology cities	0.30
			technology parks	0.45
			creativity centers	0.25
Governance framework	decentralization	0.30	field management	0.10
			administrative decentralization	0.45
			Regional administrative decentralization	0.15
	institutional capacity	0.20	diverse funding sources	0.15

		intercom	0.15
		training	0.15
Puplic participating	0.35	inclusion and diversity	0.12
		timing and accessibility	0.64
		incorporating ideas into policies	0.11
flexibility	0.15	raising awareness	0.15
		proactive planning	0.65
		sustainable managment	0.20

The relative weights of the previous activities and dimensions were calculated through the following factors:

- The simulation model was applied to the 6th of October city .Relying on time series data for decision variables through the data of the Central Agency for Mobilization and Statistics and similar previous studies and mathematical models that were similar in some decision variables for the current study.
- Factor the time period and consider time as one of the most important resources for sustainable transformation.
- Risks, issues and obstacles that hinder the implementation of those frameworks and activities.
- The degree of importance of each activity and stage, as well as the importance of each of the six dimensions. The limitations of the issue can be summarized as follows:

(1) Prices of material and human resources change with time. (2) The quantity of materials required for implementation is limited. (3) Fluctuations in resource prices.(4) The implementation period ranges from 15 to 20 years.(5)Implementation of simultaneous stage activities.

A sustainable transformation project consists of a number of stages, let it be n, and we have R the number or percentage of resources needed to complete the project (material - human - time) and let the total quantity required of one of these resources Qmax.

3. Experiments

The six frames can be grouped into one equation, as shown in Table (2), the objective function of the simulation model: [9].

$$max\ S = n\sum f(k) = \alpha 1X + \alpha 2Y + \alpha 3Z + \alpha 4E + \alpha 5T + \alpha 6G \qquad (1)$$

• (S): The expected return is limited to achieving sustainable development and achieving maximum benefit from sustainable transformation

• (n): the number of tracks and dimensions of the project (the project consists of 6 tracks).
• ($\sum f(k)$): the sum total of the decision variables
• (X) indicates the urban framework
• (Z) indicates the environmental framework
• (T) indicates the technical framework
• (α) is the degree of importance and return for each of the six dimensions
• (Y) indicates the social framework
• (E) indicates the economic framework
• (G) indicates the governance framework

Model constraints:

$$R1X + R2Y + R3Z + R4E + R5T + R6G \leq Qi1 \qquad (2)$$

$$R1X + R2Y + R3Z + R4E + R5T + R6G \leq Qi1 \qquad (3)$$

$$.....+......+.......+.....+.....+.... \leq.... \qquad (4)$$

$$RiX + RiY + RiZ + RiE + RiT + RiG \leq Qij \qquad (5)$$

$$X, Y, Z, E, T, G, Q \geq 0 \qquad (6)$$

3.1. *Specified framework path*

The figures (2&3) shows the path of the (urban, environmental, social, economic, technology, governance) frameworks after its application by the modelling program (Win qsb).

Variable -->	X1	X2	X3	X4	Direction	R. H. S.
Maximize	0.25	0.25	0.2	0.3		
C1	0.1	0.15	0.12	0.15	<=	6.5
C2	0.45	0.15	0.64	0.75	<=	2.8
C3	0.15	0.15	0.11	0.10	<=	1.1
C4	0.3	0.55	0.13	0	<=	1.6
LowerBound	0	0	0	0		
UpperBound	M	M	M	M		
VariableType	Continuous	Continuous	Continuous	Continuous		

21:58:57		Sunday	November	03	2019		
Decision Variable	Solution Value	Unit Cost or Profit c(j)	Total Contribution	Reduced Cost	Basis Status	Allowable Min. c(j)	Allowable Max. c(j)
X1	0	0.2500	0	-0.0336	at bound	-M	0.2836
X2	2.9091	0.2500	0.7273	0	basic	0.1883	M
X3	0	0.2000	0	-0.1009	at bound	-M	0.3009
X4	3.1515	0.3000	0.9455	0	basic	0.2315	1.2500
Objective	Function	(Max.) =	1.6727				
Constraint	Left Hand Side	Direction	Right Hand Side	Slack or Surplus	Shadow Price	Allowable Min. RHS	Allowable Max. RHS
C1	0.9091	<=	6.5000	5.5909	0	0.9091	M
C2	2.8000	<=	2.8000	0	0.4000	0.4364	5.4136
C3	0.7515	<=	1.1000	0.3485	0	0.7515	M
C4	1.6000	<=	1.6000	0	0.3455	0	3.0744

Figure 2. the path of the urban framework [9].

Variable -->	X1	X2	X3	X4	Direction	R. H. S.
Maximize	0.28	0.38	.24	.1		
C1	.31	.08	.145	.09	<=	1
C2	.18	.62	.51	.765	<=	8.1
C3	.24	.19	.157	.05	<=	1.5
C4	.27	.11	.188	.095	<=	1.4
LowerBound	0	0	0	0		
UpperBound	M	M	M	M		
VariableType	Continuous	Continuous	Continuous	Continuous		

22:53:49		Sunday	November	03	2019		
Decision Variable	Solution Value	Unit Cost or Profit c(j)	Total Contribution	Reduced Cost	Basis Status	Allowable Min. c(j)	Allowable Max. c(j)
X1	0	0.2800	0	-0.2000	at bound	-M	0.4800
X2	7.8947	0.3800	3.0000	0	basic	0.3800	M
X3	0	0.2400	0	-0.0740	at bound	-M	0.3140
X4	0	0.1000	0	0	at bound	-M	0.1000
Objective	Function	(Max.) =	3.0000	(Note:	Alternate	Solution	Exists!!)
Constraint	Left Hand Side	Direction	Right Hand Side	Slack or Surplus	Shadow Price	Allowable Min. RHS	Allowable Max. RHS
C1	0.6316	<=	1.0000	0.3684	0	0.6316	M
C2	4.8947	<=	8.1000	3.2053	0	4.8947	M
C3	1.5000	<=	1.5000	0	2.0000	0	2.3750
C4	0.8684	<=	1.4000	0.5316	0	0.8684	M

Figure 3. the path of the environmental framework [9].

3.2. *Overall model within min and max*

After finding the expected values within the lower bound and upper bound, the program will be fed with the distributions of decision variables within the lower bound and upper bound, as 'figures 4&5 [9].

23:57:14		Sunday	November	03	2019		
Decision Variable	Solution Value	Unit Cost or Profit c(j)	Total Contribution	Reduced Cost	Basis Status	Allowable Min. c(j)	Allowable Max. c(j)
X1	0	1.6727	0	-0.7072	at bound	-M	2.3799
X2	1.0615	4.1726	4.4294	0	basic	3.8042	6.7500
X3	0	3.0000	0	-0.5431	at bound	-M	3.5431
X4	0	2.5000	0	-0.5294	at bound	-M	3.0294
X5	5.7297	0.6000	3.4378	0	basic	0.4295	0.7249
X6	0	2.8137	0	-1.7345	at bound	-M	4.5482
Objective	Function	(Max.) =	7.8672				
Constraint	Left Hand Side	Direction	Right Hand Side	Slack or Surplus	Shadow Price	Allowable Min. RHS	Allowable Max. RHS
C1	5.1770	<=	5.1770	0	0.6450	3.1417	9.4192
C2	7.0688	<=	7.0688	0	0.6406	3.8852	11.6483
C3	5.4659	<=	11.4560	5.9901	0	5.4659	M
C4	0.9119	<=	12.1376	11.2257	0	0.9119	M

Figure 4. Overall model within the minimum Min.

00:26:52		Monday	November	04	2019		
Decision Variable	Solution Value	Unit Cost or Profit c(j)	Total Contribution	Reduced Cost	Basis Status	Allowable Min. c(j)	Allowable Max. c(j)
X1	0	1.6727	0	-1.3015	at bound	-M	2.9742
X2	0.0020	4.1726	0.0082	0	basic	2.8123	5.0657
X3	0	3.0000	0	-1.5152	at bound	-M	4.5152
X4	0	2.5000	0	-1.0675	at bound	-M	3.5675
X5	85.6683	0.6000	51.4010	0	basic	0.5096	1.9430
X6	0	2.8137	0	-4.1402	at bound	-M	6.9539
Objective	Function	(Max.) =	51.4091				
Constraint	Left Hand Side	Direction	Right Hand Side	Slack or Surplus	Shadow Price	Allowable Min. RHS	Allowable Max. RHS
C1	0.0039	<=	53.1400	53.1361	0	0.0039	M
C2	45.6700	<=	45.6700	0	0.8595	45.6634	74.9675
C3	34.2732	<=	52.8800	18.6068	0	34.2732	M
C4	34.2690	<=	34.2690	0	0.3547	8.7179	34.2739

Figure 5. Overall model within the minimum Max.

4. Conclusion

4.1. *Future forecast results*

From the results of the future forecast of the previous models, the following indicators can be summarized:

(1) The environmental path came first as a priority for implementation, by reducing carbon emissions and polluting gases and replacing them with less energy in the pollution rate, and that the implementation of the transformation plan will reduce the pollution rate by 31.8 percent, and that the development of the environmental situation of cities will see a difference for the better. About the current situation 300 percent of the current situation.

(2) The second priority for implementation was the social path, which includes the development of the health sector and the educational sector, and the implementation of the transformation plan by adopting the social path as a path with the highest priority will develop education and health 4.17 times over the current situation.
(3) The governance path came in third place, and the management of the transformation process needs an advanced management model from the current situation, as the current situation is considered below average, and the results indicate that the model for managing the transformation process needs to be developed by 280 percent from the current situation.
(4) The economic framework came fourth, as the sustainable transformation will improve the economic situation in the cities by 250 percent over the current situation during the transformation period (20-25) years.
(5) The Urban Framework Fifthly, the urban situation will improve through urban planning and green spaces, by 167 percent over the current situation.
(6) The technical framework: It will include axes and sub-stages, including communications and electronics systems, at a rate of 60 percent from the current situation.

4.2. Resource and Cost Indicators (Sensitivity Analysis)
The resource and cost indicators include the following elements: [9].

- Environmental framework: The rate of change in the prices of the material and human resources needed to implement this stage ranges from 302 to 357 percent, which will be accompanied by a change in the number of units required, the required quantities, the number of working days, etc., at a rate ranging from 7.00 to 45.7.
- The social framework: the rate of change in the prices of material and human resources needed to implement this stage ranges from 500 to 675 percent, which will be accompanied by a percentage change in the number of units required, the required quantities, the number of working days, etc., at a rate ranging from 5.17 to 53.
- Governance framework: the rate of change in the prices of the material and human resources needed to implement this stage ranges from 455 to 655 percent, which will be accompanied by a change in the number of units required, the required quantities, the number of working days, etc., at a rate ranging from 5.46 to 11.47.
- The economic framework: the rate of change in the prices of material and human resources needed to implement this stage ranges from 354 to 415 percent, which will be accompanied by a rate of change in the number of units required, the required quantities, the number of working days, etc., at a rate ranging from 0.91 to 12.14.
- The urban framework: the rate of change in the prices of the material and human resources needed to implement this phase ranges from 238 to 297 percent, which will be accompanied by a change in the number of units required, the required quantities, the number of working days, etc., at a rate ranging from 34 to 52.
- Technical framework: The rate of change in the prices of the material and human resources needed to implement this stage ranges from 72 to 194 percent, which will be accompanied by a change in the number of units required, the required quantities, the number of working days, etc., at a rate ranging from 35 to 47.

From the above it is clear to us that the six recommended areas for sustainability are (Urban-Social/Human-Environment, Economics, Technology, Public Policy and Governance. Moreover, these areas should be the organizational principles of urban management, urban design and planning, and growth management Urban, regional and urban sustainable development.

References

[1] Derk Loorbach and other, (2015). Sustainable urban transformation and sustainability transitions; Conceptual framework and case study, Journal of Cleaner Production 112, Pages 2–3.
[2] McCormick, K., Anderberg, S., Coenen, L. & Neij, L., (2013). Advancing Sustainable Urban Transformation, journal of cleaner production, Journal of cleaner production, Volume 50, 1 July 2013, Pages 7–8.
[3] Yan Yang, (2010). sustainable urban transformation, driving forces, indicators and processes, PHD dissertation, ETH Zurich -Swiss Federal Institute of Technology, Switzerland.
[4] R. E. Shannon, (1992). Introduction to simulation, Winter Simulation Conference, Arlington, Virginia, USA.
[5] S. Abar, K. G. Theodoropoulos, P. Lemarinier, and M. P. G. O'Hare, Agent Based Modeling and Simulation tools: A review of the state- of-art software, Computer Science Review, 2017.
[6] Daniel Houston, Raymond J. Madachy, (2017). What Every Engineer Should Know About Modeling and Simulation, CRC Press.
[7] Levent Yilmaz, (2015). A Tribute to Tuncer Ören, Concepts and Methodologies for Modeling and Simulation, Springer.
[8] Catherine M. Banks, John A. Sokolowski, (2010). Theoretical Underpinnings and Practical Domains : modeling and Simulation Fundamentals.
[9] Mohamad Farag Fayad, (2021). Managing the Egyptian City's Transformations towards Sustainable Ecological Urbanism, unpublished Ph.D. thesis, Department of Urban Planning, Faculty of Engineering, Al-Azhar University.
[10] Mohamad Farag Fayad, (2020). Managing of city transition toward sustainable Eco urbanism (Proposed paradigm). 13th International Conference on Civil and Architecture Engineering, ICCAE-13-2020, Proceedings of the 13th ICCAE-13 Conference, 7-9July, 2020, Military Technical College, Cairo, Egypt.
[11] Mohamad Farag Fayad, (2020). Transition governance as a model for city transformation toward sustainable future, 13th International Conference on Civil and Architecture Engineering, ICCAE-13-2020, Proceedings of the 13th ICCAE-13 Conference, 7-9July, 2020, Military Technical, Cairo, Egypt.
[12] Steffen Lehmann, (2010). Green Urbanism: Formulating a Series of Holistic Principles, S.A.P.I.EN.S., Institute Veolia Environment, OpenEdition Journal , ISSN : 1993-3819, Aubervilliers, Seine Saint Denis department, Paris, France.
[13] Wilhelm Hofmeister and others, (2014). Eco-cities sharing european and asian best practices and experiences ,EU-Asia Dialogue, regional programme political dialogue Asia, Singapore.
[14] United Nations Conference on Housing and Sustainable Urban Development (Habitat III), (2016), The New Urban Agenda, Quito, Ecuador.
[15] Nadine Chahine Bitar, (2016). Innovative solutions for sustainable cities, Envirocities E-Magazine, issue 13(Jan.2016),enviromental center of arab towns (ecat), Dubai, UAE.
[16] Ibrahim Abdel Gelil, (2017), Financing sustainable development in arab countries, Ch. 2 : environmental policy framework, AFED 2018 Conference , The 11th annual report, Beirut, Lebanon.
[17] United nations, economic commission for latin america and the caribbean, (2019). The 2030 Agenda and the Sustainable Development Goals, An opportunity for Latin America and the Caribbean, United Nations ECLAC, Available online at: https://repositorio.cepal.org/bitstream/handle/11362/40156/25/S1801140_en.pdf
[18] Wang X, Salat S, Painter D ,2018, Urban sustainability framework (USF), global platform for sustainable cities (GPSC) , first edition (English), World Bank Group, Washington, DC.

Mapping street vendors and informal economy effects in public spaces

Mohamed A. Elsayed[1, *], Ayman Ashour[1], Abeer Elshater[1], Mohamed A. Elfayoumi[1] and Shehab Mahmoud[2]

[1] Department of Urban Design and Planning, Ain Shams University, Cairo, Egypt
[2] Al-Azhar University, Cairo, Egypt

[*] Corresponding Author Email: alih980@yahoo.com

Abstract. Economic and urban planning research differ in determining the emergence of the informal economy. The literature describes the informal economy as the hidden economy in various terms. When it comes to the development of the informal sector, economic and urban planning studies are not on equal terms. It's common to link the rise of the informal sector to the increase of organised economic activity in the formally organised economy, such as tax systems and competitiveness. This study aims to develop a method for assessing the impact of street sellers on the visual appearance of public spaces. This research used mixed methods comprised of Geographical Information System (GIS) and semi-structured interviews with vendors in the selected cases. Mapping four case studies in Cairo show that street vendors occupied public places, walkways, and site coordination features. The final observations shed some light on the necessity of further investigation into street vendors in diverse situations in various locations in Cairo.

Keywords: Urban planning, street vendors, GIS, public places.

1. Introduction

Economic and urban planning research differ in determining the emergence of the informal economy [1]. There is a tendency to link the emergence of the informal economy with the emergence of organised actions of economic activities in the formally organised economy, such as tax systems and competitiveness [2]. Research recognizes another trend by linking its emergence to trade doctrine and the informal economy to the formal economy with development [3, 4]. Since the 1960s, the prosecution of street vendors in Egypt has been carried out and even deported anyone who does not have an identification card showing his home resident cities [5]. However, this action from the Egyptian governates has not eliminated the street vendors' presence in Egyptian streets. Since the mid-1970s, with the increase in unemployment, the prevalence of street vendors has increased and has become a problem facing all major Egyptian cities from Alexandria to Aswan [6]. The treatment of the phenomenon of street vendors has been adopted for decades on the security permission. Yet, these permission and hasty decisions have not addressed the phenomenon that disappears for days and returns more widespread and robust.

Previous studies discussed social security for street vendors and social safety with their families' health conditions, especially their young children who sleep next to them on sidewalks, whether the weather is mild or cold [7, 8]. Other studies have focused on their situation, specifically in the case of bad weather [9, 10]. Research the challenges these vendors might face in the knowledge-based actions towards street vendors [11, 12]. Other pieces of research discuss the competent authorities to enrol these vendors' children in schools for free that sellers are easy to pay. However, the humanity of the place was not discussed in terms of exploiting the elements of the site furniture. Building upon the case of vendors in Egypt, the exploitation of vendors in the seating areas faces challenges [13, 14]. Besides, the literature showed that street vendors occupy parking areas, cars, trees, and lamppost sellers to display [15]. Research showed that vendors are vending their goods in public places ending with a chaotic scene because the site was not designed for this purpose [16]. The literature describes the essential role of the surrounding context in creating a pleasant environment instead of a random display of low-quality products [17, 18, 19].

This research tackles the problem that many public spaces in Cairo face challenges in their humanity because of the presence of vendors randomly and unregulated on sidewalks [14, 20]. Streets and mattresses such as seats, fences, and lighting poles. In line with other pieces of research [21, 22], this negatively affects the loss of humanity for both the site visitors and vendors themselves and its lack of the most specific needs of movement and society's social life. Based on the daily trips of the selected study areas, it was found that there was the exploitation of sidewalks and benches for the walk. These benches can serve the elderly in a large proportion, with visual deformation in places exploited by vendors resulting from the presence of garbage, tree exploitation and lampposts to display their products.

- What are the changes happening in the street context because of the emergence of street vendors, and how do they affect the city's visual image?
- What is the urban policy confronting vendors?

This research aims to provide a tool to measure the confusion affecting the visual image of public spaces due to the vendors' presence and their informal economic activities. In fluffing the research aim, this article used a direct spatial survey of areas containing the informal economy in four regions of Cairo (Abbasiya Square, Al-Weili; Ahmed Tayseer Street, Heliopolis; and the health insurance area, Nasr City; and Ismail Abaza Street, Cairo Down Town). The spatial survey adapted the GIS application and mapped places occupied by street vendors in the case study areas. The spatial survey also extended to the coverage area served by these vendors. The selection of the types of land use varied from residential, administrative, or educational areas these vendors are interspersed with. The contribution of the present researcher is in mapping the visual challenges caused by vendor presence in the selected cases. The added value here is in discussing the rates of encroachment of street vendors on the elements of site coordination in the areas chosen.

2. Literature Review

The literature describes the informal economy by various terms like the hidden economy [23, 24], the shadow economy, the parallel economy [25], the unregulated economy, or the black economy [26]. All these terms try to find a comprehensive description of this phenomenon and have found it challenging to find a specific general definition of this phenomenon [26]. The International Labour Organization (ILO) defines the shadow economy as the "hidden economy," "informal economy," or "parallel economy," which includes not only illegal activities but also unreported forms of income derived from the production of legitimate goods and services [25]. The International Monetary Fund (IMF) has also defined the informal economy and divided it into legitimate and illegal activities. In the informal economy, not registered enterprises have a low level of organisation and productivity and have limited access to markets, credit facilities, formal training, and public services. They have small or unstable buildings, do not receive recognition, support, or regulation by public authorities, and do not comply with social protection regulations and employment legislation.

The informal economy has also been known to cover all economic activities, workers and economic units that are not covered by formal arrangements and that do not fall into the field of law enforcement [25]. Informal activities also mean that these workers are not protected in practice and are active outside the law [14]. The reason behind that is that the vendors, in some cases, do not follow the laws. The Office African Union Commission suggested that the term informal economy be used "instead of the informal sector. The good side of the coin is that workers and institutions in that economy were active in several sectors rather than a single sector of economic activity, which was characteristic of the definition of the African Union Commission. Literature has also discussed the positive and negative effects of the informal economy. Building on its importance as a complex economic system, the urban policy can benefit from the workforce of the informal financial sector [27]. This part of the present work clarifies by reviewing relevant literature on how the informal economy affects economic, social, and environmental policies. This urban policy can address the adverse effects and the positive impact in economic, social, environmental, and institutional dimensions.

3. Materials and Methods

3.1 Case studies
Cairo was chosen as a case study of street vendors illustrating school situations in densely populated areas and passers-by from various places. The cases selected suffer from the phenomenon of street vendors, residents, or visitors coexisting in the same space. This study investigated the services with broad interests by determining the volume of gathering points for street vendors in the city. The case studies include four types of land uses: administrative areas represented in Abbasiya area, which is located in Abbasiya Square, where the study area is about 35 acres, and vendors flock to it because of the presence of many government interests in the region in addition to the fact of the Abbasiya Metro Station allowing many passers-by as well as the faculties of Ain Shams University, which makes the area very vital and this is required for the activities of new vendors. The second area is the movement areas such as Saad Zaghloul Metro Station, which overlooks Ismail Abaza Street, where vendors occupy this street profusely. The size of the study is about 11 acres. The motivations of the vendors were the presence of Saad Zaghloul Metro Station and the componence of many departments, which makes the area active with passers-by and employees.

The third land-use type is named the health insurance areas Al-Tayaran Street in Nasr City. This site has an area of about 10 acres. One of the reasons for the presence of vendors in the area is health insurance and the increase in the number of arrivals from different places. As for the fourth and final type, this research tackles the educational areas, and the street was chosen Ahmed Tayseer in Heliopolis disrict. The area has a metro station named Girls College. The Girls' College, Ain Shams University, was selected as an area of study, about 13 acres. The initial observation of this case study location showed that the vendors targeted many female students. Having the target audience of purchasers and metro users helped increase movement on the site, which was what was meant for the street vendors. Figure 1 shows the four selected areas in Cairo.

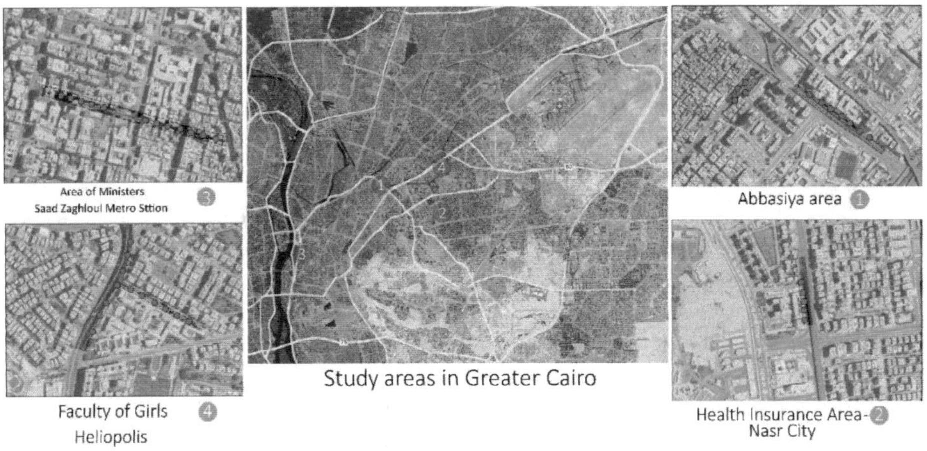

Figure 1. The aerial images of selected case studies sites.

3.2. Tools and data analysis
Cairo is chosen to study street vendors as an example of the school situation in densely populated areas that suffer from the issue of street vendors to identify gathering points for street vendors in the city. It is clear from the state of the city study that street vendors are concentrated mainly in four types of areas of the city. The example of the street selections criteria and places included administrative and commercial work such as Abbasiya area and areas surrounding institutions overcrowded, including entrances and exits of metro stations and tunnels such as the Area of Ismail Abaza Street in front of the

General Authority for Urban Planning and areas surrounding educational institutions such as schools and universities such as the area of the Ain Shams College of Girls.

Imaging tools and urban surveying of street vendors and site coordination elements (e.g., pedestrian corridors, sidewalks, trees, fences, and parking spaces). The imaginary tools took place on different days during the week and at other times (i.e., noon and mid-day and even at sunset) to study the impact of this phenomenon on these elements. This research investigated the effect of vendors on planning features located on the site was analysed for each area individually. This study aimed at determining the extent to which these street vendors and their work rate of site coordination elements. The measurement, monitoring and statistical analysis of the factors were used individually using GIS and the excel program to determine the occupancy rate for site coordination elements such as walking places. Car walks, trees, lampposts, fences, and entrances for institutions or residential buildings.

A formula was established by the researchers of the present research for each of the elements mentioned above to calculate the occupancy rate. In this context n is the variable that can estimate the walking places, the area of work of the vendors, car roads and parking areas.

$$n = \frac{\mathrm{e} * 100}{TOT}$$

Where:
n= Percentage of occupancy
e= area of elements
TOT= Total area of the vendors occupy

In the case of estimating the percentage of the lampposts, entrances and trees using the following equation.

$$nv = \frac{\mathrm{NO.} * 100}{TOT.}$$

Where:
nv = Percentage of vendors who occupied the vertical elements
NO. = number of vertical elements occupied by the vendors' activities
TOT = Total number of steep spots

Based on the above, a comparison has been reached between the selected study areas and the impact of each location on the elements of the coordination of the site. This research would expose to in the results obtained. The parameters include the occupancy rate for walking paths and sidewalks ranges from 60% to 90%, and the locations, such as trees, lamp posts, entrances, and fences, ranging from 40% to 70%. These high occupancy rates indicate that the population living in these areas, frequented, and others suffer the most from the presence of these vendors, who occupy the simplest things that a person needs to use on his typical day. Humanity has led to the lack of social aspects in many urban areas because of random and irregular work for street vendors on sidewalks. Streets and site brushes led to the lack of humanity for the most basic needs of movement, walkers, and a social life that everyone demands.

4. Result and Discussion

This research scanned and documented four case studies, focusing on the vendors and the activities that happen in study areas. The results show that the actions on the site comprised of outdoor vending that was hosted along the fences and parked cars. The observation also figured out the presence of vendors around the public building, such as governmental institutions. Figure 2 shows where vendors and their activities are found on the sites. The results showed that the timeframe of their presentence in the areas is directedly connected to the purchasers as the target audience of these kinds of goods. The target audience was mainly working or even dealing with these public institutions. Besides, the locations of vending interests were in incontinent places, including sidewalks, parked cars, building facades and fences.

The evidence yielded from the case of Ismail Abaza. The results show that Ismail Abaza Street had about 90% of the places that should be for pedestrians. However, an area of 3780 square meters was occupied by vendors using an area of 3402 square meters. For car places about 3150-meter square, the sellers occupied about 2205-meter square by 70%. At the same time, the parking spaces in the study area were about 3150 square meters, and sellers occupied 2520 square meters by 80%. In addition, the sellers used several trees (10 out of 20 trees). Besides, the lighting pillars were about 24 pillars, and the vendors occupied ten columns (41.6%). The field measurement showed that 16 public building entrances out of 25 entries were also used to display their goods.

In the case study of Nasr City Health Insurance, the pedestrian area was 600 square meters. In this area, vendors occupied about 480 square meters by 80%. Similar to the previous case, the location of car places about 2000 square meters occupied by vendors, of which about 80% and the area of car places is about 2000 square meters. One thousand two hundred square meters by 60%, while the parking spaces were about 1700 square meters, sellers occupied about 340 square meters by 20%. Similarly, the number of trees in the chosen area was about 20 vendors. Besides, about ten trees were 50%, while the lampposts were about ten columns, of which sellers occupied about six by 60%.

As for the transport and movement areas in Abbasya, the sidewalks are 7200 square meters. The vendors occupied about 5,760 square meters by 80%, and the traffic location was approximately 18,000 square meters occupied by vendors. Ten thousand eight hundred square meters were 60%, while the area of parking spaces was about 4,500 square meters, of which sellers settled about 2,700 square meters by 60%. At the same time, the trees were about 30 trees, of which sellers occupied about 15 trees by 50%, in addition to exploiting 16 lighting columns from about 36 columns by 44.44%, in addition to occupying the entrances of institutions, which reached about eight entries filled by vendors, of which about seven entrances by 87.5%.

As for the educational areas in Girls College area, the location of walking places was about 3000 square meters. Vendors occupied about 1800 square meters by 60%. The sites of cars reached about 5100 meters square, and sellers occupied about 1530 square meters by 30%. The parking spaces were filled by 30% where the area was about 2125 square meters occupied by them, of which the area of 850 square meters. The number of trees was about 50 trees, of which only ten trees were filled by 20%, so the last element is the exploitation of the walls in the area, which reached a length of 422 meters, of which the vendors occupied about 211 meters by 50%.

The concluded remarks show that the ratios of informal vending were substantial. The phenomenon of casual vending causes a loss of humanity in the places that occupy the walking positions and rests. The observation also shows that the vendors' activities on the selected case studies did pay attention to the older adults who need to rest while walking. The results also show that vendors did not occupy places designed and planned for vehicles movements or parking. Their presence in streets and obstacles traffic flow and made congestion. Besides, the field measurement figured out that the informal vending activities cause visual deformity and are derived by vending goods around trees and lampposts.

Furthermore, the exploitation of vendors for the entrances of residential buildings and administrative institutions caused quarrels between sellers and residents of facilities or employees of one of the departments. The exploitation offences were among the elements of the city's visual image, like other elements mentioned above.

The first area of Abssaya	The second area of Nasr City	The third case study of Downton Cairo	The fourth area in Heliopsis

Figure 2. The vendors' presences around the governmental and educational buildings.

The results yielded from the semi-structured interview with vendors figured out that most of the respondents mentioned that their need to provide places for sale that are close to government interests as for consumer availability and that was a large percentage of the total number of respondents, and to provide specialized job opportunities for educated youth, as well as the necessity of offering a pension and insurance

One of the sellers expressed this by hoping to find decent work regarding healthy and social issues. In this regard, respondents' answers figured out that they work with dignity, not every time someone threatens us, and they run like a burglar. In this context, another interviewee mentioned:

> "We want to work and raise our children without problems." Another interviewee said,
> "We are ready to pay, but they provide us with a suitable place. We do not want to annoy someone, or someone upset with us"

Many sellers stressed the need to provide spacious, safe, and close places to this area. Likewise, a vendor in the Ministries District on Ismail Abaza Street stated that the authority should recognize the vendors' needs in the development plans. Besides, a place close to here because the employees represent

for us the most important thing to find a customer, they don't want us at all

Two customers in Ahmed Tayseer mentioned that they wish to a place where we know where to stand, and no one bothers us; what should we do? Another interviewee said:

> "We want a place where we only eat and live, even if they make us booths, even if they take money from us, but we stand with our dignity."

In addition to the need to provide them with protection from the bullying of police secretaries and other local authorities

> "The police, nor the residents who are upset, nor the municipality, we are the same person who hears us, but they do not treat us like criminals beat and insulted and threw the goods to you, and we sat running and pushed them to take it back.".

5. Conclusion

This study investigated the definitions of the informal economy, street vendors and their relationship to the informal economy and their positive and negative effects on them with clarification. This study chose four cases from Cairo. The results show that street vendors occupied public areas, walkways, and site coordination elements. Their presence on the site varied based on the proportions depending on its commodity. The percentages of the activities that took place by vendors and their purchasers and how the elements of the coordination of the site were classified into several segments. These elements consist of the occupancy of vendors for the walkways and the places of cars and parking spaces and sidewalks and lampposts and trees and fences and entrances to buildings and entrances of institutions in strategic sites. The results also show that the vendors' presences were in public spaces between educational and community residential, administrative and health activities. These selected spaces have a high frequency in the movement of visitors and the community. This requirement is an opportunity for promising visitors/buyers. Street vendors intelligently read these strategic spaces as a chosen location for their activities and use the space as an activity place to sell their goods.

According to street vendors in the study areas, access was easy given traffic lanes and public transport availability. This facilitates the accessibility found in the public spaces between residential/student educational and community activities and increases the ease of movement of student people.

It turns out that street vendors are very good at selecting convenient places for vending their goods, in addition to taking advantage of the space to trade. In addition, public transport is available and turned, making it easier for street vendors to access the site from their place of residence and easy access to the source of goods. In addition, easy access by street vendors can make it easy for consumers to access.

The concluded remarks regarding the public spaces indicated that the elements of site coordination were calculated, where the occupancy rate of street vendors for pedestrian paths in each region was calculated, as well as car traffic areas, in addition to parking spaces in each of the selected cases. The results provide evidence of how vendors' occupancy rates for fences and light poles were studied, as well as entrances to buildings and institutions, and comparisons between them to find out the highest percentages of each element of the site coordination in the selected areas.

The concluded remarks shed light on the essential need to conduct in-depth research on street vendors in other cases for future studies. The extension on conducting such research can reveal the real problems and benefit the informal economy using development programs. Future research can also strengthen the role of non-governmental organizations in protecting the rights of street vendors, meeting their basic needs, and providing social security opportunities. In addition to meeting design requirements that preserve the spaces in general and the street vendors, it may facilitate the state's role in legalizing the status of street vendors.

References

[1] L. Kaizhi, Street Vendors in Chinese Cities Since Economic Reform:A Case Study of Guangzhou, Hong Kong: The University of Hong Kong, 2014.

[2] H. Abusaada and A. Elshaterb, "Competitiveness, distinctiveness and singularity in urban design: A systematic review and framework for smart cities," *Sustainable Cities and Society,* vol. 68, pp. 1-12, 2021.

[3] S. Rorver and C. Skinner, "Street vendors and cities," *Environment and Urbanization,* vol. 28, no. 2, pp. 359-374, 2016.

[4] R. Bromley, "Street Vending and Public Policy: A Global Review," *International Journal of Sociology and Social Policy,* vol. 20, p. 1–28, 2000.

[5] N. Ahmed, A. Elshater and S. Afifi, "The Community Participation in the Design Process of Livable Streets," in *Innovations and Interdisciplinary Solutions for Underserved Areas*, vol. 296, Cairo, Springer, 2019, pp. 144-157.

[6] I. Al-Mirghani, "hrdo center," Idea, 5 August 2014. [Online]. Available: https://hrdoegypt.org/?p=382. [Accessed 21 October 2020].

[7] R. Widjajanti and M. Damayanti, "Space Compatibility Based on Spatial Behavior of Street Vendors in Urban Public Space in Chinatown, Semarang," in *The 1st International Conference on Urban Design and Planning*, 2020.

[8] R. Oranratmanee and V. Sachakul, "Streets as Public Spaces in Southeast Asia: Case Studies of Thai Pedestrian Streets," *Journal of Urban Design,* vol. 19, no. 2, pp. 211-229, 2014.

[9] S. K. Bhowmik, "National Policy for Street Vendors," *Economic and Political Weekly,* pp. 1543-1546, 2003.

[10] T. K. Ko, J. Dickson-Gomez, G. Yasmeen, W. W. Han, K. Quinn, K. Beyer and L. Glasman, "Informal workplaces and their comparative effects on the health of street vendors and home-based garment workers in Yangon, Myanmar: a qualitative study," *BMC Public Health,* vol. 20, pp. 1-14, 2020.

[11] S. Bhowmik, Street vendors in the Global urban economy, London: Routledge, 2019.

[12] H. Abusaada and A. Elshater, "Knowledge-Based Urban Design in the Architectural Academic Field," in *Knowledge-Based Urban Development in the Middle East*, Hershey, IGI Global, 2018, pp. 204-227.

[13] P. Carrieri and I. Murta, "Cleaning up the City: A Study on the Removal of Street Vendors from Downtown Belo Horizonte, Brazi," *Canadian Journal of Administrative Sciences,* vol. 28, no. 2, p. 217–225, 2011.

[14] A. R. Arisha and N. M. A. El-Moneim, "Space Syntax Beyond Cairo Street Markets: An Application of a Theory on Socio-Spatial Contemporary Practices," in *Handbook of Research on Digital Research Methods and Architectural Tools in Urban Planning and Design*, Hershey, Pennsylvania, 2019.

[15] N. Eidse, S. Turner and N. Oswin, "Contesting Street Spaces in a Socialist City: Itinerant Vending-Scapes and the Everyday Politics of Mobility in Hanoi, Vietnam," *Annals of the American Association of Geographers,* vol. 106, no. 2, p. 340–349, 2016.

[16] L. Muñoz, "Latino/a Immigrant Street Vendors in Los Angeles: Photo-Documenting Sidewalks from 'Back-Home'," *Sociological Research Online,* vol. 17, no. 2, pp. 1-3, 2012.

[17] D. Mahadevia, A. Brown, S. Vayas and T. Patel, "Inclusive design for street vendors in India," Centre for Urban Equity, CEPT University, Ahmedabad, 2014.

[18] H. Abusaada and A. Elshater, "Effects of Urban Atmospheres on Changing Attitudes of Crowded Public Places: An Action Plan," *International Journal of Community Well-Being,* vol. 3, p. 109–159, 2020.

[19] A. Elshater, "Food consumption in the everyday life of liveable cities: design implications for conviviality Authors," *Journal of Urbanism: International Research on Placemaking and Urban Sustainability,* vol. 13, no. 1, pp. 68-96 , 2020.

[20] H. Abusaada, "Strengthening the affectivity of atmospheres in urban environments: the toolkit of multi-sensory experience," *Archnet-IJAR,* vol. 14, no. 3, pp. 379-392, 2020.

[21] H. Abusaada and A. Elshater, "Urban design assessment tools: a model for exploring atmospheres and situations," *Proceedings of the Institution of Civil Engineers-Urban Design and Planning,* vol. 173, no. 8, pp. 238-255, 2020.

[22] H. Abusaada and A. Elshater, "Improving visitor satisfaction in Egypt's Heliopolis historical district," *Journal of Engineering and Applied Science,* vol. 68, no. 1, pp. 1-22, 2021.

[23] B. S. Frey and W. W. Pommerehne, "The hidden economy: state and prospects for measurement," *The Review of Income and Wealth,* vol. 30, no. 1, pp. 1-23, 1984.

[24] N. A. H. N. Muhammad, The hidden economy "what it is - its historical establishment - its types and characteristics - its illegal crimes", cairo: Dar Al-Wafaa for the world of printing and publishing, 2020.

[25] M. El Fayoumi, "Street Vendors' Roles in Main Squares Utilization as a Type of Tactical Urbanism Application in GCR "A Criticism Review on Ramses Square"," in *Procedia Environmental Sciences*, 2017.

[26] D. A. Roy, "The hidden economy "what it is - its historical establishment - its types and characteristics - its illegal crimes"," *Middle Eastern Studies,* vol. 28, no. 4, pp. 689-711, 1992.

[27] A. Elshater, "Tactical urbanism: A method of community empowerment in Cairo neighborhoods," in *World Barcelona SB14*, Barcelona, 2014.

Evaluation of Open Spaces in Relocated Communities Using Affordance Theory: Al-Asmarat Community as a Case Study

S Soliman[1, *], S Attia[1] and H Khalil[1]

[1] Architecture Engineering Department, Cairo University, Cairo, Egypt

[*] Corresponding Author Email: salma.0.j@eng-st.cu.edu.eg

Abstract. In 2016, the urban development facility (UDF), which was previously named as informal settlement development fund (ISDF), conducted several development projects as part of the goals of Egypt 2030's vision that supports the sustainable development goals (SDGs). UDF applies several development strategies to provide residents with better quality of life (QOL), such as relocation projects as residents move from unsafe areas to a new community where they are provided with adequate housing and basic infrastructure services. However, one major issue is that open spaces of relocated projects, despite being an important component in improving livelihoods, don't fulfill the residents' needs. Affordance theory provides the link between human behaviour and the function of space. Therefore, this paper aims to evaluate open spaces in relocated communities relying on the affordance theory. Accordingly, Al-Asmarat district, one of the relocated projects in Cairo, is investigated as a case study to evaluate open spaces and livelihood in the area. This qualitative study employed 10 in-depth interviews to obtain primary data from samples taken from relocated residents. The study ends with needed indicators to ensure that open spaces in relocated communities can afford socioeconomic activities for the residents to achieve more inclusive urban development.

Keywords: Affordance Theory, Public Spaces, Relocated Communities, Evaluation indicators, Informal Settlements, Urban development

1. Introduction

Informal areas are considered "areas that have emerged in violation of laws and planning regulations" [1]. According to UDF, since 2008, informal areas have been classified into two main categories: unsafe and unplanned areas [2]. Informal areas upgrading is considered the keystone of achieving goal No. 11" *Sustainable cities and communities*" of the international Sustainable Development Goals (SDGs). Besides, it is linked to goals No. 1, "*No Poverty*", and No. 8, "*Decent Work and Economic Growth*", emphasizing inclusive and sustainable economic growth and providing appropriate work opportunities for everyone [3]. Simultaneously, establishing proper sanitary infrastructure and sewage systems is linked to goal No. 6, "*Clean Water and Sanitation*", to ensure the availability of clean water and sewage systems for everyone [4]. Therefore, UDF work emphasizes that the development techniques could achieve SDGs and, consequently, Egypt Vision 2030. Not only it can provide safe and healthy residence, but it can also upgrade the QOL and decrease poverty for the residents of these areas [5]. According to Seddik, head of UDF, 130 informal areas have already been developed. "*The plan is to eradicate slums by 2030*," he says [6].

Moreover, UDF is dealing with informal areas through different development policies. Accordingly, the authors will focus on the "Relocation in a new community" policy, where UDF displaces the residents within areas that need to be replaced. Then, the residents move to a new community where services and infrastructure are available [7]. Furthermore, they were deprived of adequate open spaces or green spaces before they were relocated from their original unsafe areas [8]. Hence, the paper aims to extract the needed indicators to evaluate open spaces in relocation projects through the affordance approach.

---

[1] Architecture Department, Faulty of Engineering, Cairo University, Cairo University Rd, 12613, Giza Governorate, Egypt

2. Background on Open Spaces

Due to the rapid urban growth that has characterized the last decade, urban open spaces are becoming increasingly important in urban development [9]. Cities and their neighborhoods can use open space development to improve their appeal to residents. The ecological and environmental impact of open spaces with high green coverage rates enhance the city's climate and reduce environmental risks. Open spaces, due to their social significance, can assist residents in adopting a healthy lifestyle. In addition, they improve the built-up character of cities due to their aesthetic significance [10]. Several open spaces are the product of local self-help and self-management processes in informal areas, often since their inception [11]. Open spaces are owned and accessed by the public; however, they are used and attached to the entire community [12]. As a result, they play an important role in the settlement's physical and social dynamics; yet, the improvement and consolidation of such places may take several years [13]. As stated in "*The Right to Public Space in The Egyptian Constitution*" [14], the right to open space refers to everyone's right to gain access to and use open space for individual expression and recreation, as well as the state's responsibility to make these open spaces available. Therefore, regardless of age, gender, class, or religion, everyone has the right to be treated equally in open spaces. Furthermore, urban security should be a part of open space design, but it should secure rather than restrict people's rights to use open space. Several articles in Egypt constitution promote fundamental rights required for the unrestricted use of open spaces. Open, accessible, safe open spaces in which residents can express themselves and their culture are essential for developing dynamic communities in Egypt and building civic society [13]. At a local scale, municipalities must focus on restoring and managing open space for urban environmental sustainability and improving the QOL of city residents.

3. The Concept of Affordance

Gibson introduced the Affordance theory in 1966 in his book 'The Senses Considered as Perceptual Systems'. Later, the concept was further developed in another book, The Ecological Approach to Visual Perception (1979/1986), in which it was defined as the following: 'The affordances of the environment are what it offers the animal, what it provides or furnishes. The verb to afford is found in the dictionary, but the noun affordance is not. I mean by it something that refers to both the environment and the animal in a way that no existing term does. It implies the complementarity of the animal and the environment.' *[15]*. Affordance is described as the properties of the physical environment perceived by the individual to promote an action [16]. Thus, it is a complementary relation between the environment and the individual, as the environment offers the individual properties that s/he can perceive with respect to his/her qualities [17]. Affordances were categorized as features of objects and layouts that are available to the perceptual system [13]. Even though Gibson didn't give much importance to categorizing affordances, he mentioned some features: medium, substances, surfaces and their layouts, attached and detached objects, places, and finally, events that refer to changes caused by other people and animals. From the point of view of learning and development in the environment, the role of the social context is fundamental, as affordances cannot be separated from the social context [17]. The nature of affordances is social [18]; almost all affordances have social dimensions. The richest affordances are those offered by other people in the environmental setting. "When you touch them, they touch you back; in short, they are interacting with the perceiver and with each other. Behavior affords behavior ..." [15,p153]. Action is essential for actualizing affordances, from perceiving to utilizing or shaping. It reveals information about the environment and helps the child realize and develop information about him/herself [15]. Finally, it is to mention that affordance connects physical and socioeconomic activities. For example, the field visits should observe a particular behavior that residents often perform in open spaces, so the action is to modify the open space's set-up to accommodate such behavior; hence affordance can be achieved. Accordingly, this research uses the concept of affordances of user activity, as the relations between objects and humans, and explains individual differences in activities in open space [14].

4. Methods and Limitations

To address the paper's objective, a double-fold analysis is used. First, the study investigated the literature of affordance theory and open spaces. Second, the paper extracted related indicators of socioeconomic activities in open spaces in relocated communities. As the study is qualitative, it used ethnographic methods such as in-depth interviews and field observations to collect primary data from residents. A sample of 10 interviewees was chosen randomly and based on various purposeful factors, including age, marriage status, residency, duration of stay in Al-Asmarat, and people who were relocated from informal areas. The researchers conducted the field observations and interviews in October 2021 (*morning, night, on holiday, and working days*). This review allowed the study to export results clarified in the needed indicators for designing open spaces that can afford social and economic activities in relocation projects. However, the study faced some limitations. First, the literature review only focuses on similar projects which evaluate open space in relocated communities. Second, some people in the investigated area refused to respond to the interviews. Third, there is no available historical data about the situation before and after relocation to validate the responses.

5. Al-Asmarat District Urban and Social Demographics Characteristics

Al-Asmarat area is 70 fd, located in Al-Muqattam district in Cairo governorate, at the edge of its hill surrounded by Nasr City and the Ring Road. This project is categorized under the relocation of a new community policy for the development of informal areas by the Cairo governorate and the UDF [19]. The majority of Al-Asmarat residents are from Manshyat Nasser, old Cairo, and other neighborhoods [20]. The residents were immediately assigned to their apartments; however, accommodating the elderly within the ground floor in each phase was considered. Therefore, the first phase was accommodated, then the second and third phases, respectively, according to the arrival priorities. The majority of the relocated families in Al-Asmarat are nucleus families (a husband and wife with children) 93.3% or extended families (including children and grandchildren) 5.1%, and the remaining are only a low percentage of composite families (including family relatives) [20]. Al-Asmarat district is considered a young community, as the residents' average age is 23 years old, with 29.3% of its total population aged between 18-35 years [21]. Therefore, investing in youth is one of Al-Asmarat's development mechanisms. The data has also indicated that the percentage of married couples of marriageable ages is about 72% of the total population, evenly distributed among males and females [20].

Several open spaces are designed in Al-Asmarat, as shown in figure 1 to ensure QOL for relocated residents. Al-Asmarat is characterized by its wide streets and several open spaces. It is to mention that the area is located near the youth center of Al-Asmarat, which residents can use with an annual membership subscription. Within Al-Asmarat, several open spaces serve the area's residents, with an estimated area of around 11,262 m^2 as stated from the urban survey and GIS analysis, accordingly making the individual share 0.6 m^2, complying with the planning standards guide for services, which is between 0.5 and 4 m^2 [22].

6. Design of the Interviews

Based on the above observations, there was a concern about some topics which needed more clarification from the residents. One of these concerns is whether the open space can afford residents' activities. The questions of the interviews were designed from the understanding of literature review and international criteria. Moreover, the three international criteria that were analyzed ensured whether the open space could afford socioeconomic activities in similar relocated communities in the Egyptian context of developing countries. Therefore, they stated several needed indicators to evaluate the open space performance. Furthermore, the indicator of smart urban quality (ISUQ) issued in Italy aims to determine public intervention priories and identify gaps to assess the compatibility of policies with local context [18]. Project for public space framework (PPS) also found that to achieve successful public space in the city; they should share four qualities which are (*accessible, activities, comfortable and sociable*), that affect places to afford social and economic activities successfully within its design setting [13].
Lastly, public space quality criteria created three categories of criteria: protection, comfort, and enjoyment to evaluate the quality of pubic space in cities [17]. Those indicators were studied carefully to facilitate the design of the interviews with the residents that respond to their needs to achieve a well-designed public space that ensures QOL.

Accordingly, 10 interviews were conducted to respond to these concerns. They include different categories of the residents according to culture, age and work, considering all groups to achieve inclusiveness (*women, children and youth, individuals with special needs*). Accordingly, they aim to collect qualitative data that reflect the residents' socioeconomic activities in open spaces after the relocation. Figure 2 shows the questions used in the interview.

(a) Semi-private Spaces:
2 in front of religious buildings.

(b) Open Spaces:
1 in front of the district building.

(c) Communal Spaces: 42 Spaces.

(d) Shaded Spaces: 22 Shaded Spaces.

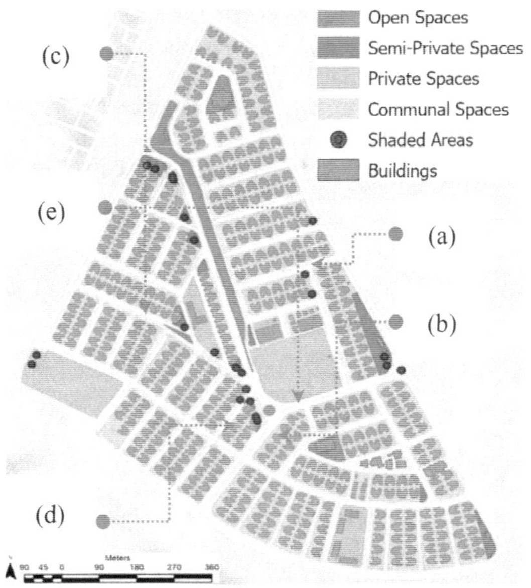

(e) Private Spaces (Fenced Playground): 5 Spaces.

Figure 1. Typologies of open spaces in Al-Asmarat, authors' photography and map analysis.

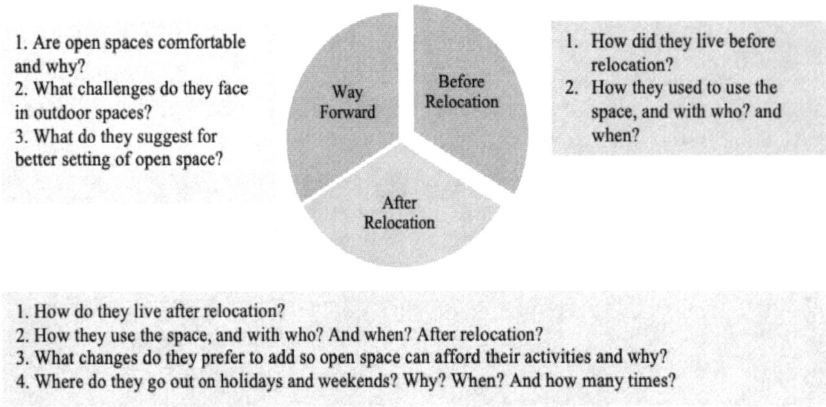

Figure 2. Design of the interview form, authors.

7. Results

The Findings of the study are revealed through field observation in addition to interviews. Interestingly, all the interviewed residents stated that they are in better living conditions after their relocation. In addition, the community offers them better infrastructure services and open spaces that enhance their QOL. Most of the residents agreed that they feel safe in open spaces in the new community where there is control and well **governance** from the district all the time.

Field observations showed that despite the availability of open spaces, green areas, and some communal spaces between the residential buildings and being **aesthetically** well-designed, there is minimal practiced activities by the residents. The limited number of wooden shaded spaces are used for picnics and sitting in the afternoon on weekdays or weekends as stated by residents [21].

The **environmental performance** within the communal spaces between the residential buildings is adequate, as they easily allow air movement. During field investigation, residents stated that the temperatures in summer within the open spaces in Al-Asmarat are comfortable. However, in winter, the temperatures decrease, especially afternoon and at night. Therefore, it is difficult for the residents to use the open spaces properly and prefer to stay home where it is warmer than outside. Regular **maintenance** of the public spaces since the project's operation was also observed. It is also recommended that designing defined spaces helps achieve a sense of **security** and limit the entry of strangers.

Some of the different responses regarding the area and open spaces asserted the following; "…. *You can see what the government has done to us now...I have access to a better housing, access to better infrastructure of potable water or sewage, environment & am provided with comfortable open space for the kids, too…. the district offers help whenever needed... where it provides a waste collection plan & that is why the community is still clean & well maintained through good control… We are lucky to have well designed * friendly open spaces that make us full of dignity.* *" (45 years old, Female).* The interviewed women also assured the existence of livelihood in open spaces. Some of them sit in the **friendly** shaded spaces to socialize and do handcrafts to improve their economic family conditions and ensure **social cohesion** between them. One of the women stated that: "…. *Almost all of us in this place depend on "our daily earnings", since we are not salaried workers but small entrepreneurs... I sometimes come to the shaded space near my house &* ***work on the needlework*** *that I sell." (50 years old, Female).* The same woman pointed to the **safety** and protection topic. Some residents prefer not to stay far outside their houses as they don't feel familiar walking around or sitting alone. The woman stated that*: "….other times, I just sit & chill with my neighbors during summer nights within the shaded space near my house … My sister's house is not near shaded spaces, so she prefers to come & sit with me by the shaded space near my house to feel* ***safer*** *& to be provided with more* ***privacy****." (50 years old, Female).* However, there are some cases of gender-based opinions; some women are prevented by their husbands or fathers from staying outside alone because they are afraid they could be harassed or attacked by strangers. The interviewed residents reacted thus: "…. *My husband prevented me from staying outside in the open spaces after sunset as he believes that most rape incidents happen at night....I simply avoid going out at night or walking around alone in isolated places….. The space is* ***monitored*** *during the day, but it is hard to feel secure at night since the* *workers & employees are not around after working hours." (23 years old, Female).* Also, it should be mentioned that some women -during their interviews- pointed out that they are benefitting more from the open spaces than men, as they sit with their neighbours. They usually meet during the afternoon to chat while watching their children play, which helps strengthen their interaction and **social cohesion**. In fact, at night, the open space (square in front of the district buildings) is considered to be **diverse** and **inclusive.** It is full of families and children who gather in the space after sunset until midnight to enjoy the environment. Some residents revealed: "*.. Around two times a week, I & my family come afternoon to this square & sit to midnight. We let our children* ***play*** *while watching them & eat some cooked food. This allows us to enjoy the* ***sensory*** *experience as it is provided with* ***lighting*** *fixtures." (32 years old, Female).* Moreover, most of the respondents stated that the new community provides the residents with spaces that can be used for entertainment and available youth centers without entrance fees. While others resort to recreational areas outside the area. Some interviewed youth stated that there are better opportunities to play football and chill in the afternoon in the new community. However, misuse of spaces, and damage in children's playing equipment during usage, led to their closure by the district. The interviews with children stated the proximity of playgrounds to their houses without exceeding 15 minutes and their participation in various sports. From the results of the interviews, it can be noted that most of the residents are satisfied with the open space design attributes. However, many residents reported that a more effective focus needs to be given to bicycle lanes, furnishing, and maintaining the open space to enhance the quality of open spaces. It is to mention that the residents were deprived of adequate open spaces or green spaces before they were relocated from their original unsafe areas. Although they are aware that they are provided with better QOL, they have suggestions to improve the not-so-well equipped open spaces.

8. Discussion

After conducting interviews with Al-Asmarat residents, the research deduces an evaluation that shows the indicators list of the open spaces in Al-Asmarat shown in table (1).

Table 1. Evaluation Indicators List to ensure open space afford socio-economic activities, authors.

Physical Design	Stay Friendly	Accessible Seating & Shaded Areas
	Urban Aesthetic Quality	Available Identity Character in Spaces Design
	Open Space Lighting	Area of Illuminated Space
	Clean	Waste Containers in The Streets & Parks.
		Distance between Garbage Containers on the Streets.
		Rate of Garbage Collection from Streets & Open Spaces.
	Quality Of Urban Landscape	Availability of Green Areas
		Air Pollutant Level
	Privacy	Usage of Natural/Industrial Elements
		Select Suitable Construction Materials to Isolate Noise.
	Environmental Performance	Temperatures in Outdoor Spaces (In Degree Celsius)
		Presence of Shadows & Air Movement in Outdoor Spaces.
Social & Economic Activities	Diverse & Inclusive	Presence of Women-Children-Ppl with Disabilities &Elderly
	Social Cohesion	Opportunities to Activities in Open Spaces
	Safety	Provide Semi-Closed Spaces to Prevent Strangers' Entry.
Accessibility & Services	Pedestrian Accessibility	Presence of Pedestrian Crossing Areas in the Open Space.
	Access To Potable Water/Sewage	Presence of Potable Water Taps in Open Space
		Presence of Toilets Near the Open Spaces
	Disabilities Access	Presence of Disabilities Equipment
	Bicycle Lanes	Presence of Bicycle Lanes & its Parking in Open Space's Zone
Governance	Waste Collection	Presence of Waste Management Plan
	Urban Maintenance Services	Presence of Urban Maintenance Plan
	Greenery Maintenance	Presence of Greenery Maintenance Plan
	Security Perception	Crime's Protection in Open Spaces
		Access to Police to Secure the Open Spaces

IOP Conf. Series: Earth and Environmental Science 1056 (2022) 012044 doi:10.1088/1755-1315/1056/1/012044

The list is extracted from interviews, field observations, literature review and readings of similar criteria. The study stated that social affordance concerns the social and recreational services available to the residents and their satisfaction and feeling of **security** within the design of open spaces after relocation. It is worth noting that residents perceive open spaces as an extension of the home, allowing for casual and planned encounters. As a result, the urban transformation process must include urban design to maximize the benefits of public spaces. Well-designed public spaces can improve the enjoyment of community life by supporting a greater density of activities, uses, and users. All the residents interviewed stated that there is a difference in using open spaces before and after relocation. The nature of using the open space in Al-Asmarat was very different, where the **governance** from the district provided control on the use of the space. The different typologies of open spaces in Al-Asmarat vary in their design to suit the residents' usages and activities. However, other factors do not allow residents to use the open space, such as the governance, the daily practices of the residents, and the conflict between the residents who come from different areas. Some of the spaces do not offer shading, making it difficult for residents to use during daytime. **Greenery** is not adequately offered in the area. Street and **open spaces lighting** is well offered at night, which gives a sense of **security** in the area. Open spaces are **clean** since there is a periodical system for garbage collection that reaches three times per day. The study found that Al-Asmarat has a low crime rate in open spaces due to **security** services. Additionally, the continuous monitoring, the availability of physical elements on the streets of the area, and the design that allow **disabilities access** ensure the residents' safety, as there are adequate street and **open spaces lighting**, **pedestrian paths**, and traffic safety. However, residents prefer to stay in groups to feel more secure.
Table 1 illustrates the ***18 evaluation indicators*** meant to evaluate the open spaces within the *physical design*, *socio-economic, accessibility and services, and governance* components in relocation projects. Those indicators are derived from the interviews conducted in the field survey. The proposed checklist's comprehensiveness demonstrates the multidimensional nature of open spaces in relocation projects. It also focuses on urban issues in open spaces, which are common in Egypt's current practices. These indicators must be considered when implementing open spaces in a relocation project to maintain an adequate standard of human well-being to allow for their activities in open spaces.

9. Conclusion

This paper investigated UDF efforts to respond to informal areas in Egypt that have emerged in violations of laws and planning regulations [23]. Moreover, it stated that open spaces are designed to ensure QOL for relocated residents from unsafe informal areas. Through the field observations in AL-Asmarat, it is noticed that open spaces lack proper utilization for activities practiced by the residents. Interviews were conducted further to understand the nature of open spaces in Al-Asmarat. The residents stated that they lived in better conditions, especially for proper open spaces, before relocating to Al-Asmarat. Additionally, some residents stated that they go to open spaces inside Al-Asmarat for recreation. However, others stated that they go outside Al-Asmarat for recreation at weekends with families. Finally, the study presented a list of 18 indicators to evaluate the affordance of open spaces within the physical design, social, economic and governance components. Furthermore, the most significant recommendation for improving affordance of open spaces in relocation projects is to consider the attachment of families during relocation to ensure the continuity of social relations and cohesion in surrounding open spaces [١٢]. In addition, upon designing a new community for relocation, the rates and standards of all open spaces and playgrounds services must be considered to provide the residents with more opportunities to enhance their livelihoods and their QOL [22].

References

[1] H. A. ElFouly and N. A. El Aziz, "Physical Quality of Life Benchmark for Unsafe Slums in Egypt," Sustain. Environ., vol. 2, no. 2, p. 258, May 2017, doi: 10.22158/se.v2n2p258.
[2] S. Attia and H. Khalil, "Urban Metabolism and Quality of Life in Informal Areas," 2015.

[3] N. M. Abdel-Moneim, H. A. E. Khalil, and R. R. Kamel, "Developing QOL Index for Resettlement Projects of Unsafe Areas in Egypt," Urban Forum, vol. 32, no. 3, pp. 349–371, Sep. 2021, doi: 10.1007/s12132-021-09419-7.

[4] A. Hoeltl, R. Brandtweiner, R. Bates, and T. Berger, "The Interactions of Sustainable Development Goals: The Case of Urban Informal Settlements in Ethiopia," Int. J. Sustain. Dev. Plan., vol. 15, no. 3, pp. 287–294, May 2020, doi: 10.18280/ijsdp.150304.

[5] N. El-Megharbel, "Sustainable Development Strategy: Egypt's vision 2030 And Planning Reform." 2015.

[6] R. Leila, "Safe relocation: Egypt's plan to eradicate slums by 2030," Ahram Online, p. 1, 2019.

[7] S. A. S. Abouelmagd, "Anwendung einer Livelihood-Perspektive zur Bewertung bestehender Politik- und UmsetzungsansätzeThe rehabilitation of slums and informal settlements in greater Cairo: applying a livelihood perspective to evaluate existing policy and implementation approaches," 2020, doi: 10.14279/DEPOSITONCE-10111.

[8] F. L. Amao, "Housing Quality in Informal Settlements and Urban Upgrading in Ibadan, Nigeria [A Case Study of Apete in Ibadan]," vol. 2, p. 13, 2012.

[9] P. I. Balogh and D. Takács, "The significance of urban open spaces and green areas in urban property developments," p. 12.

[10] P. Akkah and U. USP, "Public space as a generator of growth in african cities," Blantyre, 2015. [Online]. Available: https://issuu.com/uclgcglu/docs/public_space_as_a_generator_of_grow

[11] A. Alja'afreh, R. Al Tal, and A. Ibrahim, "The social role of open spaces in informal settlements in Jordan," Open House Int., vol. ahead-of-print, no. ahead-of-print, Jul. 2021, doi: 10.1108/OHI-02-2021-0023.

[12] P. M. Ngau and Z. Blanco, "The Production of Space and Place in Informal Settlements:," p. 13, 2019.

[13] "What is Placemaking?," Proj. Public Spaces, p. 1, 2007.

[14] "The Right to Public Space," Tadamun, 2013, [Online]. Available: http://www.tadamun.co/the-right-to-public-space/?lang=en#.YfBAVVhBxQI

[15] J. J. Gibson, The ecological approach to visual perception: classic edition. New York London: Psychology Press, Taylor & Francis Group, 2015.

[16] B. Vetter, "Perceiving Potentiality: A Metaphysics for Affordances," Topoi, vol. 39, no. 5, pp. 1177–1191, Nov. 2020, doi: 10.1007/s11245-018-9618-5.

[17] UN-Habitat), "Global Public Space Toolkit From Global Principles to Local Policies and Practice," 2015. [Online]. Available: https://unhabitat.org/sites/default/files/2019/05/global_public_space_toolkit.pdf

[18] C. Garau and V. Pavan, "Evaluating Urban Quality: Indicators and Assessment Tools for Smart Sustainable Cities," Sustainability, vol. 10, no. 3, p. 575, Feb. 2018, doi: 10.3390/su10030575.

[19] I. K. I. M. Ashmawy, "Stakeholder involvement in community resilience: evidence from Egypt," Environ. Dev. Sustain., vol. 23, no. 5, pp. 7996–8011, May 2021, doi: 10.1007/s10668-020-00894-9.

[20] AbdlRahman, "Needs & Priorities for Relocated Families toAl-Asmarat," Ministry of Social Solidarity, 2019.

[21] S. Attia, "Outcome Evaluation Study of Three Completed Informal Settlements Development Projects-Second Phase-Al-Asmarat District," Associated Consultant Company & Urban Development Fund for the African Development Bank., Cairo, Egypt, 2020.

[22] "Central Agency for Public Mobilization & Statistics." Housing and Establishment Census., 2017.

[23] "Egyptian Planning Standards Guide for Services." 2015.

[24] M. A. Khalifa, "Redefining slums in Egypt: Unplanned versus unsafe areas," Habitat Int., p. 10, 2011.

[25] S. Attia, S. Shabka, Z. Shafik, and A. Ibrahim, Eds., Dynamics and Resilience of Informal Areas. Cham: Springer International Publishing, 2016. doi: 10.1007/978-3-319-29948-8.

[26] B. El Maabady, "Reviewing the ISDF strategy for dealing with slums in Egypt," Master's Thesis, American University in Cairo, AUC Knowledge Fountain., 2015. [Online]. Available: https://fount.aucegypt.edu/etds/1110.

Implications for new cities' landscape design using place-making: A Comparative analysis study

Rehab A Zayed[1,*], Germin El-Gohary[2], Sherouk M Abdelrehim[1]

[1] Architecture Engineering Department, Faculty of Engineering, the British university in Egypt, Cairo, Egypt
[2] Urban Design & Planning department, Ain Shams University, Cairo, Egypt

*Corresponding Author Email: rehab169335@bue.edu.eg

Abstract. Despite the remarkable evolution of the new Cairo cities and the development of the quality of life, some citizens of Old Cairo districts still lack the motivation to change their residence to the new cities. The landscape design in New Cairo cities mainly manifests large green areas with water features and contemporary building exterior design to create an ordered, calm, and comforting scene. However, the landscape design in Old Cairo districts creates a significantly opposite scene. In most Old Cairo regions, green areas of large old trees occupy tiny spaces, buildings' exterior designs are bold and classic, and the whole scene manifests great chaos, crowd, and noise. Cairo is one of the ten least resilient cities around the world. The city is vulnerable because of its inadequate infrastructure, degrading environment, and vulnerable climate. According to the World Health Organization standards for green open spaces, Cairo's green open spaces provide weekly liveability. This research demonstrates theoretical and practical techniques to understand the psychological effect of the different landscape designs on Cairo citizens.

Keywords: Cairo, Cities, Place-making, Landscape, Vulnerability.

1. Introduction

Thousands of years ago, Egyptians have built their cities around the Nile River. However, Cairo, being the capital of Egypt, has experienced significant population growth special in the past 50 years as shown in figure (1), which led to urban expansion away from the Nile. The new cities of Cairo are all located in the desert, which introduces Cairo citizens to a new ecological typology that they are not familiar with. Heliopolis was the first desert oasis, a patch of planted and inhabited land surrounded by deserts. After Heliopolis, Nasr City and Mokattam were built. New Cairo region was the last desert oasis built [1]. Given the historical background, the research problem is that Old Cairo citizens lack the resilience to change their residence to new cities despite the significant crowd, pollution, and chaos in their living areas which negatively influence their overall physical and mental health. This research aims to provide recommendations for developing Cairo citizens' sense of belonging in the new cities by suggesting approaches for applying the placemaking theories in the landscape design of the new cities.

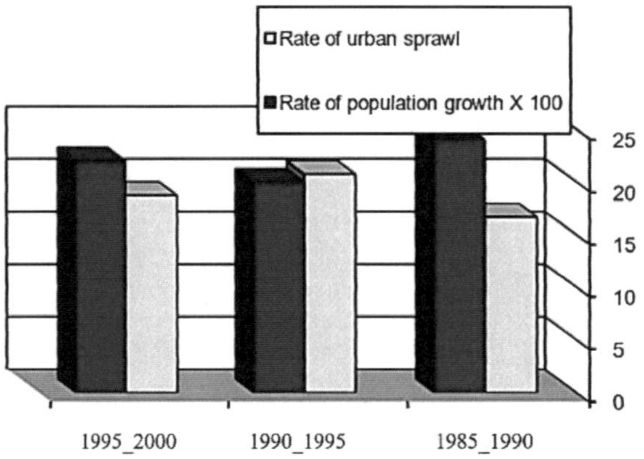

Figure 1. The population growth in Egypt and Cairo City through the period from 1985 to 2035.

2. Background

Due to excessive population growth in Cairo, it was declared one of the world's least resilient cities. The Population continued to increase the buildings' heights and number and limit the green landscape open spaces to accommodate all the citizens, which led to significant Issues. These issues have deteriorated infrastructure and inequality in the different sectors. [2]. Climate change is another issue: significantly increased temperature, severe heat waves, and decreased rainfall rates [3]. Consequently, Cairo experiences a significant increase in air and water pollution levels as. Shown in figure 2. [4]. However, the greatest issue is that the citizens are not aware of the great environmental issues they are facing [5]. The presented issues significantly affect the quality of Life in Old Cairo districts and the population's physical and mental health [6]. Accordingly, Old Cairo citizens are in a significant need to transfer their residence to new cities to give a chance for the old cities to develop. Moreover, Landscape designers need to apply the place-making theory to help in increasing the population's ability to accept the new cities as their new homes.

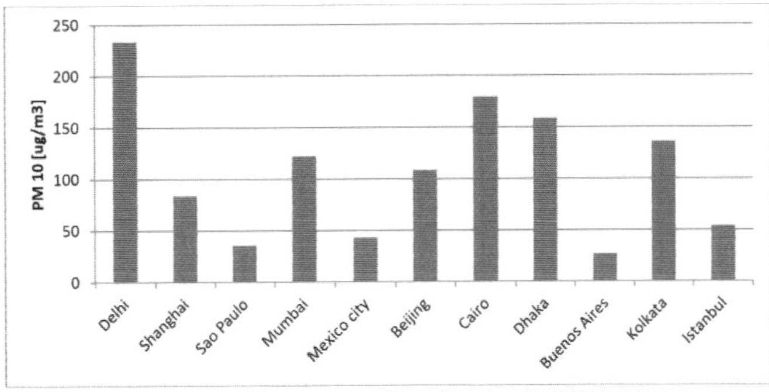

Figure 2. The most polluted cities in 2016 [7].

3. Research Problem Statement

Old Cairo Region is experiencing significant environmental changes such as temperature increases, infrastructure deterioration, and water and air pollution. These changes are due to overcrowding and few and small landscaped open areas. The World Health Organization declared that a city needs a minimum of 9msq / capita of open green spaces to be liveable. The green open spaces in Cairo only occupy 3msq / capita, which proves the city's weak life as shown in figure (3). Thousands of years ago, Egyptians have built their cities around the Nile River. However, Cairo, being the capital of Egypt, has experienced significant population growth, which led to urban expansion away from the Nile. The new cities of Cairo are all located in the desert, which introduces Cairo citizens to a new ecological typology that they are not familiar with. Heliopolis was the first desert oasis, a patch of planted and inhabited land surrounded by deserts. After Heliopolis, Nasr City and Mokattam were built. New Cairo region was the last desert oasis built the city [8]. Moreover, Old Cairo's current circumstances negatively affect the quality of life in the city and the citizens' physical and mental health [6]. However, the population in Cairo lack awareness of the tremendous ecological problem and choose to remain in their living areas where they are rooted. Moreover, the landscape design of the new cities still needs to be developed to make the Old Cairo citizens shift their residence to new cities without feeling that they are detached from their home, roots, and culture. This research aims to analyse and apply the place-making theories to suggest recommendations for designing a landscape in the new Cairo cities.

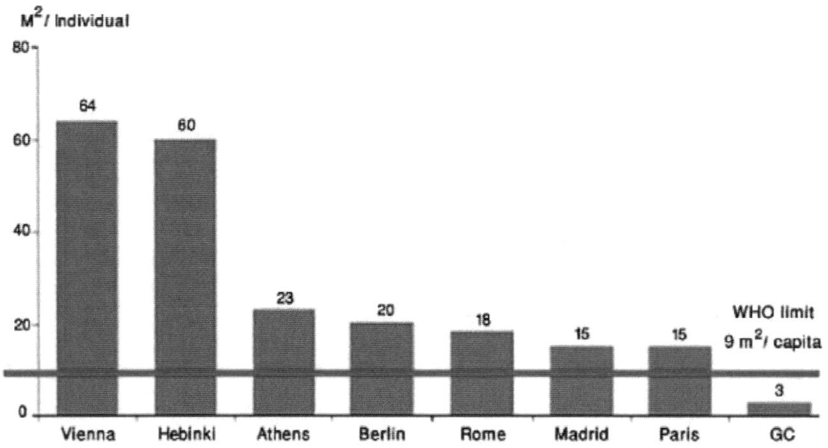

Figure 3. Share of green areas of different cities [9].

4. Research Aim and Objectives

This research aims to provide recommendations for designing new cities' landscape using the place-making theories for the sake of the following objectives:

- Collecting data of landscape elements, components, and types and their psychological effect on the population.
- Analyzing and comparing the current landscape design condition in the old Cairo districts and the new Cairo cities.
- Studying the definition, aim, characteristics, and meaning of the placemaking theories in landscape design.

5. Research Methodology

This research methodology consists of two sections: data collection and data analysis. The data collection method involves reviewing the literature and studying the previous case studies. The data analysis method is a qualitative method which involves analyzing, classifying, and comparing the collected data. Moreover, the data collection and data analysis outcomes were linked to achieve the research aim of presenting recommendations for designing new cities' landscape using the placemaking theories.

6. Landscape elements, components, and their effects on psychological and physical health

The world's continuous urbanization caused numerous adverse effects on landscape design generally. These effects are low maintenance of public green areas and neglected landscape resources [10]. Previous studies have shown that landscape elements, scenes, and characteristics have proven to have developing visual quality and health restoration in urban cities [11]. For instance, softscape elements such as water features, different vegetation, open lawn areas, and brightly colored flowers are beneficial for health and psychology. However, hardscape elements such as paved plazas and roadways considered from the damaging factors [12].

IOP Conf. Series: Earth and Environmental Science 1056 (2022) 012045 doi:10.1088/1755-1315/1056/1/012045

Table 1. Landscape elements and landscape components psychological effects (by author).

	Landscape Elements	**Landscape Components**	**Psychological Effect**	**References**
Softscape	Water Features 	Artificial lakes, fountains	It was proven that water was the most beneficial for psychological and mental health.	[13]
	Vegetation 	Old trees, green areas	Water and vegetation maintain a comfortable environment through developing the micro-climate.	[14]
	Topography 	Hills, mounts, plateau, mountains, terrain.	The convenient design of topography creates novel and diverse spatial experience and perception through bringing different sights of visitors through different levels in vertical areas.	[15]
Hardscape	Landscape constructions 	Roads and sidewalks, galleries, pavilions	Contemporary artificial constructions were claimed to have unfavorable effects on the visual quality of landscape.	[16]
	Cultural related constructions 	Poetry walls and corridors, memorials	Although culture-related components are artificial components, cultural heritage is significantly helpful in psychological restoration.	[17]

Finally, it is suggested that the effects of landscape components differed according to the environment they were used, indicating that using the same landscape elements and components can have different effects if spaces are organized in different ways [18].

7. Landscape types and psychological effects

As landscape design contains various elements and components, landscape design meanings are created and changed every day through how people perceive themselves within their environment. [19]. To understand the meaning of landscape, one is required to commit to studying a place's local information. Therefore, the meaning of landscape to a specific place is connected to the past, constructed, and supported by a social practice's network [20]. Moreover, it is required to immensely study the place's character's interpretations and their relation to creating, supporting, and maintaining cultural identity [21].

Table 2. Landscape types, definition, and psychological effects (by author).

Landscape Types	Definition	Psychological Effect	References
Familiar Landscape	The familiar landscape is related to the psychological effects of seeing a specific landscape that increases people's attachment values and meaning.	Landscape familiarity is a significant factor in perception studies as perception subconsciously affects subjective judgement. The effects can form responses to and feelings about specific landscape features and developments. For that reason, the most valued areas are those present in people's daily lives.	[22]
Aesthetic Landscape	Psychological responses to the aesthetic landscape are summarized in three main categories: coherence, fascination, and hominess.	Coherence is the analytical judgment according to the extent of space organization. Fascination is the extent of a space's visual richness and exploratory encouragement. Hominess is the warmth and coziness feelings and the judgments of a space's naturalness.	[23]
Cultural Landscape	As landscape is supposed to be beautiful, beauty is perceived through the people's cultural, social, and psychological traditions (Lothian, 2009).	The population's aesthetic preferences change rapidly as people's value for varied types of the landscape can shift according to environmental awareness, educational variation, or long-term cultural and historical changes.	[24]

In conclusion, positive psychological responses to landscape design are derivatives of landscape totality, not only from specific features but also from the embedded elements [25].

8. The transformation of the housing and landscape design sector in Cairo

In the past 25 years, the housing sector in Cairo has experienced a significant transformation on a social, cultural, and economic basis [26] New cities were built with different typologies of urban development from the original Cairo districts [27]. In these new cities, plant species used in the landscape design are selected to provide shelter and act as a natural habitat for the citizens. These plants can reduce the side effects of climatic changes. Moreover, diverse plant species increase the aesthetic value to produce a naturalistic landscape design [28].

Table 3. Comparison of landscape design between old Cairo districts and new Cairo cities (by author).

Points of Comparison	Old Cairo Districts	New Cairo Cities	References
Open Green Spaces	Very few open green spaces	Spacious green areas	[8]
Serenity	Very low level of serenity	Very calm and serene	[6]
Natural water features	The Nile River is the main water feature	No natural water feature just artificial ones	[29]
Pollution	Very high levels of pollution	Low levels of pollution	[4] [30]
Aesthetic value	Major chaos and minor aesthetic value considerations	Major aesthetic value considerations	[31]
Cultural values	Very high cultural value	Very low cultural value	[26] [27]
Climate	Tolerable climate despite the major changes due to the high pollution rates	Less tolerable due to the naturally severe conditions of the desert climate	[1]

9. Conclusions

9.1. The Place-making in urban planning and landscape design

The metropolitan Cairo districts have very high cultural value and tolerable climate. However, they have very few green spaces, a low level of serenity, and a high level of pollution. In contrast, the new Cairo cities have very spacious green areas, low pollution levels, and high aesthetic considerations. Characteristics of an effective place-making process is its central focus on the community's potential, inspiration, and assets. Moreover, the process must also shape public spaces that promote people's happiness, health, and wellbeing [32]. It was found that place-making is a deeply valued and essential process for emotionally connected to their living spaces. The place-making process aids people to see the power of their collective vision, to re-imagine their everyday areas, and to see new potentials in them [32]. According to the data gathered and analysed, it was possible to obtain several place-making approaches and processes that can be used to increase the people resilience to change their habitat to new cities.

Table 4. General place-making theories and the place-making theories and processes that can be applied in the new cities of Cairo (by author).

General place-making Theories	Place-making and processes that can be applied in the new cities of Cairo
Landscape affects the place where people choose to live and work as well as their feeling of belonging, place, and well-being [33].	Increase the familiarity of the spaces in the new cities to the old district's citizens.
47.5% of Cairo citizens still recognize the Nile River as a crucial symbolic natural landscape element that is not present in the new cities [29].	Compensate for the attachment of old Cairo districts' citizens to the Nile Valley.
Diverse plant species used in new cities are used to increase the aesthetic value and reduce climatic side effects to produce a naturalistic landscape design [28].	a) Increase the positive Aesthetic psychological responses in new cities. b) Improve the desert typology of the new cities.
Landscape should enhance cultural, social, and psychological traditions [34].	Promoting culture related landscape components in new cities.

In conclusion, the analyzed and compared literature review and the obtained theories and processes can be used to provide recommendations for new ideas and methods for improving the sense of belonging of the population in the new Cairo cities.

9.2. Recommendations

- Considering the arrangement of the landscape components is as significant as its types and variety in promoting psychological health in the new cities as shown in figure (3).

Figure 4. Mountain View Compound.

- Providing a familiar and cultural scene and landscape design in the new cities needs attention and promotion as the aesthetic scene.

Figure 5. The Unknown Soldier Memorial in Nasr City.

- Landscape design in new cities needs to improve, avoid, and control the severe climatic conditions and pollution levels.

Figure 6. Madinaty City.

References

[1] J. Schechla, (2015). "The Right to the City: Cairo. The Land and Its People: Civil Society," *Housing and Land Rights Network,* p. 133_144.

[2] R. Barkham, (2014). "Resilient cities. A Grosvenor Research Report," Grosvenor Group, U.K.

[3] UNDP, (2007). "Human Development Report 2007/8: Fighting Climate Change - Human Solidarity in a Divided World," New York. [Online]. Available: http://hdr.undp.org/en/content/human.

[4] Larsen, Bjorn, (2019). "Egypt: Cost of Environmental Degradation: Air and Water Pollution," The World Bank.

[5] H. N. E. A. G. N. Hassan, (2005). "National Strategy for Environmental Communication (NSEC) Strategy Development," Ministry of State for Environmental Affairs Egyptian Environmental Affairs Agency.

[6] W. Anwar, (2003). "Environmental health in Egypt," *Int. J. Hyg. Environ. Health 206 (4_5),* p. 339_350.

[7] A. Shirley 2016 "Which are the world's most polluted cities?," World Economic Forum. [Online]. Available: https://www.weforum.org/agenda/2016/05/which-are-the-world-s-most-polluted-cities/.

[8] N Kafafy, Y Betawi, (2011). "Urban green space benefits and the pivotal role of conservation, Cairo's case - Egypt," in *In: The Second Conference on Conservation of Architecture, Urban Areas, Nature and Landscape: Towards a Sustainable Survival of Cultural Landscape*, Zarqa.

[9] Gopp, UN-Habitat and UNDP, (2012). "Greater Cairo (GC) Urban Development Strategy," *Part 1: Future Vision and Strategic Direction; Gopp: Cairo, Egypt.*

[10] T.C. Daniel, (2001). "Whither scenic beauty? Visual landscape quality assessment in the 21st century," *Landsc. Urban Plan,* vol. 54, p. 267–281.

[11] X Wang, (2019). "Landscape Notes - Nature, Culture. Design," SDX Joint Publishing Company, Shanghai, China.

[12] Li Deng et al., (2020). "Empirical study of landscape types, landscape elements and landscape components of the urban park promoting physiological and psychological restoration," *Urban Forestry & Urban Greening,* vol. 48, no. 126488.

[13] H Nordh, C Alalouch, T Hartig (2011). "Assessing restorative components of small urban parks using conjoint methodology," *Urban For. Urban Green,* vol. 10, pp. 95-103.

[14] D Kendal, K.J.H Williams, N.S.G. Williams (2012). " Plant traits link people's plant preferences to the composition of their gardens," *Landsc. Urban Plan,* vol. 105, p. 34–42.

[15] M Arriaza, J.F. Cañas-Ortega, J.A. Cañas-Madueño, P Ruiz-Aviles (2004). "Assessing the visual quality of rural landscapes," *Landsc. Urban Plan,* vol. 69, p. 115_125.

[16] C Acar, B.C. Kurdoglu, O Kurdoglu, H Acar (2006). "Public preferences for visual quality and management in the Kackar Mountains National Park (Turkey)," *Int. J. Sustain. Dev. World Ecol,* vol. 13, p. 499–512.

[17] J Packer, N Bond (2010). "Museums as restorative environments," *Curator Mus. J,* p. 421–436.

[18] R. Parsons, (1995). "Conflict between ecological sustainability and environmental aesthetics: conundrum, canärd or curiosity," *Landsc. Urban Plan,* vol. 32, p. 227–244.

[19] T Greider, L Garkovich 1995 "Landscapes: the social construction of nature and the environment," *Rural Sociol,* vol. 59 (1), p. 1_24.

[20] W. D.R, (2004). "Environmental psychology: human responses and relationships to natural landscapes," in *In: Manfredo, M.J. (Ed.), Society and Natural Resources: A Summary of Knowledge*, Modern Litho, Jefferson City: MO, p. 337–348.

[21] J. Stephenson, (2008). "The Cultural Values Model: an integrated approach to values in landscape," *Landsc. Urban Plan,* vol. 84 (2), p. 127–139.

[22] S. Tapsell (1995). "River restoration: what are we restoring to? a case study of the Ravensbourne river, London.," *Landsc. Res,* vol. 20 (3), p. 98–111.

[23] B Adam. A,B Weinberger , P Alexander. A Christensen, A.C. Alexander Coburn, Anjan Chatterjee (2021). "Psychological responses to buildings and natural landscapes," *Journal of Environmental Psychology,* vol. 77, no. 101676.

[24] A. jorgensen, (2011). "Beyond the view: future directions in landscape aesthetics research," *Landsc. Urban Plan,* vol. 100, p. 353–355.

[25] J. Appleton (1996). The Experience of Landscape, Wiley, Chichester.

[26] D Singerman and P Amar, (2006). "Cairo as Neoliberal Capital? From Walled City to Gated Communities," *Cairo : The American University in Cairo Press,* p. 47_61.

[27] D. Sims, (2012) " Understanding Cairo: The Logic of a City out of Control," *Cairo: The American University in Cairo Press.*

[28] B Alizadeh, J Hitchmough, (2019). " A review of urban landscape adaptation to the challenge of climate change," *Int J Clim Change Strategies Manage.*

[29] M Keleg, G Butina Watson, M Salheen 2020 "Public perception of environmental change in rapidly growing cities: the case of Cairo, Egypt," SHAPING URBAN CHANGE – Livable City Regions for the 21st Century. Proceedings of REAL CORP 2020, 25th International Conference on Urban Development, Regional Planning and Information Society. [Online]. Available: https://doi.org/10.48494/REALCORP2020.8108.

[30] Filho, Leal and Walter, (2019). "Encyclopedia of Sustainability in Higher Education," *2019 Edition.*

[31] N. Alparslan 2017 "Between the Urban and the Natural Green: Marketing of Istanbul's Gated Community Projects," *Journal of Urban Studies,* vol. 8, p. 396–422.

[32] PPS, (2020). "“What Is Placemaking?” Project for Public Spaces," [Online]. Available: https://www.pps.org/article/what-is-placemaking.

[33] A. Scott, (2002) "Assessing public perception of landscape: the LANDMAP experience," *Landsc. Res,* vol. 27 (3), p. 271–295.

[34] A. Lothian, (1999). "Landscape and the philosophy of aesthetics: is landscape quality inherent in the landscape or in the eye of the beholder?," *Landsc. Urban Plan,* vol. 44 (4), p. 177–198.

Implementation of Secondary and Tertiary Treated Wastewater in Concrete Manufacturing

A Micheal[1, *], H Abdel Salam[2] and G M Ghanem[3]

[1] Centre for advanced Materials, the British University in Egypt, Cairo, Egypt
[2] Faculty of Engineering, the British University in Egypt, Cairo, Egypt
[3] Civil Engineering Department, Faculty of Engineering, Helwan University, Cairo, Egypt

[*] Corresbonding Author Email: amany.micheal@bue.edu.eg

Abstract. Water scarcity is a problem that faces many regions. In a developing and arid country like Egypt, the need for freshwater is exponentially increasing. This research is an evaluation of the recycled wastewater to be used in concrete manufacturing. The treated wastewater is blended with potable water with percentages of 25%, 50%, and 100%. The mixture of the two types of water is used in concrete mixing and curing. Control specimens with pure potable water are cast for comparison. The setting time for Ordinary Portland Cement is tested using the different water mixes. The hardened concrete is tested in compression and flexural at age of 7 days and 28 days. The results show that the treated wastewater may delay both the initial and final setting. The phenomenon is more pronounced in secondary treated wastewater. For the compressive strength, both treatment regimens can be safely used. The flexural strength is reduced at late ages.

Key words: treated wastewater, mixing water, curing.

1. Introduction

The construction industry responsible for the consumption of huge amount of fresh water. Approximately 150 litres water is required for every 1m^3 of concrete. The use of water in construction industry is not limited to concrete mixing but it extends to cleaning of equipment, aggregate washing, and concrete curing. The quality of mixing and curing water is a vital factor in producing concrete with high strength and good durability. Different specifications are set for mixing water qualities [1], [2] & [3]. An overview on the quality of concrete mixing water according to different codes can be found in the work of Babu et all [4].

One of the promising areas for saving water is to use treated wastewater in concrete manufacturing. Many researchers contribution in this field are available in literature. The published experimental programs range between checking the effect of using wastewater on cement paste properties, fresh concrete properties and hardened concrete properties mainly strength. The effect on cement setting can be found in the work of Babu and Raman [4]. Samples of cement pastes mixed with tab water and designated as control specimen, Primary treated water (PTW), Secondary Treated water (STW), and Tertiary treated water (TTW) gave initial setting time of +70, +47, and +38 minutes respectively compared to that of control, where they give +30 and +75, and +45 minutes respectively for final setting time. Different chemical compounds are observed in the XRD test compared with the compounds observed in the samples mixed with distilled water. Same observation of increased initial and final setting time is mentioned in the work of Shaikh and Inamdar [6].

The amount of soluble sulfates in the mixing water is of great importance that will affect the C3A hydration [7] & [8]. For cement soundness, Tang and Gartner [7] stated that no significant

IOP Conf. Series: Earth and Environmental Science 1056 (2022) 012046 doi:10.1088/1755-1315/1056/1/012046

differences between using PTW, STW, and control in the context of the maximum expansion in Le Chatelier test.

For properties of fresh concrete that implement treated wastewater, Tang and Gartner [7] carried out an experimental investigation on the slump of concrete mixed using PTW, STW and tab water. The results reveal that using PTW and STW leads to a decrease in slump by 4 and 1cm respectively, compared to the tab water control mix. This is attributed to the fact that PTW and STW have a higher amount of total dissolved solids that may lead to decrease in the concrete slump value.

For hardened concrete properties, focus is drawn on the compressive strength. according to ASTM C94 [1], the compressive strength at 7 days of standard mortar cube mixed with treated water should not be less than 90 percent than that of cubes constructed with tap water. As for the flexural strength, it does not have any requirements in the standards. In the work of Shekarchi M. et al [9], when STW was used, a decrease of 10 percent occurred to the compressive strength at age of 7 days and flexural strength at 28 days. On the other hand, and when TTW is used, the compressive strength at 7 days and the flexural strength at 28 days increase by 15 percent.

To help in water saving efforts in arid country as Jordan, Ghrair and Al-Mashaqbeh [10] investigated the use of Primary and Secondary Treated Wastewater (PTW, STW) in concrete manufacturing. They reported that STW is a better alternative for concrete and mortar as no negative effect on the compressive strength and setting time is observed. Conversely, using PTW as mixing water decreased the compressive strength up to 19.6%. Same comment can be found in the work of Swami et al [11]. Al Jabri et al [12] investigated the effect of using wastewater collected from car washing stations in concrete mixing. They concluded that there is no significant effect on the strength and absorption but it is suspected to have long term effect on concrete durability and steel corrosion.

El Nawawy and Ahmed [13] investigated implementation of treated wastewater from El Doha water treatment plant in concrete mixing. The chemical analysis if the water effluent revealed that it is not compatible most codes. They suggest mixing this water with tab water to improve their properties. Their results show that the treated water impacted negatively the concrete strength and cannot be used unless it is blended with tab water with percentage around 40%.

2. Water treatment methods

Wastewater is one of the types of polluted water resulting from different human activities and their multiple uses of water for many purposes. Wastewater is classified into two major categories; black water and grey water. Black water is discharged from toilet and kitchen water and it contains mainly organic matter, Nitrogen, and Phosphorous. Grey water results from recreational water, washing water and laundry water and it occupies 60-80% of sewage water.

The treatment of wastewater starts by the primary treatment process (PTW). The main objective of PTW is to remove large size of floating matter. It signifies the split-up of liquid and solids by the means of gravity settling. After the Primary Treatment about 40%-60% of the suspended solids and 30%-40% of the biological matters are omitted [14]. The key features of the Secondary Treatment stage (STW) are the use of biological processes to oxidize the organic matters and to dispose the oxidized materials using sedimentation. Purification using Ozon or Chloride is also a part of this process. The resulting water is 80-90% clear from the solid particles and 90 to 99% from harmful bacteria. Nowadays many countries adopt the Tertiary Treatment on wastewater (TTW) which involves more purification of Secondary Treated water to reduce the content of biological content and suspended matters. The Tertiary or advanced treatment results in more efficient water that can be used in purposes that need more efficient water than that results from Secondary Treated. Whatever the adopted treatment regimen, the quality of resulting water is assessed in the context of the content of the total dissolved solids (TDS), total suspended solids (TSS), biological oxygen demand (BODS), chemical oxygen demand (COD), total phosphorous (T-P), Ammonia (NH3), Chloride (Ci^{-}), and, finally, Escherichia coli bacteria.

3. Methodology

In Egypt, the water scarcity is becoming an urgent matter nowadays [15]. On the other hand, the water treatment plants are spreading in Egypt exponentially. In this research an attempt is made to implement wastewater effluent in concrete manufacturing. The wastewater is obtained from a water treatment plant located in El Shorouk City located on Cairo-Suez Desert Road. The effluent is obtained after the completion of both secondary and tertiary treatment processes. Properties of the used secondary treated wastewater effluent is shown in Table 1.

Table 1. Chemical composition of secondary treated water.

Day	PH	Temp	TS	TSS	TDS	COD	BOD5	D.O.
28-11-2020	7.4	23.2	580	25	555	37	-	4.5
29-11-2020	7.44	21	437	32	405	43	29	4.8
30-11-2020	7.35	20.2	436	26	410	37	-	4.7
1-12-2020	7.39	21.5	478	18	460	45	-	5.0
2-12-2020	7.33	21.3	475	25	450	43	-	5.2

Where:

BOD5: Biological oxygen demand
TS : Total solids
COD: Chemical oxygen demand
TDS : Total dissolved solids
TSS: Total suspended solids

One of the most significant parameters is the pH value which most of codes require to be not less than 7 to ensure suitable alkalinity of concrete medium. This is satisfied in the effluent used in concrete mixing and curing according to Table 1. Several tests are conducted on concrete made using both secondary and tertiary treated wastewater. The results are compared with concrete mixes made with tab water. The cement content in the mix is 350 kg/m^3 and the water/cement ratio is 0.45. The ratio of coarse aggregate to fine aggregate is 2:1. The coarse aggregate is lime stone with NMA of 19mm and the fine aggregate is sand.

Three percentage to replace tab water with either secondary treated or tertiary treated wastewater; 25%, 50%, and 100%. The type of water used for mixing is used for curing as well. For example, concrete mixed with water of 50% secondary treated wastewater is cured in the same mix of tab and treated water. For concrete compressive strength determination, samples of cubes of dimensions 15.8x15.8x15.8 cm according to the Egyptian Standards ECP 203 (2017) are used. The compressive strength test is conducted at age 7 and 28 days. Another test is performed to determine the concrete flexural strength. The specimens are concrete beams with length 600mm, span 500mm, section 100mmx100mm. The failure load is determined and the flexural strength is calculated using the relation:

4. Results

4.1 Cement setting time

The initial and final setting time tests are performed using the different blending percentages stated above. The test is conducted according to ESS 373/1991 [16]. The setting time are shown in figures 1 and 2 for initial and final setting time respectively.

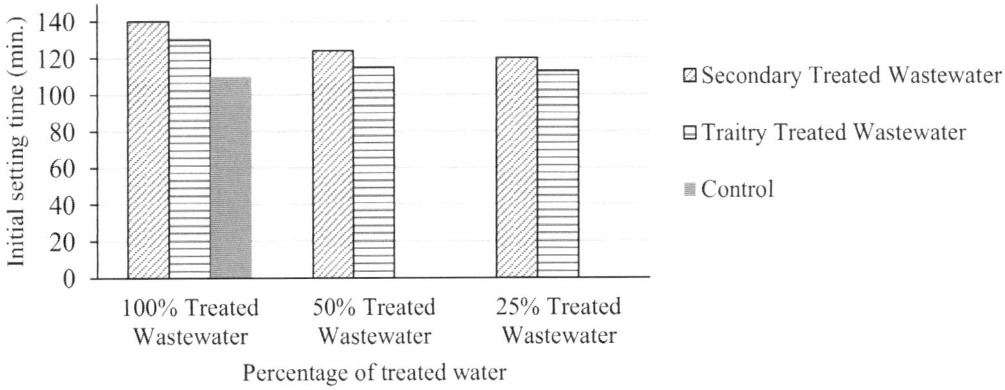

Figure 1. Initial setting time for different wastewater percentages.

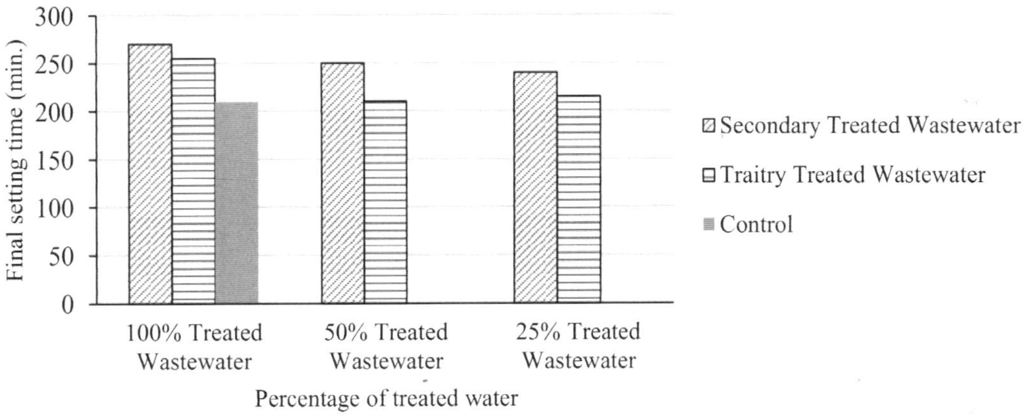

Figure 2. Final setting time for different wastewater percentages.

The trend of the results shows that there is an increase in the setting time for both secondary and tertiary waste water. The increase in more pronounced for the secondary treated water. The percentage increase in the setting time are shown in Table 2.

Table 2. Percentage increase in initial and final setting for different percentage of wastewater.

% Blending	STW		TTW	
	Initial Setting	Final Setting	Initial Setting	Final Setting
0 % (Control)	----	----	----	----
100% wastewater	+27.27	+28.57	+18.18	+21.43
50% wastewater	+12.73	+19.05	+4.55	+0.00
25% wastewater	+9.09	+14.29	+2.73	+2.38

4.2 Compressive strength

Figure 3 shows the comparison of compressive strength for different concrete mixes at age of 7 days and using the curing regimen stated before. It can be concluded that no significant change or reduction in the strength of concrete using blended water of both tab water and treated water. Both the STTW and TTTW shows good results. The same comment can be said about the compressive strength after 28 days as shown in figure 4.

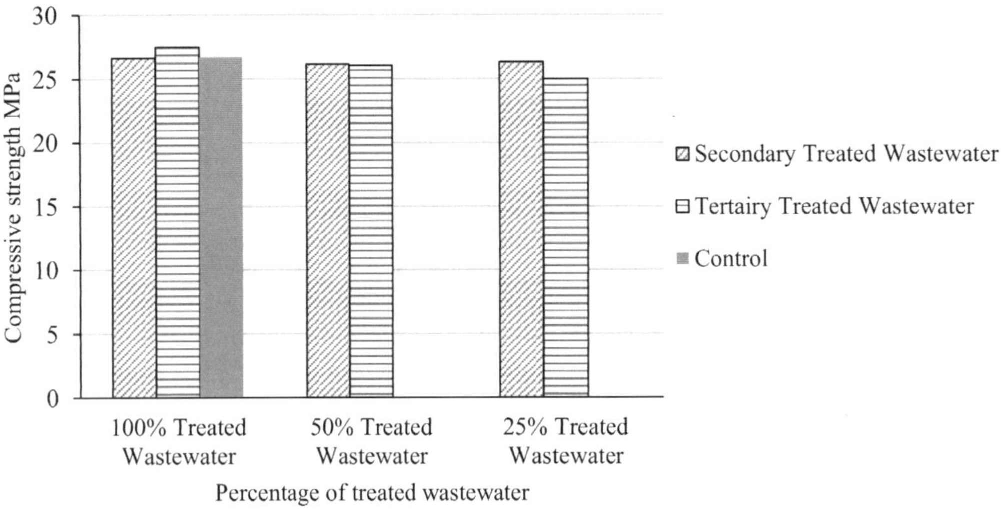

Figure 3. Compressive strength of concrete cube specimens at age of 7 days.

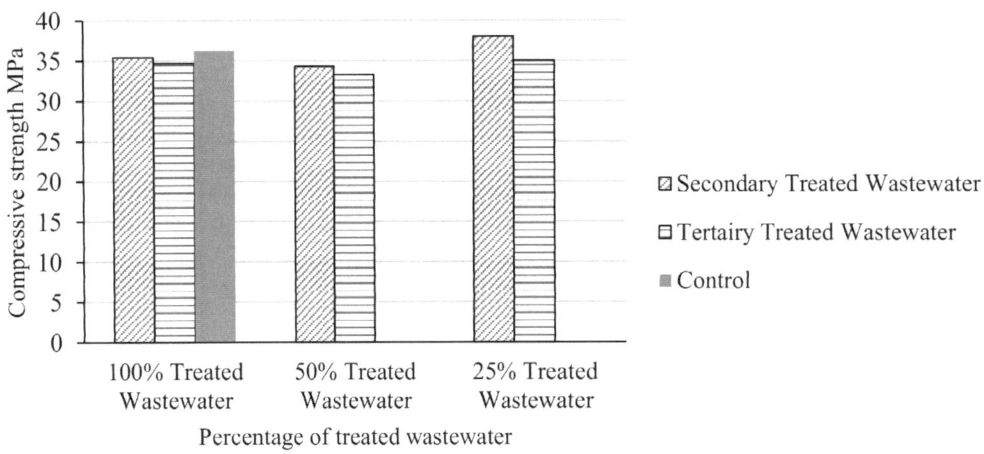

Figure 4. Compressive strength of concrete cube specimens at age of 28 days.

4.3 Flexural strength

The other mechanical test conducted in this experimental program is the three-point flexural strength test. The results are shown in figures 5 and 6 for 7 and 28 days respectively. For early ages, the tertiary treated wastewater show better performance but as hardening goes on and at age of 28 days, the control specimens takes the lead. The reduction in the flexural strength for STW is 41.8% , 29.5% and 18.9% for blending percentages of 100%, 50% and 25%. For the TTW the reduction is 40.6%, 47% and 34.2% for same blending percentages.

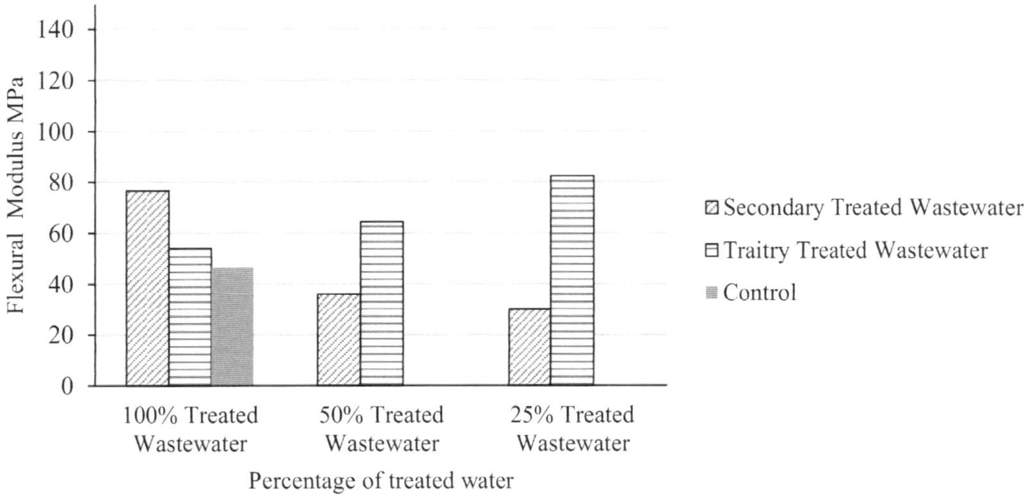

Figure 5. Flexural modulus at age of 7 days.

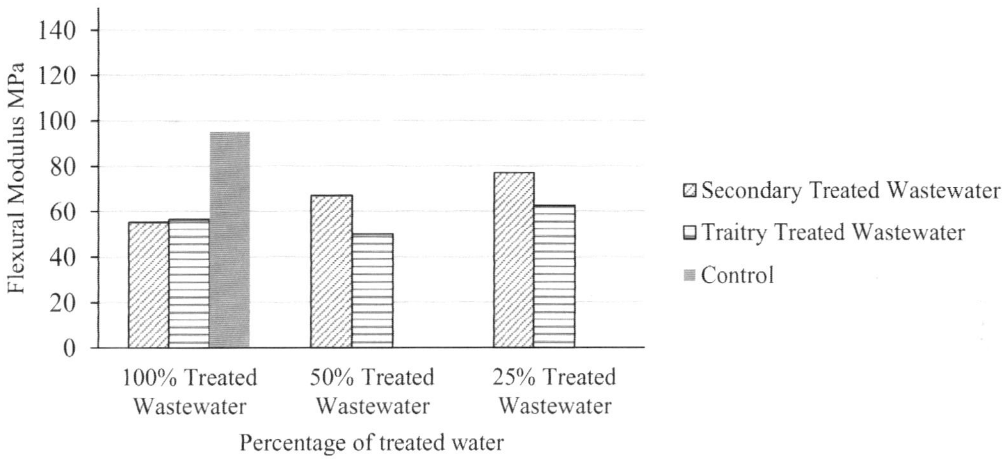

Figure 6. Flexural modulus at age of 28 days.

5. Conclusion

Implementation of treated wastewater in concrete mixing and curing is tested. Two water treatment regimens are used; secondary treated and tertiary treated water. The mixing and treatment water is a mix between tab water and treated water with percentages of 25%, 50% and 100%. The results are compared with control specimens of 100% tab water. The tests include the initial and final setting time, compressive strength and flexural strength. The results show increase in both initial and final setting time which is more pronounced in the secondary treated water. For the strength, no significant reduction can be observed in the compressive strength. For the flexural strength, the treated water shows some enhancement for the early age but for late ages, the control specimen shows better performance.

References

[1] ASTM C94 / C94M – 21 *Standard Specification for Ready-Mixed Concrete.*

[2] BS EN 1008 2002 *Mixing Water for Concrete - Specification for Sampling, Testing and Assessing the Suitability of Water, Including Water Recovered from Processes in the Concrete Industry, as Mixing Water for Concrete.*

[3] ECP 203 2017 *Egyptian Building Code for the Design and Construction of Concrete Structures*.

[4] Babu G R., Madhusudana Reddy B and Ramana N V 2018 Quality of mixing water in cement concrete "a review" *Materials Today: Proc.* **5** 1 pp 1313-20.

[5] Babu G R and Ramana N V 2018 Feasibility of wastewater as mixing water in cement *Materials Today: Proc.* **5** 1 pp 1607-14.

[6] Shaikh A R, Inamdar V M 2016 Study of Utilization of Waste Water in Concrete. *IOSR J. of Mechanical and Civil Engineering (IOSR-JMCE)* **13** pp 105-8.

[7] Tang F J and Gartner E M 1988 Influence of Sulfate Source on Portland Cement Hydration. *Adv. Cem. Res.* **1** pp 67–74.

[8] Lawrence C D 1966 Changes in Composition of the Aqueous Phase during Hydration of Cement Pastes and Suspensions *Highw. Res. Board Spec. Rep.* **90** pp 378–91.

[9] Shekarchi M, Yazdian M and Mehrdadi N 2012 Use of biologically treated wastewater in concrete. *Kuwait J. Sci. Eng.* **39** pp 97–111.

[10] Ghrair Ayoup M and Al-Mashaqbeh O 2016 Domestic Wastewater Reuse in Concrete Using Bench-Scale Testing and Full-Scale Implementation. *Water* **8** p 366.

[11] Swami D, Sarkar K and Bhattacharjee B 2015 Use of treated domestic effluent as mixing water for concrete: Effect on strength and water penetration at 28 days. *Indian Concrete J.* **89**.

[12] Al-Jabri K S, AL-Saidy A H, Taha R and AL-Kemyani A J 2011 Effect of using Wastewater on the Properties of High Strength Concrete *Procedia Engineering* **14** pp 370-6.

[13] El-Nawawy O A and Ahmad Sh 1991 Use of treated effluent in concrete mixing in an arid climate *Cement and Concrete Composites* **13** Issue 2 pp 137-41.

[14] The Egyptian Code for Design and Construction of Sanitary Water Treatment Plants 2017 *ECOP-WWTP.*

[15] Abdel-Shafy H 2002 Water issue in Egypt: Resources, pollution and protection endeavours. *Central European J. of Medicine* **8** pp 1-21.

[16] ESS 373/1991 *Egyptian Standard Specifications for Ordinary Portland Cement.*

IOP Conf. Series: Earth and Environmental Science 1056 (2022) 012047 doi:10.1088/1755-1315/1056/1/012047

Morphological changes of Cairo West power plant intake while using submerged vanes

Mohamed Eizeldin[1,2], **Ahmed Yasser**[2,*]

[1] Civil Engineering Department, Faculty of Engineering of Materia, Helwan University, Cairo, Egypt

[2] Civil Engineering Department, Faculty of Engineering, The British University in Egypt, Cairo, Egypt

*Corresponding Author Email: Ahmed163707@bue.edu.eg

Abstract. Lateral intake structures suffer from sediment deposition and instability of flow path which may have a negative influence on the efficiency of power plants that use river intakes as a primary source for their cooling system. This study aims to experiment the usage of submerged vanes as a mitigation method and determine their effectiveness in decreasing sediment deposition. A numerical model was constructed using DELFT3D of a reach that is located in the Nile River including the intake structure. Several scenarios were tested to optimise several parameters such as the usage of one or two rows of vanes and the distance of the vanes from the lateral intake. The results generated from the model showed that using two rows of vanes provided a 30% reduction of erosion in front of the intake compared to the base case and a 33% reduction in the volume of sediment entering the intake. Moreover, a 36% reduction of sediment volume was observed compared to the base case when the vanes were placed at a distance that is equivalent to 0.3 the intake's opening. The output of the model showed that it is more optimal to use two rows of submerged vanes at a distance that is equivalent to 0.3 the intake's opening.

Keywords: Intake Structure, Nile River, Sedimentation, Submerged vanes, DELFT3D and Numerical Model.

1. Introduction

Numerical modelling can save time, cost and could be used to monitor specific natural events such as climate changes. However, it could not represent some phenomena such as the scour around bridge piles because it is hard to calibrate certain parameters such as currents or waves [1]. In this case, it is better to use physical models as it could represent complex topographies. But physical models could consume more time and money. DELFT-3D is a powerful tool that provides the ability to model morphology, sediment transport, hydrodynamics and water quality for rivers or coasts. The FLOW module is used frequently and that is since it is used to calculate non-steady flow and transport phenomena that occurs from tidal and meteorological forcing on a rectilinear or a curvilinear, boundary fitted grid [2]. It is important to study sediment transport to enhance the balance between geomorphological entities or compartments from the sources to the sinks [3]. Moreover, studying sediment transport may help to determine the natural phenomena that occur in rivers such as erosion and sedimentation and how they could affect river structures.

2. Literature Review

The Meyer-Peter and Muller is one of the most popular sediment transport equations that is widely used in rivers with gravel beds. However, it will not provide the same reliable output on non-cohesive bed rivers such as the Nile River. To overcome this problem, a study was conducted to determine a correction factor that could be added to the primary equation to provide better results when representing the Nile River. The outcome of the study showed that the modified equation with a correction factor that was given a value of 1.5 provided better results when compared to the original equation and other equations such as the formulas of Frijlink, Wong and Parker, van Rijn and Huang [4]. Submerged vane is a structure that generates a secondary circulation within the flow, and this leads to a change in the direction and the extent of the bed shear stresses. As a result, the dispersal of depth, velocity and sediment transport may vary in the area that the vanes have an influence on [5]. A study was conducted to determine the optimum parameters of the vane [6]. The study recommended that the vane's angle should be 30 degrees to ensure maximum vorticity. The height of the vane should have a value of 0.4 the flow's depth and the length of the vane should have a range of 2.5 to 3 the vane's height. The study also proposed that the transverse spacing should be twice or triple the height of the vane and that the longitudinal spacing may cause a 20% reduction in bed level change when it is given a value that ranges from 15 to 20 the height. The usage of submerged vanes in front of lateral intakes could lead to a 40% to a 75% reduction in sediment deposition [5]. A previous study of the Intake structure of the Cairo West Power Plant was conducted by [7] to experiment the usage of the Baffle column structure as a sedimentation Mitigation measure. The outcome of the study showed that the baffle columns were effective in decreasing the amount of sediment entering the intake by affecting the vortex activity in front of the Intake.

3. Methodology

3.1. *Hydrodynamic model setup*

The numerical model covers an area about 700 square kilometres from the Nile River (1.3 Km in the flow direction & 0.55 Km in the cross direction). The Grid was created to cover the modelled reach. It is important to follow the ranges mentioned in the manual regarding orthogonality and smoothness. A surveying team was used to gather the bathymetric readings that were used in the model. Using interpolation and diffusion, the bathymetry file was created. The next step was to create the flow file. Grid files were added and dry points were used to represent some land areas at the bank. Courant number was used to determine the time step of the model which will affect the simulation run time. The time step used was 0.01 minute which provides a courant number that is less than 10. The surveying team also measured discharge of water and water level which will be used in initial and boundary conditions. In the initial conditions, the maximum water level was added. For the boundary conditions, the upstream of the model was entered as a discharge and the downstream was entered as a water level. This step ensures that the model keeps water level and discharge constant through the model. The physical parameters such as gravity, eddy viscosity and water density are entered. The inlet was represented by 6 uniform point discharges and the outlet was represented by one discharge point. Finally, the sediment transport file was added in addition to the local weir file.

3.2. *Calibration*

The surveying team measured the velocity and water level at a certain cross section in the reach. These measures will be compared with the model's output to ensure that the model is reliable. This is done by modifying the roughness file of the reach by adjusting the manning's coefficient. To increase velocity, the roughness coefficient should decrease and vice versa. The following figure is the velocity comparison between the model and the measured data. The figure shows that the model's velocity is nearly the same as the measured data.

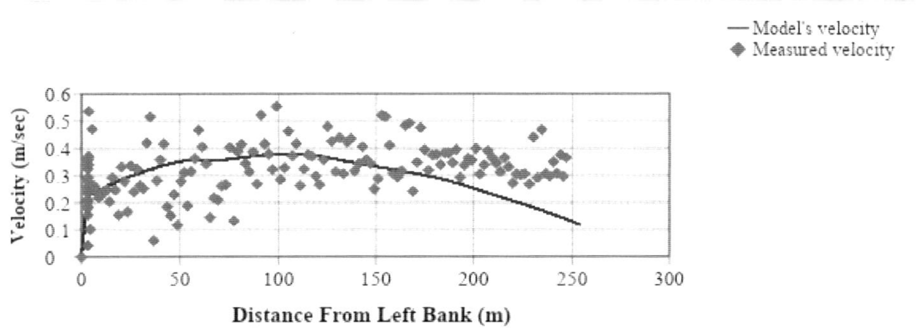

Figure 1. Calibration of model using velocity

3.3. *Scenarios of Numerical model*

The following parameters of submerged vanes are kept constant through all scenarios. The depth of the flow was 3.5 metres and hence, the height of the vane was given a value of 1.5 metres. The length of the vane was 4.5 metres and the angle was 30 degrees. The transverse spacing was 3 metres and the longitudinal spacing was 15 metres so that the vanes could be within the effective area. The base case scenario which does not include any mitigation method was the first scenario to be executed. It was used as a reference case in the comparison with other scenarios. The first parameter that was tested is the usage of one or two rows of submerged vanes. All other parameters are kept constant in both scenarios and the spacing of the submerged vanes from the intake was 10 metres. The following figures show the alignment of submerged vanes in the two cases that were compared with the base case.

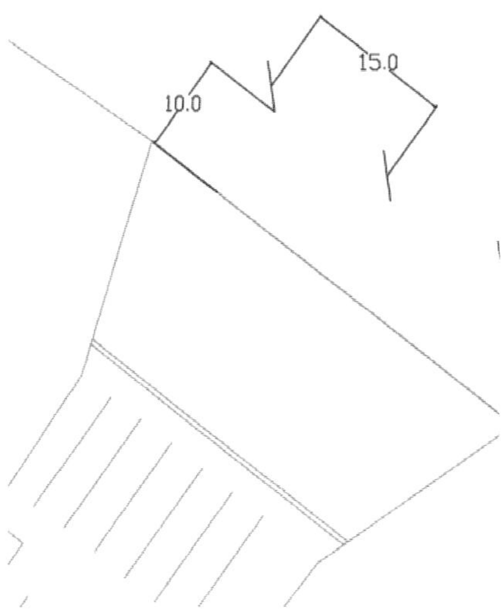

Figure 2. Vane's arrangement in scenario 1

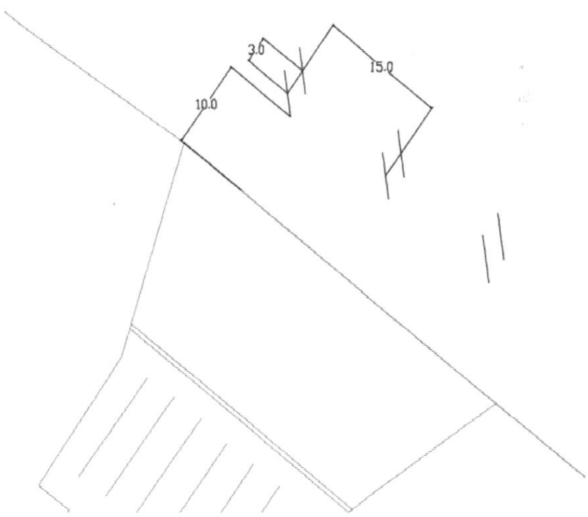

Figure 3. Vane's arrangement in scenario 2

After comparing the three scenarios, the second scenario provided better results. Therefore, this scenario was compared to two other scenarios to test the second parameter which was the distance between the vanes and the intake. The vanes of the first scenario were at a 5 metres' distance from the intake. The vanes of the second scenario were at a 10 metres' distance as mentioned before while in the third scenario they were at 15 metres. The following figures show the arrangement of the first and third scenario.

Figure 4. Vane's arrangement in scenario 1

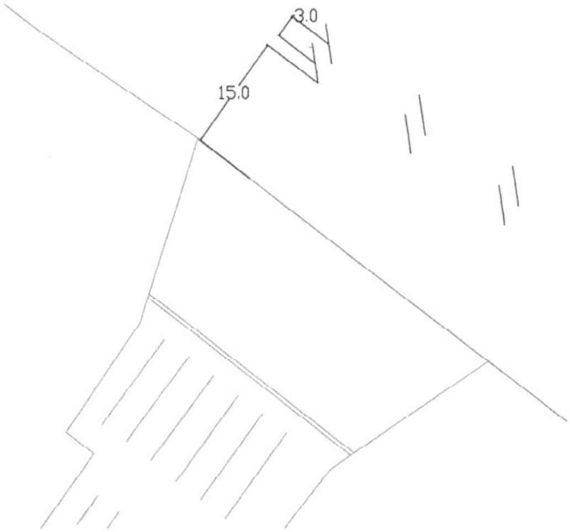

Figure 5. Vane's arrangement in scenario 3

4. Results

4.1. Results of first parameter

After comparing the two scenarios with the base scenario, a decrease in erosion was observed when using submerged vanes. The following figure shows that the first scenario faced a 17% reduction in erosion when compared to the base case. More reduction in erosion was observed when using two rows of vanes where the reduction reached 28%.

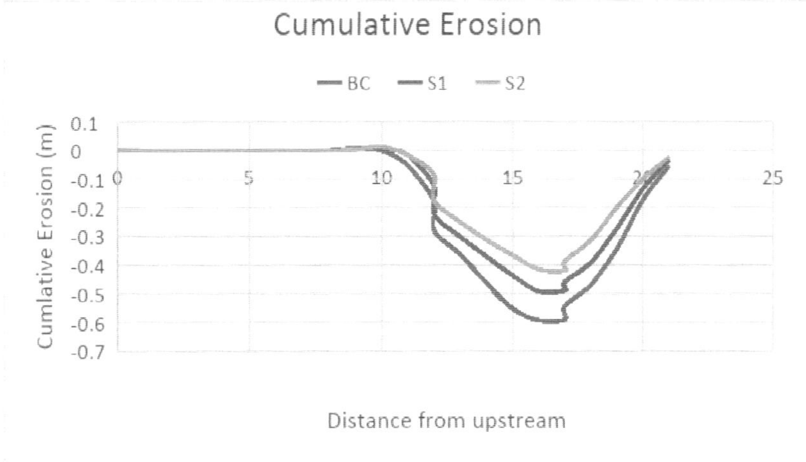

Figure 6. Erosion comparison (First parameter).

For the volume of sediment entering the intake, using two rows of vanes provided the most reduction in sediment volume. The following figure shows that scenario one had a 17% reduction in sediment volume compared to the base case while the second scenario had a 32% reduction.

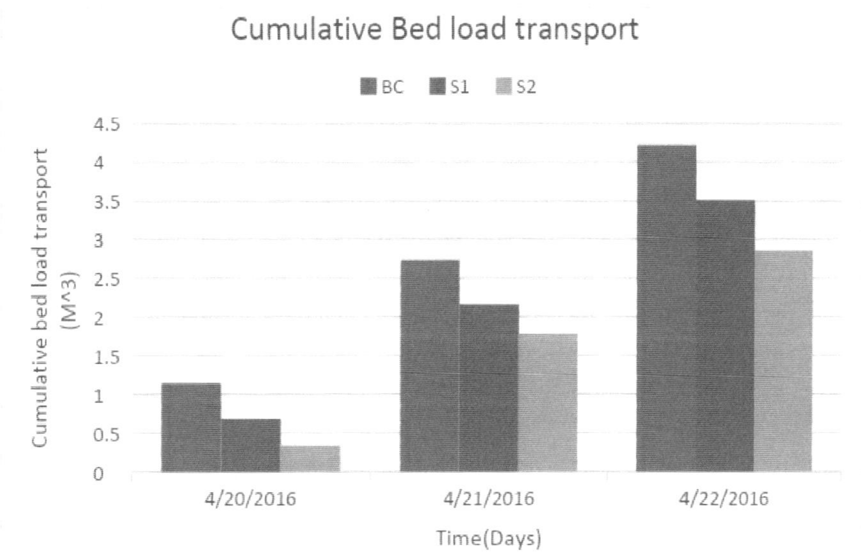

Figure 7. Sediment volume comparison (First parameter)

4.2. *Results of second parameter*

Using submerged vanes at several distances from the intake was tested. The proposed distances were 5, 10 and 15 metres from the intake. For the erosion, an increase in the percentage of reduction was observed as the distance from the intake increased. The second scenario expressed a 14% reduction while the third scenario showed a 31% decrease compared to the first scenario.

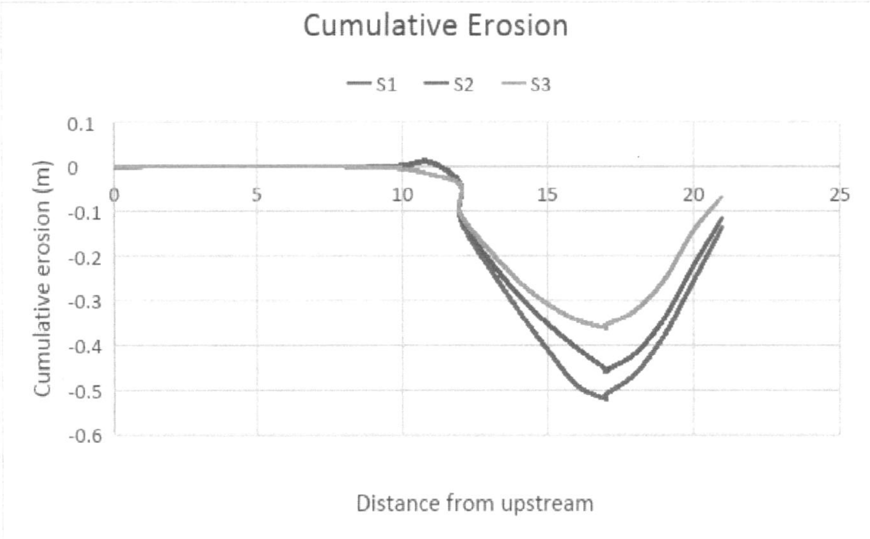

Figure 8. Erosion comparison (Second parameter)

For the sediment volume, the three scenarios were compared to the base case scenario. A reduction was observed in the three scenarios and the maximum reduction was observed in scenario 3. Compared to the base case, scenario one provided a 24% reduction while scenarios 2 and 3 provided 32% and 36% respectively as shown in the following figure.

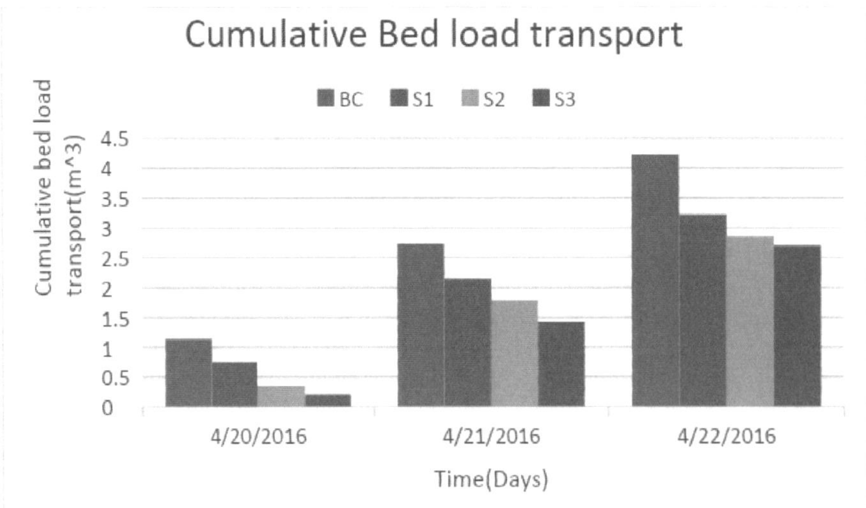

Figure 9. Sediment volume comparison (Second parameter).

5. Conclusion and recommendation for future work

Intake structures may suffer from sediment deposition or the instability of flow path. Submerged vanes are a mitigation method that is placed at river bends and intakes. They proved to be effective in decreasing sediment volume and erosion. The results of the first parameter optimization was the proposal of two rows of vanes providing better results in reducing sediment than using one row. Moreover, the outcome of the second parameter showed that placing the vanes at a distance 15 metres from the intake provided better results in terms of decreasing sediment deposition in front of intake structures. Two rows of vanes at a 15 metres' distance from the intake provided the best results. The

DELFT3D was an effective tool that provided several programs and modules that helped in creating the numerical model. The program provided various tools that were used to represent the natural phenomenon as accurately as possible. Based on these conclusions, the following should be considered by practitioners:

- Parameters such as longitudinal and transverse spacing are not mentioned enough in literature and need more testing.
- To represent submerged vanes in DELFT3D the Local weir structure was used. Other features could be used such as Modified Bathymetry and Porous plate.
- The nesting feature could be used to model the intake separately from the entire reach and hence, provide better results.
- Submerged vanes could be compared to other mitigation methods or it could also be monitored when used with other mitigation methods.

References

[1] Antunes, J. S. (2020). Physical Modelling vs. Numerical Modelling: Complementarity and Learning, 6-8.
[2] Deltares. (2020). FLOW User Manual (No. 2020). Deltares, 41-133.
[3] Ouillon, S. (2018). Why and How Do We Study Sediment Transport? Sediment transport in Coastal waters, 10(4), 1–34.
[4] Abdelhaleem, F. S. (2020). Adaption of a formula for simulating bed load transport in the Nile. Journal of Soils and sediments, 5-8
[5] El, E., & Helal, D. Y. (2017). Sediment Control at River Intakes Using a Single Row of Vanes. Ain Shams engineering journal, 4-6.
[6] Sharma, H., Jain, B., & Ahmad, Z. (2016). Optimization of submerged vane parameters. Sadhana - Academy Proceedings in Engineering Sciences, 41(3), 327–336.
[7] Saeed, R. M. (2020). *Improving Baffle Column Design to Enhance Its Effect on Sedimentation Inside River Intakes. Journal of Scientific and Engineering Research, 2020, 7 (8): 96-104*

AUTHOR INDEX